KEEPING FAITH WITH NATURE

Keeping

ECOSYSTEMS,

Faith with

DEMOCRACY, & AMERICA'S

Nature

PUBLIC LANDS

Robert B. Keiter

Yale University Press   New Haven and London

Published with assistance from the Louis Stern Memorial Fund.

Set in Adobe Garamond type by The Composing Room of Michigan, Inc. Printed in the United States of America by Vail-Ballou Press, Binghamton, New York.

Library of Congress Cataloging-in-Publication Data
Keiter, Robert B., 1946–
Keeping faith with nature : ecosystems, democracy, and America's public lands / Robert B. Keiter.
p.   cm.
Includes bibliographical references and index.
ISBN 0-300-09273-3 (alk. paper)
1. Environmental policy—United States.   2. Public lands—United States.
3. Conservation of natural resources—United States.   I. Title.
GE180.K45 2003
333.7'2'0973—dc21
2003009884

A catalogue record for this book is available from the British Library.

The paper in this book meets the guidelines for permanence and durability of the Committee on Production Guidelines for Book Longevity of the Council on Library Resources.

10   9   8   7   6   5   4   3   2   1

To Douglas, Annie, and Jackson

CONTENTS

Preface   ix
Acknowledgments   xiii

1
Introduction   1

2
Policy and Power on the Public Domain   15

3
Ecology and the Public Domain   47

4
Ecology Triumphant?
Spotted Owls and Ecosystem Management   79

5
Making Amends with the Past
Ecological Restoration and Public Lands   127

6
Shaping a New Heritage
Preservation in the Age of Ecology   171

7
Collaborative Conservation
Building Sustainable Communities   219

8
Toward a New Order
Ecosystems and Democracy   273

9
Keeping Faith with Nature   311

Acronyms and Abbreviations   329
Notes   333
Index   421

As the twenty-first century dawns, the public lands in the western United States are in an obvious period of transition. Of course, change has come regularly to the nation's public lands, often provoking acrimony and resistance. During its early years, the United States gave its public land away to encourage settlement on the western frontier. But with the arrival of the twentieth century, the country abruptly changed direction and began retaining large chunks of the public domain, putting the land and its resources to productive use. Public land policy over the next half century was the story of a growing struggle over how our timber, minerals, water, forage, and wildlife resources should be administered. Though never far removed, neither environmental nor preservation concerns were foremost in those discussions. By the mid-1970s, however, wilderness had gained official sanction and a plethora of stringent environmental protection laws were added to the mix. Since then, public land policy has increasingly focused on how these laws apply to the public domain and what portions of these lands we should dedicate to preservation. That policy debate has expanded into different venues: it is as likely to occur in a federal courtroom or a local library as on the floor of Congress or the Department of the Interior. And the debate is no longer about individual resources, nor is it constrained by the sacrosanct boundary line; it now addresses the management and restoration of entire ecosystems and thus transcends the linear boundaries that have traditionally defined the public landscape.

In this book, I identify, explain, and analyze the contemporary public land policy debates with a view toward bringing some perspective and coherence to this newly emerging era. Unlike most books on the public lands that focus on either individual agencies or resources, this one addresses public land policy in its entirety, focusing on the interconnections between the diverse lands, resources, agencies, and communities that occupy so much of the western landscape. Indeed, in this age of ecology, we have increasingly come to view the public lands as an integrated ecological entity as well as a key biodiversity stronghold. Ecology has taught us about obvious and not-so-obvious interrelationships; the alarming rate of species decline has pushed biodiversity conservation into the limelight; and the amorphous concept of ecosystem management has taken hold within the federal bureaucracy. Long-standing preservation notions—whether of entire landscapes, river corridors, wetlands, or species—have assumed new urgency in the face of the growing extinction crisis, burgeoning urban sprawl, and widespread environmental deterioration. These developments have profoundly influenced public land policy and shifted its focus toward ecological management, preservation, and ecosystem restoration. The momentum these changes have created has allowed us to begin thinking in terms of a new era in public land policy. But they are also being resisted—contested at every turn and quite tentative in many locations. (In fact, there is still much disagreement over the term *ecosystem management* itself, but I nonetheless use it as a convenient shorthand reference for the array of ecologically oriented management policies that are fundamentally reshaping public land policy.)

My task has been to chronicle the changes that forecast a new direction in public land policy, to examine the institutional forces driving (or retarding) those changes, and to offer tentative observations on what the future may hold. To do so, I have highlighted key examples of the new ecological management movement: the Pacific Northwest's spotted owl controversy; the Yellowstone wolf reintroduction; fire as an agent of ecological change; the new wilderness debates; the transformation of southern Utah's Colorado Plateau; and the Quincy Library Group's forest management initiative. Drawing upon these examples, the book focuses on the ideas, forces, and institutions that are effecting—or resisting—change on the public lands. Although Congress has the final say in how the public domain is managed, the public land agencies and the courts are playing major roles in the transformation to an ecological management regime. The general public, too, is making its will known, through a plethora of interest groups as well as new public involvement processes that have opened the federal agencies to a broader assortment of views than was true in the past. But though the general public may own the public lands and have a stake in their future, the myriad communities situated in and around them have always had a voice in how these lands are managed, just as the western states have long taken a keen interest in how the federal government administers these important assets. That is certainly true today, as federalism is alive and well in the American West, perhaps as much so as anywhere in the nation. In fact, a new homegrown movement of collaborative governance has emerged, through which diverse constituencies are playing a more direct role in managing our natural resources. By illuminating the interplay between these institutions and forces, we may better comprehend the future of public lands, even as specific controversies may evolve in sometimes unexpected directions.

Ecosystem management and restoration of the public lands is a work in progress. There is an emerging coherence to what the concepts mean and how they apply on the ground, yet no major overhaul of the laws governing the public lands has been undertaken. This is not surprising. Change has usually come slowly to the public lands, often percolating through the agencies and their various constituencies before Congress finally embeds any new notion into law. But across society, change is coming faster than in the past, driven by new social and economic realities. Not only is the polity more diverse than at any previous time, but the prevailing social norms place a high premium on quality of life concerns and a healthy environment. The nation's economy is also in a state of flux, shifting perceptibly from an industrial to a technological foundation, while also expanding into the global marketplace. Scarcity, too, has always been a driving factor in the nation's natural resources policy, and scarcity is today spelled in such terms as wilderness, endangered species, and solitude. All of this has profound ramifications for the public lands, both how we value them and how we choose to manage them.

ACKNOWLEDGMENTS

I would be remiss not to acknowledge the many people who made this book possible. The library staff at the University of Utah's S. J. Quinney College of Law—Barbara Swatt, Marsha Thomas, Felice Thorpe, John Bevan, Suzanne Miner, and Laurie Ngai—provided me with extraordinary service throughout this project, promptly responding to my endless requests for obscure materials and references. The College of Law's Faculty Research Fund supported my work over several summers, enabling me to explore fully the complex issues and controversies underlying contemporary public land policy. My colleague and friend David Williams read early drafts of the manuscript and offered much-appreciated encouragement as well as critical commentary before his untimely death; I only wish he were here to continue pursuing his progressive vision for the public lands. Andrea Hansen, Elizabeth Kirschen, Jim Holbrook, Steve Bloch, Bill Chaloupka, Sydney Cook, and the students in my spring 2002 Ecosystems and Community seminar had a hand in the project too. Many others graciously shared their time and knowledge with me during interviews; their contributions are referenced in the relevant notes. Barry Biediger, Adam Sobek, and Scott Bridwell from the university's Digit Lab provided first-rate assistance in preparing the maps found throughout the book. Jean Thomson Black and Nancy Moore Brochin at Yale University Press provided valuable encouragement and direction throughout the production process. Laramie artist Bob Seabeck graciously agreed to illustrate the book with his exquisite sketches, which do much to enrich its appearance. And my ever-understanding wife, Linda Keiter, lent encouragement and support even as the project stretched across the years. I am truly grateful to all.

ONE

Introduction

When we try to pick out anything by itself, we find it hitched

to everything else in the universe.

JOHN MUIR (1869)

The ecological ethic of interdependence may be, in fact,

a moral truth.

DONALD WORSTER (1977)

I s a new era dawning on the western public domain? With the twenty-first century upon us, the nation's policy for public lands and natural resources appears to be in a state of flux. How else to explain the near-mythical status that the previously unknown northern spotted owl has attained, or the gray wolf's transformation from the beast of destruction to a key missing ecological link? How else to explain the demise of the revered Smokey Bear image and the reintroduction of fire as an important component of the landscape? And how else to explain the heralded advent of ecological management and restoration policies within all of the principal federal agencies responsible for the public lands? The precise contours of this new era may not yet be fully defined or understood, but there is little doubt that new societal values and the ecological sciences are forcing major changes in how we view and manage natural resources across the public domain. And there is little doubt that our existing institutions, laws, and policies are being sorely tested to accommodate the emerging age of ecology on public lands.[1]

No single event accounts for the changes that are afoot. But then, no single event signaled the beginning of a new policy era during the waning years of the nineteenth century when a new utilitarian conservation philosophy first surfaced. That conservation philosophy, of course, has dominated natural resource management throughout the past century. Indeed, deeply ingrained utilitarian beliefs were originally invoked to discount the northern spotted owl in the Pacific Northwest logging controversy and to justify eradication of the wolf throughout the western states. These same beliefs also taught us that fire is bad and must be extinguished promptly and that preservation sentiments—if valid at all—must be constricted to narrowly delineated reserves, preferably composed of lands otherwise of little economic value. But these long-standing beliefs are under attack today as never before, and they are giving way—however grudgingly—to a new ecological order on the public domain. The transition, as is true of virtually all major policy transitions, is neither orderly nor complete. It is, nonetheless, being played out in various legislative, administrative, and judicial venues and, most important, on the ground.

Before 1990, few people other than a handful of scientists had heard of the northern spotted owl. Fewer people yet would have been prepared to predict that the Pacific Northwest's powerful timber industry would soon be brought to its knees by a small cadre of environmental activists and their lawyers acting on behalf of the spotted owl. Yet that is precisely what happened. On May 23, 1991, Seattle federal court judge William Dwyer made headlines across the nation when he issued an injunction forbidding all commercial logging on the Pacific Northwest's expansive public timberlands. Judge Dwyer's ruling, which culminated a decade-long struggle over federal timber policy in the region's diverse ancient forest ecosystems, triggered a storm of protest from the powerful logging industry as well as the rural communities that depended on public timber to maintain scores of local mills. Environmental advocates, drawing upon a welter of environmental laws, persuaded the judge that accelerated federal timber harvesting policies had imperiled the northern spotted owl, a reclusive bird that had come to dominate a rancorous debate over the fate of the region's timberlands as well as the nation's endangered species legislation. Following the ruling, the public land agencies were forced to acknowledge that business as usual was no longer the order of the day. The future suddenly appeared very uncertain on the region's federal forests.[2]

Another three and a half years would pass before Judge Dwyer was finally convinced to lift his injunction and allow logging to resume—at a decidedly reduced scale—on the nation's most productive public timberlands. By then, Congress had proven itself hopelessly deadlocked over the matter, an unprecedented presidential conference had been convened to defuse the regional crisis, and a high-profile team of federal scientists had developed a new ecology-based forest management plan. The proposed regional forest plan encompassed over 24 million acres of public land in three different states, extended legal protection to over 400 different species found throughout the region's ancient forests, and placed nearly 80 percent of the forest acreage off limits for timber harvesting. According to its authors, the plan contemplated a comprehensive ecosystem management strategy that first protected species diversity and ecological processes, and then permitted logging only where it would not endanger forest ecosystems. Judge Dwyer, after reviewing the plan, grudgingly affirmed its legality and lifted his injunction, admonishing that "given the condition of the forests, there is no way the agencies could comply with the environmental laws without planning on an ecosystem basis." The primary lesson from the controversy was clear: federal land managers must be prepared to address and protect ecosystems on the nation's public lands. The secondary lesson was also clear: science and litigation have the power to reshape the natural resource policy agenda on public lands.[3]

Nearly a thousand miles to the east, officials at Yellowstone National Park confronted a different dilemma. Having once rid the park of dreaded wolves, park officials were now poised to restore the extirpated gray wolf to the park ecosystem. During the 1920s, as part of a national predator extermination campaign, Yellowstone rangers killed the last park wolf in what Park Service officials now regarded as a misguided effort to protect other wildlife then perceived as more valuable. Over the intervening 60 years, the wolf's image underwent a gradual transformation, shifting from the beast of destruction to an important ecological cog. In 1974, the U.S. Fish and Wildlife Service included the Rocky Mountain gray wolf as one of the first animals listed on the new endangered species registry, and proceeded to develop a recovery plan calling for wolf restoration in Yellowstone National Park and other remote western wilderness areas. Local ranchers objected vehemently to the wolf restoration proposal and enlisted their congressional allies to block the proposal for more than a decade through a series of budget riders, congressional studies, and other political maneuvers. But in 1994, utilizing a new statutory provision that enabled ranchers to protect their livestock against depredating wolves, federal officials authorized the introduction of seven pairs of Canadian gray wolves into the Yellowstone backcountry as an experimental population. After years of frustrating setbacks, the path was suddenly cleared to reestablish historic predator–prey relationships that had not existed in the park for over 50 years.[4]

Since then, the wolf population has grown at a rapid pace. By 2000, more than 115 wolves divided into 11 different packs were roaming the Greater Yellowstone region. Although isolated livestock depredation incidents have occurred outside the park, the wolves have concentrated their attention and legendary hunger on the park's abundant elk population. As predicted, park visitation numbers increased in the aftermath of the wolf reintroduction, as both old and new visitors sought the opportunity to observe the wolves in their new domain. However, faced with a legal challenge from local ranchers, a Wyoming federal judge ruled that the entire wolf reintroduction program violated the Endangered Species Act's experimental population provision. Finding that naturally occurring wolves were already present in the park, the judge held that the statute precluded an experimental population release of new wolves. But after ordering removal of the reintroduced wolves, the judge stayed his order pending an appeal of the matter. Although the ruling placed the wolf reintroduction program in legal limbo, the wolves flourished, reclaiming their historic role in the region's ecosystem. Two years later, when an appellate court finally overturned the wolf removal order, the notion of restoring extirpated predators and other species to public lands became a firm reality.[5]

The image of wildfire may evoke even more fear than the wolf. Over the past century, federal and state officials have worked assiduously to control fire on public lands in order to protect precious timber and range resources as well as surrounding private property and scenic vistas. No less an icon than Smokey Bear carried the message that a charred landscape is a ruined landscape. The lonely fire lookout and the heroic smoke jumper represented Smokey's real-life counterparts in this annual battle against the forces of nature. Each year, as spring unfolded into the summer fire season, agency scientists took careful measure of precipitation patterns and regularly forecast the likelihood of devastating conflagrations. Fire managers paid close heed to these forecasts and prepared themselves to deploy legions of firefighters at the first sign of trouble. The goal was simple and clear: suppress all fires to protect the region's vulnerable resources and the livelihoods of those who depend on them. For the most part, the responsible agencies were remarkably successful at extinguishing fires before they could spread across the landscape.[6]

The summer of 1988, however, exploded several fire myths. With fully one-third of Yellowstone National Park burning, the nation watched transfixed as a phalanx of firefighters proved helpless in the face of the advancing flames that imperiled such venerable landmarks as Old Faithful Lodge and the historic park headquarters at Mammoth Hot Springs. Spurred by tinder-dry conditions and unrelenting winds, the Yellowstone fires jumped hastily constructed fire lines, deep riverine canyons, and administrative boundaries, burning randomly through the park's dense lodgepole pine forests until an early fall snowstorm finally brought them under control. As the Yellowstone drama was unfolding, the Park Service confirmed that its policy was to allow lightning-ignited fires to burn unchecked in the park's backcountry in an effort to emulate historic ecological processes. Scientists explained that the West's high-elevation forests historically had experienced high-intensity fires at 200–400-year intervals, which was nature's way of regenerating these coniferous ecosystems. In the investigation that followed, agency officials widely acknowledged that a century of fire suppression had converted the West's forests into potential tinderboxes and rendered them more susceptible to insect infestations and other diseases. With the forest's plight exposed, the real problem was clarified: How to restore the damaged landscape? For some the answer was to accelerate timber harvest levels to both mimic historical fire regimes and salvage valuable timber; others viewed such proposals with suspicion, fearing another large-scale logging assault on the public forests. But nearly everyone concurred that fire represents an important ecological process, and that it must be accorded a larger role in the public domain.[7]

Biodiversity conservation and ecological restoration cannot take place in a vacuum. Over the years, we have created nature reserves—national parks, wildlife refuges, and designated wilderness areas—as sanctuaries to nourish and protect wildlife resources, scenic landscapes, and other natural features. Ours has been an enclave strategy, based on drawing sacrosanct boundary lines to preserve the enclosed natural bounty. But neither the spotted owl nor the wolf nor fire respects conventional boundaries or the ownership expectations they represent. In fact, the legal boundaries we have drawn on maps to delineate protected reserves, multiple-use lands, private property, and the like are essentially irrelevant from an ecological perspective. Yet our political and legal systems, committed to the need for certainty and stability in property relationships, attach ownership rights and responsibilities based on these lines; landowners and managers are expected—even obligated—to manage their property in conformance with carefully crafted legal mandates and expectations. The Forest Service is expected to harvest timber on its multiple-use lands, the Park Service is expected to preserve its holdings in a primitive state, and most private landowners expect to do just as they please on their own lands. However, a growing national commitment to ecological preservation and restoration calls these expectations into question, just as it calls the current enclave strategy of nature conservation into question.[8]

The answer, according to ecological scientists, is to significantly expand and interconnect our nature reserve system to encompass a full array of ecosystems and to enable natural processes to unfold with minimal human intervention. Although an expansive nature reserve system with strategic linkage corridors might reduce future spotted owl or gray wolf controversies, the proposition is both highly provocative and profoundly difficult. It provokes intense opposition because it would expand the protected land base by removing productive federal lands from multiple-use management and perhaps even limit development options on adjacent private lands. It raises difficult management concerns because ecosystems are not static locations but dynamic and constantly changing entities, which makes it difficult to delineate meaningful permanent boundaries. One answer, therefore, is to better coordinate management of our existing reserve system with adjacent public and private lands. A more controversial answer is to acknowledge that the only effective means to protect biodiversity and ecological health is to expand the current reserve system by adding more lands to our existing national parks, wilderness areas, and wildlife refuges. In either event, another lingering—and hotly contested—question is whether management of these lands requires active human intervention or whether nature should be allowed to take its own course. In short, to meet ecology's challenges, we must begin rethinking our enclave-based nature preservation system as well as related interventionist management policies.[9]

Public lands are one of the West's defining characteristics. Approximately 663 million acres, or 29 percent, of the nation's land is owned by the federal government. The vast majority of this acreage is concentrated in the 11 western states (Arizona, California, Colorado, Idaho, Montana, Nevada, New Mexico, Oregon, Utah, Washington, and Wyoming), where federal ownership encompasses more than half the land base. Among these western states, the federal estate covers 82 percent of Nevada, 49 percent of Wyoming, and 28 percent of Montana and Washington. In Alaska, the federal government owns nearly 68 percent of the state's land, and the total federal acreage dwarfs that located within any of the lower 48 states. Outside these states, federal lands account for less than 5 percent of the land base, with the federal government owning less than 0.5 percent of Iowa and Connecticut. Given the sheer size of the federal presence in the West, the public lands and accompanying policies are dominant forces in the fabric of the region's political, economic, and social life.[10]

The western public domain is a virtual treasure trove of natural resources and stunning landscapes. Historically, public lands and resources were exploited and developed to promote western settlement, and they still play a key economic role in numerous western communities. Minerals and the public domain are almost synonymous; public lands contain substantial deposits of gold, silver, copper, nickel, and other important minerals, vast coal reserves that account for one quarter of the nation's domestic production, and sizable crude oil and natural gas deposits. The region's forest reserves contain some of the nation's few remaining old growth timber stands, vital wildlife habitat for both charismatic and not-so-charismatic species, and critical watersheds that nourish the region's precious water resources. The public rangelands that attracted early settlers to the West still sustain a locally important ranching industry, which annually grazes over 270 million federal acres. An ever-growing recreation and tourism industry draws directly upon the rugged splendor and open spaces of public lands to promote a wide array of outdoor recreational opportunities, ranging from off-road jeeping and big game hunting to downhill skiing and mountain biking. And for well over a century, the region's scenic vistas and uncluttered landscapes have been drawing visitors seeking respite, solitude, and restoration.[11]

The federal government, with its vast landholdings and sprawling bureaucracy, has established a significant presence across the West. To administer the federal estate, Congress has created four principal land management agencies and invested each with discrete management obligations. The U.S. Forest Service is responsible for the 192-million-acre national forest system, which consists mainly of forested lands that were reserved from early settlement to protect valuable timber and water resources from overexploitation. The national forests are managed under a multiple-use standard that opens them to mineral development, timber harvesting, livestock grazing, wildlife and fish, and recreational opportunities. The National Park Service is charged with managing the 83-million-acre national park system, consisting of 385 separate units that include national parks, monuments, recreation areas, seashores, and historic battlefields and buildings. The Park Service is legally obligated to protect park resources and to provide for public enjoyment while leaving the resources "unimpaired for the enjoyment of future generations." The U.S. Fish and Wildlife Service (FWS) is responsible for the 509 national wildlife refuges that cover over 94 million acres of federal land; it manages the refuges to conserve fish, wildlife, plants, and their habitats, while also providing compatible outdoor recreation opportunities. The Bureau of Land Management (BLM) administers the remaining unreserved public lands, which comprise roughly 264 million acres in the 11 western states; it, too, manages these lands under a multiple-use mandate that contemplates an array of resource development and protection activities. Given their diverse mandates and quite different origins, the agencies have developed distinctively different personalities, cultures, and traditions despite their shared obligation to oversee the public domain.[12]

With the emergence of new constituencies and a widely embraced environmental ethic, the agencies have been forced to reevaluate their traditional missions and priorities. For most of their histories, despite the rhetoric of multiple-use and dual mandates, the agencies actually pursued single resource management agendas with a vengeance, often to the exclusion of environmental protection and other responsibilities: in the best utilitarian tradition, the Forest Service focused on timber production; the BLM on livestock and mining; the FWS on hunter harvest rates; and the Park Service on visitor days. Of course, laws like the General Mining Law of 1872, the Mineral Leasing Act of 1920, and the Taylor Grazing Act of 1934 addressed only single resources, which thus promoted myopic agency management priorities. Recent statutory reforms, including the comprehensive and integrated planning obligations in the National Forest Management Act, the Federal Land Policy and Management Act, and the National Wildlife Refuge System Improvement Act of 1997, have sought to unshackle the land management agencies from these historical patterns. And other comprehensive statutes—primarily the Na-

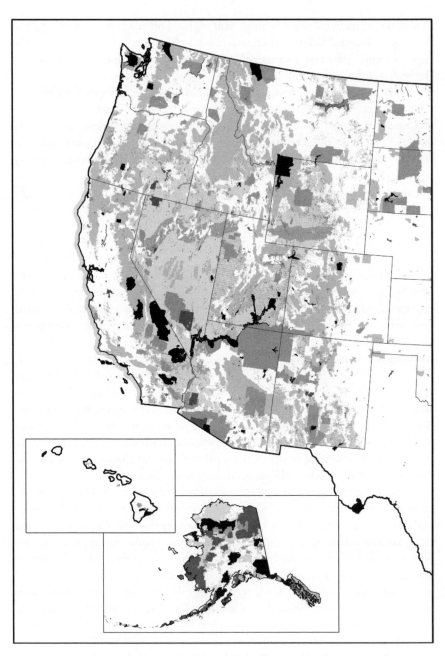

Map 1. United States Public Lands. Although federally owned lands are scattered across
the United States, the federal public lands (shaded areas) are heavily concentrated
in the 11 western states.

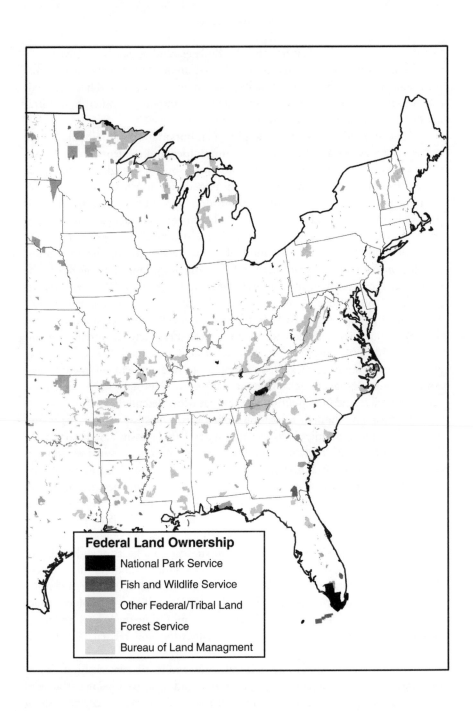

Federal Land Ownership

- National Park Service
- Fish and Wildlife Service
- Other Federal/Tribal Land
- Forest Service
- Bureau of Land Managment

tional Environmental Policy Act and the Endangered Species Act—have established important environmental protection requirements that now apply across the public domain. Thus, despite their quite different missions and histories, federal land management agencies must abide by a common yet sometimes contradictory set of guiding legal mandates.[13]

Federal public lands have long played an important economic role for the western states as well as local counties, communities, and residents. In the beginning, the lure of virtually free public land and resources brought settlers to the western frontier. While some federal lands passed into private ownership, the remaining public lands served as an economic development bank and spawned important local industries and expectations. That history is evident across the western landscape: remnant mining towns remind us that mining once dominated the economies of Montana, Nevada, and Colorado; ranches dotting the open countryside from Montana to Arizona remind us of ranching's long-standing role in the regional economy; and the ubiquitous sawmills scattered across Washington, Oregon, Idaho, and western Montana speak to the role of timber in the region's economic life. But these extractive industries are not merely historical anachronisms; they continue to represent key sectors in most western state economies. The investment and employment opportunities associated with public land mining, logging, and ranching still play a critically important role in the life of many a western community. State and local governments derive important tax receipts from these industries. And local political officials staunchly defend them.[14]

Increasingly, though, recreation and tourism are assuming a larger and larger role in the economies of the western states. Whole towns have been remade and transformed into tourism meccas; others that have long relied on tourism continue to grow at a breakneck pace. (Take Moab, Utah, and Jackson, Wyoming, as just two of many such examples.) Myriad service industries, ranging from mountain bike shops to up-scale restaurants, have been spawned, creating new local business and employment opportunities. Sales taxes and related revenues are generated by the tourists who visit nearby national parks and public lands. Some visitors have even opted to stay, citing the proximity of public lands and quality of life concerns as major reasons for their decision. And there is another rub: much of this new amenity-based economy depends on the unsullied beauty and ready availability of the nearby public lands, which puts it at odds with the older extractive industries. It has bred local political conflict, embroiling the federal land management agencies in contentious resource allocation and related policy decisions.[15]

Despite these changes, the political landscape is still remarkably familiar. Because public lands are central to the economic and social well-being of the western states, a palpable federal–state tension has persisted ever since the time of statehood. The federal government's pervasive presence—an absentee landlord, according to some—has triggered resentment and recurrent states' rights sloganeering, manifested in such movements as the Sagebrush Rebellion and the Wise

Use campaign. The fact of federal ownership removes public lands from local property tax ledgers, thus depriving the states of revenues that otherwise would be available were the lands privately owned. Local officials regularly bemoan the expenses they incur maintaining roads on public lands, providing search and rescue services, and the like. The checkerboarded land ownership pattern that prevails across much of the West, particularly the interspersed state school trust land sections, is a further setup for conflict. In many locations, the management goals of adjacent federal, state, and private landowners simply do not coincide and can work at cross-purposes to one another. It is not surprising, therefore, that federal land managers have regularly found themselves in conflict with state or local officials over appropriate land use policy, given the significant consequences that can attach to even routine resource management decisions.[16]

At bottom, most federal–state public land conflicts are grounded in disputes concerning who has authority over specific land use or resource development decisions. Under the U.S. Constitution, Congress has the "power to dispose of and make all needful rules and regulations respecting the territory or other property belonging to the United States." This provision, according to the U.S. Supreme Court, gives the federal government plenary authority over public lands and the resources located on them; it can act in both its sovereign and proprietary capacities, which means that federal power is essentially "without limitations." But Congress generally has not vested the federal land management agencies with exclusive authority over public lands. With the exception of the national parks, most of the West's public lands are subject to the concurrent jurisdictional authority of both the federal and state governments. Absent an overt conflict between federal and state law, the states can exercise regulatory authority on public lands. In fact, the states historically have assumed primary management responsibility for water, oil and gas, and wildlife resources, while federal law has focused on mineral, timber, and range resources. The states also have concurrent taxing power, which they have used to secure tax revenues from mining and other activities occurring on federal public land.[17]

These historic jurisdictional patterns, however, are breaking down and intensifying federal–state tensions on public lands in the process. As Congress has adopted new and more prescriptive environmental legislation, the federal regulatory presence has gradually expanded into areas that have traditionally been the states' domain and decidedly altered venerable public land policies and practices. The Endangered Species Act, for example, has given federal officials responsibility for wildlife species that historically were subject to state oversight, and it has created new federal regulatory limitations on timber harvesting and other economic activities long the staple fare of many rural western communities. Its prohibition on taking (or killing) a listed species by habitat modification has even been extended to privately owned lands, generating heated disputes over private property rights. Similarly, the Clean Water Act's wetlands protection and permitting provisions have displaced traditional state prerogatives and further expanded the federal regulatory presence. Although such controversies are not new on the public domain, they have plainly escalated in frequency and intensity. At the worst, federal land management offices have been bombed, federal agency employees have been threatened, timber sales have been blockaded, and trees have been spiked and logging equipment sabotaged. More often, the opponents in these controversies have turned to Congress and the courts for relief, and increasingly they are seeking to address their disagreements through new collaborative decision processes in an effort to reduce the harsh rhetoric and to avoid debilitating confrontations.[18]

One point is clear, though: public lands are both a natural and a political landscape. Some commentators have gone so far as to proclaim that the public lands are really political lands. Inasmuch as the public lands are owned by all the nation's citizens, basic policies concerning their development and protection are forged in the crucible of national and local politics. That has been true historically, and it is true today. At its core, federal public land and natural resource policy simply cannot be separated from political considerations. Without popular support, even the most enlightened theory of natural resource management will fail or wither away. This observation, of course, speaks directly to the ecological policy regime that is beginning to take hold on public lands despite sometimes stubborn local resistance. To endure, any new public land policy must be linked to a corollary political theory and broad political consensus. Whether the new ecological policies can meet this test remains to be seen.[19]

The full implications of the age of ecology for the public lands are becoming evident. Controversies like those involving the spotted owl, wolves, and fire have profoundly shaken the most basic assumptions and institutional arrangements that have dominated federal natural resource policy for over a century. Indeed, these controversies embody the fundamental reorientation now unfolding in natural resource policy on the western public domain. An array of powerful interests, ranging from loggers, miners, and ranchers to federal bureaucrats, ecological scientists, and environmental activists, are vying for supremacy in the formulation and implementation of government policy. In this clash of traditional foes, the usual agenda for doing business on the public domain is undeniably giving way to creatures and natural phenomena that have no apparent economic value and that have long been despised as the embodiment of evil incarnate. Similar stories are playing out across the western landscape: salmon are forcing an overhaul in management of the mighty Columbia River system; grizzly bears are expanding their range throughout the northern Rockies; floods are coming to be viewed as a positive force for natural regeneration; and the twin doctrines of ecosystem management and restoration are evident throughout all layers of the federal resource bureaucracy. With variations, the same story keeps emerging whether the lands are preserved as national parks, managed for multiple-use purposes, or even held in private ownership.

What accounts for this apparent sea change in natural resource management philosophy? Simply put, powerful scientific, political, legal, economic, and social forces are inexorably reshaping the public domain and the agencies that manage it. The irrefutable logic of ecological science has taken hold in the face of mounting evidence that human development has profoundly altered natural systems and imperiled critical biological resources. As environmental awareness has grown, Congress has responded by passing laws extending protection to the environment and to species facing extinction, and it has opened the federal courts to environmental litigants. As part of the nation's renewed commitment to democratic egalitarianism, administrative processes have been made accessible to citizens and advocacy organizations, and the federal courts have proven surprisingly receptive to complaints of environmental harm. The western states also have undergone major social and economic transformations; urbanization and the public's intense interest in outdoor recreation have put a premium on the environmental and scenic qualities of public lands. And the American public, confronted with urban congestion and mounting evidence of environmental degradation, no longer views the public domain merely as a development bank but rather attaches important biological, spiritual, and aesthetic values to it.

To be sure, it would be premature and presumptuous to conclude that each of these changes is now firmly rooted in natural resource policy. Powerful counterpressures are afoot. The fates of the spotted owl and the wolf are not certain. Fire policy revisions have proceeded at a slow pace. Opponents of protection for endangered species continue pressing for reform, arguing that the law is too harsh and ignores economic concerns. Western politicians and their constituents have repeatedly resorted to budget riders to block administrative reforms to mining, timber, and endangered species policies. Regulatory limitations are being challenged as violations of constitutionally protected property rights under the revitalized takings doctrine. At the same time, the scientific foundations for new ecosystem management proposals are being attacked, and the resource economists are extolling the virtues of market-based incentives in lieu of regulatory limitations for allocating access to public lands and resources. Indeed, these contrary forces—most often representing the extractive industries and resource-dependent communities—have successfully blocked major legislative and administrative reform efforts, and they seem intent on rewriting the principal environmental laws governing management of the public domain.[20]

Despite the acrimony that has accompanied the spotted owl, wolf, and fire controversies, change is coming to the West, and it is generating an unexpected degree of collaboration and commitment to resolving thorny local issues. Congress has enshrined key natural resource management compromises in legislation such as the Valles Caldera Preservation Act, Quincy Library Group bill, Grand Canyon Protection Act, and Columbia River Gorge National Scenic Area bill. Major regional administrative initiatives—including the Northwest Forest Plan, Greater Yellowstone Coordinating Committee, Sierra Nevada Ecosystem Project, and Interior Columbia Basin Ecosystem Management Project—have begun ex-

ploring how ecology-based management approaches can be effectively institution-
alized on a landscape scale. At the local level, myriad collaborative processes have
taken hold, ranging from the Malpai Borderlands Group in the desert Southwest
to the Quincy Library Group in California's northern Sierra timber country. The
overriding message from these diverse efforts to address long-standing conflicts is
that often-ignored environmental values must be accounted for, that degraded
ecological conditions must be restored, and that equitable solutions sensitive to all
affected parties must be devised.[21]

A new era is dawning on the western public domain. We appear poised on
the brink of the age of ecology in public land policy. How this state of affairs has
evolved, what it means, and where it may lead is a fascinating study in the power
of ideas, science, law, and institutions. It offers important insights into how natural resource policy is formulated and implemented in our democratic system. It
also provides an opportunity to reexamine the policies, laws, and institutions that
govern western public lands and resources. But to realize fully what the future may
hold, we must first link our important new ecological insights with a corresponding political consensus and theory. Such has been the path of reform historically,
and the same holds true today. Once that linkage is established, we may begin to
understand the full ramifications of the emerging age of ecology in federal public
land policy.

TWO

Policy and Power on the Public Domain

The man of common sense sees a wonderful future for this land.
Hard is the heart, dull is the mind, and weak is the will of the man
who does not strive to secure wise institutions for the developing
world of America.

JOHN WESLEY POWELL (1890)

Conservation cannot be considered simply as a public policy, but, far
more significantly, as an integral part of the evolution of the political
structure of the modern United States.

SAMUEL P. HAYS (1969)

ontroversy and change are nothing new on the public lands. From the beginning, the public domain has been a contested landscape. Early controversies pitted settlers against the region's native inhabitants and rival claimants against one another for nature's bounty. More recent controversies have pitted local citizens against federal land managers, and entrenched economic interests against newly ascendant environmental advocates. The partisans in these conflicts are well aware that public land policy involves inherently political questions. They have all repeatedly enlisted our governmental institutions to advance their respective agendas, invoking a spectrum of ideas, laws, scientific studies, and policies to support their positions. Over time, once revered ideas have been displaced by newer ones more responsive to contemporary concerns; once favored laws and policies have given way to newer ones with greater currency and legitimacy. Rarely, if ever, have these changes occurred without acrimony and controversy.

Nothing is static about this process. As controversy has begotten change on the western landscape, federal natural resource policy has evolved through distinctive phases marked by major revisions in the governing laws and policies to accommodate the felt necessities of the day. Throughout this process, a few key ideas have dominated the public land policy debates, and these same ideas are still central in the current debate over new ecological management principles. In each instance, our governmental institutions—Congress, the executive branch, the courts, and the states—have played key roles in converting these ideas into viable laws and policies. The principal federal land management agencies—the Forest Service, Bureau of Land Management, National Park Service, and U.S. Fish and Wildlife Service—have also been central to this process, periodically translating new ideas and laws into viable natural resource policies. But these institutions rarely act on their own accord; they are each responsive to outside forces and pressures, mainly from the myriad constituencies seeking to shape and influence public land policy. Only by understanding these institutions and the ideas that have shaped them can we begin to address the present forces that are influencing the ecological policy debates.

Understanding the Past: Ideas that Matter

Public land and natural resource policy has not wanted for ideas about how best to manage the federal estate. These ideas are deeply entwined with the history of the nation's westward expansion and settlement. This history can be roughly divided into three eras, each reflecting major shifts in federal policy based on changing societal values, preferences, and needs. Briefly stated, these eras are: 1862–91, the settlement period when federal disposal policies dominated public land policy; 1891–1964, the retention and management period when utilitarian conservation policies held sway across the public domain; and 1964–present, the nature preser-

vation period, characterized by a growing federal commitment to protecting public lands and resources from development pressures. Ever sensitive to powerful
federalism concerns, we have also experimented with various constituency-based
governance models on the public domain, making devolution—or devolved
power—another cogent idea in the natural resource policy arena. These same basic ideas, closely linked to the past yet also adapted over time to new circumstances, still dominate natural resource policy discussions, and they are influencing the current ecological policy debates.[1]

DISPOSAL: OWNERSHIP AND INCENTIVES

During the nineteenth century, following the initial acquisition and exploration of the western frontier, the United States committed to a federal disposal
policy designed to promote settlement and development in the new territories
by transferring public land and resources into private ownership. The accepted
wisdom was that the West and its resources—land, minerals, timber, grass, and
wildlife—represented a vast and seemingly inexhaustible source of potential
wealth. (Never mind the dearth of water or the region's Native American inhabitants.) Driven by the prevailing national commitment to laissez-faire capitalism,
federal policy viewed private ownership and initiative as an essential element of social progress. Congress, using laws like the Homestead Act of 1862, the Desert
Lands Act of 1877, and the General Mining Law of 1872, sought to attract prospec-
tive settlers and entrepreneurs to the western frontier with the enticement of virtually free land and minerals. The new states followed suit, employing a similar
disposal technique to administer the region's scarce water resources. Even without
acquiring actual title, the early settlers enjoyed largely unrestricted access to the region's abundant public timber, grass, and wildlife, usually without any charge or
oversight. To be sure, the legendary explorer John Wesley Powell and a few other
astute observers cautioned early on that the region's unusually dry climate and arid
landscape were ill-suited to traditional agricultural practices. But the boosters ignored these warnings and wholeheartedly embraced the disposal policies that had
worked so well farther eastward.[2]

These laissez-faire policies relied upon private initiative, fostered by the
promise of ownership and its accompanying security. The basic idea was that private ownership opportunities would attract new settlers and encourage them to
develop the region into a civilized society. Ownership, of course, presupposed
property rights, which were allocated and acknowledged on a priority basis, using
the "first in time, first in right" principle. The homesteader who first arrived and
developed his land received a federal land grant; the miner who first located a valuable claim was entitled to a mineral patent; and the individual who first diverted
water from a stream for beneficial use received an appropriative water right giving him priority over subsequent diverters. Under these laws, which have been
tellingly referred to as the "Lords of Yesterday," major chunks of the federal estate

passed into private ownership; public minerals were transformed into personal fortunes for $5 or less per acre; and early irrigators secured a vested legal right to the region's precious water resources. In short, the laissez-faire disposal policy created an elaborate set of property rights and expectations confirmed by laws that continue yet today to govern access to mineral and water resources.[3]

As the disposal era unfolded, the federal government's largesse knew few bounds on the western frontier. To encourage settlement, federal military forces were dispatched to subdue the hostile Indian tribes and to move them onto reservations where they would not interfere with the new settlers. To speed the settlement process, Congress made generous land grants to the railroads, which created an all-important transportation infrastructure as well as a quiltlike land ownership pattern along the early rail lines. To encourage new statehood applications, Congress returned several sections of public land in each township to every newly minted western state, further fragmenting the land ownership pattern across the region. By most measures, despite some colorful episodes of fraud and chicanery, these disposal policies worked quite well. By 1890, in less than two generations, the western frontier was essentially closed. Most of the territories had secured statehood, the Indians were relegated to marginal reservations, and the nation was entering the Industrial Age. The disposal era was drawing to an end on the public domain.[4]

These federal disposal policies, we now know, had a negative side too. The laws conveying free land and minerals may have been designed to benefit individual settlers (Thomas Jefferson's yeoman farmer), but they were also heavily exploited by wealthy corporations, which acquired valuable public land and resources for a pittance. These same corporations often succeeded in leveraging their holdings into raw and unyielding local political power. Witness Anaconda Copper's nearly 70-year stranglehold on Montana politics and media. The idea of inexhaustible abundance also proved false. By the turn of the century, when the Great Barbeque had mostly run its course, the region's accessible timber resources were badly depleted and its wildlife populations were decimated. The patchwork statehood and railroad land grants fragmented ownership of the public domain, creating serious management conflicts, access problems, and habitat blockages that persist yet today. In addition, the disposal era policies helped create enduring and unrealistic expectations that have proved a major obstacle to later reform efforts. Miners, ranchers, hunters, and hikers alike have each come to expect free (or nearly free) and unregulated access to the public lands. All these constituencies have consistently resisted increased federal regulatory efforts, fought hard to retain existing subsidies, and opposed various fee proposals.[5]

Federal disposal policy was synonymous with privatization and economic efficiency—terms that surface regularly in any natural resource policy discussion. Private ownership, according to its proponents, is much more economically efficient than public ownership: whether the goal is to produce more timber or to enhance recreational opportunities, private ownership and initiative can do it better and more cheaply. Privatization proponents argue that adherence to basic market principles ensures that the public receives what it truly wants, as reflected by what it is willing to pay for. Pointing to the unwieldy nature of comprehensive planning processes and unlimited administrative appeal opportunities, they assert that private ownership would eliminate these inefficiencies. They also note that private ownership would render the now tax-exempt public lands subject to state and local property taxes, and significantly reduce the federal regulatory presence across the western landscape. Moreover, private ownership would eliminate the hidden governmental subsidies that public land users enjoy in the form of timber road construction credits, discounted grazing fees, and free recreational access.[6]

Privatization opponents discount these economic arguments and focus instead on the benefits of public ownership. With the West's public lands in federal ownership, they note, all citizens enjoy equal access to the federal estate and can participate in determining how it is managed. Rather than seeing the public domain as a vast resource bank, privatization opponents recite its spiritual, aesthetic, and environmental values to justify continued public ownership. They fear that private ownership will mean limited access available only to a few privileged persons who can afford to purchase the land outright or to pay hefty entrance fees. Environmental advocates note that privatization proponents have yet to explain how the marketplace will account for noncommodity values, such as endangered species, open space, and other critical biological resources. They also fear that privatization will mean further fragmentation of the landscapes, creating additional wildlife migration obstacles and public access problems. Strikingly, many westerners, long accustomed to freely accessing the public lands for various personal purposes, fiercely oppose most privatization proposals.[7]

UTILITARIANISM: SCIENTIFIC MANAGEMENT

The excesses accompanying nineteenth-century laissez-faire capitalism and related public land disposal policies gave weight to the idea that the public interest might be better served by retaining natural resources in government ownership and employing scientific management techniques. By the end of the century, the West's land and resources were rapidly dissipating, relentlessly exploited by successive waves of explorers, speculators, and settlers. Boom and bust mining cycles had already left visible scars on the landscape: abandoned mining sites, raw tailings piles, and decaying ghost towns. Much of the region's readily accessible timber was cutover, first by the railroads and then by the ensuing settlers. Forests and watersheds were badly deteriorated; some sites were eerily reminiscent of the massive devastation visited upon the upper Midwest's prime forest lands by a prior generation of loggers. Once abundant wildlife populations had shrunk to a fraction of their earlier size; some like the plains bison were verging on extinction. Even westerners themselves had begun to decry the damage. It was time for a change.[8]

As the new century dawned, the Progressive political movement burst onto the national scene, setting the stage for a major shift in federal public land policy.

The Progressives, alarmed over corporate abuses and armed with a new faith in government, sought to use the government as guardian of the public welfare. Congress responded by crafting an entirely new federal natural resource policy designed to retain and actively manage parts of the public domain, thus launching a new conservation era. In 1891, Congress gave the president authority to designate forest reserves, which Presidents Harrison and Cleveland promptly used to place over 17 million acres in the nation's first forest reserves. Six years later, Congress passed the Organic Administration Act of 1897, which directed that the new forest reserves were to be managed for productive purposes. In 1909, President Taft utilized his inherent executive power to withdraw over 3 million acres of public land in California and Wyoming from operation of the mining laws to prevent the nation's critical oil reserves from falling into private hands. The Mineral Leasing Act of 1920 endorsed this decision, removing all energy minerals from the mining laws and instituting a federal leasing and royalty system to ensure continued government control over these strategically valuable resources. In 1934, faced with the misery of the Dust Bowl crisis and a severely depleted western range, Congress adopted the Taylor Grazing Act and brought the public rangelands under federal control, effectively withdrawing them from further homestead entry under the disposal laws. These laws, with the important exception of hardrock mining, completely reversed the earlier federal disposal policy, replacing it with a new commitment to government ownership and management.[9]

The indomitable Gifford Pinchot, first chief of the U.S. Forest Service, is generally regarded as the father of utilitarian conservation policy. Drawn from widely accepted Progressive Era tenets, the essential components of this new conservation policy were public ownership of natural resources, development and use without waste to promote the common good, reliance upon scientific management techniques, and the avoidance of monopoly power. Pinchot and his allies argued that public officials, using their technical training and expertise, should be given responsibility for managing public resources in an efficient and sustainable manner. In his seminal 1905 letter of instructions, still regarded as the Forest Service's basic charter, Pinchot explained:

> In the administration of the forest reserves . . . all land is to be devoted to its most productive use for the permanent good of the whole people, and not for the temporary benefit of individuals or companies. All the resources of the reserves are for *use* . . . under such restrictions only as will insure the permanence of the resources. . . . In the management of each reserve local questions will be decided upon local grounds . . . and where conflicting interests must be reconciled the question will always be decided from the standpoint of the greatest good of the greatest number in the long run. . . . The administration of each reserve is left very largely in the hands of local officers, under the eye of thoroughly trained and competent inspectors.

Heavily influenced by his earlier experiences in Germany, where the state intensively managed its forests as a harvestable crop, Pinchot believed that the requisite technical skills to manage nature could be learned and then employed for human benefit. He was convinced that neither the private marketplace nor the political process could be trusted to meet the public's needs while sustaining a viable natural resource base. The goal was to remove natural resource decisions from the temptations of the market and the corrupting influence of politics; it was accomplished by vesting power in expert agencies charged with employing scientific management principles to serve the public interest. Pinchot even coined the term *conservation* to describe this new natural resource policy.[10]

In relatively short order, Pinchot's progressive vision was mostly achieved as utilitarian conservation principles were incorporated into both federal and state natural resource policy. At the federal level, Congress created a smorgasbord of new agencies to manage federal public lands and resources: the Forest Service, National Park Service, Bureau of Reclamation, and Bureau of Mines. Under Pinchot's leadership, the new Forest Service was molded into a cadre of professionally trained managers with decentralized authority over the national forests. The agency was soon regarded as the quintessential example of an efficient natural resource bureaucracy. Within a few years, forestry was recognized as a profession; forestry schools were created to train students (i.e., future Forest Service employees) in silviculture and other technical disciplines; and a professional forestry organization came into being. Employing the forestry model, other natural resource professions were established in range, wildlife, and hydrology—all based on a commitment to scientific training and management. Many states emulated these developments, creating specialized natural resource agencies to oversee their water, wildlife, and other resources. By mid-century, the utilitarian conservation model was ascendant, broadly incorporated into federal policy and extended across most of the public domain.[11]

Despite widespread acceptance of the scientific utilitarianism model, many of the assumptions underlying it proved questionable. Although science provided vitally important information about the productivity and health of natural resource systems, countervailing political considerations—budgetary pressures, oversight hearings, local employment concerns, and the like—often subverted managers' professional judgments. The utilitarian commitment to productivity and use regularly caused managers to undervalue or ignore nonconsumptive resources and to resist popular preservation efforts. The multiple-use agencies, dominated by professional specialists, proved adept at managing single resources to meet production targets, but not necessarily at addressing the full spectrum of resources and related environmental concerns. The Forest Service, for example, committed itself to intensive timber management to the exclusion of other considerations. As public confidence in government waned during the post–World War Two period and as increasingly diverse constituencies sought access to government decision processes, agency officials were forced to acknowledge that resource allocation decisions were as much a question of public values as of technical judgment. The once omnipotent scientific manager was no longer fully trusted, nor could she rely upon claims of professional expertise to sustain management decisions subject to political oversight and public scrutiny.[12]

Faced with these sobering realities, once-sacred utilitarian conservation policies have been gradually revised to accommodate newly emerging environmental values and other concerns. In 1960, Congress adopted the Multiple Use–Sustained Yield Act, officially expanding the Forest Service's responsibilities to include outdoor recreation, range, fish, and wildlife. In 1964, the Wilderness Act formally established a new wilderness preservation system. In 1976, Congress passed the National Forest Management Act (NFMA), mandating comprehensive, interdisciplinary planning for individual forests and imposing significant environmental constraints—including a biological diversity conservation obligation—on the agency for the first time. The Federal Land Policy and Management Act of 1976 (FLPMA) not only gave the Bureau of Land Management its first organic charter but also subjected the public domain lands to an integrated, multiple-use planning system, and obligated the BLM to protect its lands against undue degradation. In addition, Congress passed several major pieces of environmental protection legislation that had a direct impact on public land management practices. The National Environmental Policy Act of 1970 (or NEPA) obligated land management agencies to assess the environmental consequences of their decisions before taking any action, while also providing a vehicle for public involvement in agency decision processes and a basis for judicial review of these decisions. The Endangered Species Act of 1973 imposed species conservation obligations on all federal agencies, and gave the U.S. Fish and Wildlife Service an effective veto power over any agency decision that might jeopardize a federally protected species. The Clean Water and Clean Air Acts created elaborate regulatory structures designed to reduce pollution nationwide, with provisions that apply to pollution problems on the public lands. In short, under the specter of judicial intervention, important new environmental protection standards were injected into the utilitarian calculus on the public lands, and agency officials were made broadly accountable for their decisions.[13]

PRESERVATION: SAVING WILD NATURE

The idea of preserving nature from human exploitation has deep roots in American thought and law. Fueled by the early-nineteenth-century Romantic Movement and Henry David Thoreau's transcendentalism, nature preservation emerged as a tangible political concept in 1872, when Congress created Yellowstone National Park as "a public park or pleasuring-ground for the benefit and enjoyment of the people."[14] Because this was the height of the disposal era, the immediate goal of the Yellowstone designation was to prevent the region's unique and vulnerable resources from falling into private hands. It represented the first time ever that a nation had legally removed such a large block of undeveloped public land—nearly 2 million acres—from settlement or development, and then opened it for public enjoyment. (Ironically, that same year Congress adopted the general mining law, the quintessential disposal law.) Shortly thereafter, California's John Muir embraced the nature preservation cause and placed it on the national political agenda. According to Muir, whose prolific writings helped popularize the national park idea, wild nature offered vital spiritual, recreational, and recuperative values that were otherwise unavailable in civilized society. For Muir, along with the fledgling Sierra Club he helped establish, the goal was to make nature preservation a vital part of federal public land policy.[15]

The new nature preservation cause soon collided head-on with competing utilitarian beliefs in a major test of political strength. The collision involved the oft-told Hetch Hetchy affair, which saw the preservation-minded Muir square off against the utilitarian-oriented Pinchot over the damming of the Tuolomne River inside Yosemite National Park. In the end, Pinchot prevailed: the dam was constructed to provide water for the growing city of San Francisco, and the Hetch Hetchy Valley, which according to Muir rivaled nearby Yosemite Valley in beauty, was permanently inundated with several hundred feet of water. The conflict, however, convinced preservationists that the new national parks needed more comprehensive federal protection to ward off local developers and their political allies. It also precipitated a schism between preservationists and utilitarians that persists today in conservation circles.[16]

As the twentieth century unfolded, the nascent preservation movement rapidly gained added political legitimacy. Following the Yellowstone designation, Congress created several more national parks to protect such scenic venues as the Grand Canyon, Crater Lake, Mount Rainier, and Glacier from falling into private hands and being exploited for profit. In 1903, President Theodore Roosevelt lent his authority to the preservation cause by creating, through administrative fiat, the nation's first wildlife refuge at Pelican Island, which was soon followed by similar designations elsewhere across the country. In 1906, to safeguard the nation's disappearing archaeological treasures, Congress adopted the Antiquities Act and empowered the president to unilaterally create national monuments on the public lands to protect unique scientific and historic objects. In 1916, in the aftermath of the Hetch Hetchy controversy, Congress passed the National Parks Organic Act, establishing the National Park Service to administer the fledgling national park system and giving the new agency an explicit preservationist mission. In 1918, Congress adopted the Migratory Bird Treaty Act, vesting the federal government—rather than the states—with responsibility for protecting migratory waterfowl. Less than two decades into the twentieth century, the preservation concept was well embedded in law, having taken its place beside privatization and scientific utilitarianism as a mainstream natural resource policy.[17]

The preservation model rested upon the idea that America's unique and scenic wonders should be retained in public ownership and managed by the government as nature reserves. Preservationists, like their utilitarian counterparts, strongly opposed private ownership of public lands and resources, believing it would promote shortsighted development and careless exploitation of valuable landscapes and resources. But unlike the utilitarians, preservationists were committed to protecting America's scenic marvels from logging, mining, and other industrial activity through national park and similar designations. For the most part, the early preservationists were driven by aesthetic concerns; they conceived of national parks as natural monuments that preserved the nation's frontier heritage and as a distinctly American cultural institution. Protected in their original state, the national parks would serve as contemplative retreats and provide the public with outdoor recreational opportunities in an unspoiled setting. Ironically, though, there was a distinctly utilitarian side to early preservation policies. Visitation was strongly promoted in the national parks, and the parks were managed intensively for the visiting public's convenience: lodges, roads, and other facilities were constructed, wildlife was put on display, and predators were systematically eliminated. Although this utilitarian approach to park management has gradually eroded in the face of new ecological preservation policies, park managers still struggle to reconcile competing nature preservation and visitor service responsibilities.[18]

Over time, the preservation idea has gained an expanded political foothold that extends well beyond the national park setting. In 1964, Congress passed the Wilderness Act, which formally endorsed wilderness as a legitimate use of public lands, and established a national wilderness preservation system. The Wilderness Act instantly created over 7 million acres of national forest wilderness, thus making preservation part of the Forest Service's multiple-use obligations. The Land and Water Conservation Fund Act of 1965 established a permanent federal funding source for the land management agencies to purchase sensitive private lands for preservation and recreation purposes. In 1966, Congress passed the National Wildlife Refuge Administration Act, which consolidated the national wildlife refuges into a unitary system and established a uniform compatibility standard to govern activities on refuge lands. In 1968, Congress adopted the Wild and Scenic Rivers Act, which was designed to preserve designated rivers in a natural, free flowing state and to protect sensitive river corridors. In 1976, in the Federal Land Policy and Management Act, Congress imposed preservation responsibilities on the Bureau of Land Management, requiring the agency to inventory its lands for wilderness potential and to manage those that qualified to preserve their wilderness characteristics. Moreover, the Endangered Species Act of 1973 extended federal preservation responsibilities beyond the public lands, enlisting all federal agencies in species conservation efforts and vesting the U.S. Fish and Wildlife Service with extensive regulatory powers on the public lands and elsewhere. In short, federal preservation policy has been extended across the entire public domain and into the realm of water and wildlife protection.[19]

In recent years, wildlife and wilderness protection have been linked under the rubric of biodiversity conservation, which has become the next battleground in the politics of preservation. The federal public lands, already laced with an extensive system of preserved lands, are the logical focal point for developing a national biodiversity conservation policy. According to conservation biologists, the only effective way to safeguard and restore the nation's vanishing biological heritage is to establish a network of large, interconnected reserves consisting of protected core areas, carefully regulated adjacent lands, and strategic linkage corridors to ensure secure habitat and opportunities for genetic interchange. To meet these objectives, scientists have endorsed two related strategies: first, the adoption of ecosystem management techniques to manage across jurisdictional boundaries to coordinate biodiversity conservation and restoration efforts; and second, the significant expansion of wilderness and related land preservation efforts through such proposals as the Northern Rockies Ecosystem Protection Act, the Wildlands Project, and the Yellowstone to Yukon initiative. Some preservation proponents have even advocated extending legal rights to natural objects in order to reframe preservation policy debates in explicitly biocentric rather than anthropocentric terms. These proposals mark a dramatic evolution in preservation policy—one that is giving ecological science a major new role in defining resource priorities and management strategies on the public domain.[20]

DEVOLUTION: CONSTITUENCY-BASED GOVERNANCE

The traditional model of public land governance is a centralized one based on federal supremacy. As owner of the public lands, the federal government has generally used its authority to define and enforce the rules governing these lands. With authority consolidated in the land management agencies, uniform standards are ordinarily applied across the landscape, thus promoting equity among user groups and minimizing interstate discrimination problems. But central authority has also exacerbated federal, state, and local tensions over public land policy. While such conflicts are not surprising (and should even be expected in our federal system), they can create unnecessary management inefficiencies, frustrate legitimate local interests, and promote jurisdictional fragmentation. Indeed, the greatest weakness of the centralized federal governance model may be its inability to accommodate both the ecological and social diversity prevalent across the public lands, whether rooted in different ecosystem structures, economic expectations, or social traditions. In the age of ecology, the question is whether an alternative model of devolved governance might prove more compatible with new ecological management policies.[21]

Devolution can be understood in two separate ways in our federal system. In its strongest sense, devolution means the transfer of governmental power from the federal level to the state or local level, either the full or partial surrender of jurisdictional authority. Alternatively, in a weaker sense, devolution means the sharing of decision making responsibility, either among different levels of government or with various constituencies. These arrangements can range from creating an integrated federal-state-local decision body to simply providing opportunities for public participation in the federal decision process. Public land policy, as we shall see, has experience with both of these devolution approaches.

From the beginning, local citizens and communities have played a major role in shaping public land policy. In the years before the national government solidified its control over the western public lands, state law and local customs governed mining, water, grazing, and other activities on the public domain. Even as the federal presence grew, Congress expressly incorporated state and local legal norms into such laws as the General Mining Law of 1872 and the Desert Lands Act of 1877. When Congress authorized creation of national forest reserves at the turn of the century, it did not displace state authority, but rather left federal forest officials and the states sharing concurrent jurisdiction over the new forest reserves. During the Forest Service's early days, Gifford Pinchot specifically instructed his new managers to heed local interests and needs in managing the new reserves. These early managers, many of whom came from nearby communities, usually developed strong ties with local citizens and shaped their management policies to meet local concerns. According to one prominent historian, western interests generally succeeded in resisting the centralizing tendencies of the Progressive Era conservation programs, retaining substantial local control over natural resource decisions.[22]

During the mid-1930s, Congress gave local public land constituents formal responsibility for the public rangelands. In the Taylor Grazing Act of 1934, Congress created local grazing advisory boards, composed of public land ranchers, to administer the federally owned rangelands. Dubbed "democracy on the range," this experiment in devolved governance can appropriately be viewed as a distinctive public land management model. The theory underlying the grazing advisory boards was that local livestock growers, being more familiar with range resources and conditions than anyone else, could be entrusted to protect against overuse because their own livelihoods were at stake. Because the Taylor Grazing Act also granted preferential lease rights to existing ranchers and limited participation on the advisory boards to public land livestock producers, this early experiment in public land democracy was actually more exclusive than inclusive. It was also clearly motivated by the ranchers' desire to avoid any federally imposed regulatory limits on the public range. The grazing advisory boards are now widely regarded as a classic example of local "capture" of the public land agencies.[23]

More recently, devolution has taken the form of constituency-based involvement in federal decision processes, which has become an important dimension of modern public land management. Several factors account for the significantly expanded role that the public now plays in natural resource policy. In the aftermath of World War Two, as public confidence in government waned under the weight of a burgeoning federal bureaucracy, important administrative law reforms opened federal agency decision processes to public participation and judicial scrutiny. The Administrative Procedures Act of 1946 served as the principal vehicle for injecting this new transparency into federal administrative policy. In the environmental arena, passage of NEPA in 1970, with its judicially enforceable procedural standards, required officials to solicit public comment on agency actions and assured that these concerns would be taken seriously. NEPA not only introduced environmental considerations into public land decisions but also provided interested constituents and others not previously involved in these matters a vehicle to participate. The NEPA procedures also revealed that most controversial resource allocation decisions tend to involve conflicts over values and interests as much as technical disagreements over scientific data. Subsequent comprehensive planning legislation, namely the NFMA and the FLPMA, has given added legitimacy to public involvement in natural resource decision processes by requiring federal land managers to comply with NEPA and to coordinate with local governments, tribes, and other federal agencies. In fact, the FLPMA obligates BLM managers to ensure consistency between federal resource plans and local land use plans.[24]

Not surprisingly, the public participation model of decision making has not simplified public land management. Whereas 30 years ago, public land managers dealt primarily with a relatively homogeneous constituency of ranchers, loggers, and miners, that is no longer the case. The West is now the nation's most urban region, and a couple of generations have now been educated in the importance of environmental protection. As a result, an increasingly diverse constituency has made its various voices heard over an array of issues, ranging from livestock grazing in riparian corridors to timber sale proposals. The traditional public land constituencies—finding their voice and influence challenged by members of a growing environmental movement, urban recreationists, and others—responded by seeking to reassert local control over the public lands. During the late 1970s, a disparate group of ranchers and western politicians promoted the short-lived Sagebrush Rebellion, entreating the federal government to relinquish ownership of the public lands to the states—the ultimate devolution policy. More recently, these same forces rejoined ranks under the banner of the Wise Use movement, which unsuccessfully advanced an extreme local governance model, claiming that western counties are entitled to dictate resource priorities on adjacent federal lands. Confronted with these competing claims and visions, public land managers found themselves struggling to determine whether affected local interests should enjoy greater weight than more distant voices in establishing priorities and resolving access questions.[25]

As these conflicts intensified, the constituent-based governance model—in the form of collaborative processes and partnerships—has emerged as a means to reduce acrimony and to promote understanding among competing interests. Recognizing that most natural resource controversies transcend traditional jurisdictional boundaries and can have far-reaching ecological implications, the public land agencies began informally convening the diverse affected interests in an effort to address common concerns. Numerous ad hoc citizen-initiated groups also sprang up as forums for discussing and resolving persistent conflicts outside traditional agency and judicial venues. Noteworthy examples include the Quincy Library Group in the northern Sierras, the Applegate Partnership in southwestern Oregon, and the Malpai Borderlands Group in southern Arizona. At a more formal level, the Department of the Interior established state resource advisory councils to replace the old grazing advisory boards in setting local range policy. Significantly, the regulations governing these resource advisory councils provide for an expanded and balanced membership that includes environmental and wildlife advocates as well as local ranchers. While none of these experiments in constituency-based democracy can yet be labeled an unqualified success, their proliferation lends credibility and relevance to this devolved management model.[26]

The policy of devolution—whether framed in local governance or public participation terms—tracks broader contemporary trends in political theory. As we shall see, federalism principles are resurgent at the national level, following a spate of Supreme Court decisions rekindling long dormant state sovereignty notions. The political theory of civic republicanism also has enjoyed a revival, sparking renewed interest in civic dialogue forums to address thorny public policy issues. In addition, we now have experience with diverse models of cooperative federalism—ranging from dual regulation regimes to collaborative management arrangements—that might be integrated into public land policy. None of these models offers an obvious panacea to the federal–state–local tensions that have long pervaded public land policy, but each has strengths that might be incorporated into a new ecological management regime. Indeed, the devolved governance model may provide the critical underlying political theory necessary to give ecological policy on the public lands traction in the all-important political arena.[27]

The Locus of Power: An Institutional View

Law and policy are framed in the complex milieu of our existing political institutions, and thus reflects the prevailing beliefs of the day. At the federal level, Congress, the president, and the courts are the three principal institutions responsible for public land and natural resource policy. None of them operates in a vacuum; a complex system of constitutional checks and balances gives each branch of our government a prominent yet interdependent role in formulating and implementing policy. Moreover, our federal structure ensures a dynamic interplay between these federal institutions and the states, which enjoy substantial sovereign prerogatives as part of their reserved police powers. All these governmental institutions and tensions are at work in the current movement toward ecosystem management.

CONGRESS AND THE LAW

Congress is the legislative branch of the federal government, constitutionally vested with responsibility to make law. Consisting of both the Senate and

House of Representatives, Congress is politically accountable to the populace through regular elections. Under the Constitution, Congress has been granted specific enumerated powers designed to enable it to address national problems, while the states retain a residual police power over local matters. Over the years, however, Congress has gradually expanded the sweep of federal power, employing its interstate commerce, taxing, spending, and other powers to extend the federal regulatory presence into economic and social realms previously regarded as the domain of the states. Although critics have regularly objected that Congress has ridden roughshod over state sovereign prerogatives, supporters have convincingly argued that the modern, interconnected economy and such pressing problems as discrimination, pollution, and the like can be addressed only through a uniform federal response. Since the mid-1930s, the Supreme Court has sustained congressional assertions of federal power against countervailing state sovereignty arguments, noting that the inherent local sensitivities of state congressional delegations generally serve as a political check against federal overreaching. By the 1990s, however, the Court had begun revitalizing long-dormant federalism principles in a determined effort to begin limiting the sweep of federal power.[28]

The Constitution vests Congress with broad authority over public lands and resources. The property clause is the principal source of this power: it authorizes Congress to create "needful rules and regulations respecting the . . . property belonging to the United States." In a seminal 1897 decision, the Supreme Court interpreted Congress's property clause powers quite broadly, ruling that the federal government's "power over its own property [is] analogous to the police power of the several states, and . . . [is] measured by the exigencies of the particular case." Since then, the Court has consistently reaffirmed that view, holding that the property clause power places the federal government in the role of both proprietor and sovereign on the public lands, and that "the power over the public land thus entrusted to Congress is without limitation."[29] Invoking these principles, the Court ruled, in *Kleppe v. New Mexico,* that Congress could regulate feral horses and burros on the public lands, despite New Mexico's assertion that they were subject only to state regulation. The courts have also ruled that the property clause power is broad enough so federal legislative authority can be extended beyond the boundaries of the public lands to regulate matters that affect public land or resources. In *Minnesota v. Block,* for example, the court ruled that Congress could regulate motorized activities on state-owned lands and waters within the federally designated wilderness area to protect critical natural values. Moreover, the oft-ignored enclave clause gives Congress exclusive power over public lands that were purchased by the United States with the consent of the affected state, which includes many of the national parks that are therefore subject only to federal authority.[30]

Several other constitutional provisions also give Congress power over public lands and resources. Under the commerce clause, Congress has broad power to regulate private activity that affects interstate commerce, which effectively extends federal power to any private commercial activity that has an impact on public land or resources. The Endangered Species Act provisions regulating the taking of any protected species, even on private lands, are an example of the breadth of the federal commerce power. Using its treaty powers, Congress can address and regulate any matter that is subject to treaty obligations. In the case of the Migratory Bird Treaty Act, Congress has regulated the hunting and management of migratory waterfowl, even though the states traditionally were responsible for wildlife within their borders. Furthermore, Congress regularly uses its taxing and spending powers to establish or influence public land policy. It does this directly by appropriating funds to support particular programs or policies; it does it indirectly by inserting budgetary riders into general spending bills to address policy issues, as it did on several occasions during the spotted owl controversy. With these manifold powers, Congress has the constitutional authority to establish ecological management policies on the public lands and to otherwise employ federal regulatory power to promote ecological objectives.[31]

Political considerations, however, invariably temper Congress's legislative activities and its use of federal power. In fact, the congressional political process effectively enables regional interests to exert considerable influence over any legislation with predominately local consequences. In the case of the public lands, western senators and representatives traditionally have held a near "veto" power over any reform proposals. Westerners generally dominate the congressional committees primarily responsible for public lands legislation, namely the Senate Committee on Energy and Natural Resources and the House Resources Committee (formerly Interior and Insular Affairs). They usually hold the powerful chair position on these committees, which carries the ability to control the flow of legislation. In the Senate, hoary traditions and privileges commonly enable one or a few determined senators to block unwelcome legislative initiatives, particularly if the matter is of uniquely local consequence. From this power base, western delegations have regularly used oversight hearings, budget negotiations, appropriations riders, and other institutional prerogatives to shape and influence public land policies. And based on election results since the 1980s, the western delegations, especially from the interior West, are increasingly dominated by conservative Republicans aligned with the traditional extractive industries and strongly predisposed against additional wilderness designations or more federal environmental regulation.[32]

Given these political realities, it is perhaps no surprise that Congress has not been overly receptive to new ecological policy proposals involving the public lands. General ecosystem management legislation has been introduced in Congress several times, but Congress has not broadly endorsed ecosystem management on the public lands, though it has passed place-based bills—the Quincy Library Group bill and the Grand Canyon Protection Act, for example—with obvious ecological policy overtones. In the past, of course, preservation-oriented legislation involving public lands has often divided Congress along an East–West fault line. The eastern congressional delegations have usually supported expansive national park and wilderness designation proposals, while their western colleagues have objected to the "lock-up" of productive lands. That regional fault line still exists today, though its edges are gradually becoming blurred, in part owing to population growth and urbanization in the West, particularly among the coastal states. But beyond regionalism, Congress has shown a renewed interest in limiting the sweep of federal regulatory power, which could undermine any general ecosystem-based legislation that threatens traditional state or private prerogatives. Yet even without an explicit congressional endorsement, other governmental institutions have embraced ecosystem management principles and worked assiduously to legitimize them as a viable natural resources policy.[33]

THE PRESIDENT AND EXECUTIVE AUTHORITY

The executive branch of the federal government is constitutionally empowered to implement and enforce the law. The Constitution specifically vests executive power in the president and obligates him to "take care that the laws be faithfully executed."[34] As the nation's chief executive officer, the president sets the overall policy direction for his administration within the prevailing legal framework. In this era of the bureaucratic state, the president relies upon a plethora of cabinet offices and administrative agencies to fulfill his executive responsibilities. The Constitution grants him the authority, subject to Senate confirmation, to appoint (and remove) the cabinet and sub-cabinet officials who oversee these administrative agencies. Ordinarily, he will appoint executive officials sympathetic with his own political views and then entrust them to develop and pursue his broader policy agenda. For the public lands, this means that the president sets a general policy tone, the secretaries of interior and agriculture develop specific policy agendas, and the public land agencies are then responsible for implementing these agendas.[35]

The executive branch agencies have generally been accorded considerable discretionary authority over administrative policy. Congress, with its legislative power, is constitutionally responsible for creating the executive branch agencies and for defining each agency's authority, which is usually done through an organic act. These organic acts typically delegate extensive rule making and adjudicative powers to the agencies, which must be executed consistent with specified legal standards and procedural requirements. But when defining the legal standards governing agency actions, Congress ordinarily does not legislate with much specificity in order to accommodate diverse local conditions as well as unforeseen developments. Among the public land agencies, the Forest Service's multiple-use provision is a classic example of a broad organic mandate; the courts have even opined that it "breathes discretion at every pore."[36] Because most statutory delegations of authority are framed so broadly, the federal public land agencies have been able to rely upon them to justify major policy adjustments, including the current shift toward new ecosystem-based management. But discretionary authority represents a double-edged sword: the same broad statutory standards might also be used to support a shift away from ecosystem management. In sum, the executive branch agencies have substantial leeway to define and meet their statutory responsibilities, but always with an eye on the president's overall policy objectives.[37]

Virtually all federal agency decisions are subject to political accountability and judicial review. Congress regularly employs such devices as budget appropriations, funding riders, and oversight hearings to channel administrative policies in a desired political direction and to express displeasure over contrary agency policy decisions. The Administrative Procedures Act (APA) and related laws have opened agency proceedings to public involvement and scrutiny, through such devices as notice and comment rule making, information disclosure obligations, and sunshine requirements—all of which are designed to ensure that agency officials take public concerns into account when framing new policies or making other decisions. The APA also provides for judicial review of agency decisions to ensure that agency officials do not violate either the legal standards or procedures governing their authority. The APA's judicial review standards, however, are rather amorphous: arbitrariness, capriciousness, or lack of substantial evidence. In applying these standards, the courts have regularly deferred to agency discretion, particularly when the agency is perceived to have technical expertise that the court may lack. And the Supreme Court, also invoking an agency expertise rationale, has established the so-called *Chevron* doctrine, which holds that federal courts ordinarily should defer to agency legal interpretations unless Congress clearly intended something different. Thus, even with political and judicial oversight, the federal agencies still have rather broad discretionary power to set and pursue their own policy agendas.[38]

Beyond the federal bureaucracy, the president possesses important powers that he may employ to shape public land policy. Some of these powers are constitutional in origin, others have been legislatively delegated to him. In a venerable line of cases, the Supreme Court has ruled that the United States holds the public lands as both a sovereign and proprietor, which gives it the same ownership rights over the public domain as a private landowner enjoys over his property. As the government's chief executive officer, the president is the embodiment of the federal property owner. Historically, the courts used this fact to bestow the president with an inherent withdrawal authority, which he could exercise to protect public lands or resources. President Taft used this executive withdrawal power to prevent the nation's strategic petroleum resources from passing into private ownership; President Theodore Roosevelt used it to create the first national wildlife refuges. Although Congress has twice passed legislation terminating the president's inherent withdrawal powers, it has also simultaneously granted him an explicit—albeit circumscribed—withdrawal power, recognizing that unexpected developments may necessitate prompt presidential action to protect the public interest. Notwithstanding these congressional retraction provisions, one court has suggested that the president still retains an inherent executive withdrawal power.[39]

Congress has also delegated specific withdrawal and reservation powers to the president, which have often been used aggressively to reshape the public land map. Such legislation as the General Revision Act of 1891, the Antiquities Act of 1906, the Pickett Act of 1910, and the FLPMA of 1976 have each created presidential withdrawal powers. At the turn of the twentieth century, Presidents Harrison, Cleveland, and Roosevelt used reservation powers found in the General Revision Act to place over 150 million acres of public land into federal forest reserves, which were eventually consolidated into the national forest system. Presidents have also used the withdrawal powers found in the Antiquities Act to protect sensitive sites and resources as national monuments. In fact, several presidents have used their Antiquities Act withdrawal power to create large national monuments in pursuit of overtly preservationist objectives: Franklin Roosevelt protected the Grand Teton mountain range; Jimmy Carter safeguarded the Alaska National Interest Lands; and Bill Clinton protected sensitive BLM acreage in southern Utah and elsewhere. Moreover, the FLPMA grants the president express withdrawal powers, which President Clinton, for example, used to withdraw national forest lands adjacent to Yellowstone National Park from further mining activity. In short, the president can avail himself of an array of executive withdrawal powers to pursue an ecosystem management agenda through protective public land designations.[40]

THE COURTS AND JUDICIAL REVIEW

The judiciary is responsible for interpreting the law. Under Article III of the Constitution, the federal judicial power extends to cases or controversies that, among other matters, involve federal questions or the federal government as a party. Under venerable constitutional principles first enunciated in the landmark *Marbury v. Madison* decision, the courts have the power to interpret the law and to enforce their interpretations with legal and equitable remedies. That judicial role, over the past half century, has been steadily extended into the arena of citizen-initiated public interest litigation, which has become a staple of the federal judicial docket. The Supreme Court, in a series of decisions dating from the early 1970s, has expanded the amorphous concept of standing that governs access to federal courts, ruling that a party suffering environmental injuries may seek judicial redress for those injuries. In a related development, the Court has adopted the "hard look" doctrine to review administrative decisions, which has given lower federal courts clear authority to scrutinize challenged agency actions closely. And ever since the watershed *Brown v. Board of Education* desegregation decision, the federal courts have grown increasingly willing to exercise their remedial equitable powers to oversee, often in great detail, government compliance with the law.[41]

With these developments, the judicial role in relation to the public lands has been transformed. Much of the early litigation tested the scope of federal power over public land and resources. In a remarkable set of early-twentieth-century decisions, the Supreme Court sustained the Forest Service's new organic mandate, endorsed its administrative rule making authority, and ruled that the president had an inherent executive withdrawal power to protect public lands against private appropriation under the mining laws. Over the past few decades, however, the courts have utilized their expanded judicial review powers to restrain federal agency authority. During the early 1970s, in the first wave of environmental litigation, the federal courts enjoined the Forest Service's clearcutting practices, stopped construction of the Alaska oil pipeline, and forced the BLM to revise its national grazing program. In more recent decisions, the federal courts have aggressively used NEPA, the Endangered Species Act (ESA), and other environmental laws to review and enjoin myriad logging, grazing, mineral leasing, and other resource decisions affecting the public lands. Of course, under these same liberalized standing and judicial review principles, the courts have also entertained litigation from industry and local community groups challenging public land decisions they perceived as too environmentally friendly. By any measure, the federal courts have become a major institutional presence on public land and natural resource issues.[42]

From its inception, the judicial review power has been controversial. Critics have long charged that the federal courts are predisposed toward making rather than interpreting the law. In the administrative law context, a common complaint is that judges regularly ignore or disregard congressional intent when interpreting statutes in order to promote their own favored policy goals. When courts do this, according to the critics, they are behaving in an activist or results-oriented manner that is inappropriate for a federal court, which should conduct itself in a more passive or restrained manner. This concern may be overstated, however. Judicial decisions are always subject to appellate review, and the Supreme Court has never been shy about overturning lower court rulings that disregarded clear statutory language or congressional intent. In fact, in a remarkable cession of power to the executive branch, the Supreme Court has ruled that federal courts ordinarily should defer to an agency's interpretation of its own statutory mandate. This statutory interpretation principle (also known as the *Chevron* doctrine) was intended to deter inappropriate judicial lawmaking tendencies. Moreover, what may appear as judicial activism to one observer is often legitimate judicial behavior to another, depending entirely on one's own philosophical predispositions and policy preferences.[43]

In any event, the courts have played a decisive role in promoting ecological management policies on the public domain. Despite the absence of an express ecosystem management statute, the courts have used existing laws to foster ecosystem-based policy initiatives. Relying upon the increasingly prescriptive standards that Congress has incorporated into legislation like the NFMA, ESA, and NEPA, the courts have ordered land management agencies to take account of ecological criteria in both their planning and project-level decision processes. That is the overwhelming message from the landmark spotted owl litigation, where a federal court used its judicial review powers to prod agency officials to prepare a comprehensive ecosystem management plan for the region's federal timberlands. With this precedent and the well-recognized principle of judicial deference, the agencies have considerable latitude in defining new ecological management policies for the public lands. That is, unless Congress—the lawmaking body in our governmental structure—says otherwise.

FEDERALISM AND THE STATES

Even though the public lands are under the federal government's immediate dominion, the western states are also involved in establishing basic priorities and policies. Across the 11 western states, federal public lands account for nearly 50 percent of the land base, giving the region a distinctive jurisdictional appearance. These states, as well as the communities scattered throughout them, have a keen interest in how public lands are managed. Economic opportunities, tax revenues, and well-established cultural traditions have long linked public lands to the political tides that sweep the region. In recent years, the Sagebrush Rebellion, Wise Use movement, and numerous other homegrown initiatives have emerged as indigenous western political movements with origins directly connected to federal management of public lands. Former Secretary of the Interior James Watt hailed from this background, and most western politicians are acutely sensitive to the political muscle that the traditional ranching, mining, and logging industries can wield. Although the West's urbanizing populace has reduced the influence of these industries in some locations, they remain dominant presences in several states and local politicians rarely overlook them.

Despite the oft-repeated claim that federal land managers are oblivious to state and local concerns, the western states actually exert considerable power and influence over public land policy. In the national political arena, western congressional delegations have regularly stymied efforts to reform the anachronistic General Mining Law of 1872 and fended off recurrent calls for grazing reform. Anti-wilderness forces have successfully blocked efforts to designate additional wilderness areas in the Montana and Idaho national forests, while wilderness proponents are engaged in a lengthy, acrimonious battle to protect spectacular BLM landscapes on Utah's Colorado Plateau. The states, exercising well-established legal prerogatives, are primarily responsible for wildlife on public lands; they can also claim dominion over water. And consistent with constitutional Supremacy Clause principles, the states can regulate many activities occurring on federal lands, so long as they do not contradict basic congressional policies.[44]

Through a convergence of factors, western state and local governments are now well-positioned to exercise perhaps even greater influence over public land policy. The national political mood has been growing more and more disenchanted with an expansive federal regulatory state. Congress has adopted the unfunded federal mandates bill that relieves states of their obligation to comply with federal regulatory standards absent reciprocal federal funding, and it has toyed with legislation that would return responsibility to the states in such areas as water quality, mining regulation, and grazing policy. The Clinton administration, in order to lighten the regulatory burden of laws like the Endangered Species Act and thus relieve related political pressures, consciously enlisted state agencies to oversee habitat management in areas like the California coastal sage shrub. At the judicial level, the Supreme Court has delivered several opinions pointedly reining in federal authority and reinforcing state sovereignty principles. At the same time, the states have begun to flex their political muscle to recapture traditional local responsibilities, ranging from welfare and education to environmental regulation, that the federal government had assumed. The Western Governors' Association, which during the late 1990s called for a Convention of the States, has adopted a much-publicized set of governing principles (known as Enlibra) designed to promote the local resolution of natural resource controversies on public lands. The clear message emerging from these seemingly disparate developments is a pronounced shift in power, responsibility, and focus from the federal to the state and local levels. Some observers have even suggested that this shift represents the collapse of the federal welfare state and the dawning of a new era in federal–state relations.[45]

The philosophical basis for this emerging shift in political power and emphasis is captured in the tenets of civic republicanism, as further refined by related communitarian doctrines. Numerous political and legal commentators have promoted reinvigorating the Jeffersonian ideal that public policy should be framed through civic dialogue at the level closest to those who will be affected by it. Based on the town meeting model, civic republicanism posits that government decisions made through open, local deliberative processes will tend inherently to accentuate public rather than private interests, and thus result in more public-spirited and better-accepted policies. Critics, however, have objected that the civic republican model presupposes a largely homogeneous community of interests and that it is impossible to subordinate private concerns to a greater public interest. Daniel Kemmis, in his provocative and influential book *Community and the Politics of Place,* makes a strong case for employing civic republican principles to address the West's natural resource controversies.[46] In a rough sense, the public participation processes that have become such an integral part of public land and natural resource decision making are a manifestation of the civic republican ideal. More to the point, the recent proliferation of local collaborative initiatives committed to finding common ground to resolve complex natural resource controversies can be viewed as exercises in civic dialogue. An institution like the new BLM Resource Advisory Councils, with their diverse local memberships, may also be regarded as an institutional manifestation of this impulse to address and resolve problems at a level close to those who must abide by the outcome of the process. Of course, just because a decision is reached at the state or local level rather than at the federal level does not ensure that it will be either more public-spirited or more representative and enduring.[47]

Any massive shift in power would have a pronounced impact on public land policy in the West. Most important, it would give state and local officials and citizens a critically enhanced role in formulating and implementing policy. Already the Wise Use movement, through its now discredited county supremacy legal arguments, has advocated the position that federal land management policy must be consistent with local land use policies, which generally have been framed in terms of protecting the traditional ranching, mining, and logging industries. Although environmental organizations have a secure presence in the West's urban centers, these same organizations have made only modest inroads in the region's smaller, rural communities. In some western states, environmentalists can influence policy at the state level, but they generally lack that ability at the county and local levels, except perhaps in Missoula, Telluride, Jackson, Moab, and other scattered locations that have been transformed by a changing economy and an influx of newcomers. As state and local governments begin to flex their newfound political muscle on the public domain, the evolving concept of ecosystem management will be sorely tested. But while local involvement could undermine the concept entirely, it also could provide a solid political foundation for strengthening ecologically sensitive public land policies.[48]

The Administrative Landscape: Agencies, Mandates, and Traditions

Four federal agencies are primarily responsible for managing the federal public domain. The Forest Service, Bureau of Land Management, National Park Service, and U.S. Fish and Wildlife Service administer respectively the national forests, public domain lands, national parks, and national wildlife refuges. The national forests and public domain lands are managed under multiple-use principles, and have historically been devoted to producing natural resource commodities, such as timber, minerals, and forage for livestock. In contrast, the national parks and national wildlife refuges are regarded as preserved lands; they are managed protectively, though visitation and recreation are encouraged in both locations. The U.S. Fish and Wildlife Service is also responsible for administering the Endangered Species Act, which gives it an important regulatory role overseeing resource decisions on all public lands. Despite quite different missions, traditions, and cultures, the public land agencies have each taken notable steps toward a new ecological policy agenda.

FOREST SERVICE

The U.S. Forest Service oversees the 192-million-acre national forest system, which contains myriad valuable resources, including old growth timber, oil and gas deposits, livestock forage, big game habitat, recreational ski areas, and protected wilderness lands. The Forest Service administers the national forests under a set of laws mandating multiple-use management and integrated resource planning. Created in 1906 as the prodigy of the reform-minded Progressive Era conservation movement, the Forest Service has long pursued an overtly utilitarian conservation agenda that was conceived by Gifford Pinchot, the agency's legendary first chief. According to Pinchot, the forest reserves were to be managed for productive purposes by professional managers using scientific management principles. The goal was to insulate the professional forest managers from the vicissitudes of politics and the marketplace. That has not worked, however, as the Forest Service has been unable to escape either political oversight or industrial pressures. In fact, although the agency has long been regarded as the epitome of an efficient bureaucracy, its reputation has been badly tarnished in recent years.[49]

For most of its first 50 years, the Forest Service performed a largely custodial management role. In the sparsely populated rural West, timber harvesting was not a major enterprise in the national forests. Recreation and other pressures were also relatively light. That changed, however, in the aftermath of World War Two, when the Forest Service confronted a virtual assault on its resources. The postwar nation needed timber to meet new housing demands, energy for a rapidly growing West, and additional outdoor recreational opportunities. To answer these escalating demands, the agency fashioned a new clearcut logging policy and dramatically accelerated timber harvest levels. The Forest Service, though, soon found both its policies and professional judgment under attack. Controversy over the agency's unsightly clearcutting practices quickly boiled over, fomenting court challenges in Alaska and West Virginia and a congressional inquiry in Montana. In 1975, a West Virginia federal court examined the Forest Service's organic mandate and found that Congress had effectively outlawed clearcutting by requiring individual trees to be marked before sale. For all practical purposes, the era of the Forest Service as Pinchot's omniscient manager of the nation's public forestlands was at an end. Neither Congress nor the public was prepared to trust the agency with unfettered discretion any longer.[50]

Congress responded by adopting the National Forest Management Act of 1976, which imposed, for the first time, explicit environmental limitations on the agency's timber management program and a new biodiversity conservation planning requirement. On its face, the NFMA should have set the Forest Service on a new course—one that de-emphasized timber production and paid greater attention to other forest resources and values. But neither Congress nor the White House was ready for such a drastic change. Nor was the Forest Service with its utilitarian heritage and workforce dominated by silviculturists. It took another decade and the bitter spotted owl controversy to finally blunt the agency's timber-first agenda. But before the twentieth century expired, the Forest Service's timber sales had dropped by more than 50 percent, timber road construction appropriations were subject to an annual budget battle, and even Congress was beginning to acknowledge that ecological considerations should be incorporated into the agency's policy agenda. To regain its own institutional footing, the Forest Service undertook a painful reexamination of its own managerial priorities and embraced new ecosystem management policies that would make ecological sustainability its central multiple-use goal. A century of multiple-use tradition may soon be replaced with a new ecological stewardship ethic.[51]

BUREAU OF LAND MANAGEMENT

The Bureau of Land Management (BLM) is responsible for managing approximately 264 million acres of unreserved public land located primarily in the 11 western states and Alaska. Sometimes denominated the "lands nobody wanted," the BLM public lands consist of those federally owned lands that have not been otherwise reserved or designated for another specific purpose. Until the mid-twentieth century, the unreserved public lands were available for disposal to encourage western settlement and development. The unclaimed lands remained open to local ranchers and miners who enjoyed virtually free access to what amounted to a public commons. Even after Congress passed the Taylor Grazing Act of 1934 in an effort to begin regulating the public grazing lands, western ranchers still effectively controlled these lands through the Grazing Advisory Boards, which they dominated. Given this history, it is not surprising that the BLM and its predecessor agencies—the General Land Office and the Grazing Services—have long been regarded as classically captive agencies—captured by the western ranchers and mining companies who were the principal beneficiaries of these unreserved lands.[52]

Modern land use planning and resource management finally arrived on the BLM's public domain lands when Congress passed the Federal Land Policy and Management Act of 1976, which gave the BLM its first real organic charter. Besides providing for permanent retention of the public domain lands, the FLPMA established an explicit multiple-use standard to govern the BLM's administration of public lands and created a comprehensive land planning process. It also gave the BLM express resource protection responsibilities, including the power to designate areas of critical environmental concern (ACECs) and to review its land holdings for possible wilderness designation.[53] The new FLPMA management regime was not well received by the BLM's traditional ranching and mining constituencies; they responded with the Sagebrush Rebellion in a failed divestment effort, and then actively supported the Reagan administration's pro-development agenda on public lands.[54] During the Clinton presidency, however, the BLM dramatically shifted its policy direction. Secretary of the Interior Bruce Babbitt sought to reposition the BLM: he promulgated new rangeland reform regulations, pursued administrative reforms of the General Mining Law, expanded the BLM's wilderness review authority, and vested the BLM—for the first time ever—with responsibility for managing national monuments. These policy reforms have injected environmental values squarely into the BLM's traditional ranching and mining agenda, given the BLM important new preservationist responsibilities, and thus opened the door for ecological policies to take hold.[55]

The National Park Service is responsible for overseeing the 385-unit national park system, which at 83 million acres represents a major federal commitment to preserving the nation's natural and cultural heritage. Established in 1916 in the aftermath of the Hetch Hetchy controversy, the Park Service was charged with managing the nation's fledgling park system to "conserve the scenery and the natural and historic objects and the wildlife therein and to provide for the enjoyment of the same in such manner and by such means as will leave them unimpaired for the enjoyment of future generations."[56] Over the years, the Park Service has struggled to reconcile its twin preservation and public use mandates. Early on, the agency subordinated its statutory preservation obligation to its public use responsibility: it actively encouraged visitation to the remote park areas, constructed tourist facilities and roads, eliminated wolves and other predators, suppressed fires throughout the system, introduced exotic game fish species, and promoted wildlife spectacles by feeding bears at garbage dump sites. These early management policies really represented a variation on utilitarianism; they amounted, according to a prominent Park Service historian, to facade management.[57]

During the 1960s, however, the Park Service elevated scientific management to a more prominent position on its policy agenda, setting the stage for an overtly ecological approach to park resource management. The real breakthrough came with the 1963 Leopold Report, which recommended that "the biotic associations within each park be maintained, or where necessary re-created, as nearly as possible in the condition that prevailed when the area was first visited by the white man."[58] Since then, despite sometimes hostile criticism, the Park Service has defined its preservation responsibilities in terms of maintaining and restoring native species and processes, while minimizing human intervention into natural ecological processes.[59] But in less than two decades, the national parks faced another troubling reality: external development that imperiled park resources. According to the 1980 State of the Parks Report, the parks were no longer isolated islands surrounded by undeveloped federal lands; rather, they were besieged by an array of external threats, many of which originated on adjacent public and private lands. Although Congress eventually incorporated a new science mandate into the Park Service's organic responsibilities that could strengthen its natural resource management policies, agency officials still face a bewildering array of challenges, ranging from continued diversification of the park system, further increases in visitation numbers, demands for new recreational opportunities, and continued pressure from external development. Because the national parks occupy a central role in any national biodiversity conservation strategy, the Park Service stands to be a principal beneficiary of emerging ecosystem management policies, but it has not aggressively promoted these new regionalism concepts.[60]

U.S. FISH AND WILDLIFE SERVICE

The U.S. Fish and Wildlife Service (FWS) occupies two critical public land management roles: it administers the national wildlife refuge system, and it is responsible for implementing the Endangered Species Act. The national wildlife refuge system dates from 1903, when President Theodore Roosevelt summarily established the first federal wildlife sanctuary. It has since grown into a far-flung, 509-unit refuge system covering 94 million acres of public lands located in all 50 states. Throughout their history, the refuges have been whipsawed between two competing visions: preservationists have viewed them as inviolate wildlife sanctuaries, while hunters and others have sought to open them for recreation, energy development, and other uses. The FWS, which first emerged as a separate agency in 1940, has long struggled to reconcile these diverse visions and to satisfy its disparate constituencies.[61] But in 1949, when it opened the refuges to hunting, the FWS embraced a patently utilitarian management policy and soon found itself subjected to additional recreation and other access pressures. Congress eventually responded by directing the FWS to permit refuge uses compatible with its wildlife conservation responsibilities, but neglected to give the agency a full-fledged organic charter.[62] After a series of reports documented deteriorating refuge conditions and mounting external threat problems, Congress finally passed the National Wildlife Refuge System Improvement Act of 1997, giving the FWS a long-awaited refuge system organic act. The new legislation directs the FWS to manage refuges for "the conservation, management, and where appropriate, restoration of the fish, wildlife, and plant resources and their habitats . . . for the benefit of present and future generations." It also clarifies the compatibility management standard, imposes new biodiversity conservation obligations, and creates a comprehensive conservation planning process. With this legislation, Congress has essentially obligated the FWS to engage in ecological planning and management on the nation's wildlife refuges.[63]

Besides the refuge system, the FWS is also responsible for administering the Endangered Species Act, which gives it a major regulatory presence that extends across public lands. In 1978, when the U.S. Supreme Court ruled that the ESA could empower the FWS to block any federal action that might jeopardize a listed species, the ESA became a major consideration in all agency decision processes, and the FWS was instantly elevated to a real position of power over public land policy. Since then, the FWS has been regularly criticized by business and development interests who complain it is using the ESA to block economically productive activities. Environmentalists, on the other hand, accuse the FWS of inadequately protecting at-risk species and thus failing to meet its legal-biological obligations. Dogged by these criticisms, the FWS has not employed its substantial ESA powers to promote a comprehensive federal biodiversity conservation policy on the public domain.[64] That may explain why the Clinton administration, seeking to bolster the federal government's biodiversity conservation role, bypassed the FWS and created a new National Biological Survey (NBS) agency to oversee federal biological research activities and to inventory and monitor biological resources across the country. (When the Clinton administration also transferred FWS research scientists and monies to the new agency, it noticeably diminished the FWS's technical capabilities as well as its already meager budget.) By any standard, the NBS reorganization reflected a lack of confidence in the FWS, further hindering its search for a clear identity and diluting its role in federal biological conservation efforts. Nonetheless, with its new refuge management mandate and its ESA implementation responsibilities intact, the FWS still has a real opportunity to shape ecological policy on public lands.[65]

The Constituencies: Confrontation and Collaboration

The federal government, of course, does not exist in a vacuum; it is a representative institution—one that ultimately reflects the values, views, and aspirations of its diverse constituencies. That is as true on the western public domain as elsewhere. Indeed, for more than a century, various public land constituencies have engaged in a tug-of-war for supremacy over federal policy, creating philosophical and political schisms that continue to surface regularly in any debate over future policy directions. Interest groups, corporate entities, and individual citizens all play a critical role in shaping the legislative agendas, administrative policies, and even judicial rulings that govern public lands. Any effort to identify, let alone categorize, the diverse interests connected to the western public domain is bound to be incomplete and cursory. One generalization, however, might be advanced: contemporary controversies recurrently find industry, other development interests, and motorized recreationists seeking to preserve traditional access rights to public resources aligned against environmental groups seeking to curtail those rights, either to preserve the land or to give priority to low-impact recreation and other amenity values. State and local government officials can be found on both sides of these issues, though few politicians, even in the rapidly urbanizing West, are prepared to write off the region's traditional economic interests or rural constituencies.

INDUSTRY AND ITS ALLIES

The traditional resource development or extractive industries—ranching, mining, and logging—are united by common economic concerns and a shared utilitarian perspective toward natural resources. These industries tend to view the public domain as a vast storehouse, where valuable timber, minerals, and forage commodities will go to waste if not harvested and put to productive economic use.

They share a common interest in maintaining historic access rights to public resources, minimizing the regulatory limitations governing these rights, and protecting existing subsidies. They also are united in their desire for certainty and stability in federal policy in order to facilitate rational planning and investment strategies. Extractive public land users range from large, multinational corporations, such as Weyerhauser and Phelps-Dodge, to independent loggers and solitary ranchers. More often than not, motorized recreational interests—groups like the Blue Ribbon Coalition and the Sahara Club—have joined extractive users to protect and promote road access on public lands. Within each industry, various trade organizations pursue lobbying, litigation, and other strategies at the national, state, and local levels. These trade organizations range from the American Forest and Paper Association and the National Mining Association to the Montana Farm Bureau and the Wyoming Stockgrowers Association. The extractive industries have also spawned a cadre of ideologically driven, nonprofit organizations committed to advancing a conservative political agenda focused on property rights and free market principles. Employing test case litigation strategies similar to those pioneered by the earlier civil rights and environmental movements, organizations like the Mountain States Legal Foundation, Stewards of the Range, and Defenders of Property Rights have vigorously asserted property rights and takings claims, and also challenged protective land designations and various environmental regulatory programs.[66]

These organizations have joined with disaffected rural westerners to form a loose coalition that has become known as the Wise Use movement. Conceived in 1988 to challenge the environmental movement and its purportedly "anti-people ideology," the Wise Use movement has aggressively promoted a political agenda based on developing public land resources, curtailing preservationist initiatives, and protecting private property. Although heavily funded with corporate contributions and conservative foundation grants, the Wise Use movement has a strong populist appeal, particularly attracting small-town, working-class citizens with livelihoods attached to the public lands as well as recreational vehicle enthusiasts, property rights proponents, and gun advocates. Unlike the predecessor Sagebrush Rebellion, which sought to transfer public lands from federal ownership, the Wise Use movement would retain the federal estate subject to a radical overhaul of the governing laws and policies: it has advocated opening national parks and wilderness areas to mineral development, guaranteeing access to national forest timber, protecting grazing permits as property rights, and ensuring vehicular access to most locations. It has also promoted local control over public land planning decisions, championing county land use ordinances that would require public land managers to defer to local priorities in managing their lands. In an early show of strength, the movement is credited with helping derail the Greater Yellowstone Vision Document, which emphasized protective management for the region's public lands. By most measures, the alliance between wealthy corporate entities and an extensive network of rural citizen-activists has been a potent political force on public land policy, providing both parties with clout and credibility they individually may not otherwise have. If not able to advance its own reform agenda, the Wise Use movement has at least mounted a serious challenge to the new ecological policy agenda.[67]

The traditional development industries, however, should not be perceived solely as a monolithic, single-minded entity on public land policy matters. The major logging, mining, and energy companies have vast financial and informational resources at their disposal that are not available to their smaller counterparts. With these resources, the large companies can generally meet new regulatory requirements and thus readily adjust their business operations to reform initiatives without serious financial difficulty. Ever sensitive to shareholder expectations and the financial bottom line, they often can—and will—outbid their smaller counterparts for timber sales and other public resources in times of scarcity. Corporate landowners and large ranches with extensive private land holdings are not as directly dependent on public resources as are non-landowners or small ranchers, whose very existence often depends on federal timber or range being available. Although corporate interests will typically align with those of their employees, unions, and suppliers, this is not always the case. Large timber companies, for example, have routinely closed mills and laid off workers rather than seek more distant and expensive timber sources just to maintain a local payroll. The ranching community, frequently viewed as a well-organized entity, has been shaken in several locations by the arrival of newcomers to the livestock business, who are open to new grazing techniques and may even choose to run buffalo rather than cattle.[68]

THE ENVIRONMENTAL COMMUNITY

The environmental community also defies easy characterization. Myriad national, regional, and local organizations, all fundamentally committed to protecting environmental amenities and preserving public resources, serve as watchdogs over public land policy. The major national organizations, including the Sierra Club, Natural Resources Defense Council, National Wildlife Federation, National Audubon Society, and Wilderness Society, devote substantial resources to congressional lobbying and other inside-Washington maneuvers. Some national organizations also maintain a local presence through regional offices. Various state-based environmental organizations, like the Wyoming Outdoor Council, Oregon Natural Resources Council, and Southern Utah Wilderness Alliance, focus their energies primarily on local issues, enlisting grassroots citizen-activists in wilderness designation, forest protection, and similar campaigns. A new generation of regional environmental organizations, exemplified by the Greater Yellowstone Coalition and Grand Canyon Trust, have begun promoting ecosystem-based strategies to accomplish regional environmental protection objectives. Several groups, including the Natural Resources Defense Council and National Wildlife

Federation, have assembled in-house legal staffs and use litigation aggressively to shape environmental policy; others rely on nonprofit environmental law firms, such as Earthjustice and the LAW Fund, to pursue judicial appeals in select cases. Yet other organizations, namely The Nature Conservancy and similar land trust groups, are focused on raising funds and public support to acquire and protect environmentally sensitive lands.[69]

Not surprisingly, given this diversity in organizational focus and goals, a similar diversity in strategies has surfaced. Although these groups often work cooperatively in loose coalitions on specific issues, they have also been known to disagree among themselves over tactics. In the case of the Northwest Forest Plan proposal, for example, most national organizations were receptive to a Washington-negotiated agreement on the spotted owl controversy, while local environmentalists decided to pursue a hard-line, no-compromise strategy.[70] The traditional adversarial model of political change, which relies primarily upon legislative lobbying and litigation, has long been employed by most of the national organizations as well as the more aggressive local groups. For the most part, this strategy ignores local communities and politicians, seeking instead to accomplish environmental objectives through federal institutions, either the land management bureaucracy, Congress, or the judiciary. As a political strategy, it relies heavily on mobilizing citizen-activists to bring pressure upon elected federal officials, key political appointees, and other agency policy makers. The environmental movement, with its enduring linkage to John Muir's early preservationist advocacy, has a rich radical-amateur tradition of effective political activism that has scored significant victories over the past century. Some observers, however, worry that the major national environmental organizations have lost touch with this grassroots tradition, directing too much attention to Washington-based strategies and too readily endorsing compromise solutions that shortchange important local concerns. Other observers believe that the environmental community could expand its populist appeal by expressly acknowledging and addressing local economic concerns as part of its ecological reform agenda. They fear that the environmental community has unwittingly pushed the rural working-class constituency into the Wise Use movement, which has cleverly exploited economic fears to create an effective backlash against recent ecological policy initiatives.[71]

Several environmental organizations, however, are experimenting with different tactics to promote ecological conservation and address related economic concerns. At one level, rather than ignore the local populace in public land communities, these groups are attempting to work with local officials and organizations to find common ground. Examples include the Grand Canyon Trust's Colorado Plateau Forum, the Greater Yellowstone Coalition's sustainability blueprint, and the Sonoran Institute's sustainable communities program. At another level, some organizations have borrowed from the resource economists' market-oriented philosophy and are using financial incentives to secure local cooperation for conservation initiatives. A prime example is the Defenders of Wildlife's wolf depredation compensation fund, and its companion wolf breeding reward fund, which compensates private landowners when a wolf successfully breeds on their land. Such diverse strategies have always been a necessary staple of maturing social movements, and may be even more of a necessity in today's changing political climate.[72]

Within the ranks of both industry and environmental organizations, quite different perspectives and approaches can be found on individual issues, often creating unusual alliances and collaborative opportunities. Ranchers and environmentalists have feuded for decades over federal range policy but have regularly joined forces to resist new mining proposals that threaten vital local water supplies or new ski resort proposals that could transform a peaceful mountain valley. Several organizations, weary from recurrent confrontations and frustrated over stalemated policy initiatives, have begun pursuing more collaborative solutions rather than relying solely on litigation and lobbying strategies. In the Northern Rockies, for example, the timber industry has worked with environmental groups to return grizzly bears to the Selway-Bitterroot country under the auspices of a unique citizen management committee. Some ranchers, too, have joined—even convened—collaborative citizen management teams to review and adjust local grazing practices. Other industry and environmental organizations, however, tend to regard every new policy proposal as a matter of major precedential importance and reflexively adopt an uncompromising position against change. Not only does such a position undermine the search for cooperative solutions to local problems, but it inevitably reinforces the tendency to escalate conflicts to the next highest level, ensuring that either Congress or the judiciary will eventually be brought into the fray and have the final word. While that may be necessary to establish the basic parameters of ecosystem management policy on public lands, it may not be the most effective way to cultivate grassroots political support for the countless on-the-ground initiatives that are critical components of any new policy.[73]

All of these institutions and forces are playing prominent roles in the myriad initiatives and controversies defining the new ecosystem management agenda on the western public domain. The ecosystem imperative has not only been vigorously endorsed by the environmental community, but also embraced by the primary federal institutions charged with managing public lands. While this imperative is still meeting resistance by many at the congressional, state, and local levels, there is begrudging recognition that the traditional way of managing resources is not working and that broader, transboundary collaboration among the involved governmental entities and parties is necessary. This recognition of the need for local cooperation corresponds to the growing national movement toward decentralization of federal authority. It also may provide the basis for addressing the strikingly diverse ecological and economic conditions found across the western landscape. If so, the emerging ecosystem management and restoration initiatives may yet be linked with a corresponding political theory and thus achieve full legitimacy as a new public land policy.

THREE

Ecology and the Public Domain

The biotic idea is thus an extension of the idea of management. . . .

Both stem from ecology. The biotic idea merely translates ecology

for purposes of guiding land-use.

ALDO LEOPOLD (1942)

Physically and socially, the West does not remain the same from decade

to decade any more than other places do. If anything, it changes faster.

WALLACE STEGNER (1980)

Federal public land policy is undergoing profound changes. Whereas natural resources were historically viewed as plunder for the taking and only those things with obvious economic utility were valued, society is now attaching value to resources and species that previously were viewed as worthless. Whereas public lands have long been perceived as the nation's development bank, these same lands are now being regarded as a critical biodiversity stronghold. Whereas land managers traditionally focused on single resources, they are now attempting to manage entire ecosystems in an integrated fashion. Whereas policy decisions ordinarily were based on existing political boundary lines, the public land agencies are now seeking to transcend these boundaries to manage on a landscape or regional scale. And whereas the legal system ascribed rights only to individuals and corporate entities, it is now beginning to consider that natural objects also might enjoy enforceable rights. In short, ecological facts that no longer can be denied are refashioning how society perceives the public lands and inexorably moving natural resource policy toward new management priorities.

How this has occurred and what it means are reflected in converging scientific, philosophical, socioeconomic, legal, and political developments. Current scientific and philosophical thought generally acknowledges that sustaining interconnected ecological components and processes is vitally important to human welfare, for utilitarian as well as nonutilitarian purposes. Both disciplines, committed to a holistic view of nature, endorse the related proposition that natural resources should be managed on an ecosystem scale. Broader societal trends, ranging from rising affluence and increased urbanization to a new environmental awareness, are helping propel these ideas into the policy arena. Because federal lands provide important habitat for many species and feature a large system of preserved areas, the public domain is serving as an early battleground over how a functional ecosystem management policy might be fashioned and implemented. Several working definitions of ecosystem management have been proposed, and agency officials are gradually developing them into viable planning and management strategies. Ecosystem management, however, has not been universally well received. Strong criticisms have been leveled against the concept, and these criticisms must be addressed for ecosystem management to be fully recognized as a viable natural resource policy. While the details of what management on an ecosystem scale actually means and how it should be translated into law and policy remain to be worked out, the fundamentals are sufficiently established to acknowledge the beginnings of a new era.

Origins of a New Idea

Powerful new scientific and philosophical insights have catapulted the ecosystem management concept to center stage on the public domain and elsewhere. Using the ecosystem concept, scientists have dramatically increased their knowledge of the intricate relationships connecting species to one another and to their habitats. With this knowledge, scientists have become alarmed at the tremendous impact contemporary civilization and modern technology are having on the natural environment, particularly the accelerating pace of species extinctions. Drawing upon these scientific insights, philosophers have extracted new moral imperatives to protect valuable ecological systems from human destruction. Concurrent social and cultural changes have enhanced the public commitment to environmental protection and engendered critical political support for preserving species and their habitats. The traditional anthropocentric view of nature as merely an assortment of individual commodities to be exploited for human benefit is being challenged by an emerging biocentric perspective that sees nature as an integrated organic entity entitled to respect and protection in its own right.

ECOSYSTEM SCIENCE COMES OF AGE

During the twentieth century, the science of ecology emerged from the shadows, immensely broadening society's understanding of the natural world's complexities and humanity's impact on it. Ecology first surfaced in the late 1800s as an offshoot discipline from biology designed to study species in their native surroundings. Rather than viewing species simply as entities unto themselves, ecologists conceived nature as an interdependent whole, composed of organisms as well as the surrounding physical environment. This holistic approach to biological study has become a hallmark of ecology, helping propel the discipline to the forefront in contemporary environmental policy debates. It is the one science that views land, resources, and natural processes as interrelated systems, which has profound management implications.

As a science, ecology has evolved throughout the past century, reflecting an increasingly sophisticated comprehension of the natural environment. During the early twentieth century, Professor Frederic Clements from the University of Nebraska was perhaps the preeminent ecologist of his time. To better explain the relationship between species and their environment, Clements developed the theory of ecological succession, which posited that natural systems evolved through various successional stages toward a climax state of equilibrium. Although Clements's own work focused primarily on plant communities, his successional-equilibrium model dominated early ecology. It essentially confirmed the widely accepted "balance of nature" metaphor, which viewed nature as inherently stable and predictable. Clements's successional model is still useful for understanding the dynamic—if sometimes random and unpredictable—processes that shape natural systems, but his equilibrium ideas have been largely discredited.[1]

The term *ecosystem* was introduced into ecology in 1935 as a theoretical construct. A British ecologist, Sir Alfred George Tansley, expanded upon Clements's ideas by proposing the ecosystem concept as a means of better understanding ecological succession. Tansley used the ecosystem as a model to integrate both the biotic and abiotic components of natural systems, positing that the entire system evolved, not just its organic components. But Tansley, like Clements, also adhered to the equilibrium principle: "In an ecosystem the organisms and the inorganic factors alike are *components* which are in relatively stable dynamic equilibrium."[2] Nonetheless, the ecosystem provided an extremely useful conceptual device for studying natural systems and, as it turned out, for promoting broader public understanding of environmental relationships and complexities.[3]

Other ecologists soon embraced Tansley's ecosystem concept and gave it some practical meaning. University of Minnesota ecologist Raymond Lindemann employed the concept to study and explain how a Minnesota lake ecosystem functioned and evolved. Lindemann's study provided the foundation for his landmark energy pyramid theory, which demonstrated that "ecosystems develop through ecological succession and are tied to the energy dynamics of the system and the concept that nutrient cycling, as food cycling, is linked to wider biogeochemical cycles coupling one ecosystem with another."[4] Lindemann's then-revolutionary ideas helped transform the ecosystem concept into an operational scientific tool, while also highlighting the dynamic, energy-derived nature of ecosystem processes. At least one historian, in retrospect, gives him major credit for providing the conceptual foundations for ecosystem ecology as a distinct discipline.[5] But Lindemann's ideas did little to popularize the ecosystem concept, which remained cloaked in obscurity, little known or understood beyond a narrow circle of ecologists.

After World War Two, with the public consciousness beginning to turn toward environmental protection, ecosystem science emerged from obscurity. In 1953, University of Georgia ecologist Eugene Odums published a highly successful textbook entitled *Fundamentals of Ecology*, which introduced ecological concepts (including the ecosystem) into popular discourse. In his book, Odums eloquently elaborated a series of basic ecological principles, giving a coherence to ecosystem science that previously had been missing. The fact that Odums incorporated environmental management issues into his discussion of ecosystem ecology highlighted the new discipline's practical applications and rendered it more accessible to the lay public. Describing Odums's contributions, historian Frank Golley observes: "Ecosystem ecology provided an integrated way to view the environment and therefore was quickly adopted by the public."[6] It helped, of course, that the terminology associated with ecosystem ecology was linked to engineering, economics, and cybernetics, and thus easily adapted to the lexicon of environmental policy and management.

In recent decades, the ecosystem concept has transcended the ecological sciences and taken on a major role in the public debate over environmental protection and species conservation. The term *ecosystem* has been defined as a "unit made up of all the living and nonliving components of a particular area that interact and exchange materials with each other."[7] As a scientific concept, the ecosystem provides a comprehensive model for understanding and describing the complex relationships that exist between the physical environment and biological species. These relationships, we now understand, are anything but static; influenced by multiple environmental factors, ecosystems are perpetually changing and evolving. Historically, this process usually involved gradual, incremental change produced by such natural phenomena as climate change, fires, and floods, though catastrophic (or stochastic) disturbances periodically altered evolutionary history in unanticipated ways. In the modern world, however, human activities have greatly accelerated the rate of environmental change and upset long-standing evolutionary patterns—a fact illustrated graphically by the increased rate at which species are now being lost to extinction. According to Harvard University biologist Edward O. Wilson, "the rate of extinction is now about 400 times that recorded through recent geological time and is accelerating rapidly."[8] Ecosystem preservation has, therefore, become a vitally important dimension of species conservation.[9]

Faced with what scientists widely regard as an "extinction crisis," ecosystem ecologists have focused their attention on the relationship between species and the surrounding ecosystems. An important breakthrough occurred in 1967, when Wilson and his Princeton colleague, Robert MacArthur, propounded the theory of island biogeography to explain the problem of species extinction. Employing data derived from the often less complex ecosystem structures and processes found on islands, island biogeography theory has enabled scientists to better understand and model the relationship between species and their surrounding ecosystems in the ever more complex environments that prevail on the continents. According to prevalent theories, changes in the structure of even large, mature ecosystems can have severe, destabilizing environmental effects, destroying or fragmenting habitat and thus isolating species in "habitat islands" inadequate to sustain viable populations. The initial impact is ordinarily felt by sensitive, top-of-the-food-chain species—mainly carnivores such as wolves, bears, and bald eagles—which also are the species with the greatest natural impact on the ecosystem. Most of the western North American national parks, according to one widely cited study, lost wildlife species through extinction during the twentieth century because their legal boundaries did not embrace sufficient habitat to maintain viable populations. Although no single factor alone will ordinarily destabilize a large ecosystem, the aggregate impact of multiple human intrusions can markedly alter prevailing environmental conditions and upset evolutionary patterns.[10]

In recent years, basic ecological theory has embraced a new nonequilibrium model of ecosystem ecology—a profound theoretical change with important ramifications for species conservation efforts and related laws and policies. Ecologists now acknowledge that dynamic and often unpredictable change—rather than equilibrium conditions—are an essential characteristic of most ecological systems. Clements's equilibrium paradigm stands largely rejected, replaced by a nonequilibrium model based upon dynamic and sometimes chaotic change. As a result, ecology is no longer regarded as a deterministic science; rather, it is better understood as a probabilistic science. For biodiversity conservation purposes, this means that ecosystem-level management efforts must protect resident species from both recurrent disturbances as well as unanticipated events. Ecologists express this goal in terms of resiliency (the ability of an ecosystem to recover from disturbance), meaning that ecosystem conservation efforts must ensure that targeted ecosystem components and processes are able to maintain their integrity even in the face of such stochastic events as major fires, floods, and earthquakes. And because uncertainty is inherent in nonequilibrium ecological conditions, scientists have endorsed the principle of adaptive or contingent management in order to address unexpected biological consequences. In short, the nonequilibrium paradigm has expanded the spatial and temporal scale of ecological conservation efforts and introduced adaptive management concepts into these same efforts.[11]

Species conservation is now expressed in terms of biodiversity conservation, which focuses both on preserving individual species and ensuring ecosystem integrity or resiliency. Biodiversity refers to "the variety and variability among living organisms and the ecological complexes in which they occur."[12] A multidimensional concept, biodiversity is perhaps best understood in hierarchical terms: it embraces genetic diversity (focused on individual organisms), species- or population-level diversity (focused on herds or groups of animals), ecosystem or community-level diversity (embracing the assemblage of animals and plants in a discrete area), and regional or landscape-level diversity (which includes multiple ecosystems). Ecosystems are similarly complex, consisting of components (inhabiting organisms), structures (physical life form patterns), and functions (energy flows). An effective biodiversity conservation strategy must address all these interrelated dimensions of biodiversity and ecosystem processes. Scientists generally support two interrelated and complementary strategies: the "coarse filter" approach, which seeks to conserve representative habitats (or ecosystems) based on the assumption that ecosystem preservation will protect native species; and the "fine filter" approach, which focuses on species-level conservation to ensure the survival of individual organisms and populations. In addition, scientists have admonished that biodiversity "hot spots"—areas of high species richness or endemism—deserve special attention. The ultimate goal is to ensure adequate habitat to maintain viable species population levels.[13]

The "coarse filter," or ecosystem-based, conservation strategy involves establishing and maintaining protected areas where human disturbances are minimized. It requires extending protection to the full array of representative ecosystem types to preserve the various species associated with these different environments. Size requirements for protected areas, depending upon the target species, can range from multimillion-acre wilderness-like complexes to safeguard wide-ranging species like grizzly bears to reserves of only a few acres to protect a particularly vulnerable endemic species like the desert pupfish. Focused on ecosystem preservation, the coarse filter strategy is generally a more efficient approach to biodiversity conservation than species-by-species strategies. It can also provide scientists with a benchmark or reference point for measuring ecological health and the impact of human-initiated changes in the ecosystem, particularly when large blocks of undisturbed habitat or wilderness are preserved. In today's heavily populated world, however, it is often not feasible to include large ecosystems within protected area networks. This can leave wide-ranging and endemic species with specific habitat needs inadequately protected and vulnerable to disturbances.[14]

The "fine filter," or species-based, conservation strategy is designed to address this problem. Focused on individual species, such as grizzly bears or spotted owls, the fine filter strategy essentially creates a safety net for the target species. It is designed to optimize the likelihood of species survival by ensuring that the specific habitat and other needs of individual species facing an uncertain future are addressed. A fine filter strategy can be costly, however, especially when extensive resources must be targeted on one or a few species. Moreover, if various species are in competition with one another or have different habitat requirements, the fine filter strategy can prove counterproductive for those species with ecological requirements different from those of the targeted ones. But scientists often have not identified all the species present in particular locations, so the coarse and fine filter strategies will usually complement each other and maximize the opportunities for survival.[15]

With modern technology, scientists can now design and test large-scale ecological conservation strategies. Tremendous advances in computers, satellite imagery, and related technology have significantly enhanced our ability to understand ecosystem dynamics and to develop management responses. Through the use of modern computer technology, scientists can conduct population viability analyses on particular species by assembling detailed population data and then running simulated models to predict population trends. The same computer technology can be used to construct other useful ecological models, such as predator–prey dynamics and habitat requirements. With access to satellite technology, scientists have developed extensive geographical information systems (GIS) and related mapping capabilities, including large-volume data storage capacities and spatial and statistical analysis abilities. GIS mapping techniques can be used, for example, to gather landscape information relating to natural vegetative cover and associated human use patterns, which can then be overlaid with species distribution maps to assess the location and potential causes of biodiversity gaps. Satellite-based remote sensing technology can be used to track species dispersal and distribution patterns, and thus establish critical habitat requirements as well as genetic interchange possibilities. In short, these scientific and technological advances provide sophisticated tools to assess population data and ecosystem trends—absolutely vital information for devising effective conservation strategies.[16]

The scientific community, concerned over deepening biodiversity losses, has begun mobilizing to promote ecological conservation. In 1991, the Ecological Society of America released its Sustainable Biosphere Initiative proposal, calling for more basic research to acquire the ecological knowledge necessary to solve contemporary environmental problems. The proposal, noting that the growing human population is "exerting tremendous pressure on Earth's life support capacity," gives priority to research on global change, biodiversity, and sustainable ecological systems to promote "a fundamental understanding of the ways in which the natural systems of Earth are affected by human activities." Five years later, a committee of scientists convened by the Ecological Society of America expressly endorsed ecosystem management as a means of promoting sustainable ecosystems, while noting the important role for scientists in defining appropriate spatial and temporal management scales and facilitating adaptive management strategies. A Society of American Foresters task force also endorsed ecosystem management to achieve "long term forest health and productivity." Their report recommends "managing [public and private] forests cooperatively across ownerships in large landscapes so that goods and services for human use, and ecosystem conditions such as biological diversity and ecosystem integrity are ensured in a multi-generational time frame."[17]

A new discipline—conservation biology—has been created to address the extinction crisis. Originated during the late 1970s by a group of concerned scientists, the Society for Conservation Biology has the goal of developing "principles and tools for preserving biological diversity." Conservation biologists, many of whom have traditional wildlife, forestry, and related backgrounds, believe these resource-oriented disciplines are too narrowly focused on primary species and too oriented toward utilitarian concerns to address the larger issues entwined with contemporary ecological problems. For them, conservation biology represents "a new stage in the application of science to conservation problems, address[ing] the biology of species, communities, and ecosystems that are perturbed, either directly or indirectly, by human activities or other agents." Based on "the fundamental value assumption that biodiversity is good and ought to be preserved," conservation biology is unabashedly "mission-oriented" and "more value-laden" than most other sciences. Because conservation efforts are inherently multifaceted, conservation biologists have embraced other disciplines, including philosophy, economics, sociology, geography, law, and political science, to promote a transdisciplinary conservation agenda. Several conservation biologists have spearheaded major ecosystem-based conservation proposals, including the Wildlands Project, the Northern Rockies Ecosystem Protection Act, and the Yellowstone to Yukon initiative. Critics, however, have questioned the wisdom of scientists embroiling themselves in conservation policy debates and thereby potentially undermining their professional objectivity. Conservation biology proponents respond that the extinction crisis is quite real, demanding more public attention and concerted political action.[18]

PHILOSOPHY AND ECOLOGY

Throughout history, scientific advances have regularly triggered corresponding advances in philosophical thinking. That is as true in the biological sciences as in theoretical physics, medicine, and other fields. For biologists and philosophers, the fundamental question that has occupied their attention is how people should relate to nature. Until recently, that question was answered almost exclusively in anthropocentric terms by measuring nature's value solely in terms of its usefulness to human society. But a new and competing biocentric view has crept into popular discourse. This biocentric perspective transcends a narrow utilitarianism, valuing nature for both instrumental and intrinsic reasons. By seeking to reshape how people interrelate with the natural world, biocentrism portends major changes in how natural resources are viewed and managed.[19]

During much of the nineteenth century, the human relationship with nature was defined primarily in accordance with Christian religious tenets, which meant human beings and nature were regarded as distinct and separate entities. Humankind, with its consciousness, capacity to reason, and religious instincts, occupied a higher position on earth than other animals or plants. Christian theological doctrine, premised on this overtly dualistic notion, taught that nature was to serve humanity. Not being a part of nature, humans were entitled to exploit the natural environment for their own purposes without restraint. With these utilitarian beliefs as the order of the day, wild nature was a place to be conquered and subjugated.[20]

By the mid-nineteenth century, this orthodox view of the human relationship with nature was coming under fire. The infidel was Charles Darwin and his revolutionary theory on the origin of species. In fact, Darwin's revelation that people evolved from other species basically repudiated Christianity's dualistic view of humans and nature as separate entities. Darwin's insights did not, however, displace conventional anthropocentric thinking or related utilitarian beliefs. Humans were still regarded as occupying a higher position than other animals, and nature was still treated as a commodity to be exploited for human use and consumption. This utilitarian view of nature, when linked with contemporaneous industrial and technological advances that enhanced human exploitative capabilities, continued to set humanity apart from nature and imposed few if any moral injunctions on human behavior toward the natural world. The popular attitude toward wilderness also remained largely unchanged; it was still regarded as a fearsome place, certainly in the United States where western settlement was still unfolding.[21]

Some inroads, though, were beginning to be made. Shortly after Darwin's revelations, Vermonter George Perkins Marsh published *Man and Nature,* where he chronicled civilization's harsh impact on the natural environment and propounded the concept of stewardship. Marsh observed that mankind "with stationary life . . . at once commences an almost indiscriminate warfare upon all forms of animal and vegetable existence around him, and as he advances in civilization, he gradually eradicates or transforms every spontaneous product of the soil he occupies." To reverse these trends, Marsh asserted, people owe a duty to nature: "the earth was given to [humanity] for usufruct alone, not for consumption, still less for profligate waste."[22] This custodial stewardship idea was eventually integrated into the prevailing utilitarian view of nature through the reforms of the Progressive Era conservation movement. During the latter part of the nineteenth century, John Muir's exuberance for wild nature captured public attention and fostered a nascent preservation movement based primarily on earlier Romantic-era sentiments, aesthetic sensitivities, and spiritual connections. But as an entity, nature was accorded no rights; the human obligation was merely to exercise good stewardship to avoid unnecessarily depleting nature's bounty. Nature still existed primarily to serve humanity, which could rightfully take what it wanted and needed—sensitive, if at all, only to present and perhaps future needs.[23]

As the twentieth century unfolded, with the advent of ecology as a science, it became increasingly difficult to separate humankind and nature. Ecology made the linkages between nature and human welfare ever more explicit, nurturing the view that people owed more than mere custodial obligations to the natural world. Ecologists, upon identifying a spectrum of ecological interdependencies in the natural world, observed similar interdependencies with the human world, which convinced them that human-induced change was upsetting the prevailing natural order. With these new ecological insights, they argued that the natural world should be valued for other than immediate economic reasons, rather for its long-term instrumental value and perhaps even intrinsic worth. From here, it was only a short step from ecological fact to an ethical imperative. Indeed, as Donald Worster has observed, the history of modern ecology has deep roots in moral philosophy, which virtually invited a merger between science and philosophy.[24] That merger was accomplished by Aldo Leopold, a widely respected scientist who bridged the worlds of science and philosophy and thus laid the foundation for a new ecological perspective on natural resource policy.

The ideas set forth in Leopold's *A Sand County Almanac* essays have had a profound impact on contemporary natural resource policy. Trained as a scientist at the Yale School of Forestry, Leopold graduated in 1909 and then spent his early career employed as a new Forest Service ranger in the Southwest. During these years, Leopold was directly exposed to Gifford Pinchot's wise use philosophy, which he initially embraced with few reservations. But he also had other formative—if not seemingly contradictory—experiences in the Southwest: he actively participated in the campaign to exterminate wolves, which were then regarded as worthless predators; he expressed growing concern over the deleterious impact of overgrazing on the region's riparian zones; and he promoted the then-radical idea that wilderness preservation was a legitimate forest use. In 1933, Leopold resigned his Forest Service position to assume a professorship in game management at the University of Wisconsin, where he wrote a groundbreaking wildlife management text as well as numerous scientific articles and literary essays. He also served on the state wildlife management board and became a founder of The Wilderness Society. In addition, he purchased an abandoned and badly neglected farm in Wisconsin's run-down Sand County region, which became the focal point for his personal ecological restoration efforts and, of course, the primary subject of his famous essays. All of these activities speak to the breadth of his interest and experience in the natural world, and lend compelling credibility to his related philosophical assertions.[25]

In his writings, Leopold deftly translates his penetrating ecological insights into a series of moral observations that link human welfare to the natural world and define an appropriate environmental ethic. In his "Round River" essay, Leopold observes the intricate interconnectedness of the landscape, concluding that the "complexity of the land organism" is the "outstanding scientific discovery of the twentieth century." Asserting that "the land is one organism," he admonishes against injuring any part of it: "To keep every cog and wheel is the first precaution of intelligent tinkering." Elsewhere, noting that evolution has diversified "the biotic pyramid," Leopold then explains how humanity's use of tools has dramatically accelerated the speed of evolutionary change and thus imperiled natural diversity among wild animals and plants. And in a moving and oft-quoted passage, he describes shooting a wolf as a young man and then watching "the fierce green fire dying in her eyes," later lamenting his inability to "think like a mountain" and thus understand that the wolf's death would hasten adverse ecological changes on its flanks. Both by example and injunction, Leopold conveys his conviction that human beings have an obligation to preserve ecological diversity across the landscape.[26]

In his "Land Ethic" essay, Leopold further develops his argument for the proper human relationship to nature. Drawing upon ecological insights, Leopold places human beings in "a community of interdependent parts" that includes "soils, waters, plants, and animals," observing that "man is, in fact, only a member of a biotic team." He rejects the traditional view of man as a "conqueror of the land community," and bemoans the absence of an "ethic dealing with man's relation to land and to the animals and plants which grow upon it." Noting that the human relationship to the land is now "strictly economic, entailing privileges but not obligations," Leopold proposes a land ethic that "reflects the existence of an ecological conscience . . . a conviction of individual responsibility for the health of the land." In an eloquent and widely quoted passage, Leopold then discloses his proposed land ethic: "A thing is right when it tends to preserve the integrity, stability, and beauty of the biotic community. It is wrong when it tends otherwise." With this moral imperative, Leopold embraces an essentially biocentric philosophical principle, one that is based upon his objective understanding of modern humanity's evolving relationship with the natural world.[27] His distinctive contribution, according to one environmental philosopher, was to provide "a sound *scientific* foundation for a land or environmental ethic."[28]

Leopold's expansive ideas did not, however, attract immediate public attention or acclaim. In the realm of ecology, that role fell to Rachel Carson, a highly regarded scientist and accomplished writer deeply distressed over the deteriorating natural world. In her seminal 1962 book, *Silent Spring*, Carson exposed the dangers of agricultural pesticides, demonstrating how they were adversely altering the web of life and endangering human health. Rooted in ecological insights and concerns, Carson's moving prose chronicled the devastating effect the widespread use of DDT and other pesticides had on the entire food chain. The origin of the problem, she argued, was the changing human relationship to nature: "Only within the moment of time represented by the present century has one species—man—acquired significant power to alter the nature of his world. During the past quarter century this power has not only increased to one of disturbing magnitude but it has changed in character." She concluded with a call for humility in the human relationship with nature: "The 'control of nature' is a phrase conceived in arrogance, born of the Neanderthal age of biology and philosophy, when it was supposed that nature exists for the convenience of man." Carson's book (unlike *A Sand County Almanac* and other contemporary works on pesticides) was an immediate popular success and is widely credited with educating the general public about these important scientific concepts. Besides instilling ecological principles into the public consciousness, Carson's book spurred passage of legal restraints on toxic pollutants and laid the groundwork for later pollution control legislation.[29]

Over the years, however, Aldo Leopold's ideas have endured and inspired subsequent generations of environmental scholars and activists alike. In 1972, law professor Christopher Stone advanced the then-heretical idea of incorporating an explicitly biocentric perspective into American jurisprudence, arguing that "legal rights [be given] to forests, oceans, rivers, and other so-called 'natural objects' in the environment." Supreme Court Justice William O. Douglas, an avid outdoorsman since his childhood in the Pacific Northwest, embraced Stone's proposal and used it in his concurring opinion in *Sierra Club v. Morton,* a key case in modern environmental law. Justice Douglas, after also citing Leopold's land ethic essay, stated: "Contemporary public concern for protecting nature's ecological equilibrium should lead to the conferral of standing upon environmental objects to sue for their own preservation. . . . The river as plaintiff speaks for the ecological unit of life that is part of it." Soon thereafter, the concept of deep ecology was introduced into the philosophical lexicon. Predicated on the principle of biocentric equity, deep ecology teaches that "all things in the biosphere have an equal right to live and blossom and to reach their own individual forms of . . . self realization." More recently, conservation biology has surfaced as a new discipline intent on melding ecological science and moral philosophy into a new conservation agenda. Conservation biologists have seized upon normative values derived from Leopold's writings to translate new scientific insights into biodiversity policy prescriptions.[30]

Not surprisingly, the philosophical case for biodiversity conservation and ecosystem preservation is being made in both anthropocentric (utilitarian) and biocentric terms. As a purely utilitarian matter, biodiversity is a vital component in the production of goods and products, particularly in the agricultural, industrial, and pharmaceutical sectors. Genetic cross-breeding has enhanced crop yields and livestock productivity, and thus expanded the food supply. Chemicals derived from plants and other organisms have been used for various industrial purposes, including DNA forensic analysis and glue strengthening. Little-known plants have occasionally yielded important medicinal discoveries that have proven effective in combating disease. In fact, over 25 percent of the nation's prescription drugs are derived from plants. Beyond plants and animals, ecosystem complexes—wetlands, ancient forests, barrier islands—perform important ecological services benefiting the human population. These services include climate control, oxygen production, nutrient cycling, soil generation, flood control, and water purification. One recent estimate places the economic value of these ecosystem services at over $36 trillion annually on a worldwide basis. But scientists often do not know and cannot predict precisely which plants, animals, or ecological complexes may prove beneficial. Biodiversity conservation is therefore properly viewed as a kind of insurance policy for future generations.[31]

Biodiversity affords important aesthetic and recreational benefits too. Over the centuries, people have found beauty in nature and been moved by it. The United States has a long commitment to setting aside scenic landscapes as national parks and to establishing wildlife refuges to protect native species. Legal protection has also been extended to individual species with particular aesthetic, symbolic, or charismatic appeal, such as the bald eagle and whale. Aldo Leopold, in defining his land ethic, quite consciously included beauty as a part of his ethical imperative. Activities such as hunting, fishing, bird-watching, and wildlife viewing are time-honored pursuits long enjoyed by large segments of the population. Many people also treasure the opportunity to engage in hiking, biking, and similar recreational activities in unique ecological settings. (As a side benefit, participants in these activities usually expend funds for accommodations, food, guides, and other services, thus contributing to the local economies in locations where these activities are available.)[32]

Biodiversity is also valued for purely ethical, religious, and spiritual reasons. Aldo Leopold, observing that man is only a "plain member and citizen of [the land community]," enjoined us to "enlarge the boundaries of the community to include soils, waters, plants, and animals."[33] Numerous philosophers have argued that rights should be extended beyond human beings to animals and even inanimate objects, contending that they have an innate right to exist without regard to their economic or other utility. Throughout the ages, various religions have attached spiritual significance to nature and natural objects. In the Christian tradition, the biblical story of Noah's Ark conveys a deep respect for all species. Among Native Americans, tribal members are viewed as part of nature (not separate from it), and animals are often given special spiritual significance and incorporated into tribal mythologies. In many instances, the contemporary environmental movement casts itself in spiritual and ethical terms, with proponents making the case for nature preservation in explicitly biocentric terms. In fact, American law now awards monetary damages for destroying natural resources based upon the loss of existence value.[34]

A CHANGING ECONOMIC AND SOCIAL FABRIC

Historian Samuel Hays asserts that the period in the United States following World War Two should be labeled "the Environmental Era." To support this assertion, Hays cites several factors that have contributed to the emergence of environmentalism as a major social movement and potent political force. First, during the mid-twentieth century, the nation experienced major advances in its standard of living and level of education. Greater affluence meant more discretionary income, which enabled people to focus beyond their own immediate economic needs to quality of life concerns, including amenities such as clean water, clean air, and natural landscapes. Higher levels of educational attainment exposed larger numbers of people to new environmental and ecological ideas, and thus enhanced public understanding of how human development influences natural systems. Second, the national economy rapidly evolved into an advanced industrial economy, moving from a predominantly industrial base to a primarily technology and service base. This shift, besides diminishing the role of heavy industry in the economy, also heightened public awareness of the importance of the natural world in an advanced industrial society. Third, a dramatic growth in population, particularly in urban areas, produced a more youthful population with an interest in outdoor experiences and nature preservation. As a result, according to Hays, the public embraced an environmental ethic that became a dominant social value during the latter part of the century.[35]

In the western United States, these post–World War Two trends are fully evident, and they have had a profound impact on the region. Following the war, the West's nascent urban centers consciously embarked upon a campaign to promote regional growth and development. To fuel the expansion, local political leaders turned to the region's available resources, including its precious water, fossil fuels, and old growth timber, which were harnessed and extracted in an unprecedented fashion. With some regional variations, the development patterns were remarkably similar in the Northwest, Southwest, and interior West. Describing the Southwest's postwar boom in dam construction and power production, legal scholar Charles Wilkinson tellingly labels the period "the big build-up." At the same time, the nation was embroiled in the cold war, and it too turned to the West and its resources—uranium, coal, oil, copper, and the like—to meet important national defense needs. And when the Arab oil embargo was imposed in the mid-1970s, another frenzy of development ensued, embracing the Powder River Basin's coal, the Overthrust Belt's natural gas deposits, and even oil shale on Colorado's western slope. At each juncture, the toll on the West's public lands and resources mounted, triggering public concern over the fate of the region's fragile environment and its exhaustible resources. By the early 1970s, of course, the federal government had adopted a series of laws, including the Clean Air and Water Acts, NEPA, and the Endangered Species Act, which were designed to inject environmental concerns into the development equation. And once the energy frenzy struck, the western states responded with their own laws, a combination of industrial siting acts, environmental compliance requirements, and severance taxes in an effort to moderate development impacts and to ensure some local financial return.[36]

Long regarded as the nation's storehouse, the western public lands have been the primary domain of the extractive industries for much of the twentieth century, largely devoted to producing such commodities as minerals, timber, beef, and energy resources. In fact, historians regularly described the West as a virtual colony of eastern financial interests, noting that the region was dependent on capital from elsewhere and shipped most of its raw materials to distant markets for consumption elsewhere. The myth of multiple-use management has been well-chronicled; most observers concur that the productive industries dominated management priorities on the public lands during the past century. As well, the region has long depended on a spate of federal subsidies and tax rebates to spur development, sustain local industries, and cushion cyclical downturns. Not only have the region's extractive industries played a major role on public lands, but they have also dominated local politics, as reflected in the copper industry's historical grip on Montana politics, the timber industry's lengthy reign in the Pacific Northwest, and the coal industry's role in Wyoming politics. In part, this simply reflected the fact that the mining, timber, and energy industries provided all-important jobs for many of the region's residents, who in turn embraced these industries despite the recurrent boom-and-bust cycles embedded in them. Moreover, much of the folklore surrounding the American West can be traced to the predominant role the extractive industries have played on the public lands: the self-sufficient prospector, the stalwart lumberjack, and the freedom-loving cowboy. In short, through a combination of national need, local boosterism, and sheer tradition, the West's public lands were valued largely for their productive output and not much else.[37]

But that is all changing, the result of both international and domestic forces that transcend these historical patterns. On the international front, the global marketplace, with its cyclical price fluctuations, has long influenced corporate investment strategies on public lands, a fact vividly illustrated by the oil industry's retreat from the interior West during the 1980s as the price of oil plummeted. These same market forces have also periodically squeezed other public land users, notably cattle and sheep ranchers faced with depressed markets during much of the 1990s. Today, with the opening of new international markets and growing overseas business opportunities, domestic corporations are increasingly looking abroad for new investment opportunities. Advances in modern transportation, communications, and technology have dramatically increased access to international markets. And with the collapse of the Soviet bloc, growing democratization within the developing world, and greater international stability, America's natural resource industries are turning to developing countries to pursue new business opportunities. The trend is reflected in the increased mineral development activity in South America and Russia, oil exploration in the former Soviet republics, and timber production in Latin America and southern Asia. Faced with ever more regulatory requirements governing the public lands and workplace conditions, many industries have found foreign venues more attractive, where labor, safety, and environmental restrictions are either lacking or much less rigidly enforced. With these new international opportunities, the natural resource industry's interest in development on public lands has waned.[38]

On the domestic front, rapid population growth and new social values are changing the West as well as priorities on its public lands. The West's population growth has outstripped other sections of the country, growing from 17 million in 1946 to 35 million in 1970, and then to over 60 million in 1998. In recent years, the growth rate has been phenomenal: during the 1990s, 8 of the 10 fastest growing states were western states. Not surprisingly, the region's population boom has meant a dramatic increase in urban growth. Over 80 percent of the West's residents now live in urban settings, making it the nation's most heavily urbanized region. Just 40 years ago, Nevada was the nation's most rural state; today it is the most urbanized state. The region's cities—from Denver, Phoenix, and Las Vegas, to Salt Lake City, Boise, and Portland—continue to grow at explosive rates. But the West's growth is not confined to its urban areas. Smaller towns across the region, particularly those offering scenic attractions and other valued amenities, have also experienced rapid growth as more and more people have sought refuge from urban problems, congestion, and the like. In the Greater Yellowstone region, for example, population increased during the 1980s at a rate 33 percent higher than the population growth rate in the three surrounding states. The Colorado Plateau's resident population doubled between 1960 and 1990, and then added another 15 percent between 1990 and 1994. Although these developments are not spread evenly across the landscape, it is nonetheless true that both the urban and rural West are much different places today than they were just a few decades ago. The once empty West is rapidly filling up with people.[39]

Indeed, several observers have heralded the arrival of a "new West" that is replacing the "old West" of cowboy myths and extractive economies. They observe that the postindustrial, high-tech economy has brought new economic and cultural priorities to many western communities. As an economic matter, with the traditional natural resource and agricultural industries waning in importance, the service-sector industries (health, engineering, management consulting, financial and legal services) have grown in importance, as measured by overall employment and income figures. And what economists label "unearned" income—money from such sources as retirement benefits and investments—now represents a substantial portion of the income received by many western residents. As a cultural matter, the West's new arrivals—ranging from relocated footloose industries to "modem cowboys" attached to the new technology-based economy, and from retirees to other urban refugees and recreation-seekers—have transformed many western towns. Most of these newcomers are attracted by quality of life advantages, which include environmental and amenity values associated with the western landscape, namely its scenery, wildlife, and outdoor recreational opportunities. The changes are manifest in towns such as Jackson, Bend, Durango, Moab, Flagstaff, Bozeman, and myriad other communities situated in scenic venues, often areas with extensive public land holdings. In fact, several economists now argue that the landscape itself represents a natural asset that contributes directly to the quality of life, which both attracts and retains the businesses, investments, and intellectual capital necessary to fuel the region's continuing development and growth.[40]

Confronted with these new economic realities as well as changing perceptions in the value of public lands, traditional commodity uses are in decline on public lands, while recreation, tourism, and environmental values are growing in importance. In the national forests, timber harvest levels have dropped precipitously, from 10 to 12 billion board feet (bbf) per year during the 1980s to just 2 to 4 bbf per year in recent years, in large part because of new harvesting restrictions in the timber-rich Pacific Northwest. Mineral exploration and development is down in most categories, though mining remains an important industry in specific locations, principally in northern Nevada, where gold mining grew steadily during the 1990s owing to new heap leach ore processing techniques. Oil and gas exploration (excluding coal bed methane) is down on the public lands, though production has increased, partly owing to offshore development on federally owned leases. Coal leasing has held steady, while coal production has steadily increased, particularly in the Powder River Basin in northeastern Wyoming. Livestock grazing rates have gradually declined during the past several decades, dropping from 15.4 million animal units per month (AUMs—the standard for measuring public land grazing allotments) in 1955 to 8.4 million AUMs in 1996. Increased automation in the mining and timber industries has also noticeably decreased the number of jobs available to local residents, thus further diminishing the role of these industries in many rural locations. Moreover, federal subsidy policies have been placed on the national political agenda as part of the ongoing federal budgetary deficit debate. In an unusual alliance, environmentalists have joined fiscal conservatives to target federal subsidies and the fee structure for logging roads, grazing permits, water projects, and mining activities.[41]

Recreation is no stranger to the public lands. In the aftermath of World War Two, with its rising birthrate and growing affluence, the American public began actively seeking outdoor recreational opportunities. Among other things, this burgeoning interest in outdoor experiences fostered the National Park Service's Mission 66 program to enhance park visitor facilities, the exponential growth of the ski industry, and dramatic increases in off-road vehicle use. It was also the genesis for the 1962 Report of the Outdoor Recreation Review Committee, which helped pave the way for such legislation as the Wilderness Act of 1964, the Land and Water Conservation Fund Act of 1965, the National Trails System Act of 1968, and the Wild and Scenic Rivers Act of 1968—all designed to enhance the nation's commitment to nature preservation and outdoor recreation. Once the interstate highway system was completed, the nation's public lands became even more accessible to the growing urban populace, while new lightweight camping equipment made extended trips into remote areas ever easier. And as the postwar baby boom generation matured and embraced new environmental values, it not only heightened the demand for outdoor recreational opportunities but also generated corresponding pressures to protect pristine, undeveloped landscapes. In combination, these factors have contributed to an enormous growth in popular use and appreciation of public lands for recreational and environmental purposes. One study even suggests that recreation has now effectively displaced commodity uses as the dominant use on public lands.[42]

The growth in recreational use of public lands parallels the evolving role of tourism in the new economy, as well as the general public's steadfast commitment to protecting the environment. Whereas the extractive industries—mining, timber, and livestock—once dominated western state economies, that is no longer the case with the enormous growth of the service-based economy and the related tourism sector. A recent Western Governors' Association report found a high dependency on tourism among the western states, noting that "Tourism is relatively more important for the western United States than to the remainder of the nation."[43] Already the asserted largest private employer in 7 of the 11 western states, tourism provides many rural communities with "one of the few opportunities to enhance the local economy."[44] In fact, several communities have literally remade themselves from depressed single-industry towns to rejuvenated tourism and outdoor recreation centers. When the local lumber mill closed in Dubois, Wyoming, for example, the town shifted its focus toward wildlife tourism, winter recreation, and retirement living, actually experiencing an upsurge in population and income during the transition. On a related note, the American public's commitment to protecting the environment, including plant and animal species, is very pronounced in national polls, and thus would appear to support the expanded role that biodiversity conservation is assuming in public land policy. Not surprisingly, public support for environmental protection is strongest among the younger generation—a fact that has obvious long-term ramifications for public lands. As these new economic and social realities are absorbed into mainstream politics, management priorities and bureaucratic behaviors will increasingly reflect greater sensitivity to environmental protection and related economic benefits.[45]

These changes in the West's social fabric and economic priorities, however, are not coming easily. True to historical patterns, tensions persist between the West's urban and rural communities. The new West culture and its accompanying economic opportunities are not spread evenly across the western landscape. Many smaller, rural communities continue to depend on agriculture and extractive industries for their economic sustenance, and they often resent and fear the economic and social changes that are occurring around them. Some of this local frustration has been directed toward federal land managers and their new policy agendas, leading occasionally to confrontational incidents and even violence.

More often, it is manifested through the Wise Use movement and similar groups, who have adopted the same roster of political and legal tactics to forestall change that their environmental counterparts have used to spur change. In addition, the region regularly witnesses clashes over values between new arrivals and longtime residents—clashes that often focus on the role of extractive industries and development on the public lands. The new environmental values, which are very strong among the region's urban populace and many newcomers, are not so readily embraced by many of its longtime rural residents who still rely on public lands for their livelihoods. But recent data suggests that the region's urban and rural residents may not be so far apart: they both disfavor policies giving priority to resource development at any cost. This nascent local commitment to environmental values will only deepen with further changes in the region's populace and economy.[46]

In a related development, Native American tribes have become a force to be reckoned with in managing natural resources. During the 1970s Columbia River native fishing rights controversy, local tribes successfully asserted historic treaty rights governing the salmon harvest, thus gaining legal protection for their interests against competing claims. The controversy demonstrated that tribes are not merely interested constituents when natural resources are at stake; rather, they are sovereign entities whose treaty-guaranteed rights must be recognized, and they are entitled to equal treatment under any conservation policy. Often, reservations are located adjacent to public lands, which gives the tribes a stake in how contiguous public lands are managed and likewise gives public land managers an incentive to coordinate management policies with neighboring tribes. In some locations, tribes enjoy treaty-guaranteed access rights to public lands, as illustrated by the Blackfeet tribe's hunting, fishing, and wood gathering rights on the ceded strip in northwestern Montana, which encompasses part of the Lewis and Clark National Forest. Many tribes also have sacred sites on public lands, giving them a First Amendment interest in protecting and accessing these locations for religious purposes. And as sovereign entities, Native American tribes are now empowered to assume responsibility for establishing clean water, air, hazardous waste, and other environmental standards on reservation lands, a power that has extended tribal influence over activities occurring on adjacent lands. Thus, owing to their sovereign status, treaty rights, and other legal entitlements, the region's tribes are becoming an increasingly important force in determining public land priorities and policies.[47]

Ecosystem Management as Policy

Occasionally, a new idea will emerge and gain sufficient credibility to challenge conventional wisdom. That new idea, often denominated a paradigm shift, has the power to shake the very foundations of existing legal and institutional structures. In the case of public lands, only a few fundamental transformations—

or paradigm shifts—have occurred. One happened at the beginning of the twentieth century when scientific utilitarianism displaced the laissez-faire disposition policies that had long governed the public domain. Another has occurred more gradually, reflecting the steady absorption of preservation concepts into federal policy to complement extant utilitarian and disposal policies. These doctrines have served the nation well in their times, but a new set of ideas is now pressing for legitimacy. Its prominent features are that biodiversity conservation should take precedence on public lands, that entire ecological systems must be the focus of management, and that public participation is a vital aspect of natural resource policy. Together, these powerful biological and democratic concepts are being melded into a new paradigm in natural resource policy under the rubric of ecosystem management.

SETTING THE STAGE

Within federal public land agencies, convergence toward ecosystem management has been more evolutionary than revolutionary. Tempting as it may be to view ecosystem management as the original creation of an environmentally minded Clinton administration, that would be wrong. As we have seen, important scientific, philosophical, and socioeconomic forces have been at work reshaping natural resources thought and policy throughout the latter part of the twentieth century. Faced with this changing reality, the responsible federal agencies had little choice but to begin viewing their management obligations in a broader ecological context and taking account of the transboundary ramifications of their policy decisions. For the National Park Service and the U.S. Fish and Wildlife Service, interrelated problems concerning wildlife management, species extinction, and external threats highlighted the ecological interconnections between their protected reserves and neighboring lands. Elsewhere, the traditional logging, grazing, and mining agenda of the Forest Service and the Bureau of Land Management was under increasingly heavy fire for ignoring widespread ecological damages and related wildlife habitat needs. In each instance, the agencies responded—sometimes reluctantly and timidly—by experimenting with rudimentary ecological management concepts.

By the mid-1970s, it was evident that the national parks no longer existed as isolated, self-contained islands. Whether surrounded by other federal, tribal, state, or private lands, an increasing number of parks faced serious environmental threats, ranging from air pollution to clearcut logging, that emanated from sources beyond their boundaries. Early in the decade, the problem surfaced starkly in the case of Redwood National Park, which was suffering severe ecological degradation caused by extensive upstream logging just beyond park boundaries. After a series of court decisions ordered the Park Service to take action to protect Redwood's resources, Congress was finally persuaded to adopt legislation expanding the park and clarifying the agency's general responsibility to protect park resources. In 1980, building upon its Redwood experience, the Park Service issued a congressionally mandated State of the Parks Report identifying myriad external activities that threatened park resources, including air pollution, energy exploration, timber harvesting, and subdivision development. Similar external threat problems had also surfaced in the national wildlife refuges. At the Kesterson National Wildlife Refuge in central California, a massive bird-kill attributed to pesticide-contaminated water draining into the reserve from adjacent farms received widespread publicity. With these events confirming the need for a more comprehensive, ecologically oriented management strategy, both agencies began to move the external threats issue to the forefront of their policy agendas. But the new Reagan administration soon thwarted these efforts.[48]

Not surprisingly, the volatile Yellowstone region spawned several early ecosystem management proposals and a related high-profile interagency planning experiment. The Craighead brothers, whose pioneering studies of Yellowstone National Park's grizzly bears spanned two decades, are generally credited with linking the Greater Yellowstone ecosystem concept with wildlife management policy. After documenting that Yellowstone's bears regularly wandered beyond the park's rectilinear boundaries and noting that the bears were managed quite differently depending upon their jurisdictional location, the Craigheads argued forcefully that a single bear management policy should be adopted for the entire Greater Yellowstone ecosystem. To be sure, the Greater Yellowstone terminology had been coined as early as 1917, and others had previously used the ecosystem concept in connection with wildlife management. But the Craigheads, by employing the Greater Yellowstone ecosystem concept for managing the charismatic grizzly bear, linked a specific image to ecosystem management and forged a critical ecological tie between the national park and surrounding national forestlands. Drawing upon their work, the Greater Yellowstone Coalition and other environmental organizations soon began promoting a comprehensive ecosystem management strategy for the park's dwindling grizzly population as well as related transboundary environmental problems.[49]

During the late 1980s, acting through the Greater Yellowstone Coordinating Committee (GYCC), the Park Service and Forest Service initiated a much-heralded but ill-fated regional ecosystem management experiment. Chastened by a congressional report finding that a lack of regional management coordination placed the wide-ranging grizzly bear at risk, the GYCC began a vision document process to establish Greater Yellowstone as a "world-class model" for integrated and coordinated natural resource management. The draft vision document seemed to do just that: it called for ecosystem management and envisioned "a landscape where natural processes are operating with little hindrance on a grand scale . . . a combination of ecological processes operating with little restraint and humans moderating their activities so that they become a reasonable part of, rather than incumbrance upon, those processes."[50] But fearing a significant shift in policy, some area residents and commodity groups denounced the document and enlisted local political officials to subvert the process. Which they did. When the GYCC released the final document in 1991, it represented a pale replica of the original visionary statement, omitting any reference to ecological management and emphasizing instead the separate missions of the two federal agencies. Although the final document has been largely ignored in subsequent regional policy debates, the Greater Yellowstone vision process gave an early measure of legitimacy and enhanced exposure to new ecological management ideas within the federal bureaucracy. It also demonstrated, however, that political and legal support are essential if such new initiatives are to succeed.[51]

Meanwhile, the Forest Service was deeply embroiled in its own resource controversies with significant ecological management overtones. In 1980, just as the Forest Service was embarking on its first round of forest planning mandated by the National Forest Management Act, the Reagan administration assumed office and endorsed accelerating national forest timber harvest levels. The timber issue, as we shall see, quickly escalated into a major regional controversy in the Pacific Northwest, where the reclusive northern spotted owl became a legal surrogate for old growth forest ecosystems and a potent political symbol. Elsewhere, as national forest timber quotas escalated, critics began protesting other sales too, filing one administrative appeal after another and then turning to the courts in an effort to protect diminishing ecological values. By 1989, the Forest Service's own officials had joined the chorus: the Region One forest supervisors wrote the chief a widely publicized letter sharply criticizing agency timber targets and lamenting the environmental damage being done to the forests. Disillusioned Forest Service employees banded together to create the Association of Forest Service Employees for Environmental Ethics (AFSEEE) in an overt challenge to the agency's single-minded timber focus. Stung by the criticism and by litigation setbacks in the courts, agency officials responded first by endorsing a New Forestry timber policy and then by embracing a New Perspectives program, both designed to inject ecological factors into the agency's timber management regime. In mid-1992, the chief of the Forest Service finally announced that the agency would begin using ecosystem management techniques to guide its timber harvesting practices. By then, though, the courts were firmly in control of the Pacific Northwest forests, and the agency had few choices but to design a meaningful ecosystem management policy addressing the full array of forest values.[52]

Controversy also swirled around the BLM, which was enmeshed in its own set of disputes. By the mid-1970s, once Congress had passed the National Environmental Policy Act and then the Federal Land Policy and Management Act, the BLM found itself with a new organic multiple-use mandate, as well as related resource planning, environmental assessment, and wilderness protection obligations. Neither the agency nor its constituencies were fully prepared for this new regime. On one flank, as the BLM began to translate its new legal responsibilities into more prescriptive management policies, its traditional ranching constituency protested vehemently, launching the Sagebrush Rebellion in an effort to thwart the agency's efforts to regulate livestock grazing and other extractive activities. On the other flank, environmentalists successfully invoked the nascent NEPA environmental impact statement requirements and forced the BLM to begin addressing the environmental implications of its livestock grazing program. The environmental community also mounted legal challenges to the agency's burgeoning coal leasing and mineral exploration agenda, part of the federal response to national energy shortages. Once the Reagan administration took office and further accelerated resource development on the public lands, the litigation docket swelled even more. Like the Forest Service, the BLM was struggling with its own identity, trying to understand its new environmental protection and resource planning responsibilities while still appeasing its traditional constituents and addressing the nation's perceived energy needs. Not surprisingly, the result was contentiousness, litigation, and acrimony across the public lands.[53]

In the midst of these developments, biodiversity conservation emerged as a key element in contemporary environmental policy. Spurred by the accelerating rate of species extinctions, scientists placed the biodiversity concept squarely on the public agenda and made a compelling case for intensified conservation efforts, as reflected in passage of the powerful Endangered Species Act (ESA) and the more recent international biodiversity convention. Once passed, the ESA brought the FWS squarely into the public land resource controversies. Following the U.S. Supreme Court's 1978 ruling in the well-publicized snail darter case, the FWS—as the agency primarily responsible for implementing the ESA on public lands—was effectively cast in the role of a regulatory agency overseeing public land planning and project decisions whenever a protected species was present. With its jeopardy review authority and the related taking prohibition, the FWS holds a veto power over any agency decision that might jeopardize the continued existence of a listed species or adversely modify critical habitat. To implement its species recovery obligations, the FWS began developing recovery plans for such wide-roaming creatures as the grizzly bear, gray wolf, and bald eagle—plans that necessarily addressed land management policy from a primarily biological perspective. Grounded in science, these ESA obligations forced the land management agencies to begin viewing their responsibilities in broader ecological terms, focusing on the biological needs of protected species rather than on competing economic or political concerns.[54]

During this same period, other ecological management concepts also penetrated federal public land agencies. In the early 1970s, after endorsing UNESCO's Man and the Biosphere program, the federal government designated several national biosphere reserves, including such well-known parks as Yellowstone, Glacier, and Olympic as well as adjacent national forestlands. The biosphere reserve concept was conceived as a mechanism to protect biodiversity resources and representative ecosystems while also providing sustainable economic opportunities for nearby communities. It employed a concentric zoning strategy, with core protected areas (national parks) buffered by moderately disturbed lands (often national forests) that were surrounded by lands where intensive development activities occurred. Although the biosphere reserves did not have any legal stature under federal law, they offered an early model of how an ecological management strategy might be implemented on public lands to achieve both ecological and economic goals. By the late 1980s, the Park Service and Forest Service were actively beginning to investigate how ecosystem management concepts might apply in national park and wilderness area settings. A 1987 interdisciplinary conference at the University of Washington's Pack Creek Experimental Forest brought representatives from both agencies together for exploratory discussions. The conference culminated in publication of a book entitled *Ecosystem Management for Parks and Wilderness,* which contained the first comprehensive discussion of ecological management on the federal landscape. By the late 1980s, therefore, public land agencies had begun to identify and define ecologically oriented policies that might be used to address the increasingly contentious transboundary issues that were coming to dominate their resource management agendas.[55]

Congress, however, was plainly not prepared to address these complex ecological, social, and economic issues through legislation. Although national park protection legislation to remedy the problems concerning external threats surfaced regularly during the early 1980s, it died in the Senate just as regularly. And although a congressional committee report forecast doom for Yellowstone's grizzly bear population, local congressional delegations actively participated in undermining the GYCC's visionary interagency planning process that was designed to cure this problem. Practically every senator and representative with constituents touched by the Pacific Northwest timber harvesting controversy advanced a legislative solution to the conflict, but no one could muster a consensus for any of the proposals. Instead, Congress addressed the controversy with appropriations riders designed to maintain current logging levels, which only served to aggravate the matter. With Congress either unwilling or unable to craft any effective ecosystem management legislation, the agencies were left to address the mounting ecological controversies on their own. And in the Pacific Northwest, the courts soon demonstrated that they were prepared to play an active role in promoting an ecological management agenda.[56]

By 1993, when the Clinton administration took office, the stage was set to bring an ecological management vision to the fore within the federal bureaucracy. With court injunctions blocking all public land timber harvesting in the Pacific Northwest, with the threat of similar litigation looming elsewhere, and with major regional resource management controversies simmering in the San Francisco Bay Delta area, the Florida Everglades, and elsewhere, the new administration had few options but to chart a new course. Beginning with their ascension to positions of power, key Clinton administration officials, including Vice President Al Gore, Secretary of the Interior Bruce Babbitt, and Environmental Protection Agency administrator Carol Browner, endorsed the ecosystem management concept and began integrating it into the new administration's natural resource policy agenda.

Ecosystem management became the strategy of choice for resolving the Pacific Northwest timber controversy, and similar ecologically based strategies were put forth to address the Bay Delta, Everglades, and other long simmering controversies. To institutionalize this new approach, the administration established an Interagency Ecosystem Management Task Force to develop a coherent federal ecosystem management policy for use by all federal agencies. Since then, as we shall see, public land agencies have undertaken myriad initiatives under the rubric of ecosystem management. But before we examine what ecosystem management has meant on the ground, it is necessary to give it some content and to identify potential problems with this new natural resource policy.[57]

ON DEFINITIONS, PRINCIPLES, AND LAW

Ecosystem management is still a young and evolving concept—one that has not yet been captured in a succinct and tidy sound bite. Most commentators agree, however, on several key propositions that go to the core of emerging ecosystem management policies. With an extinction crisis looming, biodiversity conservation and ecosystem restoration are major ecosystem management concerns. But the human presence on the landscape cannot be ignored, which means that economic, social, and political concerns must be factored into any ecological management policy. To address these interrelated concerns, ecosystem management has embraced sustainability as a paramount policy goal. Sustainability is defined broadly: it includes both ecological and economic sustainability, as measured by the needs of both current and future generations. To achieve these sustainability objectives, planning and management decisions must be framed in ecologically relevant geographical and temporal terms, unconstrained by conventional boundary lines, jurisdictional jealousies, or short-term political, economic, or social considerations. Management on this enlarged scale requires collaborative, cross-jurisdictional planning protocols as well as adaptive management approaches. In combination, these propositions represent a dramatic departure from traditional natural resource management institutions and policies, and introduce controversial new complexities into the public land policy equation.[58]

Numerous general definitions of ecosystem management have been proffered, but none has yet been universally accepted or endorsed. According to the Ecological Society of America, ecosystem management is "management driven by explicit goals, executed by policies, protocols, and practices, and made adaptable by monitoring and research based on our best understanding of the ecological interactions and processes necessary to sustain ecosystem composition, structure, and function." Another comprehensive definition posits that "ecosystem management integrates scientific knowledge of ecological relationships within a complex sociopolitical values framework towards the general goal of protecting native ecosystem integrity over the long term." Yet another definition emphasizes ecological goals: "Ecosystem management focuses on the conditions of the [ecosystem] with goals of maintaining soil productivity, gene conservation, biodiversity, landscape patterns, and the array of ecological processes." Another definition, however, downplays the concept's scientific origins and goals: "the application of ecological and social information, options, and constraints to achieve desired social benefits within a defined geographic area over a specified period." Other suggested definitions have tended to emphasize either the biological or social-political aspects of the concept, depending largely on the proponent's professional affiliation or disciplinary background.[59]

Not surprisingly, a less striking diversity in views can be found in the original federal agency definitions of ecosystem management. According to the Interagency Ecosystem Management Task Force, the ecosystem approach was "a method for sustaining or restoring natural systems and their functions and values. It is goal driven, and it is based on a collaboratively developed vision of desired future conditions that integrates ecological, economic, and social factors. It is applied within a geographic framework defined primarily by ecological boundaries." The Forest Service subscribed to "an ecological approach to achieve the management of national forests and grasslands by blending the needs of people and environmental values in such a way that national forests and grasslands represent diverse, healthy, productive, and sustainable ecosystems." The Bureau of Land Management saw ecosystem management as "the integration of ecological, economic, and social principles to manage biological and physical systems in a manner that safeguards the long-term ecological sustainability, natural diversity, and productivity of the landscape." The National Park Service viewed ecosystem management as a "philosophical approach that respects all living things and seeks to sustain natural processes and the dignity of all species to ensure that common interests flourish." The U.S. Fish and Wildlife Service defined ecosystem management as "protecting or restoring the functions, structures, and species composition of an ecological ecosystem, recognizing that all components are interrelated." Thus, while public land agencies generally agreed on the goal of maintaining sustainable ecosystems, they did not all place the same emphasis on related economic concerns—a clear reflection of the divergent statutory mandates governing each agency.[60]

Because these proffered definitions are so general, the ecosystem management concept has also been defined in terms of governing principles. First, to ensure healthy natural resource systems and to address pressing species extinction concerns, a primary goal (objective) of ecosystem management is to maintain and restore biodiversity and sustainable ecosystems. Second, because people are a part of nature and human values inform any natural resource policy, ecosystem management goals must be socially defined—ordinarily through broad public participation—to incorporate ecological, economic, and social concerns into workable sustainability strategies. Third, because species and ecological processes transcend jurisdictional boundaries, ecosystem management requires coordination among federal agencies and collaboration with state, local, and tribal governments as well as opportunities for public involvement in planning and decision processes.

Fourth, given the dynamic, nonequilibrium nature of ecosystems and the unpredictability of related disturbance processes, ecosystem management requires management on broad spatial and temporal scales in order to accommodate ecological change and to address multiple rather than single resources. Fifth, given the important role science plays in understanding natural systems, ecosystem management is based on integrated, interdisciplinary, and current scientific information that can be used to address risk and uncertainty. Sixth, because ecosystem management and the accompanying science are still experimental, ecosystem management requires an adaptive management approach that includes establishing baseline conditions, monitoring, reevaluation, and adjustment to reflect changes in scientific knowledge as well as evolving human concerns.[61]

In each instance, these governing ecosystem management principles represent a stark departure from traditional natural resource management policies. Whereas ecosystem management places a premium on biodiversity conservation and ecological integrity, traditional resource management has emphasized the production of commodities and services. Whereas ecosystem management contemplates broad public involvement in defining sustainability goals, traditional resource management has generally relied on the technical expertise of agency officials to establish priorities and strategies. Whereas ecosystem management disregards administrative boundaries and promotes institutional coordination, traditional resource management has been highly deferential to jurisdictional boundaries and agency officials have jealously guarded their own managerial prerogatives. Whereas ecosystem management acknowledges that resource systems are dynamic and nonequilibrium in character, traditional management has taken a more static and deterministic view of the landscape. Whereas ecosystem management views natural resources in a holistic, interrelated, and systematic context, the traditional management approach has focused on individual resources and short-range time frames. Whereas ecosystem management employs an adaptive approach to establishing resource policies and priorities, traditional management has usually established firm production targets and resisted subsequent modifications in order to promote predictability. While not exhaustive, this comparison between the two policies demonstrates how an ecological perspective alters the narrow and rigid management approaches that have dominated natural resource policy.[62]

Although existing federal law contains few explicit references to biodiversity or ecosystems, it is still possible to construct a solid legal foundation for ecosystem management on the public domain. The primary laws covering public lands are the organic statutes governing individual land management agencies, the Endangered Species Act, and the National Environmental Policy Act. According to the NFMA and FLPMA, the Forest Service and BLM operate under broad organic multiple-use mandates; they also must adhere to various environmental protection requirements and follow detailed planning processes. The NFMA features an explicit biological diversity provision, which has been translated into a viable population requirement to ensure the forest's ecological health.[63] The NFMA and FLPMA both provide for interdisciplinary planning, public participation in the planning process, and transboundary coordination, all of which are important components of ecosystem management.[64] The science-based Endangered Species Act contains powerful species and habitat protection requirements that give protected species priority over other statutory responsibilities on public lands. Under the ESA, the FWS has a major regulatory presence on public lands with its jeopardy review power and takings oversight responsibilities. The ESA essentially creates a habitat-based zoning system to protect listed species, thus effectively legitimizing ecosystem management as an overlay on public lands.[65] Although the NEPA is a procedural statute, it also promotes ecosystem management, primarily through its rigorous environmental impact statement requirements.[66] The NEPA cumulative effects environmental analysis requirements are designed to expand the temporal and spatial dimensions of agency decisions, which means full disclosure of serial development proposals and a comprehensive examination of transboundary impacts. These same provisions also have been used to inject conservation biology principles into land management decisions. Moreover, NEPA consultation and disclosure requirements promote interagency coordination, an important dimension of ecosystem management.[67] Beyond multiple-use lands, the national parks, wilderness areas, and national wildlife refuges are governed by separate organic statutes with explicit preservation mandates that generally support ecological management objectives on these lands.[68]

But the legal case for ecosystem management on the public domain does not rest on any one statute. It is, instead, based on the aggregate impact of these laws, including the substantive ecological protection standards found in the ESA and the various organic statutes as well as the integrated planning and procedural responsibilities that attach under the NEPA, NFMA, FLPMA, and other laws. The spotted owl litigation, as we shall see, brought this point home forcefully. In sustaining the federal government's ecosystem-based management plan for the Pacific Northwest's public forests, the court first cited to these myriad laws and then observed that the responsible agencies had no legal choice but to begin planning on an ecosystem basis. In other forest planning litigation, the United States Court of Appeals for the Ninth Circuit has ruled that proposed forest plan revisions require further ESA consultation whenever a new species is listed, which has forced the Forest Service to work with the FWS in managing the region's forests as ecological entities, giving due deference to all protected species. With the courts reading such ecological management obligations into these collective statutes, the public land agencies ignore these legal responsibilities at their peril. Rather, they can—and should—rely upon these statutes and decisions as well as their own inherent discretionary authority to develop functional ecological management policies for the public lands.[69]

The concept of ecosystem management offers land managers an opportunity to break down the humanity–nature or utilitarian–preservation dichotomy that has characterized natural resource policy for the past century. Even the words *ecosystem management* imply a merging of the natural and human: *ecosystem* suggests a natural setting shaped primarily by ecological processes, while *management* contemplates a human presence and involvement in shaping the natural world. Utilitarianism or multiple-use management, with its strong commitment to efficiency and scientific planning, has gradually acknowledged that human interests in natural resources cannot be defined solely in economic terms but must also include biological, aesthetic, and other considerations. The preservationist movement, despite its long-standing commitment to aesthetics, has been influenced profoundly by recent scientific developments; it is now using the ecological sciences to protect biodiversity, natural processes, and linkage corridors in a dynamic landscape setting. Both philosophical schools have generally accepted the cogent scientific and philosophical arguments supporting the need to safeguard all species—not just charismatic megafauna—against extinction. And managers from the public land agencies—whether multiple-use or preservationist in orientation—are all well aware that resource decisions inevitably trigger repercussions beyond their immediate boundaries, which can also affect their own management options and strategies. In short, ecosystem management concepts are being embraced by both the utilitarian and preservation schools of conservation thought. Whether this new ecological perspective can finally fuse these two divergent philosophies together will depend on related political developments and how these new principles are applied on the ground.[70]

CRITICISMS AND CHALLENGES

The emerging doctrine of ecosystem management cuts profoundly against many of the long-standing assumptions undergirding natural resource law and policy. Property rights and most natural resource laws are based upon the notion of fixed boundaries, which have rarely been defined in ecological terms. The very concept of legal ownership implies certainty and stability, but the nature of ecosystems is instability and disequilibrium, requiring flexible management strategies based on adaptive experimentation. The existing legal order is generally designed to ensure prompt and tangible financial returns, while ecological management gives priority to biodiversity conservation and often requires managerial forbearance. Critics have seized upon these and other problems in an effort to discredit new ecosystem management concepts.[71]

One key point of contention is whether ecosystem management has any substantive content or whether it is merely a process. Noting that many ecosystem management definitions give priority to biodiversity conservation and ecosystem integrity goals, critics assert that ecosystem management is really just a poorly disguised effort to elevate environmental protection goals above commodity production on the public domain. They reject this transparent shift in management priorities, arguing that it violates existing legal standards governing public lands, has not been endorsed by the general public, and ignores important countervailing economic and social considerations. Other analysts, taking a more circumspect view, suggest that neither ecosystem protection nor biodiversity conservation can properly be regarded as an ecosystem management goal; instead, they view ecosystem management as a process that enables natural resource managers to identify specific production, protection, and restoration goals. These goals, however, are not inherent in ecosystem management but reflect "desired social benefits" that should be "defined by society, not scientists." Under this approach, biodiversity conservation and ecological preservation may (or may not) be a desired social benefit that emerges from ecosystem management processes.[72]

A second major point of contention focuses on the inherent difficulty in defining and using ecosystem boundaries for management purposes. Because the ecosystem concept originally was conceived as a theoretical construct for research purposes to establish experimental boundaries, critics contend that ecosystems are ill-suited for land management purposes. They note that the ecosystem concept is inherently malleable and can be defined on multiple scales and in diverse settings. If ecosystem management is applied on a large scale (which is what most proponents envisage), then critics fear that regulatory chaos will ensue. On the public lands, managers confronting transboundary issues will face an unsolvable dilemma in determining whose legal mandate should prevail in the event of a conflict over resource priorities. Private landowners, on the other hand, will face the prospect of an expanding federal regulatory presence that could limit their land use and development options. Others, also concerned about potential boundary problems, believe ecosystem management requires clearly defined, place-based boundaries, which will enable managers to address concrete resource problems.[73]

A third potential problem is the inherent vagueness and uncertainty of the ecosystem management concept and related terms. Ecosystem management definitions that do not establish definitive priorities among environmental protection and related socioeconomic concerns provide managers with little clear guidance on how the policy should be implemented on the ground. In the event of a conflict between ecological and economic objectives, should the manager authorize a timber sale or mining project on undeveloped land, or does ecosystem management dictate a negative decision in this instance? From an economist's perspective, the vague ecosystem management concept provides no evident basis for making a cost-benefit analysis or for determining appropriate trade-offs. Even if the protection of ecosystem integrity or biodiversity conservation is the highest priority, critics ask how these concepts are to be defined and measured. They similarly question whether the amorphous concept of sustainability provides any clearer guidance. Relatedly, they observe that uncertainty pervades ecological science, and that scientists frequently disagree among themselves over basic concepts and the proper interpretation of experimental results. They note that the inherent instability and unpredictability of ecological processes make it difficult to define a clear management target or goal. Given these definitional problems, critics believe, it is virtually impossible to set meaningful ecosystem management standards, which means there is no basis for measuring management performance or for holding managers accountable. From a commercial perspective, these definition problems can create intolerable uncertainty, deter financial planning, and increase transaction costs. From the government's perspective, an expanded planning regime could prove exceedingly costly and bureaucratically unwieldy.[74]

Fourth, critics have questioned whether ecosystem management can be reconciled with basic democratic principles. Several critics have charged that an ecosystem management policy based on protecting ecosystem integrity is so technical that it essentially vests scientists with ultimate management authority and effectively excludes affected parties and the general public from any meaningful role in management decisions. From this perspective, ecosystem management recalls an earlier and thoroughly discredited era when claims of scientific expertise were allowed to dictate natural resource policy without regard to competing social value preferences. One particularly harsh critic, noting that a widely cited ecosystem management proponent has advocated "avoid[ing] the democratic trap of giving equal weight to all interest groups [because] many would destroy biodiversity for economic gain," asks who would be entitled to serve as nature's proxy in determining appropriate management objectives.[75] An unduly vague ecosystem management policy that lacks clearly articulated and enforceable standards would also violate fundamental democratic principles of government accountability. Moreover, because ecosystem management contemplates interagency coordination to achieve shared ecological objectives, any ambiguity in the basic policy could undermine coordination efforts and leave participants without any clear sense of direction or accomplishment. Coordination merely for coordination's sake is the antithesis of an efficient and accountable bureaucracy.[76]

Fifth, noting that several ecosystem management models contemplate local collaborative processes to set the resource policy agenda, environmental critics fear these processes will inevitably promote commodity production over countervailing ecological goals. If ecosystem management is merely a process to establish natural resource priorities, they fear, industry and its allies in resource-dependent communities will dominate that process, as they have in the past. They note that the environmental constituency tends to be concentrated in urban areas and is thus logistically unable to participate effectively in locally based collaborative processes where the critical decisions would be made. They also are concerned that locally based collaborative processes will tend to ignore or discount important scientific evidence documenting ecological conditions in order to promote local economic interests. Eschewing local partnership arrangements, they would rather place their faith in national regulatory standards, which are designed to protect environmental values and are enforceable in courts. In fact, citing the Forest Service's 1995 proposed revisions to its NFMA regulations in the name of ecosystem management, environmental critics are concerned that ecosystem management is simply a subterfuge to return maximum discretion to the land manager or to local partnerships. Unless ecosystem management policy contains legally enforceable standards promoting clearly defined ecological preservation goals, they reject it as merely another effort to decrease environmental protection of the public lands.[77]

Finally, critics have asserted that there simply is no legal basis for ecosystem management on public lands, which makes it an entirely illegitimate policy. Not only does existing law contain no explicit ecosystem management provisions, but Congress has failed to adopt proposed ecosystem management legislation when presented with the opportunity to do so. Granted, one might stitch together legal authority supporting ecological management policies from the diverse environmental and other statutes governing public lands, but critics believe these laws do not displace the governing organic mandates, which clearly mandate commodity production and other tangible outputs as primary policy goals. As a practical matter, in the aftermath of the 1994 elections and the Republican recapture of Congress, the term *ecosystem management* itself was called into question and deleted from the working vocabulary of many Washington-based policy makers otherwise committed to instituting new ecological policies on public lands. (The term has not, however, disappeared from the policies and practices of land managers outside the nation's capital, though the George W. Bush administration is no proponent of it either.) Thus, convinced that ecosystem management lacks a strong legal or political foundation, critics argue that any administrative policies promoting ecological objectives must first meet existing multiple-use and other statutory standards governing the public domain.[78]

Nonetheless, ecosystem management has plainly taken hold on the public domain, and it is proving resilient enough to surmount these criticisms and problems. Indeed, within the federal bureaucracy, the halls of Congress, and elsewhere, an overt struggle is under way between the proponents and opponents of ecosystem management to shape the concept to fit their particular agendas and predispositions. While acknowledging that both definitional and technical problems must still be resolved, public land agencies are nevertheless engaged in myriad ecosystem-based experiments, employing new scientific insights and cooperative decision making models to reshape natural resource policy on the ground. Related ecological reforms are also evident in several recent legislative and administrative initiatives. Whether the term of art is ecosystem management, an ecosystem approach, integrated resource management, or collaborative stewardship, there is general agreement on the underlying rationale and direction for new ecological approaches to public land management. How these experimental endeavors fare and whether any universal lessons emerge from them will help further shape ecosystem management strategies and determine whether they represent a viable long-term natural resource policy. In the meantime, with the spotted owl controversy as a catalyst, federal ecosystem management policy has already gained a measure of legitimacy from the courts and developed a momentum of its own across the federal landscape.

FOUR

Ecology Triumphant?

Spotted Owls and Ecosystem Management

Implicit in the ecosystems concept is recognition that maintenance
of the ecosystem depends upon the consistency of man-made
standards, laws, and boundaries with those that have evolved
through natural processes.

LYNTON CALDWELL (1970)

It seems to me obvious that we have a de facto policy of biodiversity
protection, particularly on national forest lands. In fact, biodiversity
protection becomes an overriding objective of federal land
management.

JACK WARD THOMAS (1993)

After percolating for more than a decade in the academic journals, ecosystem management finally attained legitimacy in the rancorous spotted owl controversy. Few if any conflicts have tested our governmental institutions and environmental laws like the epic struggle over the Pacific Northwest's ancient forests. Encompassing nearly a decade of contentious litigation, congressional recalcitrance, and administrative vacillation, the old growth imbroglio starkly demonstrates the important yet often painful adjustments that the age of ecology is bringing to the public domain. The controversy was originally cast as an "owls versus jobs" fight, but most observers soon acknowledged that the northern spotted owl really was a surrogate for the region's imperiled ancient forests, a veritable ecological treasure trove as well as an important recreational and spiritual resource. As that controversy unfolded, the federal courts played a pivotal role in bringing ecosystem management principles onto the public domain. By the early 1990s, finding themselves enjoined from selling any more timber, the Forest Service and the Bureau of Land Management (BLM) were finally forced to address the needs of the region's forest-dependent species. Environmental litigants had by then gained the upper hand in the timber wars, having converted biodiversity conservation into a legally enforceable dimension of national forest policy. With Congress unable to achieve any legislative consensus, an energetic new president seized the initiative and convened a team of scientific experts to address the problem. They proceeded to conceive a bold, new forest plan, coalescing around a stunningly broad ecosystem approach to forest management—one that encompassed all the region's public forest lands, mandated biodiversity protection across the landscape, and dramatically cut timber harvest levels. When the district court endorsed the plan and lifted its injunction, it heralded the first major federal experiment with ecosystem management on public lands.

Indeed, the spotted owl controversy thrust biodiversity conservation and ecosystem management into the mainstream public land policy debates. The Northwest Forest Plan, despite some bumps along the way, restored a semblance of order to the region. Timber started flowing again though at greatly reduced levels; the agencies began working together and paying attention to the ecological ramifications of their decisions. More broadly, both the Forest Service and BLM moved to incorporate ecosystem management principles into their overall management agendas, giving biodiversity protection a prominent—if not priority—position on public lands. Key planning regulations were revised to reflect these new commitments, and the agencies undertook several high-profile regional ecosystem management initiatives designed to reorder priorities across the public landscape. Although few courts have since brandished their injunctive powers so widely, the judiciary's post–spotted owl decisions still evince a heightened sensitivity to the role of biodiversity protection and ecosystem conservation on public lands. Congress, however, has been a reluctant—and often obstructionist—participant in this new ecosystem management movement. Its 1995 timber salvage rider effectively overrode the Northwest Forest Plan's biological protections, and it regularly tried to thwart key Clinton administration reforms, including its star-crossed National Biological Survey initiative. But it has also written biodiversity conservation into new legislative mandates, extended the role of science on the public lands, and funded several innovative, regional planning initiatives. In an era steeped in the ecological sciences, pervasive environmental regulation, and public interest litigation, new biodiversity and ecosystem management concepts have plainly become forces to be reckoned with on the public domain.

The Spotted Owl Controversy

TIMBER AND THE OLD GROWTH FORESTS

The Pacific Northwest is a land of forests. Dense coniferous forests blanket the region, stretching from the Pacific seacoast up and across the rugged Cascade mountain range. Blessed with a mild climate and regular rainfall, the forests west of the Cascade range produce abundant timber, long regarded as one of the region's most important products. Approximately 55 percent of the west-side timberlands are public lands, owned by the federal government, the state, or the counties; the remainder is privately owned, and most of it has been logged for commercial purposes. Roughly 30 percent of the public land is protected by national park, wilderness, or similar protective designations, and thus unavailable for commercial timber harvesting. The remaining public lands (approximately 21 million acres) are contained in 19 national forests and 7 BLM resource districts that stretch across Washington, Oregon, and northern California, and that are open to logging. It is these lands that have been the focus of the old growth controversy.[1]

The Northwest's federal timberlands are managed by the Forest Service and the BLM. The Forest Service manages its lands under its classic multiple-use mandate, while the BLM manages its forestlands in western Oregon under a sustained yield-forest production standard. In the years following World War Two, timber dominated both agencies' agendas. With few environmental laws on the books, the question was simply when, not whether, to cut. But by the mid-1970s, a plethora of new laws—namely the National Forest Management Act (NFMA), Federal Land Policy Management Act (FLPMA), National Environmental Policy Act (NEPA), and the Endangered Species Act (ESA)—had injected environmental constraints and detailed planning obligations into the management equation. As the old growth controversy played out, these laws became the primary vehicles for bringing what had become a logging juggernaut to a standstill.[2]

Before World War Two, the nation's public forests were consigned to custodial management and spared from extensive logging. Most commercial timber was harvested from private forestlands, which were adequate to meet prewar demands. In fact, fearful of governmental competition, the timber companies had actively discouraged the Forest Service from harvesting its own trees. After the war, however, everything changed. A building boom and soaring birthrates triggered an enormous demand for timber. With the nation's private forestlands near depletion, the timber companies turned their attention to the public forests to address a projected housing shortage. Federal timber was abundant, mature, and largely intact, particularly in the western United States. In relatively short order, the nation's public forests were transformed into a timber breadbasket: timber sale receipts tripled from 1946–50, and then tripled again from 1950–56. The volume of timber harvested from the national forests jumped from an average of 1 billion board feet (bbf) annually prior to the war to a record high of 12.1 bbf in 1966.[3]

The Pacific Northwest, with its mature forests and high-quality wood, quickly became a major source of the nation's softwood lumber. Elsewhere the national forests could not match the Northwest's old growth timber: the upper midwest had been harvested during earlier eras and contained less productive second growth timber, while the Rocky Mountains were too high and dry to grow much valuable timber. By 1959, the Northwest was providing nearly 45 percent of the national forest timber harvest. The region's abundant Douglas fir stands produce large amounts of commercially valuable lumber: a single mature tree may contain over 2,000 board feet of lumber, and only seven of these trees are necessary to build an average house. With technological improvements, such as the gas-powered chainsaw and diesel-powered heavy equipment, the timber was increasingly accessible. No longer dependent on rivers to transport timber to market, loggers began bulldozing new roads into the mountains and expanding harvest operations into roadless areas and onto steeper terrain. Gradually, as the onslaught continued, the environmental toll began to mount.

By the 1980s, the timber industry had installed itself as a major component of the regional economy, and it wielded extensive political power. According to some reports, timber accounted for nearly one third of Oregon's economic base, and it played a lesser but nonetheless important role in Washington's state economy. By 1970, more than 10 percent of the regional workforce was employed in timber-related jobs, though that number suffered a slow but steady decline over the ensuing years. Even with a recession-induced slowdown, timber harvesting on the region's public forests averaged 4.5 bbf annually from 1980 to 1989. (By comparison, the entire national forest system produced approximately 11 bbf annually during these same years.) The cutting reached a fever pitch in 1989 when the total regional harvest from all land ownerships topped 15.6 bbf amid growing protests that the cut was unsustainable. Environmentalists objected that excessive logging was damaging fragile forest resources, including water quality and critical old growth species. But the politically powerful timber industry and its allies were not about to countenance any changes. Acutely sensitive to these important constituents, the region's congressional delegations used the Forest Service's annual budget to establish ever higher timber targets, known as the "allowable sale quantity" (ASQ) in forestry vernacular.[4]

In the decades following the war, small towns and families throughout the region grew increasingly dependent on what seemed an inexhaustible supply of old growth timber. The Forest Service, under a long-standing agency commitment to promote local industry, actively fostered an expectant set of timber-dependent communities across the region. The timber industry provided relatively high paying jobs for loggers and mill workers in rural locations that generally lacked other comparable employment opportunities. County governments relied on the 25 percent federal stumpage fees to support local schools and other public services. Even as increased automation forced some mills to lay off workers and even as storm clouds began brewing over the mounting environmental toll, the region's rural communities remained staunchly attached to the timber economy and its manifold local benefits. For many rural residents who had seen one generation follow another into the forest, logging represented a way of life that was disappearing elsewhere in the United States. Besides, small town living enabled them to avoid the social ills that plagued modern urban life. Most of the region's political leaders, acutely sensitive to these concerns, were not prepared to reevaluate timber's role in the region's economy to address disputed environmental concerns.[5]

THE ANCIENT FOREST ECOSYSTEM

The old growth forest has not always been viewed as a rich and diverse ecosystem. Before 1970, with the postwar logging industry in full swing, scientific studies on the Northwest's forests focused primarily on silvicultural protocols, namely how to optimize harvest levels and promote faster regeneration. Most biologists regarded the region's old growth forests as biological deserts, believing the thick coniferous canopies blocked sunlight from reaching the forest floor, thus creating an impoverished ecosystem. Environmental organizations also showed little interest in the region's old growth forests. During the 1970s and early 1980s, local activists concentrated their energies on winning wilderness designation for the region's alpine mountain ranges, many of which were eventually protected as wilderness sanctuaries. They ignored most of the valuable adjacent timberlands, which were left unprotected.[6]

By the mid-1970s, scientists were beginning to reevaluate the biological importance of the old growth forest. Eric Forsman, an Oregon State University graduate student, sounded the initial alarm. While doing research for his doctoral dissertation, Forsman grew concerned that clearcut logging practices were harming northern spotted owls in the region's forests. Not long afterward, a team of Forest Service–affiliated scientists concluded that old growth forest ecosystems are really quite different from younger forest ecosystems, particularly those represented in managed forests. By then, though, accelerated logging was rapidly converting the region's remaining old growth forests into even-aged stands and fragmenting the forest landscape into smaller and smaller patches of old growth trees.[7]

Subsequent studies revealed just how ecologically unique the Northwest's old growth forests really are. These ancient forests contain more than a dozen different species of conifers adapted to the region's topography and climate. The dense forest vegetation both shades and protects streams, thus providing an important cold water aquatic habitat and enhancing water quality. Many of these trees are very old (700–1,000 years) and quite tall (200–300 feet); they create a thick, multilayered forest canopy with a correspondingly rich understory. In the ancient forest, even dead trees perform vital ecological functions, providing snag habitat for cavity nesting birds and other animals, generating decaying matter to enhance soil nutrients, and stabilizing soil and streambanks as deadfalls. Fallen or "blow down" trees are also important, because they open the canopy and thus allow sunlight to penetrate the forest and stimulate new growth on the forest floor. In addition, scientists discovered that these unique old growth ecosystems actually support a remarkable variety of life not ordinarily found in younger forest ecosystems, including spotted owls, pine martens, and fishers. In short, the old growth forest constitutes a unique and vital ecosystem rich in biodiversity.[8]

Among the old growth species, none has achieved the notoriety of the northern spotted owl, which became both a legal icon and a potent political symbol in the struggle over the region's remaining old growth ecosystems. A nongame species with no obvious economic value, the northern spotted owl is a brown, medium-sized owl with white and mottled coloration; it is reclusive and nocturnal, though not afraid of humans. It is very habitat-sensitive, depending mainly on old growth forests in western Washington and Oregon, northwestern California, and southwestern British Columbia for its nesting, foraging, and other needs. A highly monogamous species, owl pairs typically require 3,000–5,000 acres of forested home range habitat, with a substantial portion of that range in uncut, old growth timber. By the 1980s, scientists estimated that only 3,000–4,000 owl pairs remained. They attributed the rapid owl population decline to intense timber harvesting that was destroying its necessary habitat. Other forest-dependent species— the marbled murrelet, various anadromous fish runs, northern goshawk, pine marten, and lynx—were disappearing too, a further sign of serious ecological degradation.[9]

THE CONTROVERSY CRYSTALLIZES

With these scientific disclosures, public land agencies slowly started to address owl protection concerns in their timber harvest plans. In the mid-1970s, following Forsman's original revelations, the Forest Service designated the northern spotted owl as an indicator species for assessing the ecological health of the region's old growth forests. In 1977, four years after receiving initial owl protection recommendations, the Forest Service and BLM agreed to protect 400 owl pairs by designating 300-acre set-asides around spotted owl nesting sites, but only on an interim basis until they completed their respective planning processes. Environmentalists formally entered the spotted owl debates when they appealed this administrative decision, arguing that it would not adequately protect the owl or ensure its long-term survival. In response, the Forest Service agreed to address spotted owl management concerns in its forthcoming Regional Guide, which was being prepared as part of the new NFMA forest planning process.[10]

But when the regional guidelines finally appeared in 1984, they displeased virtually everyone. With little explanation, the Regional Guide recommended a rigid (very unecological) grid pattern of 1,000-acre set-asides in old growth forests—known as spotted owl habitat areas (SOHAs)—to protect a meager 375 spotted owl pairs (later revised to 551 pairs). Another administrative appeal ensued, which the Forest Service lost. It was ordered to prepare a supplemental environmental impact statement (SEIS) to the regional guidelines that would examine a broad range of alternative spotted owl management plans and contain a biological rationale for its chosen alternative. For the first time, the agency was required to provide a scientific justification for its old growth timber management program.[11]

In 1986, the Forest Service released its draft SEIS with additional spotted owl management protections. The SEIS proposal still protected just 550 spotted owl pairs, but it increased the set-aside allowances from 1,000 to 2,200 acres for SOHAs—a significant concession of timberlands from the agency's perspective. While forecasting a 5 percent reduction in timber production, the proposal still gave the owl only a small chance of long-term survival, and it did little to stem environmental criticism. With President Ronald Reagan ensconced in the White House and his conservative administration committed to economic development on public lands, the Forest Service was determined to minimize the owl's impact on its timber harvest levels. Moreover, the Democrat-controlled Congress had a long history of setting high-level ASQ expectations in the agency's annual budget, and there was no sign it was ready to change direction. Two years later, the Forest Service's final EIS adopted basically the same position, giving environmental advocates a final decision that they could take to court. Meanwhile, the BLM refused repeated requests to undertake a similar scientific analysis of spotted owl population numbers and habitat needs on its forests, setting itself up too for a lawsuit to halt its aggressive harvesting policies.[12]

By the late 1980s, scientists and environmental organizations were gravely alarmed over the region's declining old growth forests and the agencies' continued indifference to the owl's plight. The science concerning old growth ecosystems and species population dynamics had improved markedly over the past decade, giving a real sense of urgency to the developing controversy. Nearly everyone agreed that less than 15 percent of the region's old growth forests remained, reflecting the rapid pace at which the ancient forests were being liquidated. Ecologists were also convinced that spotted owl population numbers were dangerously low and that large blocks of contiguous old growth habitat were essential to the species' long-term survival. State agencies shared these concerns. By 1988, the state of Washington had listed the owl as a state endangered species, and the Oregon-Washington Interagency Scientific Committee believed more rigorous spotted owl protections were needed than those recommended by the federal agencies. Armed with this extensive and unrefuted scientific evidence, environmental advocates were positioned to mount a litigation campaign challenging federal timber management policies.[13]

The federal agencies, however, remained committed to an administrative solution to what was quickly devolving into the "owl wars." In 1989, still seeing the controversy in purely technical terms, the federal agencies convened a panel of scientific experts—dubbed the Interagency Scientific Committee (ISC)—to develop "a scientifically credible conservation strategy for the northern spotted owl." The resulting report, which became known as the Thomas Report (after Forest Service biologist Jack Ward Thomas, who chaired the committee), concluded that "the owl is imperiled over significant portions of its range because of continuing losses of habitat from logging and natural disturbances" and that "current management strategies are inadequate to ensure its viability." Drawing upon population viability models, the Thomas Report recommended a habitat conservation area (HCA) strategy designed to protect "large blocks of [unlogged] habitat capable of supporting multiple pairs of owls and spaced closely enough to facilitate dispersal between blocks." It also advocated an adaptive management strategy that would involve regular monitoring of the owl population and management adjustments based on new information. And to address serious inconsistencies among the agencies responsible for managing spotted owl habitat, the report called for a uniform and coordinated conservation strategy. A comprehensive, landscape-level strategy, the report noted, would benefit other old growth species too.[14]

One year later, Congress was finally stirred to action. Faced with growing political fallout, Congress established a Scientific Panel on Late Successional Forest Ecosystems to report on how a network of old growth reserves would affect the region's federal timber program. The committee (which became known as the Gang of Four) consisted of four distinguished Ph.D. scientists with extensive forestry experience. Pooling their knowledge, they evaluated an assortment of ecological management strategies for the region's old growth forests. Their conclusion was unequivocal: "Increases in the probability of retaining a functional LS/OG [late-successional–old growth] forest network, viable populations of northern spotted owls, or habitats of other LS/OG species and potentially threatened fish species and stocks decrease sustainable harvest levels, which in turn result in decreased regional employment and income levels." In other words, a viable biodiversity conservation strategy would mean further reductions in timber production.[15]

Taken together, the two reports revealed that the controversy had now become a zero sum game. Any increase in spotted owl protection would come at the expense of timber production. There simply was no painless way to preserve the ecological integrity of the region's public forests while maintaining a timber harvest program at existing levels. Even more alarmingly, the reports demonstrated that a significantly expanded conservation strategy was essential to ensure the owl's survival. Perhaps these conclusions should have been obvious by then, but the reports defined the stakes with renewed clarity. Nonetheless, Congress was still paralyzed, unable to reach any consensus over an acceptable long-term solution. And the agencies, with their recalcitrance and foot dragging, had lost all credibility. This left the matter—by default—to the courts.

Law, Litigation, and Ecosystem Protection

By the late 1980s, the basic ingredients were in place for environmental advocates to mount a serious legal challenge on behalf of the spotted owl and related forest ecosystem concerns. Northwest environmental activists, with the state national forest wilderness designation campaigns winding down, were primed to refocus their attention on the remaining unprotected national forestlands, which were also the region's timber-rich old growth forests. Proven litigation tools were available: both the National Environmental Policy Act and the Endangered Species Act had been used effectively to block federal resource development projects to protect environmental values. The yet untested National Forest Management Act of 1976 also might be invoked: it imposed detailed, interdisciplinary planning responsibilities and environmental constraints, including a new biodiversity conservation obligation, on forest managers that should provide a basis for judicial review of timber sale decisions. Moreover, a little-known Massachusetts environmental group, Greenworld, had already petitioned the U.S. Fish and Wildlife Service (FWS) to list the spotted owl on the endangered species registry—an action that would soon bring that regulatory agency directly into the timber controversy. Unlike the land management agencies, the FWS was charged with making purely scientific judgments about the owl's current status, and it held a veto power over any timber cutting decision that might jeopardize the owl's existence. The stage was set for a protracted legal and political battle over the future of the spotted owl and the region's old growth forests.[16]

Spanning nearly a decade, the old growth litigation tested the limits of the nation's environmental laws, federal judicial power, and administrative authority. The bitter and seemingly endless court battles took three separate yet convergent tracks. On one track, environmental advocates successfully invoked the Endangered Species Act to protect the spotted owl, first forcing the FWS to list the owl as a threatened species and then compelling it to designate critical habitat. On a second track, environmental advocates invoked the NEPA, NFMA, and ESA against the Forest Service to force dramatic revisions in national forest timber harvest policies to protect the region's old growth ecosystems. On a third track, they used similar laws to bring the BLM's timber harvest program to heel on its Oregon and California (O and C) lands, likewise forcing a radical drop in the harvest volume.

In the course of the litigation, Congress adopted several appropriations riders to suspend operation of the environmental laws, key cases bounced back and forth between the trial and appellate courts, and the president was ultimately forced to broker a compromise. When the dust finally settled, timber production on the region's federal forests was reduced by almost 80 percent, and the federal land management agencies had embraced a comprehensive ecosystem-based plan for managing the region's federal forests.

ENDANGERED SPECIES LITIGATION

The ESA, which is administered by both the FWS and the National Marine Fisheries Service (NMFS), has been described as the "pit bull" of environmental statutes. Under this act, Congress established a powerful species preservation mandate that obligates all federal agencies to conserve and protect any "listed" endangered or threatened species from harm—a substantive obligation that takes precedence over any other statutory obligation. In a seminal 1978 decision, the Supreme Court strictly interpreted the act's preservation purposes, enjoining completion of the Tennessee Valley Authority's multimillion-dollar Tellico Dam to avoid inundating the protected snail darter's remaining habitat. To enforce its species preservation mandate, the ESA establishes an intricate array of substantive and procedural obligations: these include an initial "listing" decision, a critical habitat designation, preparation of a recovery plan, a "jeopardy" review process, a prohibition against taking any endangered species, and an elaborate exemption process, all enforceable through a broad citizen suit provision. Most of these provisions were litigated during the old growth controversy, which saw the courts enforce these obligations against a variety of recalcitrant federal officials who desperately sought to avoid the ESA's hard strictures in order to maintain their own discretionary authority.[17]

Before the ESA can come into play, a species must first be "listed" as either threatened or endangered, a decision based solely upon scientific considerations. In the case of the spotted owl, despite unrefuted evidence that its population was dwindling, the FWS initially rejected Greenworld's 1987 listing petition, concluding summarily that listing "is not warranted at this time." On appeal, however, Washington federal district judge Thomas Zilly ruled that the agency acted arbitrarily when it "disregarded all the expert opinion on population viability, including that of its own expert, that the owl is facing extinction, and instead merely asserted its expertise in support of its conclusion." The court ordered the agency to explain its nonlisting decision and to present further substantiating evidence. Not surprisingly, given the overwhelming weight of scientific opinion, the FWS reversed itself in June 1990 and listed the owl as a threatened species. With this decision, the powerful ESA was brought directly into the old growth controversy, making the FWS responsible for reviewing all timber sale plans and proposals.[18]

But the FWS was not finished equivocating. Although the ESA provides that a listing decision ordinarily should be accompanied by a critical habitat determination, the FWS announced that it could not determine critical habitat for the spotted owl. A critical habitat determination, of course, would require the FWS to develop information on spotted owl locations and on the condition of the region's old growth forests—information that both the FWS and the land management agencies feared could force a drastic reduction in timber harvesting. To explain its decision, the FWS observed that the northern spotted owl's range extended over 7 million acres of old growth forest habitat and that much of this land was fragmented into isolated habitat islands by extensive logging, concluding that it lacked information on connecting linkages between these islands and on the economic impacts accompanying a critical habitat designation. Again, however, Judge Zilly found the FWS had ignored its statutory duties, holding that habitat designation "is a central component of the legal scheme developed by Congress to prevent the permanent loss of species." He ordered the agency to prepare a critical habitat proposal within 45 days. Two years later, after innumerable delays, the FWS finally issued its critical habitat determination in January 1992, designating approximately 6.9 million acres for the northern spotted owl. Although such an expansive critical habitat designation would ordinarily forestall most development activities, by this time other laws had blocked most public land logging proposals. Besides, the timber industry's allies initially succeeded in setting aside the critical habitat designation, thus further delaying application of this important ESA protection.[19]

Beyond the spotted owl, several other imperiled species also shaped the Northwest's old growth controversy. When the FWS listed the marbled murrelet as a threatened species, this coastal forest-dwelling seabird became an important consideration in the Northwest Forest Plan, and its presence has imposed limits on private land logging practices. Faced with several listings and proposed listings of salmon runs, the Northwest Forest Plan team created a new aquatic conservation strategy in an effort to mitigate the impacts of timber harvesting, road construction, and erosion on salmon spawning grounds and other key habitat areas. Under the aquatic conservation strategy, the Forest Service and BLM are obligated to protect designated riparian reserves and to conduct watershed analyses before disturbing any riparian habitat. Once listed, the mere presence of these wide-ranging species brings the FWS and NMFS squarely into the forest planning process, giving them a significant regulatory presence as well as jeopardy review authority. Notwithstanding these ESA protections, the NFMA contains a biodiversity conservation obligation, which has been translated into a minimum viable population regulation that obligates the Forest Service to identify sensitive or potentially endangered or threatened species and to monitor population and habitat trends for each species as forest plans are implemented. As we shall see, once the agencies committed themselves in the Forest Plan to protect these species, the court warned that failure to monitor population trends could jeopardize the plan, which is precisely what happened when the agencies reneged on these obligations.[20]

NATIONAL FOREST LITIGATION

In the initial legal skirmish over national forest timber policy, environmental advocates scored an early but shortlived court victory on behalf of the spotted owl. In December 1988, after the Forest Service released its SEIS on spotted owl management, environmentalists filed a lawsuit in Washington federal district court over the proposed plan, alleging that the agency had violated the NFMA, NEPA, and Migratory Bird Treaty Act. The case was assigned to federal judge William Dwyer, who had been appointed to the federal bench by President Reagan. Judge Dwyer surprised many people when he agreed with the environmental plaintiffs and granted an initial injunction that blocked 140 pending timber sales. At the same time, the FWS announced it was proposing to list the spotted owl under the ESA. In tandem, the two decisions sent an immediate chill throughout the timber industry and the region's timber-dependent communities, who feared— correctly, as it turned out—that environmentalists might use these legal tools to force changes in federal timber policy.[21]

Congress promptly intervened in the budding controversy to protect the region's timber industry, adopting the so-called Northwest Timber Compromise as part of the 1990 Department of the Interior appropriations bill. This was not the first time Congress had employed an appropriations rider in response to litigation over timber harvesting in the Pacific Northwest: it had used a rider for three years to stop litigation against the BLM's timber sales on its spotted owl lands in Oregon, and it earlier used a rider to overturn a court order stopping logging on Oregon's Siuslaw National Forest. To appease timber proponents, the 1990 rider (known simply as Section 318) established an annual 5.8-bbf timber sale target for the region's national forests. To address environmental concerns, Section 318 protected already designated SOHAs, mandated minimum fragmentation of old growth forests, and set aside additional acres from logging. But then, in its most controversial provision, Section 318 declared that these environmental protections were sufficient to meet the legal standards involved in the pending old growth litigation (specifying the cases by name) and prohibited further judicial review of this litigation. Faced with this new statutory language, Judge Dwyer felt compelled to dissolve his preliminary injunction and to allow logging to resume. The case would eventually wend its way to the U.S. Supreme Court, where the controversial Section 318 rider was sustained over the argument that Congress had violated constitutional separation-of-powers principles by intruding into the judicial domain. But by then, the rider had expired, Congress had become more sympathetic to the environmental position, and the regional congressional delegations were unable to muster enough votes to support another rider. A final judicial showdown loomed.[22]

In this next round of litigation, environmental advocates again challenged the legality of the Forest Service's spotted owl management plan. But now the owl was listed under the ESA as a threatened species, and the ISC had issued its report, concluding that the current management approach "is unacceptable and has contributed to a high risk that spotted owls will be extirpated from significant portions of their range." In response, the Forest Service abandoned its original spotted owl management proposal (as adopted in the 1988 Forest Service EIS) and amended its forest plans to provide for timber management "in a manner not inconsistent with [the ISC recommendations]." But the agency then took the inexplicable position that, because the owl was now protected by the ESA, it no longer had any obligation under the NFMA diversity requirements to ensure minimum viable populations through its planning processes. And it did so without any opportunity for public involvement or review.[23]

Environmentalists immediately counterattacked, alleging that the Forest Service had violated both the NFMA and NEPA in adopting its revised spotted owl management policy. Again the district court agreed. Judge Dwyer clearly attached independent significance to the NFMA biodiversity requirements, ruling that "the duty to maintain viable populations of existing vertebrate species requires planning for the entire biological community—not for one species alone. It is distinct from the duty, under the ESA, to save a listed species from extinction." And he ruled that the Forest Service violated the NFMA amendment procedures by not providing for public participation. On May 23, 1991, Judge Dwyer granted an injunction barring further timber sales on the region's old growth forests, and he ordered the Forest Service to develop spotted owl management guidelines consistent with the NFMA planning procedures, including preparation of an EIS. Remarkably, the judge found that "the most recent violation of the NFMA exemplifies a deliberate and systematic refusal by the Forest Service and the FWS to comply with the laws protecting wildlife . . . [which] reflects decisions made by higher authorities in the executive branch of government." On appeal, the Ninth Circuit sustained the district court judgment, observing that the NFMA and ESA contained concurrent statutory requirements that would require coordination among the Forest Service, BLM, and FWS "to take aggressive steps to avoid a species extinction and preserve its viability."[24]

With these rulings, the courts brought biodiversity conservation to the fore in the Pacific Northwest forests. Confronted with the federal government's single-minded commitment to timber production goals, the courts had little choice but to extract a biodiversity conservation obligation from the governing environmental laws and to treat it as a counterweight to timber on the region's multiple-use lands. The decisions linked the NFMA and ESA species conservation obligations together, and then further linked these obligations to ecosystem preservation across agency boundaries. Moreover, the sweeping remedial order brought national forest timber sales in Oregon, Washington, and northern California to a standstill. When Congress failed to adopt another rider removing the controversy from this judicial forum, the courts became the primary venue for shaping new ecological policy.

Less than a year later, in an effort to resume its timber program, the Forest

Service completed a new EIS adopting the ISC spotted owl management strategy for the region's national forestlands. But once again, environmental advocates challenged the plan. They alleged a NEPA violation because the EIS failed to address relevant scientific information, and they asserted NFMA violations because it did not adequately protect other old growth–dependent species or ESA-designated critical habitat. Judge Dwyer again agreed with the environmental plaintiffs, and he enjoined old growth timber sales until the Forest Service remedied the EIS's deficiencies. Putting the controversy in perspective, Judge Dwyer noted that "a chief concern of scientists of all persuasions has been whether the owl can survive the near-term loss of another half-million acres of its habitat." He then held that the EIS failed to address credible new scientific information inconsistent with the ISC's conclusions on declining spotted owl population trends, that it failed to account for 13 timber sale exemptions the BLM had received for its adjacent spotted owl timber lands, and that it failed to address adequately the plan's potential adverse impact on 32 other old growth–dependent species. On this latter NFMA point, the court subsequently elaborated that "whatever plan is adopted, it cannot be one which the agency knows or believes will probably cause the extirpation of other native vertebrate species from the planning areas." Once again the Ninth Circuit affirmed the district court's injunction, observing that "the Forest Service's EIS rests on stale scientific evidence, incomplete discussion of environmental effects vis-à-vis other old growth–dependent species and false assumptions regarding the cooperation of other agencies and application of relevant law."[25]

Faced with these back-to-back setbacks, the Forest Service could no longer ignore its biodiversity conservation responsibilities. Under the existing environmental laws and related scientific principles, the courts agreed that multiple-use management of the national forests required protection of old growth forests and native vertebrate species through "systematic management of a biological community." In June 1992, in a last ditch effort to right his floundering agency, the Forest Service chief announced that it would henceforth follow a new ecosystem management strategy and limit clearcutting in the national forests. The judiciary had catapulted the ecosystem squarely into the forest planning and management process. And litigation had served as the catalyst.[26]

BLM FOREST LITIGATION

Although the biological considerations were the same, rather different legal issues drove the environmental litigation challenging the BLM's management of its old growth forests. The BLM manages its western Oregon and northern California timberlands under the FLPMA and the O and C Act, which establish legal standards quite different from the NFMA standards governing management of the national forests. Although both the NFMA and FLPMA are comprehensive, multiple-use planning statutes, the FLPMA does not contain a biodiversity provision that might be invoked to protect the spotted owl in the BLM planning process. Moreover, the O and C Act is a timber production rather than multiple-use statute; it directs the BLM to manage these lands "for permanent forest production" using sustained yield principles. Environmental advocates, therefore, relied primarily upon NEPA and the ESA to challenge the BLM's timber policies and sales.[27]

The initial round of BLM litigation was triggered by the agency's failure to incorporate new spotted owl scientific information into its timber sale plans. In 1987, after the region's dwindling spotted owl population had become a major concern, the BLM refused to reevaluate its existing timber management plans, which contemplated over 200 timber sales in spotted owl habitat during the next three years but failed to assess their impact on the owl. Although the Forest Service was then preparing a SEIS to evaluate new scientific information on how old growth harvesting affected the spotted owl, the BLM stubbornly refused to reexamine its timber sale plans despite repeated requests from scientists and others.[28] After environmentalists sued, challenging the BLM's refusal, Oregon federal district judge Helen Frye found that the BLM had violated NEPA by not incorporating new spotted owl information, through an SEIS, in its timber planning decisions. But the judge concluded that a recently enacted congressional appropriations rider—known as the Section 314 rider—prohibited judicial review of the BLM's timber program and denied any relief. Under the Section 314 rider, obviously adopted to block the pending lawsuit, Congress mandated that "there shall be no challenges to any existing [BLM] plan . . . solely on the basis that the plan does not incorporate information available subsequent to completion of the plan." Over the next three years, with the rider and its progeny protecting it from legal scrutiny, the BLM continued furiously harvesting its timber with little apparent regard for the spotted owl or old growth ecosystems.[29]

In late 1991, however, the Ninth Circuit Court of Appeals opened the courthouse door for judicial review of the BLM's timber management program. When Congress failed to renew the rider for fiscal year 1991, the court concluded that Section 314 had expired and that the earlier NEPA claim was therefore properly before the district court.[30] Little more than a month later, in *Portland Audubon Society v. Lujan,* Judge Frye issued an injunction blocking further BLM timber sales in spotted owl habitat. In a chilling indictment of the BLM's biological indifference, the judge noted that in the decade since its timber management plans were first formulated, the BLM had "neither addressed the effects of its timber sales on the survival of the northern spotted owl as a subspecies in a NEPA document, nor promised such a NEPA document until mid-1993 at the earliest." She concluded that "the issue of the effects of the [BLM's] timber management plans on the long-range survival of the northern spotted owl subspecies should be addressed by the agency under NEPA." And she noted that Congress, in its original Section 314 appropriations rider, had directed the BLM to ensure that its completed resource management plans met all statutory requirements—another factor requiring the BLM to prepare an updated SEIS on the spotted owl. Thirteen months later, the Court of Appeals for the Ninth Circuit affirmed the injunction.[31]

With this ruling, the BLM was finally forced to confront the ecological ramifications of its intensive timber harvesting program. The SEIS process, of course, would require public involvement in reviewing and assessing scientific information about the owl, and it would require the BLM to respond to contrary scientific opinions and to provide a rationale for its final decision. Given the difficulties the Forest Service was experiencing in its efforts to address similar spotted owl concerns, the BLM could not be sanguine about maintaining its current timber harvest levels. In addition, because the owl was now listed as a threatened species, the BLM had no choice but to begin formal consultation with the FWS as it started to prepare the necessary environmental analysis.

In fact, the ESA was already being felt on BLM lands. When the FWS announced in June 1990 that it was listing the spotted owl as a threatened species, the BLM was forced to revise its timber management policies to account for the owl's new legal status. To do so, the agency unilaterally adopted the so-called Jamison strategy (named after Bush administration BLM director Cy Jamison), which was designed to protect the owl on BLM forests while still allowing timber sales at the rate of 750 million board feet (mbf) for 1991 and 1992. Although the Jamison strategy represented a substantial change in its existing timber policy, the BLM refused to consult with the FWS over the effect of its new policy on the owl. (Curiously, the BLM simultaneously announced 170 new timber sales and proceeded to consult with the FWS on these individual sales—a process that eventually led the FWS to conclude that the Jamison strategy "does not sufficiently protect spotted owls.") Environmentalists immediately sued under the ESA, alleging that the BLM was required to consult with the FWS before implementing its new owl conservation policy. After the district court rejected this claim, the Ninth Circuit ruled on appeal that the BLM must consult with the FWS, and it enjoined any further timber sales that might affect the owl pending such consultation. As with the national forests, the courts had now taken old growth timber harvesting on the BLM forests out of the agency's hands.[32]

At this point, its timber program essentially shut down, the BLM took the extraordinary step of requesting the secretary of the interior to convene the Endangered Species Committee (the so-called God Squad) for an exemption from the ESA's legal strictures. Specifically, the BLM sought to exempt from the ESA's formal consultation requirement 44 pending timber sales that the FWS had determined would likely jeopardize the owl's survival. No longer protected from judicial review by a congressional rider and faced with extensive new scientific evidence confirming the adverse impact of old growth harvesting on spotted owls, the BLM was now invoking an infrequently used provision of the ESA to escape legal accountability for its timber policies. Under 1978 amendments to the

ESA, adopted in the wake of the Supreme Court's controversial snail darter decision, Congress established the cabinet-level ESC to review agency requests for exemptions from the statute. Until this time, the ESC had been convened only twice, and in each instance, the ESC either denied or modified the requested exemption.[33]

The spotted owl, though, was not so lucky. Faced with mounting political fallout over the court injunctions, the ESC voted 5–2 to exempt 13 of the 44 timber sales from the ESA's consultation requirement. Outraged environmentalists immediately appealed the ESC exemption decision to the Court of Appeals for the Ninth Circuit, arguing that the ESC had violated key administrative law principles, and that improper ex parte contacts had been made between Bush administration White House officials and ESC members in an effort to influence their votes. The court agreed and ordered an evidentiary hearing into the matter.[34]

By this time, however, the political dynamics surrounding the spotted owl controversy were beginning to change. In January 1993, the Clinton administration took over the executive branch, and it proceeded to withdraw the BLM's timber sale exemption requests. This decision restored the BLM's formal ESA consultation obligations and effectively canceled the exempted timber sales. Clearly, the BLM would have to meet its judicially decreed NEPA and ESA obligations before selling more timber on its forests. Like the Forest Service, the courts now controlled its destiny.

The spotted owl litigation saga contains important institutional lessons for natural resource policy. As a threshold matter, the owl's environmental advocates repeatedly made calculated strategic judgments concerning how to use the available legislative, judicial, and administrative forums to advance their biodiversity policy agenda. During the early stages of the controversy, the regular appropriations riders reconfirmed Congress's dominant policy making role on public lands, as well as its overt hostility to the environmentalists' cause. Even though Congress was responsible for the environmental laws that ultimately carried the day on behalf of biodiversity, the rider litigation showed that Congress can alter—at least temporarily—those same laws through the subterranean annual budgeting process and usually escape public scrutiny. Given Congress's legislative and fiscal powers, court victories sometimes proved ephemeral absent a corresponding legislative strategy—a lesson that was reconfirmed, as we shall see, when Congress modified the Northwest Forest Plan with its 1995 salvage logging rider. And given the agencies' intransigence over reducing logging levels, the owl's advocates had little recourse but to rely upon the courts to enforce a plethora of often untested environmental laws through litigation and injunctions. In the final analysis, with Congress hopelessly stalemated, the judiciary carried the day and forced the public land agencies to take their ecological management responsibilities seriously. Congress's own environmental laws left the courts and ultimately the agencies with no other choice.

An Ecosystem Management Plan

TOWARD AN ECOSYSTEM SOLUTION

By late 1992, the federal judiciary was essentially running the Northwest's federal forestlands, logging was at a standstill, and no congressional solution was in sight. The environmentalists' spotted owl litigation campaign had finally stopped the logging juggernaut and achieved a temporary stalemate in the Northwest woods. To be sure, the Forest Service was working feverishly to devise an acceptable logging plan and just beginning to reform its timber-first management practices, the BLM had secured tenuous interim relief from the ESC (though that decision was on appeal and unlikely to stand), and the U.S. Fish and Wildlife Service was nearing completion of a spotted owl recovery plan. Both major presidential candidates were forced to address the controversy during the 1992 election campaign. Incumbent candidate George Bush used the occasion to call for a major overhaul of the Endangered Species Act, whereas challenger Bill Clinton indicated he would find a solution that would both protect the environment and bring the region's timber industry back in line. After the Clinton election victory, the new administration confronted the crisis head-on and proceeded to develop the president's Northwest Forest Plan, which also represented the first major federal experiment with landscape-scale ecosystem management on the public domain.

On April 2, 1993, less than three months after his inauguration, President Clinton convened the Northwest Forest Conference in a Portland, Oregon, downtown hotel, with the express goal of resolving the gridlocked ancient forest controversy. Both the environmental community and logging interests viewed the conference as an opportunity to move the new administration toward an acceptable solution. Environmentalists were encouraged by President Clinton's early cabinet and subcabinet appointments, which included Bruce Babbitt as the new secretary of the interior and Jim Lyons as the assistant secretary responsible for overseeing the Forest Service in the Department of Agriculture, and by the obvious presence of Vice President Al Gore, who had championed environmental causes during his Senate tenure. Timber interests, on the other hand, were well aware that the conference sprang from a campaign promise that candidate Clinton had made to timber union workers during an earlier election year visit, that the new president was indebted to labor for its support during his presidential campaign, and that Clinton had long supported social justice and economic development programs as the governor of Arkansas. In recognition of the occasion's importance, a phalanx of newly appointed executive branch officials accompanied the president and vice president to the conference: Secretary of the Interior Bruce Babbitt, Secretary of Agriculture Mike Espy, Secretary of Labor Robert Reich, Secretary of Commerce Ron Brown, EPA administrator Carol Browner, and OMB director Alice Rivlin. Although the region's congressional delegations attended the affair, they were not invited to speak, nor were the various agency officials who had overseen the crisis. Their absence from the program highlighted the new administration's commitment to focusing the conference on solutions rather than recriminations.[35]

The Northwest Forest Conference provided the controversy's major players with a high-profile forum to air their concerns, ideas, and expectations. The 54 invited speakers—a carefully chosen amalgam of environmentalists, timber industry representatives, loggers, scientists, academics, and others—underscored just how intractable the problem had become given the condition of the region's forests. Several perhaps predictable themes emerged: environmentalists emphasized the importance of judicial oversight under the Endangered Species Act, the need to protect the remaining old growth forests, and the changes otherwise occurring in the regional economy; the timber industry and its allies repeatedly stressed the importance of the region's timber resources, the growing human suffering in rural communities, and the need to break the gridlock preventing timber sales; other speakers noted the impact of logging on the region's fisheries, changes linked to international markets and automation, and the need for a balanced, environmentally sustainable solution based on a coordinated ecosystem approach to forest management. In a telling summation, Jack Ward Thomas, the plainspoken Forest Service scientist who had overseen key studies at critical junctures during the crisis, bluntly asserted that biodiversity conservation was now de facto federal policy on public lands. He concluded that a complex ecosystem protection strategy—one that would tax the limits of current knowledge and management acumen—represented the only viable conservation strategy.[36]

By the time of the conference, it was clear that Congress—still badly divided over the region's forest wars—was incapable of forging an acceptable legislative compromise. This meant that an administrative solution was the only way to resolve the impasse and to escape the federal injunctions. Recognizing this fact, the president promised to complete a legally sufficient forest plan within 60 days, one that would meet Judge Dwyer's environmental concerns and jump-start the region's timber industry. But well aware that any plan would be subject to legal challenge and that Congress might be reluctant to shield it from such a challenge, the administration committed to follow NEPA procedures before adopting the plan. This approach, it was hoped, would protect the plan from political reprisal, while the resulting EIS could be used as a supplement to the challenged EIS presently in abeyance under Judge Dwyer's injunction. As for the plan itself, the president enumerated five governing principles: protect long-term forest health; ensure a stable timber supply; employ sound scientific and ecological knowledge; address the economic needs of loggers and timber communities; and improve federal interagency cooperation. The principles clearly anticipated—virtually dictated—an ecosystem approach to regional forest management.

Responsibility for preparing the forest plan fell to a cadre of federal agency scientists directed by the indefatigable Thomas, who the president himself enlisted for this delicate task. Thomas and his scientists—the Forest Ecosystem Management Assessment Team (FEMAT)—were instructed to "identify management alternatives that attain the greatest economic and social contribution from the forests of the region and meet the requirements of the applicable laws and regulations," and to "take an ecosystem approach to forest management . . . particularly . . . [the] maintenance and restoration of biological diversity."[37] Thomas promptly assembled more than 90 federal scientists and technical experts from the Forest Service, BLM, Environmental Protection Agency, U.S. Fish and Wildlife Service, National Park Service, National Marine Fisheries Service, and several universities. The group then began the marathon task of integrating over a decade's worth of scientific, economic, and other data into a viable regional forest management plan.[38]

As the FEMAT process unfolded, it bore little resemblance to a conventional public land planning process. First, the FEMAT team did not include any managerial personnel, but consisted exclusively of scientists, economists, social scientists, and other technical experts. Second, the FEMAT team conducted its deliberations and drafting sessions in secret, preparing the plan in a severely compressed time frame without any formal public involvement. Third, after more than a decade of study and strife, the FEMAT members had an enormous amount of scientific information and other data at their disposal. Fourth, working in the shadow of Judge Dwyer's comprehensive injunction barring any logging on the region's federal forests, the FEMAT team clearly understood the primary legal and political constraints governing its deliberations. In short, the FEMAT plan was both a legal and scientific document: prepared by scientific experts, it was designed to meet critical scientific and legal objectives necessary to start timber flowing again to the region's mills. It was anything but a political document.[39]

THE NORTHWEST FOREST PLAN

Three feverish months later, the FEMAT released the president's Forest Plan, unveiling the federal government's first serious regional ecosystem management effort on the public lands. The FEMAT selected Option 9 from among 10 alternative management proposals. Under Option 9, the plan created a complex set of land allocations and management strategies designed to achieve broad biodiversity and ecosystem conservation goals as well as a 1.2-bbf annual timber harvest during the plan's first decade. Option 9 tracked broadly accepted ecosystem management principles: it gave priority to biodiversity conservation and ecosystem protection; provided local public participation opportunities; promoted transboundary coordination; expanded planning scales and time frames; was based on integrated scientific considerations; and adopted an adaptive management strategy with extensive monitoring obligations. Though decidedly not the end product of a traditional public land planning process, Option 9 nevertheless demonstrated just how far the courts, the law, and the new ecology had brought public land policy, at least in the Pacific Northwest.[40]

In broad outline, Option 9 adopted a landscape-scale ecosystem management strategy to provide adequate habitat for old growth species and to protect old growth forest ecosystems while also permitting timber harvesting within these ecological constraints. Option 9 allocated the region's 24 million acres of federal lands into two basic categories: nearly 80 percent of the land (18.8 million acres) was protected in reserve status, and the remainder (5.6 million acres) was opened to timber harvesting and active management. In selecting the protected lands, the FEMAT sought to create an interconnected regional reserve system to preserve terrestrial species and old growth forests by facilitating dispersal among disjunct populations and by reestablishing natural disturbance regimes. The reserved lands included: congressionally designated national parks, wilderness areas, and research natural areas; riparian reserves consisting of buffer strips bordering streams, lakes, and wetlands; administratively withdrawn lands protected for recreational, scenic, and other purposes; and late-successional reserves intended to ensure a viable old growth ecosystem. Outside these protected areas, federal forestlands were open to regulated timber harvesting and fell into three different categories: managed, late-successional areas where limited logging was allowed to reduce fire hazards, diseases, and insect infestations; adaptive management areas, ranging in size from 84,000 to 400,000 acres, that were created to foster innovative local ecosystem management strategies; and matrix lands that represented the remaining federal lands and the primary areas open to logging. Within each of these land designations, Option 9 prescribed detailed management standards governing all resource decisions. These included, for example, specific silvicultural limitations, green tree and woody debris retention requirements, natural fire control provisions, and an elaborate survey and manage system designed to protect amphibians, mollusks, vascular plants, and other species.[41]

Option 9 also proposed a unique aquatic conservation strategy (ACS) designed to "restore and maintain the ecological health of watersheds and aquatic ecosystems contained within them on public lands." According to the FEMAT, extensive logging, road building, and other federal land use activities had contributed to a significant loss of aquatic habitat, including marked declines in the region's native salmon and other aquatic species populations. To reverse these trends, the ACS adopted a four-prong approach: riparian reserve designations interspersed within the matrix lands; key watershed designations to protect anadromous salmonid and bull trout populations; watershed analysis requirements; and a watershed restoration program to reclaim high-quality aquatic habitat. The plan designated 164 key watersheds across 9.1 million acres as an overlay on the other land designations covering the region's 24.5 million acres of federal land. Within these key watersheds and roadless areas, timber harvesting, most road building, and other resource activities were prohibited until a watershed analysis was completed. This analysis would be used to assess aquatic ecosystem conditions, with the information then being used in resource management decisions, including riparian reserve boundary adjustments, management prescription revisions, and monitoring programs. In short, to avoid another legal donnybrook over native

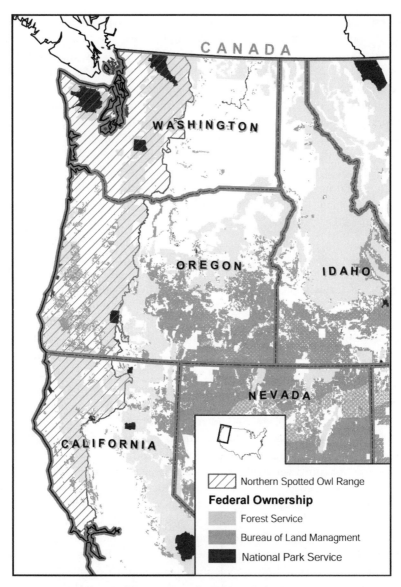

Map 2. Northwest Forest Plan Area. Northern spotted owl habitat extends across a vast acreage encompassing three states in the Pacific Northwest. To protect the owl and other forest-dependent species, the Northwest Forest Plan covers 24 million acres of federal land, including 19 national forests and 7 BLM resource districts.

aquatic species, the ACS linked terrestrial and aquatic ecosystem components together in a comprehensive ecosystem management plan to protect aquatic habitat for the region's at-risk fish species.[42]

In its economic analysis, the FEMAT plan acknowledged that this new ecosystem management approach yielded "probable timber sale levels that are substantially less than was historically sold and harvested from the federal forests in the region." Whereas annual federal timber sales averaged 4.6 bbf from 1980–89 and 2.4 bbf under the 1990–92 forest plans, Option 9 would provide only 1.2 bbf annually. More than 5,000 timber-based jobs were expected to go on the chopping block. The creation of additional employment opportunities was anticipated in recreation, tourism, and other non–timber forest resources, but these sectors would not make up for all the lost timber jobs. The region's counties also faced a dramatic decline in federal timber payments: Under Option 9, the payments were expected to drop from approximately $400 million annually in 1990–92 to approximately $100 million dollars annually. Overall economic projections, however, anticipated continued regional economic expansion (primarily in the metropolitan areas) linked, at least in part, to maintaining the Northwest's reputation for a high-quality environment.[43]

The region's timber-dependent communities could also expect myriad social problems as this new ecosystem management approach took hold. According to the FEMAT social scientists, "Most negative community effects will be concentrated in rural areas." Within communities, the potential impacts included changes in the age and class structure of the local populace, reduced tax revenues, a heightened sense of alienation from government decision processes, and a deteriorating infrastructure. For individuals, the potential impacts included a decrease in wages and standard of living, a greater sense of isolation and disenfranchisement, and increased levels of deviant social behavior, such as crime, domestic abuse, alcoholism, and truancy. Some of these negative impacts, however, were linked to larger economic changes—including increased global competition and technological developments in the timber industry—and not solely to local forest management decisions. Moreover, communities dependent upon recreation and tourism were expected to benefit from the decrease in timber harvesting, which removed some scenically attractive local landscapes from the timber inventory. To soften the blow, the FEMAT report recommended greater community involvement in natural resource decision processes and additional job retraining and educational assistance.[44]

To address these socioeconomic concerns, the Clinton administration devised the Northwest Economic Adjustment Initiative, which provided $1.2 billion in federal assistance to communities, businesses, and individuals affected by the federal timber harvest reductions. The economic initiative supported dislocated worker retraining, business diversification, community infrastructure, and a short-term Jobs in the Woods program.[45] To provide further assistance, Congress also repealed the long-standing log export tax incentive, which had subsidized raw log exports from state and private timberlands to Japan and other Pacific Rim nations, and then earmarked these funds to guarantee payments to the states.[46]

With its ecosystem-based management provisions, the Northwest Forest Plan proposal presented serious jurisdictional challenges. Without federal interagency coordination, the ambitious FEMAT scientific management plan would tax the leadership skills of even the most committed land manager. Indeed, Judge Dwyer's 1992 injunction, based largely on the Forest Service's failure to even consider the environmental impacts of timber harvesting on adjacent public and private lands, had highlighted the need for much greater interagency coordination. President Clinton had identified the same problem at the 1993 Northwest Forest Summit: "I was mortified when I began to review the legal documents surrounding this controversy to see how often the departments were at odds with each other . . . we've had our own agencies suing one another in court, often over issues which are hard to characterize as monumental."[47] But with 7 different departments and 16 different agency programs involved with forest management across the three spotted owl states, it would not be easy to overcome the divergent statutory mandates and incentives, different agency cultures, and general lack of trust that had hampered previous regional coordination efforts.

The Clinton administration, therefore, established an Interagency Coordination Working Group to promote institutional reform among the region's federal land management and regulatory agencies. The group promptly created an intricate, tiered committee structure to facilitate coordinated planning and to broaden public involvement in forest management decisions. At the regional level, a federal Regional Interagency Executive Committee, advised by a separate intergovernmental advisory committee with state, tribal, and local representation, was responsible for implementing the Forest Plan and overseeing coordination among the agencies. At the local level, the region was divided into 12 different provinces that were defined in ecological rather than political terms. Each province was administered by a separate federal Interagency Executive Committee, assisted by a Provincial Advisory Committee with both governmental and nongovernmental members who were responsible for integrating local concerns into the decision process.[48] An interim assessment, while noting obvious bureaucratic overtones and built-in inefficiencies, concludes that this structure had improved interagency coordination and intergovernmental relationships, as well as public participation in resource decisions through partnerships and similar arrangements.[49]

BACK TO COURT

The Northwest Forest Plan was not universally well received. The timber industry and the region's timber-dependent communities were visibly upset that the final plan contemplated a 75 percent drop in the annual timber sale volume, which meant the loss of another 5,000 jobs and reduced tax revenues. Although some environmentalists objected to any further old growth harvesting, most of the weary environmental community grudgingly accepted the Option 9 solution, even while recognizing that its hastily drawn provisions allowed logging in some inappropriate locales. Political reality provided them with few other options. Their congressional allies were convinced that the president's plan represented an acceptable solution to this long festering controversy, and the Clinton administration hinted that it might seek legislation insulating the Forest Plan from legal challenge—a particularly troublesome prospect given the environmental community's negative experiences with appropriations riders denying judicial review of timber decisions. Within the agencies, more than a few longtime Forest Service and BLM employees were unhappy about the plan too, lamenting the agencies' shift away from their historic timber production missions and the loss of managerial flexibility that would inevitably accompany the new ecosystem management policies.[50]

Once the Northwest Forest Plan was submitted to Judge Dwyer, the timber industry and environmental organizations unleashed a flurry of lawsuits challenging the plan and its new ecosystem management policies. In a final effort to maintain existing logging levels, the Northwest Forest Resource Council mounted a frontal assault on the plan, arguing that the agencies lacked the basic legal authority to adopt an ecosystem-based management strategy and that they had violated NEPA and other procedural statutes in preparing the plan. Various environmental organizations, still unhappy that logging would continue in the region's old growth forests, attacked the plan's projected harvest levels, its old growth habitat protections, and the scientific analysis supporting its spotted owl viability conclusions and the new aquatic conservation strategy. To expedite the judicial review process, Judge Dwyer consolidated the various cases into a single proceeding and undertook a comprehensive assessment of the plan's validity.

On December 21, 1994, Judge Dwyer rendered his verdict: the Northwest Forest Plan met all legal requirements, thus ecosystem management was the new order of the day on the region's 24 million acres of federal forestlands. In response to the Northwest Forest Resource Council's arguments, the court rejected the claim that the agencies lacked the legal authority to adopt a regional ecosystem management plan, finding that the combined force of the governing statutes—namely the NFMA, FLPMA, NEPA, and ESA—virtually compelled such an approach: "Given the condition of the forests, there is no way the agencies could comply with the environmental laws *without* planning on an ecosystem basis." In response to other substantive claims, the court sustained the NFMA viability regulation as consistent with the Forest Service's multiple-use obligations, rejected the assertion that the Oregon and California Lands Act overrides the BLM's statutory species conservation obligations under NEPA and the ESA, and found that the agencies had properly assumed that ESA-based species conservation obligations would also apply on private lands to supplement federal protection efforts. Similarly, the court rejected the industry's procedural claims, finding that the agencies were not obligated to consider a "no action" alternative given the illegal nature of previous forest management practices, and that the aquatic conservation strategy was validly included in the plan given the credible scientific information linking logging to degraded aquatic habitat. Having previously found that the timber industry and the agencies had illegally jeopardized the region's biological resources through unsustainable logging practices, Judge Dwyer was not about to conclude that the law forbid the agencies from pursuing a new ecosystem-based management approach that would now protect these same biological resources, even if it meant a drastic curtailment in federal timber production. In fact, the court bluntly observed that "any more logging sales than the plan contemplates would violate the laws."[51]

Judge Dwyer likewise rejected the environmental plaintiffs' objections to the Forest Plan. In response to the NEPA-based argument that logging should be prohibited on all spotted owl habitat land, the court found that the agencies were not required to consider this alternative, and that there was adequate scientific information to allow logging in some old growth forests. The court also sustained the aquatic conservation strategy, finding that the agencies had reviewed the relevant scientific information and acknowledged the uncertainties and risks associated with the plan. Given these uncertainties, the court emphasized the need for ongoing monitoring to assess how the plan was in fact affecting at-risk species: "Monitoring is central to the plan's validity. If it is not funded, or not done for any reason, the plan will have to be reconsidered." In addition, noting the programmatic nature of the SEIS, the court found that the agencies could postpone a cumulative effects assessment of the relationship between federal and nonfederal lands until site-specific project proposals were made. As with the timber industry's claims, having been finally presented with a cooperative, science-based, regional forest management plan, Judge Dwyer was not prepared to overturn this finely wrought compromise. Nonetheless, the court's opinion contained two clear warnings: any logging beyond the plan's limits was unacceptable, and continuous monitoring of old growth species was essential given the experimental nature of this new federal ecosystem management endeavor.[52]

The timber industry and three environmental groups immediately appealed the ruling. The Court of Appeals for the Ninth Circuit, having consistently upheld Judge Dwyer's earlier decisions limiting logging on the region's federal forests, again sustained his conclusion that the forest plan passed legal muster. In response to the timber industry's limited procedural appeal, the court of appeals ruled that the district court had properly joined the Northwest Forest Resource Council as a party in the forest plan litigation despite its efforts to challenge the plan's validity in a Washington, D.C., venue rather than before Judge Dwyer. In response to the environmentalists' NEPA claims, the court of appeals held that the agencies had considered an adequate range of alternatives in the plan and undertaken a sufficient cumulative effects analysis given their assumption that nonfederal lands would be managed to avoid harm to threatened species. The court also agreed that the plan's old growth harvest levels did not violate the NFMA viability regulation, noting that the Forest Service's statutory multiple-use obligations precluded the agency from forgoing all timber harvesting and that the viability provision itself incorporated several multiple-use references. Although the Ninth Circuit's concise opinion did not end all ancient forest litigation, it did place a definitive judicial stamp of approval on the Northwest Forest Plan. And it placed the burden squarely on the Forest Service and BLM to prove that their new ecosystem management strategy would provide both environmental and economic values from the region's multiple-use public lands.[53]

ANOTHER RIDER, MORE LITIGATION

Once Judge Dwyer blessed the president's Northwest Forest Plan, the stage was set—or so it seemed—for a serious experiment with ecosystem management. Despite the pending appeals, responsibility for the region's forests was back in the hands of the land managers, who were no longer operating under a federal injunction. While no one expected either the region's powerful timber industry or its battle-scarred environmental activists to fade from the scene, both sides were obviously weary from the long struggle and the controversy was finally quieting to a simmer. Certainly, no one was prepared for the acrimonious next chapter in the Northwest timber wars.

During 1994, however, two events conspired to propel the region's forests back into another death-struggle that once again pitted the timber industry and environmentalists against each other. Throughout the West, the summer of 1994 was hot and dry, triggering a long fire season that saw over 2 million acres lost to flames and 14 federal firefighters killed in a Colorado forest fire. The timber companies, joined by industry scientists and western politicians, asserted that long-standing federal fire suppression and timber management policies had fueled the fires by allowing a combustible mixture of dead and insect-infested timber to accumulate. They called for an aggressive salvage logging campaign to remove fallen and diseased timber from the forests. Still clinging to its timber-first mentality, the Forest Service responded by announcing a Western Forest Health Initiative that would provide an additional 4.5 bbf of salvage sale timber over the next two years. Just as the salvage logging idea was gaining momentum, the 1994 elections propelled the Republican Party into control of both houses of Congress for the first time in 40 years. As longtime Republican allies of the West's extractive industries assumed leadership positions on key congressional committees, the timber industry perceived a golden opportunity to bring its salvage logging proposal into a suddenly much friendlier legislative arena.[54]

Once again, the tool of choice was the dreaded appropriations rider. In June 1995, the new 104th Congress attached a comprehensive salvage logging rider to the Budget Recisions Act of 1995, a politically popular bill that contained funds for the Oklahoma City bombing victims and for California flood relief. After initially vetoing the bill, President Clinton acquiesced in a revised version that still contained the salvage logging provisions. The logging rider—misleadingly labeled a salvage provision—established a breathtakingly broad federal timber cutting program that extended over an 18-month period. The rider mandated both salvage and green timber sales: it sanctioned virtually unrestrained salvage logging on all national forests; it authorized the immediate sale of all previously withdrawn Option 9 timber sales; and it revived all previously unsold Section 318 sales, which covered all Pacific Northwest timber sales that had not been harvested under Congress's earlier timber riders. The rider eschewed legal accountability: it vested the secretaries of agriculture and interior with sole discretion for determining whether sales met legal environmental requirements; it contained sufficiency language obviating application of NFMA, NEPA, ESA, and other environmental laws; it prohibited all administrative appeals; and it precluded any judicial review for compliance with the environmental laws. Once again, unrestrained logging was the order of the day in the Northwest, the courts were removed from any oversight role, and the environmental laws were effectively forsaken. Congress had reinserted itself into the forest management business.[55]

The outraged environmental community valiantly rallied its supporters in yet another anti-logging campaign, hoping to save some vital old growth remnants. The Clinton administration, for its part, insisted that it would protect the Northwest Forest Plan and follow all applicable environmental laws. After signing the Recisions Act, the president directed the federal agencies "to implement their timber-related provisions in an environmentally sound manner, in accordance with my Pacific Northwest Forest Plan, other existing forest and land management policies and plans, and existing environmental laws, except those procedural actions expressly prohibited by Public Law 104–19." The agencies likewise reaffirmed their "commitment . . . to continue their compliance with the requirements of existing environmental law while carrying out the objectives of the timber salvage related activities authorized by P.L. 104–19." Nonetheless, relieved from the threat of judicial intervention and subjected to intense congressional pressure to sell the timber, neither the Forest Service nor the BLM was overly sensitive to existing environmental laws or policies. In Idaho's Boise and Payette national forests, for example, the Forest Service proceeded with the controversial Thunderbolt salvage sale despite repeated objections from the FWS and EPA that it would further degrade salmon habitat and water quality on the already overlogged south fork of the Salmon River. Similar problems surfaced in other locations: grizzly bear habitat was threatened by salvage sales in Montana's Kootenai National Forest; Mexican spotted owl habitat was compromised in Arizona's Coronado National Forest; and the soon-to-be-listed Umpqua cutthroat trout was ignored in Oregon's Umpqua National Forest. Ultimately, the Clinton administration was forced to revise its salvage sale policy with six months remaining on the rider, conceding that serious abuses had occurred, including the sale of healthy green timber under the salvage provisions.[56]

In the meantime, environmental lawyers turned grimly to the courts in a vain attempt to narrow the scope of the rider and to reassert the environmental laws. The courts, however, repeatedly refused to review alleged violations of environmental standards or even allegedly arbitrary decisions permitting clearcut logging in ecologically sensitive locations with troublesome endangered species ramifications. In the Kootenai National Forest, for example, the Forest Service ignored its own grizzly bear management standards as well as the U.S. Fish and Wildlife Service's contrary habitat degradation conclusions when it authorized a massive salvage sale in roadless grizzly bear habitat. Acknowledging that the sale flew in the face of clearly contradictory scientific evidence, the Court of Appeals for the Ninth Circuit nevertheless ruled that the salvage rider precluded any judicial review under the environmental laws, gave the Forest Service sole authority to make endangered species determinations, and permitted only "arbitrary and capricious" review for nonenvironmental claims. In the controversial Thunderbolt sale, not only did the courts refuse to intervene, but the Idaho congressional delegation blatantly pressured the Department of the Interior into retracting its scientific criticism of the sale's impact on salmon habitat and water quality. With the rider in place, the Forest Service became the sole arbiter of the salvage sale program, subject only to the intense political pressure exerted by the measure's congressional sponsors.[57]

The timber industry's lawyers, intent on optimizing this unexpected harvesting opportunity, enlisted the courts to ensure that the rider received an expansive interpretation. In the first major test of the salvage rider's scope, industry lawyers argued successfully, in *Northwest Forest Resource Council v. Glickman*, that Section 2001(k) required the Forest Service and BLM to complete all previously offered (and withdrawn) timber sales throughout Oregon and Washington under the original sale terms, even though log prices and environmental protection requirements had changed dramatically in recent years. Many of the Section 318 timber sales (referring to the earlier 1989–90 fiscal year rider) had been withdrawn by the agencies for environmental reasons, some as much as five years earlier. The Clinton administration joined environmental lawyers to argue that the rider's reference to Section 318 limited its geographic and time period coverage only to 1989–90 sales in the spotted owl forests originally covered by Section 318. However, Oregon federal district judge Michael Hogan agreed with the industry's interpretation, finding that the statutory language and accompanying legislative history supported a broader geographic and temporal interpretation. Judge Hogan ordered the Forest Service and BLM to award nearly all timber sale contracts offered in any Oregon and Washington forests between October 1, 1990, and July 27, 1995. Eventually, the courts allowed even Section 318 sales previously canceled for environmental reasons to be resurrected and reoffered, unless they were still subject to a court injunction. Thus, in patent disregard of the Northwest Forest Plan, the rider granted the timber industry a green timber windfall that included cutting on important and only recently protected spotted owl, marbled murrelet, and aquatic habitat.[58]

The Option 9 rider provision similarly gave timber companies unimpeded access to all timber sales authorized in the Northwest Forest Plan without fear of any judicial oversight. Ostensibly added to prevent environmental challenges to these sales, the Option 9 rider suspended operation of the environmental laws and limited judicial review of sale decisions, just as it did for salvage sales. The full implications of these provisions were brought home dramatically in the North Umpqua River headwaters. Even though the Forest Service's own biologist concluded that additional clearcutting would severely damage the already cut-over watershed's aquatic habitat and place sensitive fish species at risk, agency officials authorized the controversial sale under the rider. The courts rebuffed an environmental challenge to the sale, observing that the rider's "notwithstanding any other law" language left them without any substantive standard to apply in measuring the legality of the agency's actions. Under this rationale, the rider's "arbitrary and capricious" review provision was rendered meaningless: because the rider precluded judicial review under any environmental laws, there were no meaningful standards to apply in reviewing the sale for arbitrariness. Despite Judge Dwyer's admonition that the legality of the Northwest Forest Plan was contingent on maintaining the plan's logging limitations, logging was now allowed without any judicial oversight of its environmental effects. Congress was dangerously close to upsetting the carefully constructed Northwest Forest Plan and the uneasy peace that accompanied this initial foray into ecosystem management.[59]

The salvage rider controversy contains important institutional lessons concerning the future of the Northwest Forest Plan and ecosystem management. Although the Forest Plan is still basically intact, the rider illustrated once again that Congress, employing its considerable legislative and appropriations powers, has the legal authority to establish and to revise federal public land policy. This authority includes the power to override existing environmental laws, to ignore contrary administrative arrangements, and to render the courts powerless to respond. The Section 318 rider litigation established this power at the beginning of the spotted owl controversy; the salvage rider litigation reconfirmed it during the controversy's later stages. As much as the successful spotted owl litigation campaign and major advances in ecological knowledge deserve credit for propelling the Forest Service and BLM to embrace ecosystem management, the salvage rider illustrates that raw political power is also a primary ingredient in promoting lasting reform on the public domain. Congressional political pressures initially forced the agencies to implement the rider aggressively, while countervailing political pressures eventually forced the Clinton administration to revise its salvage harvest policies before the rider expired. It is telling, though, that Congress has been unable to muster sufficient support to permanently change the existing environmental laws or to remove the courts from overseeing public land decisions. As a result, with the law and the spotted owl precedent as a bulwark, public land agencies are making overt administrative forays into the realm of ecosystem management. And Congress may well be unable or disinclined to stop this new trend.

THE AFTERMATH: AN INTERIM ASSESSMENT

Despite the 1995 salvage rider and the harvesting frenzy it generated, the Northwest Forest Plan's ecosystem management strategy remains in place and governs agency decisions on public lands embracing spotted owl habitat. The obvious question is whether ecosystem management has made a significant difference on these lands. Are the agencies adhering to generally accepted ecosystem management principles? Has the trend toward ecological decline been reversed? Are the region's biological resources receiving equal treatment with its timber and other commodity resources? Have the responsible federal agencies established a more effective and coordinated management regime? Are the region's communities and inhabitants better or worse off? Has the level of controversy subsided, or is the region still quarreling over public forestlands? Although definitive answers to these questions remain elusive, the current evidence confirms the fundamental soundness of the new federal ecosystem management strategy.

The level of timber harvesting on the Northwest's public lands has dropped dramatically since the Forest Plan was adopted, even with the 1995 salvage rider. Whereas the Forest Service and BLM were cutting an average of 5 bbf annually during the 1980s, the Forest Plan set the annual harvest level at 1.2 bbf, which represented an 80 percent reduction in the cut. But the agencies did not reach even that reduced harvest level during the plan's first four years, when annual sale figures averaged approximately 687 mbf. In fact, agency officials twice revised the plan's timber harvest levels downward, projecting an annual harvest of just 746 mbf over the first decade, more than a 25 percent decrease from the original estimate. The reduction in harvest levels has been accompanied by a corresponding reduction in the quality of timber available for sale, owing to selective harvesting requirements and other restrictions on old growth cutting to meet conservation objectives. Meanwhile, expenses associated with timber harvesting under the plan have nearly doubled, highlighting the high cost of biodiversity conservation and creating an additional barrier for some logging firms. Smaller mills historically dependent on public timber have found it difficult to compete effectively with the multinational logging companies, many of which have turned to their own private landholdings for a steady timber supply. But this was not surprising: by absorbing biodiversity conservation obligations on federal lands, the plan was designed to relieve state and private lands of this burden in order to stabilize harvest levels on these lands. Rural counties, moreover, report a 35 percent reduction in federal stumpage fee payments since 1989, which has had an adverse impact on local tax revenues. After the Bush administration assumed office in 2001, though, the new Forest Service chief announced that the agency will consider administrative changes to the plan in an effort to expedite the timber sale process. But even if changes ensue, logging is no longer the dominant consideration in managing the region's forests, and the shift in priorities is being felt within adjacent communities.[60]

The Forest Plan, however, has not entirely reversed regional ecological trends. Even before the plan was contemplated, the National Marine Fisheries Service had begun listing Columbia Basin salmon runs under the ESA to protect the declining populations. Since then, despite the Forest Plan's aquatic conservation strategy, the National Marine Fisheries Service has listed several additional salmon runs as well as five steelhead runs, citing excessive logging, roading, and grazing as one factor to support these decisions. In 1998, the U.S. Fish and Wildlife Service added the landlocked bull trout (or Dolly Varden, as it was previously known) to the endangered species registry, again citing logging, roading, and grazing as the basis for the listing. In total, during the Forest Plan's first five years, there were 18 new listings, consisting of 15 fish species, 1 frog, and 2 plants, all dependent on forest ecosystems in one way or another. Agency officials also have acknowledged that northern spotted owl population trends are not clear, despite the attention this species has received in the forest planning process. The timber industry, however, has mounted a legal challenge to the northern spotted owl and marbled murrelet listings while others have challenged salmon listings, signaling a new round of endangered species litigation. Given the extensive habitat needs of the aquatic species added to the endangered species registry and the potential for renewed litigation over key listing decisions, the stage may well be set for yet another ecological maelstrom that could further test the Forest Plan, its ecosystem management strategy, and the nation's commitment to protecting its biological heritage.[61]

Moreover, the agencies have not consistently met their ecological management commitments under the plan, provoking yet another round of timber sale injunctions. In 1999, after watching the courts rebuff their early challenges to Forest Plan timber sales, environmental groups successfully sued the Forest Service and BLM for not fully implementing the Forest Plan's survey and management requirements, which were designed to identify and protect nondispersing species such as mollusks, amphibians, and voles that would not otherwise benefit from the plan's reserve system. In deciding the case, Judge Dwyer reconfirmed the importance of the Forest Plan's species monitoring requirements in meeting its biodiversity conservation objectives: "The surveys are designed to identify and locate species; if they are not done before logging starts, plants and animals listed in the ROD [Record of Decision] will face a potentially fatal loss of protection." The judge first enjoined 9 pending timber sales and then extended his injunction to 25 additional sales, prompting Congress to threaten another rider to override the plan's survey and management requirements. After settling the lawsuit, however, the agencies amended the Forest Plan to reduce their scientific planning burdens and to facilitate timber sale decisions, removing 72 species from the survey and management requirements and shifting other species within the plan's monitoring categories. And confronted with another round of survey and monitoring litigation, the Bush administration announced it would revise these requirements to give agency officials greater flexibility.[62]

In other litigation, Judge Barbara Rothstein twice invoked the ESA to enjoin controversial timber sales in the Umpqua River Basin, finding that the agencies had ignored the Forest Plan's ACS commitments. The judge ruled that the NMFS, by focusing its analysis of timber sale effects on the "endangered" Umpqua River cutthroat trout at the watershed rather than site-specific level and by employing a long-term rather than short-term time frame, had not met its ESA Section 7 consultation obligations, observing the "overwhelming evidence" of ongoing habitat degradation in the basin. After the Court of Appeals for the Ninth Circuit sustained these rulings, the Bush administration began rewriting the ACS standards; in the future they will most likely be relaxed and a broad-scale analysis of timber sale proposals permitted in order to expedite the ESA consultation process. Thus, although environmental groups have successfully enforced these key Forest Plan ecological protection provisions in court, the Forest Service now seems intent on using its administrative authority to loosen them and increase the flow of timber.[63]

The Forest Plan implementation process has also been harshly criticized for not sufficiently protecting old growth trees and aquatic ecosystems. Environmentalists have chastised the Forest Service and BLM for selling old growth timber in the designated late successional reserves and matrix lands, objecting that no additional old growth should be harvested and that roadless lands should be protected from harvesting to meet the plan's ecological objectives. They fear that any cutting of the remaining old growth diminishes the spotted owl's survival chances and also imperils other old growth–dependent species. However, land management agencies and the timber industry respond that the plan was not designed to prohibit all old growth harvesting and that the other reserved lands will regenerate adequate old growth habitat over time to replace the habitat now being cut. Although most observers agree that the Forest Plan's aquatic conservation strategy is a vital ecological management tool, critics believe it is too weak to protect aquatic species or their habitat. Specific criticisms include the lack of adequate restrictions on timber harvesting and other activities in riparian corridors, inadequate information about existing watershed conditions on public and adjacent private lands, and the agencies' reluctance to make or implement watershed analysis recommendations that would benefit aquatic habitat. As we have seen, however, the Bush administration seems intent on revising the aquatic conservation obligations in an effort to expedite timber sales. With so many aquatic species already listed on the endangered species registry, environmentalists will surely contest these revisions. By any measure, controversy continues unabated over endangered species management, scientific implementation strategies, and harvest levels, which means that yet more litigation defining federal ecosystem management obligations is on the horizon.[64]

From other quarters, however, the Forest Service and the BLM have received generally high marks for integrating basic ecosystem management principles into the Northwest Forest Plan. According to a 1999 General Accounting Office (GAO) assessment, the agencies have now begun planning at an appropriate ecosystem scale to meet their diverse legal mandates and biological obligations, eschewing traditional boundary lines that previously impaired planning on an ecologically meaningful scale. The GAO and others believe the agencies have significantly improved interagency coordination in the region: the federal regulatory agencies—FWS, NMFS, and EPA—are routinely involved earlier in the land management agencies' planning and decision processes, and the agencies have all begun to standardize their previously diverse data collection systems. The GAO was also convinced that the Provincial Advisory Committees had improved public involvement in plan implementation decisions, thus providing a vehicle to communicate local concerns and values as part of the decision process. And the GAO concluded that the plan's three-tiered monitoring program was well-designed to meet adaptive management requirements, though it did not assess the effectiveness of the agencies' actual monitoring programs. However, not everyone involved in implementing or overseeing the Forest Plan agrees with the GAO's assessments. As the scale of forest planning efforts has expanded, participants complain that it has become increasingly difficult to determine who is accountable for management policies and decisions. Both environmentalists and local community leaders complain that the Provincial Advisory Committees are cumbersome and lack any real ability to affect resource management decisions. And, of course, the agencies failed to adhere to the plan's survey and management requirements, which have now been significantly modified.[65]

The Forest Plan also contained provisions designed to buffer local communities. The plan not only funneled federal funds to local communities for worker retraining and economic development programs, but also provided residents with an opportunity to participate in resource management decisions through the Provincial Advisory Committees. Federal officials report that over $1.2 billion have been channeled to local communities, that between 1994 and 1998 more than 1,440 jobs were created in the Jobs in the Woods program, and that in 1997 more than 2,900 displaced workers were provided job retraining opportunities. Local officials, however, assert that the assistance came too late to reach many displaced workers who were forced to leave logging communities after losing their jobs. They also note that much of the money has gone toward local infrastructure development without reaching displaced workers and that many of the new jobs have been short-lived, make-work opportunities. Although the Provincial Advisory Committees were designed to provide residents and others with an opportunity to participate in resource decisions, local officials complain that these committees have been dominated by agency employees and proven too bureaucratic to make many meaningful resource decisions. Nonetheless, the regional economy has grown in the aftermath of the Forest Plan, and several of the affected communities near metropolitan areas have rebounded from mill closures. For other rural communities, though, the transition from a purely logging economy to a more diversified one has taken longer and left deeper scars.[66]

In sum, the Northwest Forest Plan's science-based ecosystem planning regime has produced mixed results. It has significantly reduced timber harvest levels, but it has not abated ecological decline as measured by endangered species listings, nor has it ended old growth harvesting. It has integrated the entire ecosystem complex into a single management framework, but this has added new layers of scientific complexity and interagency coordination to the planning and management processes. It has injected new adaptive management protocols and local advisory groups into the management process, but the agencies have been slow to embrace these new inventorying and consultation obligations. And while it restored a semblance of peace to the region, it has not eliminated controversy from the forests. In fact, with the courts now involved in refining federal ecosystem management obligations and with the Bush administration intent on revising the plan, it is apparent that the agencies are still finding it difficult to reconcile their new ecological responsibilities with countervailing political and economic pressures.

Coming of Age: Ecosystem Management on Public Lands

To say that ecosystem management has secured a foothold in public land policy is to state the obvious. Without question the spotted owl controversy elevated biodiversity conservation on public lands, legitimizing it as an important— if not priority—management goal. At the same time, ecosystem management emerged as the strategy of choice—perhaps even necessity—to integrate biodiversity conservation into public land policy. At every level and in every forum, these ideas not only resonate, but they have been translated into concrete proposals, several of which have been given the sanction of law. During the Clinton presidency, each branch of the federal government—the agencies, the president, the courts, and even Congress—endorsed various ecosystem management principles as important new elements of public land policy. Although counterexamples exist within each of these venues, the result is still clear: new laws, policies, and experimental initiatives designed to change the traditional way of managing our public resources. Even if the arrival of the Bush administration makes it premature to enshrine biodiversity conservation as the new dominant use on public lands, ecosystem management has gained a foothold across much of the public domain.

ADMINISTRATIVE INITIATIVES

In the aftermath of the spotted owl controversy, federal public land agencies made a concerted effort to incorporate an ecological perspective into their management policies. During the Clinton administration, from the president to the secretary of the interior to individual agencies, the clear direction of administrative policy was toward greater ecological sensitivity in managing public lands. Not only did Clinton administration officials use their executive authority to reform outdated policies, but they also undertook numerous initiatives to institutionalize ecological management principles. The initiatives included new regulations reinterpreting existing laws, expansive regional planning efforts, a new landscape conservation system, key leadership appointments, and even creation of a new agency. But faced with a recalcitrant Congress, the Clinton administration was forced to temper its efforts; a disgruntled Congress could undo its reforms, with either new legislation or a carefully crafted budget rider. And, of course, the new Bush administration can employ its own administrative authority to reverse its predecessor's handiwork.

Shortly after assuming office, in an effort to elevate biodiversity conservation on the national agenda, Secretary of the Interior Babbitt created the National Biological Survey (NBS) as an independent agency within the Department of the Interior responsible for inventorying and monitoring the nation's biological resources. As originally envisioned, the NBS would function much like the U.S. Geological Survey did during the settlement of the American West, surveying and compiling data on a key resource vital to the nation's future well-being. But the NBS met immediate resistance from conservative political forces. Property rights advocates launched a vigorous counterattack, fearing the new agency would intrude on private lands and bring the regulatory authority of the federal government to bear on otherwise innocent property owners. Using its budgetary powers, Congress forced Secretary Babbitt to revise his original plans by relocating the NBS as a mere division within the Geological Survey, thus downplaying both its stature and role within the department. Regardless, the mere fact that the federal commitment to biodiversity conservation merited a separate government bureau signaled a major realignment in federal priorities for natural resource management purposes.[67]

As part of its ecological management agenda, the Babbitt-led Department of the Interior implemented major Endangered Species Act reforms. Most notably, the department expanded the ESA Section 10 habitat conservation planning (HCP) process, refocusing it on ecosystem protection to safeguard the full array of biological resources in one location rather than continuing with a single species protection strategy. It also adopted a "no surprises" policy to shield private landowners harboring a "listed" species on their property from potential ESA liability if they entered into an HCP adequately protecting known species on their land. In effect, the federal government said if you will make these current commitments to protect endangered species, we will bear the burden of providing any needed future protection on public lands. Although department officials eventually acknowledged that unforeseen developments could require future HCP modifications, these changes in ESA policy helped defuse the political rhetoric of property rights proponents, undermining their well-orchestrated campaign for legislative changes that would have weakened federal endangered species policy. They also illustrate how the ESA is being used as a tool for ecosystem protection on private lands, which has given public lands a significantly enhanced role in national biodiversity conservation efforts.[68]

In the wake of the spotted owl controversy, the Clinton administration moved aggressively to instill ecosystem management principles into the federal bureaucracy. Under its Government Performance Review program, the administration identified ecosystem management as the strategy of choice for streamlining federal natural resource policy. An interagency task force then developed a blueprint for federal ecosystem management policy, defining it as "a [goal-driven] method for sustaining or restoring natural systems and their functions and values . . . based on a collaboratively developed vision of desired future conditions that integrates ecological, economic, and social factors . . . within a geographic framework defined primarily by ecological boundaries."[69] In December 1995, the public land agencies convened a major interdisciplinary conference to develop key ecological management concepts, resulting in a three-volume publication that contains the most comprehensive discussion of ecological stewardship principles to date. Since then, the federal government established a Cooperative Ecosystems Studies Unit network, announced a new unified watershed approach policy toward managing federal lands and resources, and promoted a plethora of agency-specific ecological management policies. The Clinton administration also used its appointment powers to bring reform proponents into key leadership positions in the land management agencies. Most notably, the successive appointments of Jack Ward Thomas and Mike Dombeck, both veteran biologists, to head the Forest Service symbolized the administration's commitment to this new policy direction. Not only did they break the mold of former Forest Service chiefs—all men with timber management backgrounds—but their appointments ensured that the agency would take ecological management principles seriously. Although the administration's overt ecosystem management rhetoric quieted following the 1994 Republican congressional election victory, its basic policies and reform initiatives continued to emphasize biodiversity protection and ecological sustainability on public lands.[70]

Chastened by its spotted owl experience, the Forest Service has been an aggressive proponent of new ecosystem management policies. In 1997, the Forest Service convened a Committee of Scientists to recommend revisions to its NFMA forest planning regulations. Under the new regulations, adopted during the Clinton administration's waning days, the overarching purpose of forest planning is sustainability to ensure the ecological integrity of watersheds, forests, and rangelands. As might be expected, the regulations were widely attacked by the timber industry and its congressional allies. After reviewing them, the Bush administration proposed wholesale revisions that deemphasize the agency's ecological sustainability obligations, minimize its NEPA obligations, and dilute its biodiversity conservation responsibilities. The Forest Service has also been a principal advocate of new regional planning initiatives designed to take an ecological approach to public land management. In both the Interior Columbia Basin Ecosystem Management Project (ICBEMP) and the Sierra Nevada Forest Plan Amendment project, the Forest Service dramatically expanded its planning efforts to integrate ecological concerns—biodiversity conservation, fire, and other disturbance processes—into its management agenda. In addition, under Chief Dombeck's guidance, the agency embraced cooperative stewardship as its new watchword in an effort to improve relations with its diverse constituencies and to encourage managers to work across boundaries to achieve common resource conservation objectives—other key elements of ecosystem management.[71]

Moreover, asserting that roads represent a serious threat to its species conservation and environmental protection obligations, the Clinton administration pursued two controversial national forest initiatives designed to protect sensitive wildlife habitat and roadless areas. First, the Forest Service imposed a moratorium on the construction of new roads into the national forests pending a road closure survey. Eventually, the moratorium was lifted, but only after the agency revised its road management policies to significantly curtail new road construction. Second, the Forest Service undertook a nationwide roadless area review with the express purpose of protecting the remaining unroaded national forestlands from any new incursions for ecological and other purposes. The ensuing Roadless Area Conservation EIS and related regulations encompassed over 58 million acres of national forest land and effectively eliminated commercial logging and new road construction from these areas. Although the initiatives do not preclude all timber harvesting, they make it more expensive and difficult—in the absence of new roads—to access remote sites, thus further diminishing timber's role on the agency's agenda. Opponents to the roadless area regulations, however, temporarily secured a federal court injunction preventing them from going into effect, and the Bush administration is reviewing the regulations, with all signs pointing toward a major reduction in the level of protection they afford.[72]

Long regarded as a captive of the extractive industries, the BLM has also endorsed ecosystem management and planning principles. Under Secretary Babbitt's guidance, the BLM revised its rangeland regulations, adopting ecologically oriented standards to govern livestock management on the public domain and creating new state-based Resource Advisory Councils (RACs) to replace the old rancher-dominated grazing advisory boards. The RACs, which are required to have a diverse membership including environmentalists, scientists, and other representatives, are responsible for establishing local grazing policies consistent with national rangeland health standards. These changes not only refocus the BLM's livestock management responsibilities on ecological performance standards, but also broaden the constituencies formally involved in range management issues. It is, for all practical purposes, a marriage between ecosystem science and local democracy. The Babbitt-led BLM also adopted new hardrock mining regulations that imposed, for the first time, explicit environmental performance and financial assurance standards on the mining industry; these standards were designed to protect key ecological features on the landscape, both water quality and wildlife habitat. (At the same time, the BLM asserted the legal authority to deny a mining permit to protect competing ecological or cultural resources, but the Bush administration promptly repudiated this position.) In addition, the BLM has revised its FLPMA-based resource management planning process to incorporate new regional planning guidance and adaptive management concepts that are based on ecosystem management principles. Moreover, under the Clinton administration, the BLM assumed responsibility for 14 new national monuments that have been incorporated into a new National Landscape Conservation System, giving the agency a prominent new role in ecological preservation and environmental protection. Though the rhetoric of ecosystem management may be less evident in these initiatives, the net effect has been to infuse the BLM with new ecologically oriented management responsibilities that should ultimately help reshape its on-the-ground priorities and policies.[73]

Within the preservation-oriented agencies—the National Park Service and the U.S. Fish and Wildlife Service—a commitment to ecosystem management is also manifest in several administrative initiatives. During the Clinton administration, the Park Service revised its all-important Management Policies document, which establishes fundamental park resource management policies. Noting the agency's primary resource protection mission, the revised policies are designed "to integrate parks into sustainable ecological, cultural, and socioeconomic systems," using a planning process that "considers the park in its full ecological, scenic, and cultural contexts . . . and as part of a surrounding region." Park managers are directed "to maintain all the components and processes of naturally evolving park ecosystems, including the natural abundance, diversity, and ecological integrity of the plants and animals native to those ecosystems." Drawing upon the document's strict interpretation of the Park Service's legal resource protection responsibilities, the Clinton administration concluded that recreational snowmobiling in Yellowstone National Park impaired park resources and undertook to phase it out on environmental grounds. The U.S. Fish and Wildlife Service also adopted an ecosystem approach to its wildlife conservation responsibilities, which entails "protecting the natural function, structure, and species composition of an ecosystem." Using principles derived from the federal Interagency Ecosystem Management Task Force, the FWS reorganized itself into ecosystem units along watershed boundaries, and it endorsed adaptive management and partnership strategies to support its new ecological approach. Through its ESA regulatory responsibilities, the FWS has also helped advance ecosystem management goals on public lands. Its myriad species listing decisions, which have included multiple salmon and steelhead runs, the northern and Mexican spotted owls, the lynx, and the bull trout, have had the practical effect of giving priority to the habitat needs of these protected species on large chunks of the public domain. And using its Section 10(j) reintroduction authority, the FWS is actively restoring several controversial species, most notably the wolf, whose presence has clear ecological overtones. In short, the Park Service and FWS have integrated important ecosystem management concepts into their policy agendas and their day-to-day decisions.[74]

JUDICIAL INTERVENTION AND ACQUIESCENCE

There can be little doubt that the federal judiciary deserves much credit for bringing ecosystem management to the public lands. Judge Dwyer's dogged determination to protect the spotted owl and to curb runaway logging in the Pacific Northwest forests, which rested upon the cumulative impact of the NFMA and FLPMA planning laws, the ESA, and NEPA, ultimately forced the Forest Service and BLM to adopt a new ecologically oriented approach to administering the region's public forests. Since then the public land agencies have credibly used the threat of more spotted owl–type litigation to justify incorporating basic ecosystem management principles into public land policy. If the powerful timber industry could be brought to its knees in the Northwest, then the same could happen anywhere. No one—neither the Clinton administration, the agencies, industry, nor local communities—relished the prospect of another spotted owl imbroglio, particularly when the environmental community's lawyers had a proven litigation road map available to them. Besides, federal courts elsewhere demonstrated that they too were prepared to enforce the environmental laws, even if they did not push as hard as Judge Dwyer and rejected some of the environmental arguments presented to them. On balance, the judiciary has endorsed most of the key elements of ecosystem management, finding that the existing law basically supports this new approach to public land policy. This means that the agencies have a limited range of options for meeting their stewardship responsibilities on the public domain.

Among the ecosystem management principles, the courts have struggled the most with the role of biodiversity conservation in public land management. Although the courts have aggressively enforced the Endangered Species Act and its protective provisions, they have been reluctant to translate the NFMA biodiversity provision into an affirmative agency responsibility. Once listed under the ESA, various charismatic and noncharismatic species—including the northern spotted owl, red cockaded woodpecker, Mexican spotted owl, and various salmon runs—have provided the legal ammunition necessary to reform forest and range management practices in several locales. When the FWS has dallied or obfuscated in making initial listing decisions, the courts have not hesitated to order the agency to reconsider, reminding agency officials that they are obligated to employ the best available scientific information. And when the FWS has tried to avoid designating critical habitat for key species, such as the northern spotted owl, Mexican spotted owl, and California gnatcatcher, the courts have consistently ordered the agency to make this important ecological determination, which also has potential legal ramifications under the statute's consultation provisions. When the FWS has overlooked vital population and habitat data in preparing recovery plans, as it did for the grizzly bear, the courts have forced it back to the drawing board to incorporate this data into its long-range planning. As a result, the ESA has played a vital role in elevating biodiversity concerns within the public land agencies. And as species listings and critical habitat designations mount, more and more of the public domain is being overlaid with ESA obligations, thus bringing the FWS or NMFS directly into most major public land planning and management decisions.[75]

But when the legal basis for species conservation shifts from the ESA to the NFMA's biodiversity provision, the courts have been decidedly less willing to impose affirmative species conservation obligations on public land agencies. To be sure, Judge Dwyer affirmatively used the NFMA biodiversity provision and its accompanying minimum viable population regulation to protect the spotted owl, finding that the Forest Service was required to ensure well-distributed, viable populations in its plans even after the owl was listed under the Endangered Species Act. But the Court of Appeals for the Ninth Circuit subsequently ruled that the minimum viable population regulation does not require the Forest Service to select the management alternative that provides the highest likelihood of species survival, nor does its habitat maintenance requirement impose an absolute prohibition on forest fragmentation. And elsewhere, the courts have generally given the Forest Service broad discretion in meeting its diversity obligations, primarily on the theory that the agency—not the judiciary—possesses the requisite technical expertise to make these scientific judgments. The Court of Appeals for the Seventh Circuit, for example, rejected the argument that the Forest Service was required to incorporate conservation biology principles into its biodiversity analysis for forest planning purposes, choosing instead to defer to the agency's own scientific conclusions. Other courts, including the Court of Appeals for the Tenth Circuit, which has jurisdiction over much of the inland West, have reached similar results. But under the revised NFMA forest planning regulations, the Forest Service faces more extensive biodiversity conservation obligations, including detailed new viability analysis requirements for focal and at-risk species that extend to habitat needs and ecological processes affecting these species. Thus, while the courts may not have moved biodiversity to the forefront of the agency's planning agenda, the agency may itself have achieved that result through its own administrative reforms.[76]

The courts have also prompted the land management agencies to expand the scale of their environmental analysis, basically forcing them to begin planning on an ecosystem scale. The principal legal tool has been the NEPA cumulative effects analysis requirement. In case after case, the courts have enjoined agency plans or projects that failed to consider the full spatial and temporal impacts that a proposal will have on the environment. The Forest Service, for example, is required to consider and analyze the aggregate impacts of linked or concurrent projects, such as proposed timber sales and related road construction plans, in a single NEPA document; it cannot segment its analysis of the environmental impacts. Both the Forest Service and BLM cite their broad-scale NEPA environmental analysis obligations as a primary reason for pursuing such initiatives as the ICBEMP regional assessment process. Similarly, the courts have interpreted NEPA to require the agencies to analyze the impact of proposed activities on wildlife corridors connecting impacted areas, thus also ensuring that an ecosystem-focused analysis is done before final project decisions are made. As the scale of environmental analysis expands, public land agencies are finding themselves forced to coordinate their planning and management decisions with other agencies and adjacent landowners, another key dimension of ecosystem management. Most notably, the courts have interpreted the ESA to require the Forest Service to reconsult with the FWS on its forest plans following a new species listing decision, rather than simply consulting on individual project decisions. The practical effect of this ESA reconsultation requirement on resource planning is to expand the scale of environmental analysis for biodiversity conservation purposes.[77]

The courts have also begun to enforce adaptive management principles, imposing inventorying and monitoring responsibilities on the agencies. In the climactic Northwest Forest Plan litigation, Judge Dwyer emphasized the importance of inventorying and monitoring, and then forcefully drove the point home in a subsequent ruling that enjoined further timber sales when agency officials ignored this important adaptive management responsibility. Other federal courts have made the same point. The Court of Appeals for the Eleventh Circuit, for example, has held that NFMA-based forest plans require the Forest Service to gather inventory data on ESA-proposed, ESA-listed, and sensitive species that might be affected by timber harvesting before offering the sales. In a related development, the courts have enforced resource analysis and monitoring provisions in existing forest plans, thus requiring agency officials to reassess forest conditions before making further project-level decisions. The decisions also stand for the proposition that forest plans establish legally enforceable obligations, which provides a basis to enforce other ecosystem management commitments in the plans.[78]

In addition, the courts have helped define the role and limits of public involvement in public land planning and management processes. Under existing law the agencies must provide meaningful opportunities for public participation, while ecosystem management principles contemplate that public values will play a major role in shaping long-range planning and project decisions. Confirming the importance of public opinion, several courts have sustained resource decisions predicated primarily on public values and concerns. A Montana federal district court, for example, upheld a forest supervisor's decision not to lease Rocky Mountain Front forest lands for oil and gas development primarily because of the public's well-documented view that these lands were treasured as a special place in their undisturbed state, even though the exploration impacts could be adequately mitigated. In a related development, when public land agencies seek to involve the public directly in management decision processes, the courts have begun to establish clear boundaries on such citizen participation. Agency officials, for example, cannot relinquish their authority to a citizen management committee and vest it with the power to make resource decisions, even though Congress may have expressly provided for the creation of such an advisory group. Under the Federal Advisory Committee Act, the courts have ruled that agency officials who solicit advice from persons outside the federal government must establish an official advisory committee and comply with rigorous procedural requirements—including notification, charter, balance, and open meeting standards—before using any such advice for management purposes. In short, the courts have allowed public land agencies to predicate management policies and decisions on public values, while limiting the role of nongovernmental parties in agency decision processes without adequate procedural safeguards.[79]

On a related note, the courts have generally sustained the Clinton administration's ecologically oriented reform initiatives. The courts have consistently supported the FWS's use of its experimental population reintroduction authority, thus giving the agency an important ecological restoration tool. In two separate cases, the D.C. Circuit Court of Appeals has ruled that President Clinton acted well within his authority under the Antiquities Act in designating several new national monuments, even those with expansive boundaries. The Supreme Court upheld Interior Secretary Babbitt's rangeland reform regulations, rejecting rancher arguments essentially seeking to maintain a preferred position in range policy matters. When the FWS balked at designating critical habitat for ESA-listed species, the courts admonished the agency to make the designation, citing the statutory linkage between habitat and species survival. In addition, after a lower court blocked the Forest Service from implementing the Clinton administration's controversial Roadless Area Conservation initiative, the Court of Appeals for the Ninth Circuit lifted the injunction, finding that the agency had complied with NEPA's public involvement and other requirements.[80]

To be sure, it would be a mistake to read the developing case law as an unalloyed judicial endorsement of ecosystem management on the public lands. Ready counterexamples can be found where the courts have rejected ecologically based legal arguments, and thus sanctioned business-as-usual on the public lands. Most of the developing case law focuses on the Forest Service and forest management policy; few court decisions have held the BLM to the same high ecological management standards. The Court of Appeals for the Ninth Circuit, with the grueling spotted owl litigation under its belt, has proven more receptive to ecologically focused legal arguments than its counterparts elsewhere, though it also has handled far more public lands litigation than the other circuits. No court has yet ordered the Forest Service or any other agency to engage in ecosystem management; rather, the courts have interpreted diverse legal provisions to incorporate various elements of ecosystem management policy—in essence creating a nascent common law of ecosystem management. And while the courts have rejected most of the Wise Use movement's litigation agenda, they have forced the U.S. Fish and Wildlife Service to interject economic considerations into the ESA's critical habitat analysis. Yet the trends are nonetheless remarkable. Even when the organic statutes—NFMA and FLPMA—are not read to impose enforceable ecosystem management standards on public land agencies, the universal laws that apply across public lands—like the ESA and NEPA—are being enforced in a manner consistent with ecosystem management principles. Moreover, the courts are importing very clear science-based standards and adaptive management obligations into these laws, thus highlighting the importance of good science as a key dimension of public land policy. In short, with the courts increasingly involved in resolving environmental controversies on the public domain, public land agencies ignore ecosystem management principles at their peril.[81]

CONGRESSIONAL RECALCITRANCE AND REPOSE

Rather than joining the shift toward a new ecological management regime on public lands, Congress has been anywhere from ambivalent to largely hostile toward these new developments. During the early years of the Clinton administration, Congress appeared poised to endorse the emerging ecological management regime through legislation. But since 1994, when the Republicans assumed control, Congress has been stalemated over most matters involving multiple-use public lands. Unable to muster a consensus behind reform proposals addressing the General Mining Law, the Endangered Species Act, and other public land laws, Congress has used controversial appropriations riders to achieve short-term goals aimed toward increasing commodity production activities on public lands. It has also used riders to block or delay various administrative reform proposals that establish more environmentally sensitive public land policies. And it has refused to expand significantly the existing wilderness system. Yet despite its overt hostility to many of the Clinton administration's public land initiatives, Congress has not reversed these policies, suggesting that the ongoing shift toward ecological management has long-term implications. Moreover, outside the multiple-use public lands, Congress has actually adopted legislation that promotes ecological management on national wildlife refuge and national park lands, thus linking land preservation with new biodiversity conservation policies.

In the early 1990s, confronted with the new Clinton administration and the intractable spotted owl controversy, Congress was beginning to assess what new ecosystem management legislation might entail on public lands. The responsible committees, recognizing that major social and economic changes were occurring within the public land states, held hearings to examine the long-term implications of these changes. At Congress's behest, the General Accounting Office and the Congressional Research Service prepared detailed reports explaining ecosystem management, both how it squared with existing public land policy and the changes that might be required. As the Pacific Northwest's spotted owl controversy intensified, a Republican member of the region's delegation twice introduced an ecosystem management bill that would have created a study commission to recommend future legislative changes. At the same time, several of the bills introduced to resolve the spotted owl controversy called for new ecosystem management protocols on the region's public lands. Whether any of these legislative initiatives would have fared differently without the turnover in congressional control would be sheer speculation, but the stage was plainly set for a legislative endorsement of the ecosystem management concept.[82]

After the 1994 congressional elections, however, the legislative atmosphere changed dramatically. Conservative Republicans assumed control of Congress and announced new legislative priorities. In their Campaign for America agenda, neither the environment nor ecology was mentioned, except as targets for legislative reforms promoting deregulation, property rights, and devolution. But even though the 104th Congress passed the Unfunded Mandates Reform Act of 1995 as part of its environmental reform agenda, the legislation has had little effect on public lands, other than amending the Federal Advisory Committee Act to facilitate consultation between federal managers and state and local governmental officials. Despite strong industry support for revising the Endangered Species Act and the National Forest Management Act, neither the 104th Congress nor its successors have been able to rewrite these laws, which are lynchpins in the litigation arsenal to promote more ecologically sensitive management policies. Nor has Congress been able to pass industry-supported revisions to the General Mining Law or federal grazing policy that would enhance the legal posture of these industries. Congress has also been unable to move takings legislation that would give property owners greater rights and protections. Moreover, Congress has been unwilling to reverse legislatively several controversial Clinton administration policy initiatives, including the presidential Antiquities Act national monument designation powers, livestock grazing reforms, a moratorium on mineral patents, and various Forest Service reform initiatives. In short, notwithstanding a revolutionary shift in congressional political power, the basic legislation governing public lands remains intact, which means the fundamental legal building blocks for new ecological management policies remain in place.[83]

But the Republican-controlled Congress has adeptly used appropriations riders to achieve short-term political gains in the public lands venue. As we have seen, the controversial 1995 salvage logging rider represented a major congressional effort to reshape the Clinton administration's timber policies by eliminating judicial oversight under the existing environmental laws. The fact that Congress included the limitation on judicial review was at least a tacit acknowledgment that it could not muster the necessary political support to change the governing environmental laws through direct legislation. But while the rider briefly expedited timber sales across public lands, it proved so politically controversial that even Congress acquiesced in the administration's decision to terminate the salvage program six months early. After that, Congress used other riders to block temporarily Clinton administration public land policy reforms, including new regulations imposing environmental standards for mining on BLM lands and proposed R.S. 2477 regulations governing road management on public lands. In addition, Congress used its budgetary powers to downgrade the new National Biological Survey, originally conceived as an independent agency in the Department of the Interior, to a mere division within the U.S. Geological Survey, where its stature is clearly diminished. Yet at the same time, Congress funded major Clinton administration ecosystem management initiatives, namely the Interior Columbia Basin Ecosystem Management Project, the Sierra Nevada Ecosystem Project (now the Sierra Nevada Forest Plan Amendments), the Everglades Restoration program, and the San Francisco Bay Delta ecological restoration program. These projects, most of which are still moving forward under existing law, signal a fundamentally new ecosystem-based approach to federal natural resource policy—one that Congress has explicitly endorsed with its continued appropriations. Thus, despite the sparring over riders addressing administrative policy reform initiatives, the real story of reform on public lands may prove to be Congress's acquiescence in these numerous place-specific ecological management and restoration initiatives.[84]

Congress has also made several noteworthy changes in public land law, mostly outside the controversial multiple-use lands. In 1997, Congress passed the National Wildlife Refuge System Improvement Act, which gave national wildlife refuges an organic act, established a comprehensive conservation planning process, and endorsed biodiversity conservation as a refuge purpose. With this new legislation, the national wildlife refuges joined the national forests in having biodiversity conservation as part of their organic mission. In 1998, Congress passed the National Parks Omnibus Management Act, which made scientific research an explicit part of the National Park Service's organic agenda. The legislation acknowledged that the agency's scientific research obligations would inevitably involve matters beyond park boundaries, and thus recognized that national parks are interlinked with larger ecosystems. With its new scientific research agenda, the Park Service essentially has legislative approval to begin viewing the national parks as vital biodiversity reserves containing important scientific resources and knowledge—all of which support an expanded ecological role for the parks on the public landscape. In addition, the 106th Congress established a new six-year land acquisition and conservation funding program, entitled Land Conservation, Preservation and Infrastructure Improvement, which substantially increased the amount of federal money available for wildlife habitat and recreation purchases, ESA species recovery efforts, and other conservation purposes. And it potentially reordered local public land priorities by revising the payment in lieu of taxes (PILT) system and decoupling federal payments to the counties from timber production and other commodity development activities on public lands.[85]

Congress has, in addition, unleashed its own experiments in ecosystem management and community conservation. The most obvious example is the Quincy Library Group legislation, which placed a congressional imprimatur of approval on a locally negotiated national forest management experiment. That Congress saw fit to endorse the Quincy proposal, albeit under the Forest Service's auspices and subject to the full panoply of national environmental laws, further suggests that Congress is not adverse to new ecological management approaches that integrate local economic and other concerns into the planning equation. In a notable departure from its usual acquisitions policy, Congress adopted the Valles Caldera Preservation Act to administer the ecologically significant 95,000-acre Baca Ranch lands in north central New Mexico, which are surrounded by other public lands. Rather than simply turning these newly acquired private lands over to federal management, Congress sought to allay local concerns over the growing federal estate by establishing an experimental public-private management regime consisting of an independent board of trustees charged with overseeing these lands under a sustainability mandate also designed to preserve their ecological and historic values. Although it would be premature to forecast a wholehearted congressional endorsement for local management of public lands, Congress is obviously receptive to such carefully crafted arrangements, so long as both national and local interests are adequately protected.[86]

In sum, despite its hostile rhetoric toward new ecological management policies on public lands, Congress has done remarkably little to reverse this trend. To be sure, it has regularly added controversial riders to its budget bills, some of which have temporarily stalled new administrative initiatives. During the Clinton administration, it regularly convened oversight hearings to chastise administration officials, even threatening severe budget cutbacks in an effort to derail new initiatives. But Congress has been unable to advance a contrary legislative agenda. If it really wants to curtail ecological management trends on public lands, it will have to overhaul the principal environmental laws—NFMA, FLPMA, NEPA, and the ESA—that the courts have used to sustain environmental challenges to ecologically insensitive planning and management decisions. Budget riders and other ad hoc political strategies simply will not work. Even Congress seems to recognize this new legal landscape, as reflected in its continued funding of the Sierra Nevada, ICBEMP, and other place-based initiatives incorporating new ecological policies into a regional management agenda. Moreover, Congress's willingness to import new biodiversity and scientific research obligations into the organic mandates of the preservation-oriented agencies can be seen as a further manifestation that ecological management approaches are here to stay. To change direction, Congress would have to mount a prolonged assault on popular environmental laws that have become a well-accepted part of the public landscape. This would require an enormous expenditure of political capital for dubious gains, particularly given the prevailing scientific consensus and the broader socioeconomic changes underlying the new ecological policy initiatives.

Just as an ecosystem experiences dynamic and random changes, the nation's political institutions and priorities can change too. The 2000 elections witnessed just such a shift in national politics. George Walker Bush, a conservative Republican, was elected president to succeed Bill Clinton, a liberally inclined Democrat. President Bush's key appointments to the Department of the Interior and the Department of Agriculture signaled a clear shift in emphasis away from an ecologically driven public land policy agenda to one more attuned to industry's interests and local influence. The Bush administration has begun to dismantle some Clinton era reforms, notably its mining law reforms, and to revise others, including the Forest Service's Roadless Area Conservation initiative and its new NFMA planning regulations. Other Clinton era reforms, including the National Biological Survey agency and the BLM's National Landscape Conservation System, and its national monument management policies, are at risk too. And equally important, in the aftermath of the September 11, 2001, terrorist attacks, the Bush administration has aggressively pursued a new national energy policy that would open public lands to accelerated leasing, exploration, and development while reducing environmental protections and related planning obligations. A relatively evenly balanced Congress has thus far not seemed in the mood for major policy reforms on public lands, but that may change with future elections, as the Republicans capture and lose control of Congress by turns. Without an overhaul of the governing laws, the prevailing judicial precedent virtually forces public land agencies to incorporate ecological concerns into their planning and decision processes. Though the new Bush administration may be disinclined to actively promote ecosystem management ideas, it runs substantial legal and even political risks if it blindly ignores them. Perhaps it will perceive a unique opportunity to integrate emerging ecosystem management principles with its community involvement agenda to reinforce the connection between a sustainable landscape and the people who inhabit it.

Regardless, the new ecosystem management agenda reflects a fundamental shift in how we value and manage our public lands. Rather than treating the public domain as merely a development bank, an ecological perspective puts a premium on it as a biodiversity stronghold where nature's forces must be respected. Much of the federal estate, however, bears the scars of our past policies: denuded mountainsides, depleted rangelands, an impoverished biota, abandoned mine sites, and sediment-choked streams. If the goal of ecosystem management is to conserve a full suite of species and ecological processes on public lands, then we will need a comparable commitment to undoing our most egregious legacies. It will not be enough to start trying to manage for ecosystem health; we must also begin restoring natural systems. Indeed, ecological restoration is—and must be—a critical component of any viable ecosystem management policy.

FIVE

Making Amends with the Past

Ecological Restoration and Public Lands

The task . . . is to become a co-worker with nature in the
reconstruction of the damaged fabric.

WILLIAM PERKINS MARSH (1864)

This is a day of redemption and of hope. It's a day when the limits of
what is possible have been greatly expanded because we are showing our
children that restoration is possible, that we can restore a community
to its natural state.

SECRETARY OF THE INTERIOR BRUCE BABBITT (1995)

About 200 years ago, when the Lewis and Clark Expedition traversed the American West en route to the Pacific Ocean, it encountered a largely untouched and still primitive landscape. Millions of bison crisscrossed the Great Plains, grizzly bears roamed the region, salmon choked many of the rivers, and periodic fires burned the prairies and forests. Native ecosystems, having evolved over the millennia, were shaped primarily by natural disturbance regimes. To be sure, the region's native inhabitants had a hand in the process, routinely setting fires and taking wildlife, but these impacts had not unraveled historic evolutionary patterns. Today that same western landscape looks quite different. European settlement and the persistent onslaught of modern civilization have markedly altered ecological patterns: cattle have replaced bison on the plains, only a few remnant grizzly bears remain, annual salmon runs have declined to dangerously low levels, and fires are regularly suppressed with ruthless efficiency. Intent on making the landscape safe and productive, we have eliminated entire species and disrupted natural processes on a hitherto unprecedented scale. Even on the uninhabited public lands, ecological simplification has been the order of the day.

Emerging ecological management concepts are designed to reverse this pattern and to restore extirpated species as well as historic disturbance regimes. Indeed, ecological restoration is a key component of ecosystem management and an essential element in any meaningful biodiversity conservation or sustainability strategy. On public lands, the return of the wolf and fire epitomizes our commitment to making the landscape whole again, representing a sea change in how we value the natural world. These and other restoration proposals have stirred passionate policy debates with profound economic, social, and cultural overtones that transcend the immediate environmental impacts. At the center of these policy debates are difficult questions concerning active versus passive resource management strategies, which are heavily colored by the fact that aggressive management strategies fostered the very ecological degradation that is now the subject of restoration efforts. The current law, though not framed in explicit ecological restoration terms, provides the public land agencies with sufficient authority to pursue an active restoration agenda, and they are plunging ahead into this yet uncharted realm. When presented with specific restoration proposals, Congress has generally—though sometimes reluctantly—added its stamp of approval to these projects. As these experimental restoration initiatives expand in size and scope, they are not only reshaping public land policy but also redefining the human relationship with nature.

Of Wolves and Fire: Restoring the Ecosystem

BRINGING THE WOLF BACK

For over a century, the symbolic importance of the wolf (*Canis lupus*) has transcended the biological realities of its existence. Depicted as "a mythic and blood-lustful killer," the wolf was eliminated from the West by the mid-1930s through a full-blown federal eradication effort. In both a real and symbolic sense, the success of that campaign was viewed as the triumph of civilization over the forces of nature. Nearly a century later, the wolf has reappeared, but this time it is the symbol of an ecological renaissance. The story of wolf restoration is as much a study in cultural transformation and *real politik* as it is a study in ecology. It represents a major shift in policy and power that is moving ecological restoration to center stage in public land policy.[1]

When European settlers first arrived in the West, wildlife proliferated across the vast landscape. Numerous contemporary accounts describe the plains blackened by millions of bison with wolf packs in tow, subsisting on the young and weak stragglers who could not defend themselves. The bison were the first to go, slaughtered by market hunters seeking only profit from their valuable hides. Then came the ranchers, who replaced the bison with cattle and were soon clamoring for government protection against wolves and other marauding predators. The early state and territorial legislatures needed little convincing to establish a bounty on predators, igniting another market-driven extermination campaign, with the wolf as the figurative force of evil. Even Teddy Roosevelt, later renowned as the nation's first conservation-oriented president, joined the anti-wolf chorus, labeling it "the beast of waste and desolation." By the mid-1890s, all the western states had wolf bounties, ranging from $1–$15 per hide. Some private bounties ran as high as $400 per animal. Wolves were dispatched in every conceivable manner—shot, clubbed, poisoned, burned, tortured, and "denned." But despite the local zeal for extermination, the states were unable to eliminate the wolf until the federal government finally joined the effort, bringing its superior financial and technical resources to the campaign.[2]

Once the federal government entered the picture, the wolf's days were numbered. In 1914, responding to entreaties from the western states and its livestock industry, Congress funded the Department of Agriculture's Biological Survey to initiate an all-out war against the wolf. The wolf was not safe anywhere, not even in the national parks. Yellowstone, despite its role as an early wildlife sanctuary, had permitted bounty hunters to trap wolves in the park since its inception. The National Park Service, after its establishment in 1916, also supported predator eradication in the national parks, hoping to protect more highly valued species from depredation and to curry favor with its neighbors. After the federal government actively entered the extermination campaign, it became a stunning success: by 1942, government hunters had dispatched 24,132 wolves, mostly in Colorado, Wyoming, Montana, and the Dakotas. Wolves were gone from Yellowstone National Park by 1927, and they were eliminated from the rest of Wyoming as well as most of Montana and Idaho only a few years later.[3]

Only once the wolf was gone did biologists begin to reassess its role in the ecosystem and the myths that had prompted people to drive it from the western landscape. Gradually, a different view of the wolf started to surface—one that was grounded more in science than myth and hyperbole. During the early 1930s, after the wolf had been eliminated from Yellowstone, the Park Service announced it would no longer countenance predator control in the parks, though this new policy was not strictly followed over the ensuing years. In the mid-1940s, deeply contrite over an earlier wolf shooting experience that helped shape his ethical views, Aldo Leopold suggested that wolves might be "used to restock Yellowstone." Soon thereafter, separate pioneering studies by biologists Sigurd Olson, Paul Errington, and Adolph Murie provided new insights into predator–prey relationships and helped dispel the view that wolves would inevitably destroy big game populations. In the mid-1960s, following publication of the influential Leopold Report, the Park Service radically altered its wildlife management policies to provide for "maintaining, and when necessary reestablishing, indigenous plant and animal life."[4] At the same time, another Leopold Report on federal predator control policies concluded that all native animals should be valued and that local control efforts should be strictly limited to troublesome individuals and not undertaken indiscriminately. In 1970, biologist David Mech published his groundbreaking study of wolf ecology, which provided a comprehensive scientific analysis of the wolf and its habits. The ancient negative myths that had plagued the wolf were gradually being dispelled, restoring the animal to biological if not political respectability.[5]

After passage of the Endangered Species Act of 1973, wolf recovery became official federal policy. Under the ESA, which is administered by the U.S. Fish and Wildlife Service (FWS), the secretary of the interior is required to "list" all endangered and threatened species on the endangered species registry and to promote their recovery. The wolf, extirpated throughout most of the United States, was listed in 1974 as an endangered species everywhere except in northern Minnesota, where it was treated as a threatened species. Once wolves had been listed, all federal agencies were obligated to take affirmative steps to conserve them, and they received maximum legal protection. Moreover, the FWS was required to prepare a wolf recovery plan, detailing how the species would be protected, managed, and eventually recovered. To promote the wolf's recovery, the FWS could avail itself of the ESA's new experimental population reintroduction provision, which Congress had added in 1982 to create management flexibility and facilitate controversial species restoration programs. Under this provision, known simply as Section 10(j), the secretary of the interior can designate a reintroduced species as either an essential or nonessential population; if designated nonessential, the species receives significantly reduced legal protection and can be taken in defined circumstances. As expected, the FWS's wolf recovery plan recommended using the Section 10(j) experimental population designation for the Yellowstone wolf recovery effort; it relied on natural recolonization to recover wolves in Idaho and Montana. But not even the powerful ESA could ensure actual wolf recovery. Still regarded in influential quarters as the beast of destruction, the wolf's fate rested with a hostile administration and a reluctant Congress, which controlled the FWS's purse strings.[6]

The next chapter in the wolf restoration saga is perhaps best understood as a classic study in biological fact, raw political power, and the enduring realities of federalism. Although the federal government was required to prepare a wolf recovery plan, top Reagan administration officials—well-attuned to western state sensitivities—adamantly opposed wolf reintroduction. They delayed completion of the revised wolf recovery plan for nearly seven years before finally releasing it in 1987. By then, wolves were a growing reality in the northern Rockies: in 1982, the so-called Magic Pack was discovered denning on the northwest periphery of Glacier National Park, and other wolves were soon dispersing from Canada into northern Montana, bringing with them the ESA's full protection as an endangered species. In response to this unexpected biological development, political positions quickly hardened throughout the region. Montana, Idaho, and Wyoming vehemently opposed any federal wolf reintroduction program, concerned that the ESA would prevent them from managing the wolves and inevitably curtail economic activity on public lands. The region's extractive industries generally shared these same concerns. Ranchers also opposed reintroduction, citing the likelihood of increased federal control over their grazing privileges, limitations on their ability to defend livestock against depredating wolves, and the absence of any federal compensation program for livestock losses. Environmental organizations uniformly supported an aggressive wolf restoration program, though some organizations feared that the ESA's experimental population provision might not adequately protect the wolves. Public opinion polls consistently indicated strong national and regional support for wolf reintroduction. Nonetheless, the region's congressional delegations stood squarely against reintroduction; they employed budgetary riders, congressionally mandated study committees, and assorted other delaying tactics to block the recovery plan as long as they could. By 1991, though, the local delegations had exhausted their stalling tactics. Confronted with numerous studies demonstrating that wolves posed little threat to domestic livestock, local big game populations, or other economic activities, Congress quietly attached a rider to the Department of the Interior appropriations bill directing the secretary of the interior to prepare an environmental impact statement (EIS) on wolf reintroduction to Yellowstone and central Idaho.[7]

In the ensuing EIS, the FWS proposed to reintroduce an experimental population of gray wolves into both Yellowstone National Park and central Idaho. Relying upon Section 10(j) of the ESA, the draft EIS recommended translocating 30 wolves annually from Canada into Yellowstone and central Idaho over a 3-year period until a wild wolf population was established in each area. The translocated wolves would be designated a "nonessential" experimental population, which would eliminate any ESA-based land use restrictions on either public or private land. Under a proposed, legally binding Section 10(j) regulation, agency officials would be responsible for moving any wolves that attacked local livestock, ranchers could kill wolves caught actually attacking livestock, and landowners could harass wolves from their property. Although the federal government would not compensate livestock operators for any wolf losses, the Defenders of Wildlife (an avid wolf reintroduction proponent) had established a Wolf Compensation Fund to cover such losses, which had already been used to address Montana livestock depredations. Recovery would occur once 3 separate wolf populations with 10 breeding pairs of wolves had inhabited the recovery area for 3 consecutive years, and delisting could ensue after the states had adequate wolf management plans in place. Once released for public review, the Wolf Reintroduction EIS generated over 160,000 comments, most of which were supportive—then the largest number of public responses received concerning any federal conservation initiative. In the end, the reintroduction proposal was approved, a special experimental population reintroduction regulation was promulgated, and the FWS began live-trapping Canadian wolves for translocation to Yellowstone and central Idaho.[8]

The controversy, however, was not over yet; another chapter remained to be written in the federal courts. The region's agricultural community, spearheaded by the Wyoming Farm Bureau Federation, had vehemently opposed wolf restoration from the outset, and it immediately turned to the judiciary to block the reintro-

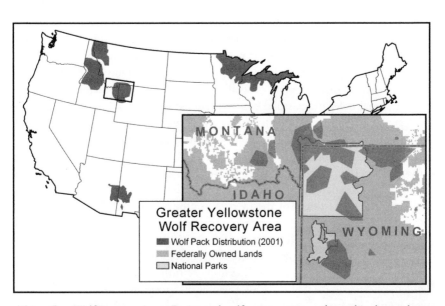

Map 3. Gray Wolf Recovery Areas. Designated wolf recovery areas are located in the northern Rockies, upper Midwest, and the Southwest. Through 2000, the various wolf populations were treated differently under the Endangered Species Act: reintroduced wolves in the Greater Yellowstone, central Idaho, and southwestern areas were governed by the act's experimental population provision; naturally occurring wolves in northern Montana, Wisconsin, and Michigan were treated as endangered species; the northern Minnesota wolf population was designated a threatened species. The Canadian wolves reintroduced in 1995 into Yellowstone National Park have dispersed widely from the central park area, as illustrated on the inset map.

duction program. Represented by the conservative Mountain States Legal Foundation law firm, the Farm Bureau sought a preliminary injunction in Wyoming federal district court to stop federal officials from bringing wolves from Canada into the United States for release into Yellowstone and central Idaho. In December 1993, Wyoming federal district judge William Downes denied the Farm Bureau's preliminary injunction request, paving the way for the wolf to reclaim more of its native habitat. In mid-January 1994, 14 wolves arrived in Yellowstone National Park and were transferred to a holding pen, where they would acclimate for 10 weeks before being released into the park backcountry in a procedure known as a "soft release" (so named because of the holding pen acclimation period). At the same time, another 15 wolves were transported to the edge of central Idaho's River of No Return Wilderness Area and given a "hard release" directly into the wild. Wolf recovery had become a reality, while ecosystem restoration gained more legitimacy on the policy agenda.[9]

But as the wolves were adapting to their new habitat, the judicial machinery ground on and soon enmeshed the entire recovery process in doubt. The Wyoming Farm Bureau Federation, still objecting to the legality of the reintroduction process, pressed its argument that the FWS could employ the Section 10(j) experimental population provision only when the proposed reintroduction would occur "outside the current range of such species." According to the Farm Bureau, because isolated wolf sightings in Yellowstone and central Idaho established the presence of naturally occurring wolves in these areas, the government could not release an experimental population. Several environmental organizations broke ranks with the other wolf recovery supporters and mounted a similar legal challenge, contending that the released central Idaho wolves could not be treated as a nonessential experimental population because that would effectively reduce the level of ESA protection enjoyed by naturally occurring wolves in the area. In response, the government argued that these Section 10(j) reintroduction limitations applied only when an established population of wolves (defined as two or more breeding pairs) was present in the release area, and that there was no evidence of such a population in either Yellowstone or central Idaho. The district court, however, read Section 10(j) differently and ruled that the reintroduction was illegal. Judge Downes interpreted the statutory language to preclude the use of the Section 10(j) reintroduction provision whenever any other members of the same species might be present, even if they did not constitute a population. In a stunning move, the judge then ordered federal officials to remove all the released wolves from both Yellowstone and central Idaho. But giving the wolves a reprieve, the judge immediately stayed his own order pending an appeal. If allowed to stand, Downes's ruling would have not only terminated a successful and popular ecological restoration program, but also undermined Section 10(j) as a viable ecological restoration strategy for extirpated species.[10]

By nearly any measure, the wolf restoration program was proving a remarkable success. As the year 2000 dawned, approximately 360 wolves were roaming the Greater Yellowstone and central Idaho recovery areas, with more than 60 pups born in each area during 1999. The wolves had formed 23 separate packs averaging 10 wolves per pack, with home ranges covering roughly 350 square miles. To monitor the wolves, the FWS had placed radio collars on more than 80 individuals. Livestock depredation rates were relatively low, but not inconsequential: approximately 250 cattle and sheep had been confirmed killed by wolves, and Defenders of Wildlife had paid over $60,000 in compensation. As predicted, visitation to Yellowstone National Park increased perceptibly in the aftermath of the reintroduction, confirming earlier projections that wolves would add a net economic benefit of $7–$10 million per year to local economies. Some hunting organizations, however, remained concerned that the wolves were reducing elk and other big game numbers, though there was no scientific evidence that this was the case. Moreover, neither Idaho nor Wyoming, still smarting over the Section 10 (j) reintroduction program, had yet taken any significant steps to prepare itself to assume legal responsibility for wolves once the recovery goals were met. In fact, the Idaho legislature's obdurate intransigence toward the recovery program had eventually prompted the FWS to vest the Nez Perce tribe—rather than the state—with responsibility for monitoring the central Idaho wolves.[11]

Perhaps the Court of Appeals for the Tenth Circuit took note that the reintroduced wolves were thriving in their new habitats, because it reversed Judge Downes's removal decision and sustained the FWS's wolf reintroduction program. In January 2000, after more than three years of legal limbo, the Tenth Circuit ruled that the reintroduced wolves would remain, concluding that the presence of solitary naturally occurring wolves did not preclude an experimental population reintroduction. The court was persuaded that Congress intended under Section 10(j) to give the secretary of the interior considerable flexibility to reestablish extirpated endangered species in locations where such reintroductions may prove controversial, which was plainly the case with wolves. Applying well-established principles of judicial deference to agency legal interpretations, the court sustained the FWS's interpretation of the Section 10(j) requirement that reintroduced wolves be "wholly separate geographically from nonexperimental populations" as meaning that the new wolves must be separate from other distinct wolf populations, not from lone dispersing wolves. The court also found that the reintroduction program, which treated all wolves in the recovery area as members of the experimental population regardless of their actual origin, was properly designed to avoid law enforcement problems that otherwise would occur if different levels of ESA protection were created for the area's wolves. These legal conclusions confirmed important biological and political realities: because wolves often roam many hundreds of miles, it is virtually impossible to segregate individual wolves by origin; because predator reintroduction proposals are almost always locally controversial, administrative flexibility is the key to using the experimental population reintroduction provision. In short, by deferring to the FWS's superior technical competence in making complex species management decisions under the

ESA, the court preserved the Section 10(j) provision as a meaningful and flexible biological restoration tool.[12]

Since then, the wolf recovery program has proceeded apace, and the federal government is poised to begin the delisting process. At the end of 2001, roughly 480 wolves were located in the Yellowstone and Idaho recovery areas, and they were increasing at a 30 percent annual rate. The FWS verified at least 34 wolf packs in the northern Rockies wolf recovery area, which met the recovery goals for the second consecutive year. Livestock depredation numbers were still relatively low; 40 cattle and 138 sheep were lost to wolves during the year. Since its inception, the Defenders of Wildlife compensation fund has paid about $215,000 to 188 different ranches. Well-known dispersers, the wolves continue to spread farther afield, raising legal questions about how roaming wolves should be treated outside the three immediate recovery states. Significantly, in a challenge to existing wolf control policies, an Idaho federal court ruled that wolves take precedence over livestock in the Sawtooth National Recreation Area, owing primarily to the area's special legal status. Despite continued resistance, the Idaho legislature has approved a state wolf management plan that is under federal review, while Montana and Wyoming have released draft wolf management plans. From the states' perspective, a major concern continues to be the costs involved in assuming responsibility for a recovered wolf population, which they estimate at $800,000 per year.[13]

The ecological and cultural lessons from the wolf restoration program are readily apparent. As a biological matter, the wolf's presence in Yellowstone completes the ecosystem, restoring predator–prey relationships that were ruptured over a half century ago when the wolf was exterminated from its historical range. The restoration program also demonstrates the power of scientific knowledge in the public policy arena; once the biological realities of wolf behavior were fully understood, local recovery opponents were left with few credible arguments against the reintroduction proposal other than a visceral dislike of wolves and an unsubstantiated fear of livestock depredation losses (which was effectively addressed by Defenders' compensation fund). As a cultural matter, the wolf's return to Yellowstone and the northern Rockies stands as a potent symbol of the nation's growing commitment to ecological preservation and restoration: without redrawing any wilderness boundary lines, the wolf's mere presence has de facto extended the wilderness concept wherever these wild creatures may roam. Wolf reintroduction may also mark another manifestation of a new West—one where an ecological protection ethic seems to be inexorably replacing an old West ethos based on raw utilitarianism and the conquest of nature.

The political and legal lessons from the wolf restoration controversy are more complex and still unfolding. Although the ESA plainly required the federal government to implement a wolf recovery program as part of the national commitment to endangered species recovery, the political branches were able to hinder the effort for over a decade, responding to powerful state and local opposition to the initiative. First a recalcitrant Reagan administration delayed completion of the recovery plan, and then a reluctant Congress stalled its implementation. The judiciary lent further credence to claims of federal overreaching when Judge Downes entered his wolf removal order, shrouding the entire recovery effort in doubt for another three years. When the Tenth Circuit finally lifted the legal shroud, it confirmed that Section 10(j) could be used as a flexible planning tool to reconcile federal ecological restoration goals with local economic concerns. But with wolf numbers proliferating and the recovery objectives in sight, a delisting proposal is now just a matter of time, potentially evoking further federal–state conflict. Are the states ready to assume management responsibility for the wolf? Neither Idaho nor Wyoming has a state endangered species act; Wyoming's draft wolf management plan still treats the wolf as an unwelcome predator outside the Yellowstone region wilderness areas; and Wyoming's plan is not coordinated with the Idaho and Montana plans. None of the state wolf management plans have received federal approval, and no one knows who will cover the costs of such a program. Anti-wolf sentiment still runs strong in some communities too. Yet the ESA delisting process will require firm assurances from the states that the wolf will be adequately protected before the federal government can relinquish management responsibility to them. As the delisting debate unfolds, it will undoubtedly offer another object lesson in federalism and endangered species restoration.[14]

FIRE POLICY IN FLUX

Fire is no stranger to the western landscape. Historically, lightning-ignited fires shaped the vegetative environment, which was well-adapted to periodic burning. In many forests, regular low-intensity fires would remove the understory without damaging mature trees, thus minimizing the occurrence of destructive, high-intensity forest fires. On the prairies, fires regularly consumed the dead grasses, replenishing soil nutrients and fertilizing another rich carpet of prairie grasses for the next season. Drought and wind conditions would occasionally trigger larger conflagrations—all part of a well-established fire regime in this fire-adapted environment. Native Americans also broadcast fire across the landscape, routinely using it in their agricultural, hunting, and military pursuits. The legendary explorer John Wesley Powell reported that "Native American ignited fires regularly destroyed larger or smaller districts of timber . . . and this destruction is on a scale so vast that the amount taken from the lands for industrial purposes sinks by comparison into insignificance." When the early settlers arrived, they too used fire to clear the landscape for farming and to create seasonal burn perimeters as protection from dreaded, runaway wildfires.[15]

Once the federal government began carving national forests and national parks out of the undisposed public lands, it inherited the fire problem as one of its central concerns. Nearby settlers immediately started looking to the federal government for fire protection. The newly created Forest Service was ready to oblige, assuming an early leadership role in federal fire policy that has endured throughout the past century. Under the ascendant Pinchot doctrine of utilitarian conservation, the new agency committed itself to excluding fire from the forests in order to protect valuable timber and other resources. Not everyone agreed, though. In California and the South, settlers had regularly torched the landscape for agricultural and fire control purposes. These "light burning" proponents, who saw fire as a beneficial natural process, argued against the proposed suppression policy.[16]

The 1910 fire season, however, sealed the debate's outcome and inaugurated the federal total suppression policy. In that year, more than 5 million acres of national forestland were consumed in flame, 78 firefighters died in the line of duty, and smoke trailed from the Rocky Mountains to the Great Lakes. These events reportedly "traumatized" the new Forest Service, cementing the Pinchot-inspired view of fire into place and foreclosing any further discussion of fire as a valuable ecological process. But the realities of land management limited the policy to front country venues—locations where firefighters had ready access to the forest, where valuable timber was visibly at risk, and where adjacent landowners expected federal protection. Besides, without a ready firefighting staff, federal officials had to rely on local volunteers to mount an effective campaign against threatening wildfires. As a result, fires continued to burn unabated in remote wilderness locations across western public lands.[17]

The federal agencies, however, soon expanded their fire control commitment. With the advent of the New Deal and establishment of the Civilian Conservation Corps (CCC), the Forest Service found itself with a ready supply of labor and proceeded to create a firefighting infrastructure. At the same time, the CCC road building projects began to open the backcountry, making it more accessible for firefighters. Buoyed by these developments, the Forest Service announced its legendary "10 a.m. policy," designed to ensure all new fires were controlled by 10 a.m. after any fire was first reported. After World War Two, the Forest Service further expanded its fire control efforts, eager to protect a newly mobile American public that was moving from the cities into suburbs where greater wildland fire dangers lurked. It launched its highly effective national Smokey Bear fire prevention campaign and began purchasing surplus military airplanes and other hardware to mechanize its firefighting capabilities. Yet, just as the federal government finally had the means to fight fire across the entire landscape, scientists began to question the all-out suppression policy. Besides a growing recognition of fire's ecological role, the high costs associated with the total suppression policy were hard to justify, especially when the fires were burning in remote wilderness areas and did not threaten human life or property. A particularly bad 1970 fire season, when extensive fires blazed across northern California and central Washington late into the fall, confirmed the need to further modify the total suppression policy.[18]

Already, though, the federal agencies had begun to reexamine their fire policies. During the mid-1960s, the Park Service became the first agency to acknowledge that wildfire plays a vital role in public land ecosystems and to change its fire management policies. Spurred by the seminal 1963 Leopold Report, the Park Service adopted new resource management policies allowing naturally caused fires to burn when they would promote vegetation or wildlife management objectives. After the Wilderness Act of 1964 gave the Forest Service wilderness management responsibilities, it too began reexamining the ecological, economic, and institutional costs associated with its all-embracing fire suppression policy. Within a few years, the Forest Service also announced major policy changes: it would allow "prescribed fires" to burn unabated in its wilderness backcountry areas (within certain carefully defined or "prescribed" parameters); it was dropping its much vaunted 10 a.m. policy; and it would concentrate on managing rather than controlling fires. Both agencies also authorized the use of human-ignited fires (within prescription) for resource management purposes, generally to mimic natural fires and thus restore a more natural ecological order on the landscape. In addition, recognizing that wildfires defied administrative boundaries, the federal public land agencies began coordinating through the newly established Boise Interagency Fire Center and otherwise unifying the federal response to wildfire incidents.[19]

For more than a decade, the agencies quietly pursued their nascent prescribed fire policy without incident. By the mid-1980s, they had approved more than 60 prescribed natural fire plans, but the actual acreage burned remained relatively modest. Then, in the summer of 1988, the Yellowstone fires ignited amid drought conditions, precipitating a very public and acrimonious debate over the prescribed fire policy. The Yellowstone fires, several of which were initially allowed to burn as prescribed fires and some of which were promptly fought as illegal ignitions, consumed over 1.5 million acres of national park and national forestlands, sparking myriad investigations and recriminations. The spectacle of uncontrollable wildfires captured national headlines and introduced the American public to the prescribed fire policy, though the media pejoratively relabeled it Yellowstone's "let burn" policy. Historic park structures suffered close calls from the advancing flames, while several local residents lost summer cabins, thick smoke blanketed the northern Rockies for more than a month, and the area's tourism business came to a standstill. Once the flames were finally extinguished, following an early fall snowstorm, both the Park Service and Forest Service suspended their prescribed fire policies pending further review.[20]

In the wake of the Yellowstone fires, opinions on the federal prescribed fire policy were anything but ambivalent. As an ecological event, scientists agreed that the Yellowstone fires largely mimicked the region's historic fire regime that had periodically burned its high-elevation lodgepole pine stands. They saw not a destroyed landscape, but rather an ecologically resilient and rejuvenated forest that was now opened into more diverse vegetative patterns. But the political reaction was quite different. The fires, having burned indiscriminately across administrative boundaries, confirmed the need for an ecosystem approach to fire policy, but political realities called into question the principal goal of such a management policy—the pursuit of ecological integrity through the maintenance of native ecosystem processes. Local politicians viewed the summer fire season as a disaster, not a watershed ecological event. For them, the fires had devastated the park's once beautiful native forests, unnecessarily endangered neighboring communities, disrupted summer tourism businesses, and thus represented a flawed policy experiment. Moreover, the Yellowstone fires highlighted another growing problem: even in this wilderness-like setting, communities and homes now abutted once wild public lands, placing the potential ecological benefits of fire squarely in conflict with private property rights. The ensuing congressional oversight hearings, interagency policy review panels, and scientific assessments brought all these sentiments to the fore, forcing the agencies to reassess their fire policies.[21]

As the political spotlight faded, the agencies reaffirmed the prescribed fire policy, while simultaneously faulting how the policy was implemented. The interagency Fire Management Policy Review Team concluded: "The objectives of prescribed natural fire programs in national parks and wildernesses are sound, but the policies need to be refined, strengthened, and reaffirmed."[22] To revise park and wilderness fire plans, the report recommended tightening the prescriptions governing natural fires, using hazardous fuel burning programs, and promoting public involvement in the planning process. Meanwhile, land managers were ordered to control all fires regardless of origin or location. But the planning process moved slowly: by 1992, fewer than 10 natural fire plans were approved compared to 68 active plans in 1988. As a result, prescribed fires were increasingly rare, the annual burned acreage dropped noticeably, and the corresponding fuel buildup threatened yet more catastrophic fires. And that is precisely what happened during the calamitous 1994 fire season that ravaged the drought-ridden West, claiming 14 firefighter lives in Colorado and reconfirming the urban–wildland interface problem.[23]

Once again, the agencies were forced to reexamine their fire policies and once more they endorsed fire on public lands. The 1995 Federal Wildland Fire Management Report basically adopted an ecosystem management approach to fire. First, the report recognized that wildfire is "a critical natural process [that] must be reintroduced into the ecosystem." Second, it acknowledged that fire and resource management decisions must be coordinated through appropriate landscape-scale planning processes using the best available science. Third, calling for uniform federal fire policies, it promoted interagency coordination as well as cooperation with state, local, and tribal jurisdictions. Fourth, the report addressed the urban–wildland interface problem through the use of risk assessments to protect human life, health, and property. And making an obvious devolutionary judgment, it noted that state and local governments are primarily responsible for protecting private property in this interface zone. Fifth, the report called for more public involvement in establishing fire management plans. Put simply, the agencies recommitted themselves to a prescribed fire policy driven by the ecological role of fire on public lands yet tempered by a new sensitivity to encroaching urbanization and other local concerns.[24]

Since then, federal fire management policies have come under even more intense scrutiny in the face of several explosive fire seasons. The seminal event was the May 2000, Cerro Grande fire in New Mexico, where a Park Service–initiated prescribed burn at Bandelier National Monument roared out of control, scorching 40,000 acres and destroying 240 Los Alamos homes. Overall, more than 8 million acres burned during the 2000 fire season, and more than 7 million acres burned during the drought-driven 2002 season, including over 2,000 homes and other structures. In the wake of the Cerro Grande fire, the public land agencies reviewed and reconfirmed the 1995 Fire Management Policy while acknowledging that it had not been fully implemented. Specifically, the 2000 Fire Policy Review reaffirmed the vital role of wildland fire as "an essential ecological process and natural change agent." It explained, however, that the fire hazard (owing to fuel accumulations) was much greater than previously thought, noting that 70 million federally owned acres were classified as high fire risk. The report further lamented the growing urban–wildland interface fire risk, continuing shortfalls in the federal prescribed burning program, and an ongoing lack of coordination. To address the coordination problem, top federal officials joined the western governors to announce a new collaborative approach to reducing fire risk. The agreement emphasizes the need for community fire protection as well as local involvement in fire management planning. With over 20,000 communities identified as potentially at risk, however, the urban–wildland interface fire prevention policy remains entangled in a curious paradox: many communities and homeowners have resisted prescribed burning proposals as a preventative fuel thinning technique, while environmental groups have protested alternative mechanical thinning proposals, many of which seem to be poorly disguised logging projects. Regardless of how this paradox is resolved, nearly everyone now acknowledges fire's important ecological role and the need to reduce its risks.[25]

Fire ecology varies across public lands. Whether fire is allowed to burn or not, ecosystems will change, because they are dynamic, not static, entities. But fire can both speed and alter ecological change. Dependent on ecosystem conditions, fire regimes vary from high-frequency, low-intensity surface fires to low-frequency, high-intensity stand-replacing fires. The classic southwestern Ponderosa pine forest supports regular low-intensity fires that remove the understory but do not damage mature trees. Such fires rarely alter prevailing vegetative patterns or displace resident wildlife species; they usually replenish soil nutrients without serious erosion problems. The lodgepole pine forests common in the Yellowstone region typically experience infrequent high-intensity fires that can reshape the forest. These fires will often alter vegetative structures, damage soil, create erosional problems, increase streambed siltation, destroy wildlife habitat, and even displace some species.[26]

Over the past century, federal fire suppression policies have altered ecological conditions across the western landscape. The absence of fire has changed the composition and distribution of tree and plant species, facilitated the spread of exotic species, promoted the buildup of woody debris (fuel loading), and displaced some species. Scientists generally agree that these policies have created older, denser, and less healthy forests prone to larger, more intense fires, which can present serious control problems and endanger nearby communities. In fact, some public land fires have been larger and more destructive than in the past: in the decade between 1986 and 1996, for example, more than one quarter of the Boise National Forest was consumed by wildfire. A policy designed to save the forests, ironically, is now understood to have seriously endangered them. Indeed, many scientists believe the inland West faces a massive forest health crisis. Widespread debate over what that means and how it may affect management policy is now under way. Unlike in times past, however, that debate is not over the intrinsic value of fire as an agent of ecological change; rather, it is over how to use or mimic fire in remedying the ecological conditions that now prevail in its absence—in other words, how to go about restoring these fire-shaped ecosystems.[27]

Curiously, although fire plays a major role on the western landscape, few federal laws directly address fire management on public lands. This stands in stark contrast to the extensive legal regimes covering the principal public land resources—timber, minerals, water, grass, wildlife, recreation, and aesthetics. Fire is not addressed in any of the organic acts governing the primary public land agencies, nor is there any separate statute indicating how fire should be managed on public lands. To be sure, various criminal, environmental, and other laws influence how the agencies address fire, but none establishes a definitive federal policy toward fire, leaving the matter largely to the agencies' administrative discretion. Because uncontrolled fire poses a real threat to public land resources, Congress has adopted criminal sanctions to deter people from intentionally setting fires or negligently allowing them to burn uncontrolled. Given the environmental consequences of fire and the intense public interest in it, agency fire policies should require some degree of compliance with the National Environmental Policy Act (NEPA), especially as those policies are integrated into resource management plans. And given the ubiquitous presence of smoke wherever fire is allowed to burn, federal fire policy must also address Clean Air Act requirements, which typically present problems in urban–wildland interface areas. In addition, because fire has such destructive potential, agency officials must weigh their fire policy options against the backdrop of the Federal Tort Claims Act (FTCA), which imposes liability on the federal government if its agents act negligently. But because the FTCA contains a discretionary judgment immunity provision, the negligence standard is more likely to apply when fire policy is being implemented than when it is formulated.[28] Also, the agencies must consult and adhere to state and local fire codes so long as these laws are not superseded by contrary federal laws or policies. But within this diffuse assortment of federal and local laws, the agencies plainly have the legal authority to restore fire to a more prominent role in the ecosystem, and they are beginning to do exactly that. In fact, fire—like biodiversity—is a key element on the public land ecological restoration agenda.[29]

Ecological Restoration Policy: Origins and Evolution

A LEGACY OF SIMPLIFICATION

Public lands have not often been linked with ecological restoration. The reason is obvious: for most of the twentieth century, these lands were managed primarily for utilitarian purposes. The goal was to produce timber, grass, minerals, big game, and scenic vistas for human consumption. Public land law and policy, focused as it was on specific commodities, essentially segmented the public domain into separate resource-based regimes. In the case of timber, the forest was managed to produce harvestable and preferably even-aged timber stands, which meant eliminating natural fires, suppressing pest infestations, and accelerating growth rates. In the case of grasslands, the range was managed to produce maximum forage for domestic livestock, with little concern for competing wildlife needs or other range resources. In the case of wildlife, habitat was protected for valuable big game species while other species were either ignored or actively eradicated. Even when sustainability was the goal, it was defined narrowly in sustained yield terms, not with a view toward maintaining ecosystems or species in perpetuity. In the name of productivity, vital ecological components and processes were dismissed, suppressed, or otherwise eliminated from the landscape to protect more valuable resources. The result, we now know, has been the ecological simplification of landscapes, including the loss of biodiversity, impaired ecosystem processes, and a generally degraded environment.

The legacy of unbridled utilitarianism is painfully evident. On lands administered by the Bureau of Land Management (BLM), unregulated mining activities have scarred the landscape, leaving behind gaping pits, open mine shafts, and unsecured tailing piles as silent testaments to another era. In the arid Southwest and elsewhere, the public range has been sorely abused: sensitive riparian areas have been trampled and destroyed, while overgrazing has depleted the native grasses, facilitated the spread of exotic plant species and noxious weeds, and eliminated important fish and wildlife habitat. On national forestlands, widespread clearcutting has denuded entire hillsides, triggered countless landslides, degraded once-pristine water sources, reduced vegetative diversity, and destroyed valuable wildlife habitat. Over 400,000 miles of access roads have been constructed in the national forests, further fragmenting the landscape, silting streambeds and displacing resident wildlife. And by routinely excluding fires from the landscape, the public land agencies eliminated a key ecological process and dramatically altered the composition and structure of forest and range ecosystems. This cumulative legacy of widespread environmental degradation and ecological simplification has set the stage for a new restoration policy.[30]

On occasion, however, the western public lands have played an important albeit limited ecological restoration role, primarily as nature reserves. One of the nation's first and most important wildlife restoration achievements occurred at the end of the nineteenth century when Yellowstone National Park's military caretakers imported remnant plains bison from Montana and Texas to begin recovering this once plentiful species from near extinction. Yellowstone provided the transplanted bison with a sanctuary, where they eventually flourished under the Park Service's watchful eye. Early in the twentieth century, Yellowstone also helped recover the West's badly depleted elk populations, translocating park elk to nearby states to stabilize existing herds and establish new ones. As protected sanctuaries, the early national parks and national wildlife refuges often sheltered big game and waterfowl from local hunting and poaching pressures. And in the years predating passage of the Endangered Species Act, the region's national parks and wilderness areas provided a final refuge for dwindling grizzly bear populations and other large carnivores. Of course, the national parks were not otherwise immune from the era's utilitarian values. The government's aggressive predator control campaign extended into the national parks, eliminating bears, wolves, and other "worthless" species from both public and private lands. And from its earliest days, the Park Service routinely constructed roads and visitor facilities throughout the park system with little regard for how these projects would affect wildlife habitat or for any other ecological consequences. Thus, even as the national parks and other public lands played a key role in early wildlife preservation efforts, the responsible agencies regularly pursued contradictory policies and rarely displayed any regard for the ecosystem as a whole.[31]

Elsewhere, restoration has occupied a more central role on public lands, particularly in the East and Midwest. According to one study, "The work of the Forest Service in rehabilitating the eastern national forests . . . is one of the great conservation achievements of American history." Early in the twentieth century, following a series of devastating floods and fires, Congress adopted the Weeks Act of 1911, which authorized federal funding to purchase lands "located on the headwaters of navigable streams." Armed with this new acquisition authority, the Forest Service began buying cut-over eastern forestlands beset with serious erosional problems and high wildfire risks. Once these lands were purchased, the Forest Service set about rehabilitating them with the goal of re-creating productive forests, though without restoring fire to the landscape. Remarkably, through a combination of tree replanting programs, various CCC projects, and simply letting nature (but not fire) take its course, the eastern forests were transformed and rejuvenated in little more than half a century. Buoyed by the region's moist weather and rich soils, the federal revegetation efforts were largely successful: hardwood stands now cover many of the northern forests, whereas restocked softwood pines blanket many of the southern and midwestern forests. The question is no longer how to heal these forests, but rather how to apportion their timber, watershed, recreational, and other resources among the large and diverse populace that lives nearby. Whether a similar transformation awaits the western national forests remains to be seen; many of the same forces that drove eastern forest restoration are now present on the western landscape.[32]

TOWARD A NEW RESTORATION IMPERATIVE

That the public land agencies have not yet formulated a comprehensive eco-logical restoration agenda is not surprising. Restoration ecology is still in its in-fancy, having emerged as a distinct subdiscipline of ecology only during the latter part of the twentieth century. The concept of ecological restoration has its origins in the Civilian Conservation Corps' 1935 tall-grass prairie restoration initiative at the University of Wisconsin Arboretum. The project was designed to restore the full suite of native grasses that covered the midwestern prairies before they were converted to monocultural farmlands. After several years of fruitlessly planting and replanting native grasses without discernible progress, arboretum scientists discovered that fire, which had periodically scorched the prairies before European settlement, was an essential element in maintaining these grassy ecosystems. This discovery confirmed two important and interrelated insights: a comprehensive understanding of ecosystem structure and processes was critical to any restoration initiative, and meaningful restoration efforts must focus on the ecosystem as a whole and not individual elements of it. Meanwhile, Aldo Leopold, who was by then also at the University of Wisconsin and a participant in the arboretum ex-periments, initiated his own well-chronicled prairie restoration project, seeking to recover native plant and animal communities on a piece of derelict farmland he had acquired. In tandem, these two Wisconsin initiatives are widely credited with giving birth to the concept of ecological restoration.[33]

Before the mid-1930s, there was little interest in ecological restoration and even less understanding of it. Ecology was still viewed primarily as a theoretical and descriptive science with little practical application. Most ecologists did not manipulate the natural systems they studied; they were content merely to observe, describe, and analyze them. With utilitarianism dominating natural resource pol-icy, nature manipulation was the province of the resource disciplines. It was the agriculture, range, forestry, hydrology, and wildlife scientists who were energeti-cally engineering nature to increase crop, forage, timber, water, and big game pro-ductivity. But Leopold and other ecologists were beginning to realize that the era's unbridled logging, grazing, and predator eradication policies had impoverished the landscape and upset long-standing ecological relationships. As usual, Leopold was prescient about what was needed: "The time has come for science to busy it-self with the earth itself. The first step is to reconstruct a sample of what we had to start with."[34]

Over the ensuing years, various ecological restoration concepts have at-tained independent stature, and restoration ecology has emerged as a discrete discipline. The success of the University of Wisconsin Arboretum prairie restor-ation experiments not only demonstrated the value and complexity of ecologi-cal restoration but it also spawned additional restoration initiatives. Although most early ecological restoration experiments involved relatively small-scale, site-specific projects, the next generation of projects has expanded the restoration agenda to embrace entire watersheds as well as large chunks of the public domain. Ecology as a science has evolved from a purely theoretical discipline into a practical one too: scientists now regularly manipulate ecosystems to study disturbance processes; they generally agree that ecological systems can be re-created; and they routinely employ new computer and satellite technologies to assist with these tasks. The American public, confronted with widespread environmental degradation, has demanded that public officials begin to redress this unsavory legacy by reclaiming or restoring degraded landscapes. A new generation of environmental laws, including the Clean Water Act, Endangered Species Act, Surface Mining Control and Reclamation Act, and Superfund legislation, have not only legitimized the notion of environmental restoration but compelled sometimes expensive corrective actions. In 1987, bolstered by these developments, a group of involved scientists established the Society for Ecological Restoration at the University of Wisconsin to promote "ecological restoration as a means of sustaining the diversity of life on Earth and reestablishing an ecologically healthy relationship between nature and culture."[35] Since then, numerous books and two new journals devoted to the topic of ecological restoration have been published, providing both theoretical and practical meaning to this new discipline.[36]

Ecological restoration can be defined simply as "the return of an ecosystem to a close approximation of its condition prior to disturbance." It seeks to re-create or repair "the structure, function, and integrity of indigenous ecosystems and the sustaining habitats they provide." Restoration is thus perceived as a holistic process focused on the entire ecosystem and not individual elements. Moreover, restoration frequently entails active management "to accelerate recovery of degraded ecosystems by complementing or reinforcing natural process."[37] Restoration ecologists employ several different—though often complementary—strategies to change ecological conditions toward a desired predisturbance state. Structural restoration strategies are generally directed toward reengineering an ecosystem's prevailing physical attributes, which can involve realigning stream banks, using fire to re-create habitat conditions, or removing physical barriers to fish or wildlife movement. Land use strategies can be employed to change the intensity, distribution, timing, or duration of uses affecting the landscape; these strategies can range from eliminating timber harvesting in sensitive drainages to rearranging livestock grazing patterns. Biological control strategies are designed to alter species composition within an ecosystem, either by removing undesirable species like noxious weeds or by reintroducing desirable ones like endangered wolves or black-footed ferrets. All these restoration strategies are being employed on public lands, with the choice depending on the overall goals, existing conditions, project scale, and related political, economic, and social factors.[38]

Although closely related to other remedial environmental management concepts, ecological restoration should be distinguished from rehabilitation, reclamation, and mitigation. Restoration ordinarily contemplates restoring the full range of natural conditions to an ecological setting, while rehabilitation seeks to recover biological productivity but not the full range of species diversity or ecosystem processes. Reclamation pursues even more modest goals; it is designed to reestablish only selected structural or functional elements of seriously degraded ecosystems. In choosing a remedial goal, the key question is whether the targeted site can be restored to its historic ecological condition or whether the site is so degraded that restoration is impossible. Another key consideration is whether related economic, social, or political factors will support an ambitious restoration agenda. Mitigation, in contrast, is a much broader concept; it is used to lessen or compensate for the ecological impacts of development activities or project proposals. Mitigation may involve ecological restoration efforts as well as other on- and off-site environmental protection or remedial efforts. Thus, among these broadly related concepts, ecological restoration is the most ambitious, with its emphasis on restoring ecosystem integrity to preexisting environmental conditions.[39]

Not surprisingly, the concept of ecological restoration mirrors the closely related concept of ecosystem management, both of which are driven by paramount ecological considerations. Both concepts give priority to maintaining and restoring biodiversity and ecosystem integrity. They also emphasize public involvement, interjurisdictional coordination, extended temporal and spatial planning scales, interdisciplinary scientific information, and adaptive management strategies. Restoration proponents, well aware that even degraded ecosystems can provide important economic and social benefits, emphasize the need for public involvement in setting and pursuing ecological restoration goals. And because successful restoration efforts often require landscape-level initiatives, restoration advocates usually frame restoration projects in broad spatial and temporal terms, while acknowledging the need for interjurisdictional coordination to replicate predisturbance ecosystem conditions. Given the inherent complexities and uncertainties involved in understanding ecosystems, they also rely heavily upon interdisciplinary scientific information and regularly utilize adaptive management techniques to monitor and adjust restoration strategies as new information becomes available. In short, just as ecosystem management represents a major departure from past practices, ecological restoration is also best understood as a new contingent and experimental policy—one designed paradoxically to reshape the present by re-creating the past.[40]

Ecological restoration, like its ecosystem management counterpart, has roots in both the utilitarian and preservation traditions that have long dominated natural resource policy. On the one hand, with its recognition that active management and intervention may be necessary to reestablish ecosystem components or processes, ecological restoration resembles traditional conservation policy, which has historically viewed human intervention as essential to improve nature for productive purposes. But with its focus on restoring ecosystem integrity, ecological restoration deviates from traditional conservation policy, which is ordinarily associated with the sustained use of single resources without regard for the broader implications of such use. On the other hand, committed to re-creating predisturbance natural conditions, ecological restoration emulates traditional preservation policy, which focuses on protecting undisturbed landscapes and pristine settings. Yet with its emphasis on active intervention, ecological restoration deviates from traditional preservation policy, which has generally sought to preclude or minimize human intervention into natural systems. In an effort to reconcile these seemingly contradictory traditions, restoration advocates argue that by placing people in an active and reciprocal relationship with nature, ecological restoration fosters a more pragmatic and enduring human connection with nature than is true with a purely noninterventionist preservation policy. Others worry, however, that restoration ecology should not be used to create a false dichotomy between preservation and restoration or to stoke our latent hubris over the human ability to control nature.[41]

ON MANAGING NATURE: TO INTERVENE OR NOT?

The tension between competing utilitarian and preservation approaches to conservation is evident in the often contentious debates over restoration goals and strategies. Indeed, translating ecological restoration concepts into comprehensive policy prescriptions for the public domain raises several vexing and divisive issues. First, because ecological restoration seeks to reestablish natural or historic predisturbance conditions, some consensus must be reached over how these targeted conditions are defined or what historical era should be used as the appropriate reference point. Second, with the continuing schism between utilitarian conservationists and pure preservationists, restoration policy must confront the question of intrusiveness: When is it appropriate to intervene intensively with engineering solutions, and when should nature be left to take its own course? Third, given the dynamic and complex nature of ecosystems, restoration policy must establish appropriate spatial and temporal scales for restoration initiatives. How the public land agencies resolve these issues will reveal the depth of their commitment to ecological restoration and provide a framework for addressing related political, economic, and social concerns that cannot be divorced from public land policy.

Few ecological management issues have proven as contentious as defining restoration goals by reference to historic ecological conditions. Should ecological restoration goals be framed in terms of re-creating the predisturbance conditions that prevailed immediately prior to the arrival of European settlers to a particular region? Should another specific historical date be selected as the target, and, if so, then on what basis? Or should ecological restoration goals be framed in terms of naturalness, that is, managing to re-create a natural environment? And how should the selected goal or reference point take account of the ecological impacts of Native Americans, who most environmental historians agree played a considerable role in shaping the landscape the early settlers encountered? Alternatively, given the dynamic and ever-changing nature of ecosystems, should restoration goals be framed in terms of evolutionary environmental conditions rather than attempting to re-create a snapshot in time? Do we know enough about historical conditions or trajectories of environmental change to set reliable ecological goals? While there are few easy answers to these questions, they also cannot be avoided if ecological restoration is to play a meaningful role in the public land policy agenda.

In one form or another, these same questions have dominated the rancorous debates over Yellowstone National Park's often controversial resource management policies. In the mid-1960s, the now-famous Leopold Report recommended that national parks "be maintained, or where necessary recreated, as nearly as possible in the condition that prevailed when the area was first visited by white man . . . [to] represent a vignette of Primitive America." The Park Service, squarely challenged to begin thinking in restoration terms, responded by altering its natural resource management policies to focus on maintaining and restoring a more natural and undisturbed landscape.[42] In Yellowstone, this shift in resource management policy—sometimes denominated the natural regulation policy— prompted the Park Service to endorse restoring wolves and natural fire to the Yellowstone landscape, eliminating exotic species, and otherwise protecting the full range of native species within the park. Critics charged, however, that it was impractical to choose one historical time period as the managerial goal, because ecosystems are dynamic in character and do not lend themselves to any static characterization. The same critics asserted that any attempt to re-create pre-European natural conditions must take account of Native American influences, which significantly shaped the park's precontact environment. Noting the park is not an island, these critics also observed that restored wolves and other park animals will disperse beyond park boundaries just as unchecked fire will likewise burn across these same boundaries, creating serious management problems and political conflicts with adjacent landowners. Finally, the critics argued that a Park Service management regime based on naturalness or any other imprecise standard lacks objectivity and is inherently unaccountable. Despite these criticisms, the Park Service has given ecological restoration a prominent role in its management agenda, intent on recapturing an earlier and less disturbed ecological setting.[43]

The Yellowstone experience helps frame the restoration issue elsewhere on public lands. Multiple-use public lands, unlike national parks, have been used and often abused over the past century, making restoration a real and practical necessity. Clearly, any ecological restoration policy for these already disturbed public lands must draw upon the past to establish meaningful management goals. Without some knowledge of earlier ecological conditions, there are no reference points to use in establishing restoration goals, measuring rates of change, or monitoring progress. Most observers, while acknowledging that Native American practices had significant impacts on the local environment, agree that the arrival of European settlers triggered even greater environmental changes, creating differences in kind as well as degree. While not inconsequential, Native American hunting and burning practices generally paled in comparison to the market hunting and agricultural pursuits of the early western settlers, to say nothing of how logging, grazing, mining, roading, and dam-building altered the same landscapes in later years. Whether these disturbed landscapes can be restored to a prescribed pre-European condition is less important than returning them to an ecologically sustainable condition that forestalls further degradation, while reestablishing native species and basic ecological processes whenever possible. Besides promoting ecological integrity, such restoration goals can also help return the land to a more economically productive state. The key, therefore, is to select an appropriate time period predating large-scale landscape conversions as a reference for defining restoration targets.[44]

Ecological restoration goals must also, however, take account of the dynamic or nonequilibrium character of ecosystems. While a historical reference point can provide an initial restoration target, the ultimate goal is a moving target based upon the trajectory of evolutionary change, which reflects the changing nature of environmental conditions. The primary objective is to bring the ecosystem back into a more natural evolutionary pattern, not to deny or stop environmental change. Such an approach to ecological restoration acknowledges that ecosystem changes would have occurred since Europeans came into contact with Native Americans, just as they did during the precontact period, but not at the rate or intensity that prevailed following European settlement. It also avoids the trap of defining restoration goals in static terms, simply by reference to one selected point in historical time. Because change is inevitable in nature, defining restoration goals in evolutionary ecological terms should give land managers some latitude in setting restoration targets, yet not relieve them from accountability. And it will enable restoration initiatives to proceed even in the absence of detailed historical information that often may not be readily accessible or easy to interpret. Of course, when knowledge of the precontact environment is available, agency officials can better understand and define patterns of evolutionary change, and adapt their strategies accordingly.[45]

Besides defining threshold goals, a related ecological restoration policy challenge is to identify appropriate strategies to achieve these goals. If the goal is to reestablish evolutionary patterns of change based on predisturbance conditions, can it be achieved without actively intervening in the ecosystem and thus further disturbing it? Or put differently, can meaningful ecological restoration occur under a hands-off, natural regulation management strategy? For Yellowstone's critics, the answer is plainly no. They argue that the Park Service's decision to minimize human intervention into ecological processes—principally management of the park's northern range ungulate populations—has seriously upset the park's native ecology and will require extensive intervention to restore the damaged ecosystem. Similarly, many forest restoration proponents advocate more—rather than less— logging in order to re-create more diverse forest landscapes akin to those prevailing during presettlement times. Underlying these interventionist restoration prescriptions is an abiding faith in science and in the human ability to use it properly. But earlier land managers also thought they understood the science of the time, and it was, of course, their decisions that created the current need for an expansive restoration agenda.[46]

The question of natural versus engineered restoration strategies defies an easy answer. Local ecological conditions, prevailing human values, political feasibility, and the limits of scientific knowledge are all factors in determining an appropriate restoration strategy. In some instances, where the ecosystem has not been severely damaged or where evolutionary patterns of change are not seriously askew, it may be possible to rely upon natural evolutionary processes to mend the damage. If the Park Service is correct, Yellowstone's northern range may be a case in point. But in other instances, the very nature of restoration—reestablishing long-absent pieces or processes to an ecosystem—may require active intervention to accomplish the ecological objectives in an acceptable and timely manner. To achieve its species recovery goals, for example, the FWS has used captive population breeding programs, translocation reintroduction strategies, and otherwise actively managed controversial species, even sanctioning the limited killing of reintroduced predators to reduce local opposition to its recovery programs. And where fire restoration is at issue, the public land agencies have consistently employed both lightning-ignited and human-ignited fires within predefined conditions to re-create a more natural burning regime on the landscape. Some logging designed to simulate historical fire conditions may also be necessary to restore native forest ecosystems. But whenever active intervention is the chosen strategy, humility and restraint are the order of the day. Given the complexity of ecosystems and ecosystem processes, every restoration management strategy should be regarded as contingent, subject to periodic reassessment based on what is actually occurring on the ground. In short, the restoration choice is not between interventionist or passive strategies; rather, restoration should reflect a commitment to adaptive management, an active monitoring program, and the willingness to adjust strategies as necessary.[47]

There are, moreover, critical differences among intrusive restoration strategies. It may be necessary, for example, to capture and translocate wolves in order to restore them to the Yellowstone ecosystem, but that may not justify extensive radio-collaring and monitoring of individual wolves, which could effectively eliminate any sense of wildness from the restoration effort. Ecological structures might be restored to an overgrown and deteriorating forest by either reintroducing prescribed burning or permitting salvage logging, but only fire may fully mimic historic patterns of ecological change. Similar concerns bedevil contemporary river restoration debates. In the Columbia River salmon restoration debate, for example, the question is whether removal of the lower Snake River dams or fish barging around the dams is the key to recovering native fish runs. Although both strategies are intrusive, only dam removal seems to hold any real promise of restoring the river to its historic ecological role and maintaining a viable native fishery. In each case, as the responsible agencies have wrestled with these thorny restoration strategy issues, they have invariably sought to balance ecological effectiveness, economic implications, political viability, and other considerations in making what is fundamentally a political decision framed by important scientific concerns. But despite the obvious political implications, restoration strategy must focus on reestablishing an ecologically resilient and sustainable landscape that meets not only current but also future human expectations.[48]

Ecological restoration proposals inevitably raise complex questions of scale, namely what spatial and temporal scales to use in recreating historical conditions. Because restoration goals are framed in ecological terms, the relevant spatial and temporal boundaries should be defined in similar ecological terms rather than in traditional political or commercial terms. If the restoration goal is to reintroduce a viable population of a native, wide-ranging species such as wolves or grizzly bears, then the project design must encompass a large landscape that will inevitably affect an array of public and private lands. If the restoration goal is to reestablish historical disturbance patterns, such as natural fire or flood regimes, then an evolutionary time frame must be used to assess preexisting ecological conditions and to monitor progress toward the restoration goals. An evolutionary standard is particularly necessary if the disturbance regime is marked by infrequent episodes or if a species targeted for restoration has a low reproductive rate. It is also necessary to accommodate dynamic patterns of change. For example, wolves and grizzly bears will move into new territory as their populations expand; catastrophic fires and floods are inevitable though not easily predictable occurrences. To be sure, an expansive, landscape-level ecological restoration initiative can run afoul of powerful political, economic, and other interests, which may force agency officials to modify the proposal in order to meet local concerns or other constraints. But even when this occurs, broad-scale ecological restoration goals still might be met by linking a series of smaller projects together, particularly if biodiversity goals depend upon species interchange opportunities or if predisturbance conditions can be emulated by creating a matrix of landscape patterns.[49]

Large-scale ecological restoration projects invariably create thorny planning problems, especially given the political importance of jurisdictional boundaries. The National Forest Management Act (NFMA), Federal Land Policy and Management Act (FLPMA), and related statutes limit agency planning authority to those lands immediately under the governing agency's jurisdiction, which may or may not fully encompass the relevant ecosystem for restoration purposes. But NEPA cumulative effects analysis requirements obligate agencies to assess the full spatial and temporal ramifications of project proposals, including environmental impacts involving adjacent public or private lands. Moreover, where an endangered species is at issue, the ESA is not constrained by conventional political boundary lines; its coercive provisions are defined primarily in habitat terms and extend across the landscape. In the Pacific Northwest's spotted owl controversy, the courts forced the public land agencies to expand their planning horizons to embrace the entire west-side forest ecosystem and to use a 100-year planning cycle to ensure the owl's continued existence. For these agencies, the only answer was to inaugurate a new interagency planning regime, to cooperate with adjacent non-federal landowners, and to employ an adaptive management strategy to evaluate ecological trends over the long term. The resulting regional forest plan demonstrated both the necessity and feasibility of ecosystem-level planning to achieve broad-scale restoration objectives. As we shall see, the agencies have now prepared other regional, landscape-scale assessments as part of the evolving federal commitment to ecosystem restoration on public lands. Regionalism, however, is not without political peril or potential legal vulnerability.

Setting the Restoration Agenda

Ecological restoration projects are now ubiquitous across the public landscape, stunning in both their diversity and origins. Whether the agency mission is multiple-use or preservation, restoration is now acknowledged to fit within that mission. The projects encompass an impressive array of ecological concerns and settings: rangelands, riparian corridors, forest health, native species diversity, abandoned mine lands, wetlands, river corridors and dam sites, exotic species, wildfires, seasonal floods, and other disturbance regimes. Many of the projects are small-scale in design, covering only a few acres or a single drainage; others are much larger in scope, covering entire river systems or eco-regions. Some projects have grassroots origins; others have been administratively conceived by agency officials. Several have been driven by litigation or the threat of litigation; still others have their genesis in congressional legislation. Many cut across traditional agency boundaries, creating new interagency relationships as well as myriad public–private partnership arrangements. While still uncharted territory, this budding era of ecological restoration places less emphasis on commodity development and more emphasis on safeguarding environmental amenities.[50]

Rather than impeding this new restoration agenda, the existing law can be employed to promote it. Many of the applicable laws trace their origins to the heady 1970s when Congress passed a welter of new environmental protection laws, few of which were originally viewed in ecological restoration terms. The organic statutes governing the public land agencies may not contain explicit references to ecosystem restoration, but the basic multiple-use and preservation mandates are broad enough to encompass restoration within the agencies' overall responsibilities. These laws also establish an integrated, interdisciplinary planning process that is well-suited to identifying and developing restoration opportunities. The powerful Endangered Species Act, with its explicit commitment to avert pending extinctions, essentially compels all federal agencies to integrate species conservation and recovery considerations into their planning and decision processes. Other laws also support an active public land restoration agenda: the Surface Mining Control and Reclamation Act (SMCRA) imposes explicit reclamation obligations for mine sites; the National Forest Management Act contains express biodiversity protection and tree restocking requirements; and the Clean

Water Act's unambiguous purpose is "to restore and maintain the chemical, physical, and biological integrity of the Nation's waters." And NEPA establishes a time-tested procedure to examine the environmental implications of restoration proposals and engage the public in the process. Within this legal framework, the agencies have begun integrating restoration projects into their long-term planning processes, though not without litigation and political controversy.[51]

FOREST AND RANGE RESTORATION

Although the issue is still debated in some quarters, most observers agree that the public forest and rangelands are in poor ecological condition. The national forests have been badly fragmented and their ecosystems simplified by excessive logging and related road construction, while the rangelands have declined as a result of overgrazing. Fire suppression has altered the composition of both forest and range ecosystems, even within national parks and wilderness areas. The public land agencies have begun to address these problems through various restoration initiatives, generally designed to reestablish a more natural landscape. The examples abound: a comprehensive 1996 study documented over 30 restoration projects involving western public lands, including myriad Forest Service and BLM range improvement, riparian stabilization, and forest health initiatives. While not immune to controversy, most of the local, site-specific projects have avoided the white-hot conflict and political intermeddling that have dogged larger-scale restoration initiatives.[52]

On BLM public lands, federal restoration efforts have primarily involved restoring rangelands and damaged riparian corridors. The principal culprit, of course, has been the cow: too many of them trampling the range and overconsuming its vegetation, often displacing the native grasses and inviting weedy exotics to replace them. Cattle are also notoriously hard on streams; they congregate in sensitive riparian terrain, destabilizing stream banks, fouling the water, silting streambeds, and creating erosional problems. Indeed, rangeland health has long been an issue on public lands. The heralded Taylor Grazing Act of 1934, originally intended to stabilize deteriorating rangelands, precipitated only minimal improvements to range conditions. Neither the FLPMA of 1976 nor the Public Rangelands Improvement Act (PRIA) of 1978 stemmed the continuing decline in range conditions. A 1988 General Accounting Office study reported that "many [riparian areas] are in degraded conditions" and "the number of [restoration] successes is small compared to the area still needing restoration." Unable to stir Congress to further action, the Clinton administration promulgated controversial rangeland reform regulations, establishing ecologically based range management standards and vesting statewide Resource Advisory Councils with oversight responsibility for improving range conditions. After surviving a rancher-initiated legal challenge, the new regulatory process has created an atmosphere conducive to range restoration experiments.[53]

Numerous restoration strategies are available to address overgrazing problems and to improve range conditions. The techniques include: reductions in allotment numbers; exclusionary fencing; seasonal, rotational, and other timing restrictions; re-vegetation; prescribed fire; stream bank stabilization; and even beaver
reintroductions to restore riparian functions. Some range improvement strategies
have proven more controversial than others. Ranchers have steadfastly resisted any
strategy that will reduce herd numbers or require more oversight of grazing livestock, citing diverse financial concerns for their opposition. Yet environmental
groups have long implored BLM and Forest Service officials to reduce livestock
numbers and to impose strict limitations on grazing livestock. Both sides have resorted to litigation, and the resulting court decisions confirm that the agencies have
both the duty and authority to remedy degraded range conditions. In fact, the
courts consistently have sustained federal authority to reduce livestock numbers in
order to protect or restore environmental values. In the closely watched Comb
Wash proceedings, environmental groups successfully argued that the BLM had illegally authorized grazing in a sensitive, desert riparian zone in southern Utah, and
the agency was ordered to remove the cattle to restore the canyon's vegetation and
stream corridor. Endangered Species Act litigation aimed at protecting native fish,
the Southwest willow flycatcher, and other riparian species has also prompted court
decisions ordering agencies to control livestock grazing and restore functional riparian habitat. Although environmental advocates failed in their initial effort to extend Clean Water Act permit requirements to public land livestock grazing decisions, the court decision does not preclude the agencies from relying on water
quality concerns to pursue riparian zone restoration initiatives.[54]

Against this backdrop, the BLM and area ranchers have undertaken several
cooperative restoration projects designed to improve range and riparian conditions, and ultimately forage productivity. Two well-documented examples are the
Trout Creek Mountains Working Group and the Malpai Borderlands Group. By
the late 1980s, public rangelands in the Trout Creek Mountains of southeastern
Oregon were in serious decline: creek banks were badly eroded, native grasses had
been displaced by juniper and other exotics, and the water table was dropping.
Despite some initial reluctance, local ranchers established a working group with
the BLM and concerned environmentalists, to address the obvious overgrazing
problem. Within two years, the Trout Creek Mountains Working Group had devised a long-term grazing strategy that excluded cattle from riparian zones for
three years and then opened these areas to short-duration, early-season grazing,
while also modestly reducing overall livestock numbers. The grazing plan survived
the subsequent listing of the local Lahontan cutthroat trout as a federal threatened
species, and most observers agree that the riparian areas and range conditions are
dramatically improved. Although ranchers continue to object to monitoring cattle movements, most working group members acknowledge that without these
operational modifications even greater grazing changes may have been forced
upon them.[55]

Far to the south, on the Arizona–New Mexico border, the Malpai Border-lands Group was conceived in 1994 by a coalition of local ranchers and private landowners. Its purpose was clear: to "restore and maintain the natural processes that create and protect healthy, unfragmented landscapes to support a diverse, flourishing community of human, plant and animal life in the Borderlands region." In an 800,000-acre area of mixed public and private land ownership, the group has worked with the federal land management agencies and state officials, using a science-based ecosystem management approach to reintroduce fire to the range ecosystem in order to restore native grasslands. To forestall subdivision development, the group also established a unique grassland banking system, which supports ranchers who donate conservation easements with access to a communal grass bank. Both projects have met with initial success: by 2002, more than 15,000 acres of rangeland had been intentionally burned with prescribed fires, 42,000 acres of private land were protected by conservation easements, and several grass bank arrangements were in place. Local endangered species issues have tested relations between the federal agencies and private landowners, but most ranchers continue to support the collaborative approach to ecological restoration that is evolving. Thus, while not without lingering controversy, the Trout Creek and Malpai initiatives represent a new breed of consensus-based, landscape-level resource management initiatives designed to restore range ecosystems to a more sustainable condition.[56]

On the western national forests, the signature ecological restoration issue is forest health. Heavily affected by resource development pressures and long-standing fire suppression policies, the region's forests have lost native diversity as well as much of their innate resiliency. Excessive logging has denuded mountainsides and fragmented the landscape, diminishing old growth habitat and pushing sensitive species toward extinction. The access roads that inevitably accompany logging have exacerbated the fragmentation problem and opened once secure wildlife habitat to human traffic. With fire excluded from the ecosystem, the forests have also aged: trees have died, bugs have invaded, and diseases have spread. Despite recent efforts to reintroduce prescribed fires, the amount of acreage burned remains quite small compared to the amount of land historically scorched by fires. As a result, aging forests now represent a significant fire hazard that endangers not only valuable timber and other resources but also nearby communities, many of which have expanded right up to the forest boundaries. It is a supreme irony that the current forest health controversy culminates nearly a century of active management policies that were supposed to sustain, not endanger, forest resources. And to compound the irony, the solution may entail yet more aggressive manipulation of the ecosystem in order to restore it.[57]

Although nearly everyone supports the general goal of forest restoration, major disagreements have surfaced over how to re-create presettlement forest ecosystem conditions. Environmentalists argue that the major problem is unrestrained timber harvesting, which has created a forest monoculture and done enormous environmental damage. Viewing native species biodiversity as a primary restoration goal, they want to reconnect the forest landscape, restore natural processes, and protect critical habitat. They focus on the need for reforestation, road closures, and soil stabilization projects, and they generally support prescribed burning policies as well as carefully controlled thinning near urban–wildland interface areas. The Forest Service and other agencies, while acknowledging that logging has altered the forest mosaic, see fire suppression as the major problem, arguing that overmature forests are unhealthy and represent a serious fire hazard. Although they too would reintroduce fire into the forest ecosystem, they believe the region's altered forests cannot withstand extensive burning because of excessive fuel buildup. Fearful that catastrophic fires could endanger nearby communities and irreversibly damage vital ecosystem components, they advocate immediate intervention with strategic logging to thin the forests and would limit the use of prescribed fires. Their emphasis is on restoring forest productivity, to enhance both ecological diversity and resource development opportunities. Logging sales, they note, will also produce revenues that can be used to cover restoration costs while providing residents with employment opportunities. But the environmental community, having long fought to limit timber harvesting on the public domain, is not convinced that the forests need more logging for their own ecological good.[58]

Historically, native forest restoration was not a major concern on public lands, except as a means to promote wood fiber productivity. When Congress passed the Knudsen-Vandenberg Act of 1930, it gave the Forest Service a powerful financial incentive to cut old growth timber and replace it with plantation-like tree farms. By allowing local forests to retain 75 percent of their timber sale revenues, contingent on these funds being used for reforestation, the act promoted even-aged timber management strategies at the expense of a more diverse forest landscape. Since then the National Forest Management Act of 1976 has imposed environmental constraints on the Forest Service's logging practices, but it also excepted salvage timber sales from these limitations, and its five-year tree restocking requirement has helped promote even-aged forestry. Not surprisingly, the agency has regularly invoked the NFMA's salvage exceptions to harvest timber as a supposed antidote to insect infestations, fire damage, and other forest problems. Environmentalists have challenged several of these sales under the ESA and NEPA, particularly those involving roadless lands, meeting with some success. In short, until the present forest health dilemma, ecological restoration was rarely—if ever—seen as a primary management goal.[59]

But that has now changed. The high-stakes debate over the interior West's forest health problems has moved into the political arena, triggering legislative intervention as well as administrative reform proposals. In a major display of raw political power, the 104th Congress attached the infamous timber salvage rider to the 1995 appropriations legislation. To expedite salvage sales, the rider exempted the Forest Service from complying with environmental laws, removing these sales from any meaningful judicial review. Ostensibly designed to address the forest health problem, the salvage rider unleashed an avalanche of logging across the national forests, much of it with little connection to forest health concerns. After an initial round of judicial challenges to the salvage bill proved fruitless, the Clinton administration interceded when it became apparent that the cutting was primarily benefiting the timber companies rather than the forest resource. Following the explosive 2000 and 2002 fire seasons, the Bush administration aggressively promoted the president's Health Forests Initiative, which calls for extensive forest thinning and salvage operations to minimize large-scale fires. After the 107th Congress failed to pass comprehensive forest health legislation, the agencies unveiled their own administrative reform agenda, seeking to eliminate NEPA review of forest thinning and rehabilitation projects through categorical exclusions, to expedite endangered species consultations, and to limit administrative appeal opportunities. The environmental community, still embittered by the earlier salvage logging fiasco, sees this new initiative as a frontal assault on key environmental laws and as a sop to the timber industry that will neither improve forest conditions nor reduce the fire danger. Their congressional allies seem to share this view, which means a major political and legal confrontation looms over fire policy.[60]

Meanwhile, several smaller-scale forest restoration projects have moved forward, generating an unusual degree of harmony between the Forest Service and its diverse constituencies. In northern Arizona, a local partnership has used an integrated silviculture and prescribed burning approach to restore native Ponderosa pine ecosystem conditions. Outside the town of Flagstaff, forest conditions have changed dramatically since European settlers arrived: high-grade logging practices, extensive livestock grazing, and fire suppression had all combined to create a dense forest structure where catastrophic crown fires rather than slow-burning ground fires were becoming the norm.[61] In 1996, after a summer wildfire burned out of control for more than two weeks, the Greater Flagstaff Forests Partnership was born. Spearheaded by the Grand Canyon Trust with local Forest Service support, the group is committed to protecting forest ecosystems, improving forest management practices, and reducing the likelihood of another conflagration. After initially sparring over tree removal strategies (environmentalists wanted all large-diameter trees left standing, while the Forest Service supported a more aggressive removal strategy), the partnership has begun implementing several forest thinning projects. It is also trying to foster a local forest products industry to utilize the small-diameter trees that are being cut, and also hoping to generate revenue for its restoration work. However, a local environmental organization, objecting to the logging strategy, has mounted a successful challenge to one of the early thinning proposals and remains skeptical of the plan's commercial aspects. And permanent funding beyond the initial foundation grants remains an issue. Yet despite these problems, this ambitious, partnership-based restoration project has moved forward with broad local support for restoring ecological health to the forest.[62]

RESTORING BIODIVERSITY UNDER THE ENDANGERED SPECIES ACT

Biodiversity conservation evokes widespread controversy even on public lands. Still largely undeveloped, the public domain has come to be regarded as a primary biodiversity refuge, where wide-ranging species can be accommodated without too much disruption or conflict with private landowners. But species restoration proposals regularly trigger opposition from state and local officials as well as adjacent landowners, whose concerns are typically more legal than biological in nature. This is not surprising: most federal species recovery programs are governed by the powerful Endangered Species Act, which establishes a highly prescriptive regulatory regime. The science-driven ESA gives primacy to endangered species on public lands, which means its prescriptive limitations—listing decisions, jeopardy opinions, and taking determinations—apply without regard to competing economic or social considerations. Recognizing that these statutory provisions leave little room for negotiation, Congress added the Section 10(j) experimental population reintroduction provision to create a more flexible framework for negotiating controversial species restoration initiatives with local officials and landowners. Both the promises and pitfalls surrounding the Section 10(j) provision are gradually being worked out in a series of complex public land restoration initiatives involving California condors, black-footed ferrets, wolves, and grizzly bears.[63]

The California condor and black-footed ferret reintroduction initiatives, although not without controversy, have established important principles governing the Section 10(j) species recovery process. In both cases, the U.S. Fish and Wildlife Service (FWS) captured all remaining wild condors and ferrets, then transferred them to a captive breeding program in a last-ditch effort to prevent them from sliding into extinction. Sued over its decision to capture the surviving condors, the FWS prevailed in the ensuing litigation, establishing the legal efficacy of captive breeding programs as a technique to restore wild populations of ESA-protected species. Then, overcoming local opposition, the FWS availed itself of the Section 10(j) experimental population reintroduction provision to release captive-bred condors and ferrets onto public lands, even in locations outside the species' known habitat. Although a local county challenged the California condor experimental population release in southern Utah, the lawsuit was dismissed, reinforcing the legitimacy of Section 10(j) as a biodiversity restoration strategy. Both species are gradually reestablishing themselves in their new public land habitats: more than 40 condors soar the California and Arizona skies, while several hundred ferrets occupy four different sites across the West. In tandem, the California condor and black-footed ferret reintroductions demonstrate that captive breeding programs can be combined with the experimental population release procedure as a viable endangered species restoration strategy.[64]

Predator restoration proposals regularly generate layers of opposition far exceeding what accompanied the condor and ferret restoration programs. The wolf, with its legendary status as a wide-ranging and voracious predator, is a case in point. Besides its controversial northern Rockies wolf restoration program, the FWS has used the Section 10 (j) experimental population provision to reintroduce wolves in North Carolina and New Mexico. In each instance, local landowners sued the federal government over wolf restoration; in each case, the government prevailed and established important precedents clarifying its Section 10(j) authority. After the FWS successfully reintroduced a nonessential experimental population of captive-bred red wolves into the Alligator River National Wildlife Refuge in 1987, the North Carolina legislature passed a law permitting wolves to be killed in the counties surrounding the refuge. Local landowners then sued the federal government, arguing that the new state law governed the reintroduced wolves because the FWS had no constitutional authority to regulate wildlife in the state. However, a local federal court disagreed, finding that the FWS's Section 10(j) regulations were a proper exercise of federal power under the commerce clause because wolves moved across state lines and tourists regularly visited the area to see them. In New Mexico, local landowners also challenged the FWS's Mexican wolf reintroduction program, contending that the agency's experimental population proposal had not fully considered the genetic background and depredation proclivities of the reintroduced wolves. But a local federal court, noting the extensive scientific evidence supporting the FWS's reintroduction proposal, upheld the Section 10(j) release program. These rulings, along with the Tenth Circuit decision sustaining the Rocky Mountain wolf reintroduction program, confirm the FWS's broad discretion under Section 10(j) to negotiate flexible restoration programs on public lands.[65]

Nonetheless, federal–state political tensions are manifest in the implacable local opposition that has dogged these controversial predator restoration programs. The ultimate ESA goal, of course, is to recover listed species, remove them from federal control, and return management responsibility to the states. In the recovery process, the FWS has regularly opted to use Section 10(j) to promote state and local cooperation (or at least forbearance) and thus enhance the prospects for successful on-the-ground predator reintroductions. But the states and local residents have not always responded in kind. That the North Carolina legislature blatantly countermanded the federal recovery program does not inspire confidence that the state will soon be prepared to assume management responsibility for the red wolves once recovered population targets are reached. In the Southwest, the problem is even more serious: so many of the originally released Mexican wolves were killed illegally that the FWS was forced to revise its reintroduction strategy and to move the remaining wolves to an alternative wilderness location. And in the northern Rockies, the states of Montana, Wyoming, and Idaho initially refused to help manage the reintroduced wolves; Wyoming still classifies the wolf as a predator that can be shot on sight. Clearly, the federal government's superior ESA regulatory power, counterbalanced by its willingness to negotiate local concerns under the Section 10(j) provision, has been key to bringing these important predator restoration experiments to fruition. Whether the states, given their close affinities with local landowners and their strong commitment to economic development, are prepared to assume responsibility for a restored but still vulnerable wolf population remains to be seen. But with wolf population levels growing, the federal government has proposed both delisting and reclassifying the wolf's legal status, which will put the states to the test.[66]

The long-simmering grizzly bear recovery controversy offers another powerful object lesson in the science, law, and politics of biodiversity restoration. The wide-ranging grizzly bear—a top of the food chain carnivore—is regarded as a barometer for overall ecosystem health and an unalloyed symbol of wilderness. Fewer than 1,000 grizzly bears persist in the United States, located mostly on public lands in Glacier National Park, Yellowstone National Park, and northwestern Montana's Cabinet-Yaak region, as well as a few other remote wilderness strongholds. In 1975, the grizzly bear was listed as a threatened species on the endangered species registry based upon its declining population and a shrinking habitat base. These bears present difficult recovery challenges: they are reclusive and not easy to count; they reproduce at a low rate (one or two cubs every three years); they require extremely large, undeveloped habitat areas where human encounters can be minimized; and they have been known to kill people, which injects a fear element into any recovery discussion. Because the various grizzly bear populations are distant from one another, biologists have concluded that linking corridors are important to promote genetic variability within the populations, and that such interchange opportunities are particularly important for the isolated Yellowstone population. Environmental organizations believe the bear requires even greater legal protection to ensure its long-term survival across the northern Rockies; through litigation they have forced the FWS to reassess whether the Cabinet-Yaak and nearby Selkirk grizzly populations should be reclassified as endangered rather than threatened species. State and local officials, however, fear continued federal oversight on public lands of virtually all development activity involving the grizzlies' far-flung habitat.[67]

The federal recovery program is cloaked in controversy and uncertainty. A federal Interagency Grizzly Bear Committee (IGBC) was created in 1983 with representatives from the FWS, Forest Service, National Park Service, and state game and fish departments to establish uniform federal bear management policies across the various grizzly bear ecosystems. While the FWS has not designated critical habitat for the grizzly bear, the IGBC has created an elaborate habitat zoning system intended to ensure adequate space for grizzly recovery and to safeguard the bear from conflicts with humans.[68] The ESA-based recovery program has achieved several positive results, which include halting grizzly bear hunting, cleaning up local garbage dumps that attract nuisance bears, reducing sheep grazing in bear country, and blocking several controversial development proposals in prime grizzly habitat. Nevertheless, major disputes persist over grizzly bear population trends, particularly in the Greater Yellowstone recovery area, where the FWS is interested in delisting the bear as an ESA-protected species. Most observers also believe that grizzly bear habitat is continuing to shrink on public lands owing to logging, mining, livestock grazing, recreation, and nearby residential development. In fact, the IGBC's habitat zoning system has not stopped logging and roading in the national forests, which has triggered several lawsuits ordering the Forest Service to close roads and otherwise protect bear habitat. Environmental organizations also successfully challenged the FWS's revised grizzly bear recovery plan, convincing a federal court that the plan did not adequately address habitat status and trends, failed to justify its population size measurement standards, neglected potential genetic isolation problems, and overlooked bear mortality related to livestock grazing. Federal officials, in an effort to address these concerns, have sought to expand the recovery effort, and they are beginning to prepare for an eventual delisting proposal.[69]

In March 2000, in an unprecedented concession to local sensitivities, the FWS proposed using Section 10(j) to reintroduce a nonessential experimental population of bears into the Selway-Bitterroot wilderness areas of Montana and Idaho that would be managed by a politically appointed citizen committee. The proposal, which evolved from discussions between two environmental organizations (Defenders of Wildlife and the National Wildlife Federation), local timber companies, and timber industry unions, was intended to avoid the time consuming and costly litigation that surrounded the controversial Rocky Mountain wolf reintroduction program. Biologically, the proposal would create another important population (and genetic reservoir) of bears in a remote region where conflicts with humans should be minimal and where the reintroduced bears could perhaps help link the isolated Yellowstone grizzly population with bears in the Glacier and Cabinet-Yaak recovery areas. The reintroduction proposal, however, met substantial opposition, from the state of Idaho, local residents, and even environmental organizations. Local residents feared for their safety if grizzly bears were translocated into the area, while state officials were concerned about the federal regulatory limitations that would inevitably accompany the bears even if they were treated as a nonessential experimental population. Recalcitrant environmental organizations argued that the Section 10(j) designation would not provide the sensitive bear with adequate protection and that the proposed citizen management committee, being a political rather than scientific entity, could not be trusted to oversee the bears. The Idaho congressional delegation, citing state and local opposition, successfully used a budget appropriations rider to block the reintroduction proposal for a year by prohibiting the FWS from spending any federal funds on it. As we shall see, this innovative attempt to marry biodiversity restoration with local political control raises serious institutional and legal questions concerning management authority on federal public lands.[70]

Looming behind these often bitter restoration controversies is the question of returning management responsibility for the grizzly bear, wolf, and other endangered species from the federal government to the states. In the ESA's jargon, this process is known as delisting. A pending but still unannounced delisting proposal has already provoked controversy in the case of the Yellowstone grizzly bear population, which is approaching the recovery plan's population size and dispersal goals. Because grizzly bear habitat covers expansive portions of the northern Rocky Mountain public lands, the states of Montana, Idaho, and Wyoming are interested in having the bear delisted, which would relieve local industry and residents from the stringent ESA regulatory limitations that accompany the bear's presence. Under the ESA, a species may be removed from the endangered species list if it no longer meets the requirements for listing. This entails several inquiries: the current condition of grizzly bear habitat, whether the grizzly is being overutilized for recreational or scientific purposes, the adequacy of existing regulatory mechanisms, and any other natural or human-created factors affecting its continued existence. Because the federal agencies have had such difficulty securing adequate grizzly bear habitat on public lands, the states—with their diverse wildlife management policies, limited legal authority, and entrenched political interests—will likely encounter even greater obstacles in managing the bear and its habitat.[71]

Setting aside the biological questions concerning the size and range of the current bear population, any delisting proposal must answer two difficult legal-political questions. First, given the basic principle that federal law supersedes state law, do the states have sufficient legal authority and adequate laws in place to ensure grizzly bear habitat on public and private lands throughout the Greater Yellowstone Ecosystem? Second, given the tristate nature of the recovery area and the need for secure wildlife corridors connecting the Yellowstone bear population with other populations, is there any assurance that the three states will undertake the extensive interstate coordination necessary to protect the bear population? Even if the legal authority is available to safeguard the bears, bear habitat, and dispersal corridors, there is precious little evidence that the states have the political will to impose restraints that might hamper local economic activity. Of course, with the newly reintroduced wolves proliferating across the northern Rockies, these delisting concerns may initially be addressed with wolves rather than grizzlies. In either event, a predator delisting debate would test the limits of federal–state cooperation over endangered species restoration programs on public lands.[72]

RESTORATION ON A REGIONAL SCALE: THE ICBEMP EXPERIENCE

As expansive as the grizzly bear recovery program is, it pales beside the massive restoration initiatives under way in the Pacific Northwest. On the west-side forests, as we have seen, the federal land management agencies are engaged in a major biodiversity restoration effort on behalf of the northern spotted owl and other old growth forest species—an initiative spawned by litigation, designed by scientists, and overseen by the federal courts. On the east-side public lands, the Forest Service and BLM collaborated on the Interior Columbia Basin Ecosystem Management Project (ICBEMP), a multiyear planning initiative to develop an ecological restoration policy for over 63 million acres of public land in four different states. A common thread in each of these regional restoration projects is the salmon, an anadromous fish that has long symbolized wild nature across the Pacific Northwest and that is now imperiled throughout much of its native range. Public land timber, livestock grazing, and roading policies have diminished critical salmon habitat, contributing to the species' decline. Excessive logging and past fire suppression policies have also badly frayed the east-side forests, enhancing the fire risk across the region. Any major policy modifications, of course, could presage related economic and social changes, touching myriad communities, constituencies, and individuals with interests tied to the region's public land resources. How, then, have the federal public land agencies approached regional federal restoration on this unprecedented watershed scale? The ICBEMP experience provides an early glimpse into what a truly regional restoration commitment might entail—both its promise and its perils.[73]

The ICBEMP restoration initiative has a distinctly administrative origin linked to very real ecological and legal concerns. During the 1993 Northwest Forest Summit, faced with mounting concern over the ecological health of the east-side forests, the new Clinton administration instructed the Forest Service to prepare a science-based, ecosystem management plan for the region. Once the BLM joined the planning effort, the agencies expanded it to encompass the entire Columbia River Basin east of the Cascade mountain range—an area covering 144 million acres in seven states, including eastern Oregon and Washington, western Montana, most of Idaho, and small portions of Nevada, Utah, and Wyoming. Confronted with dwindling salmon runs, deteriorating forest conditions, and the looming specter of another spotted owl litigation imbroglio, east-side public land managers saw the regional planning project as a potential lifeline to avoid the protracted controversy that had engulfed west-side forests for nearly a decade. When Congress funded the ICBEMP initiative, the project gained critical political legitimacy and became a key part of the Clinton administration's budding ecosystem management agenda. The ICBEMP strategy would provide the basis for revising management policy for 32 national forests and 16 BLM districts that stretched over millions of acres.[74]

Subsequent events have only strengthened the case for the regional ICBEMP restoration initiative. In 1992, when the National Marine Fisheries Service listed the Snake River chinook salmon as a threatened species, the decision brought the powerful ESA to bear on east-side forest, range, and other management decisions. Soon afterward, the courts temporarily enjoined all east-side timber sales until the Forest Service consulted with the NMFS over the effect of the new salmon listings on existing forest plans, proposed timber sales, and other forest management activities. Jarred into action, the agencies promptly developed interim aquatic habitat management standards (PACFISH, INFISH, and East-side Screens), thus managing to avoid the prolonged shutdown that west-side forests experienced during the spotted owl controversy. But the threat of further litigation and judicial intervention now hung over the ICBEMP initiative, especially with other ESA listing decisions pending. That threat became reality when several other salmon runs and the bull trout were added to the endangered species list. The courts, faced with more litigation over these newly listed species, began overtly prodding the agencies to complete the ICBEMP planning process and establish permanent aquatic habitat and other management standards. Moreover, the West's devastating 1994 fire season further heightened concern over regional fire risks, highlighting the need to address deteriorating forest conditions. In short, the agencies had little choice but to develop a coordinated management strategy for restoring ecological integrity to the region's public lands.[75]

The ICBEMP has evolved through distinctive phases since its inception, vividly illuminating the scientific, administrative, and political challenges involved in developing a regional restoration agenda. At the outset, confronted with a dearth of useful scientific information, ICBEMP managers established a Science Integration Team to compile data on the ecological, economic, and social conditions prevailing across the basin. The team's work culminated in a regional ecosystem assessment, which concluded that 60 percent of the basin's public lands rated low in ecological integrity and that 67 percent of the basin's population lived in counties with a high socioeconomic resiliency rating. The basin's resource-dependent rural communities, however, protested that the assessment did not fairly capture their plight, and they pressured Congress to require the agencies to develop a more detailed socioeconomic profile. (That study subsequently revealed that more than half the region's counties ranked low in economic resiliency, 30 percent of the communities were economically isolated, and 70 percent of their workforce depended on public lands and were ill-prepared for any major shift in management priorities.)[76] Using the ecosystem assessment data, the agencies next prepared two draft environmental impact statements that outlined alternative ecosystem management policies for the upper and lower basin areas and analyzed the environmental as well as socioeconomic ramifications of each alternative. The draft EISs reaffirmed that the region's terrestrial and aquatic ecosystems were badly stressed: "Native cold water fish stocks are rapidly declining, soil fertility is at risk, forest structures and patterns have changed from naturally occurring conditions, wildfire and pest attacks are becoming more severe, noxious weeds are invading rangelands, and riparian and aquatic health is deteriorating." Anxious to atone for past mistakes, the agencies endorsed a new management policy designed "to aggressively restore ecosystem health through active management, the results of which resemble endemic disturbance processes, including insects, disease, and fire."[77]

The ICBEMP's aggressive restoration proposal proved to be only the opening salvo in what was becoming an increasingly contentious regional ecosystem management experiment. The public reaction to the proposed alternative—over 83,000 comments—was uniformly negative. Many rural residents objected that the proposal's ecological restoration emphasis excluded local economic concerns. The timber industry was concerned with the lack of production goals, the use of

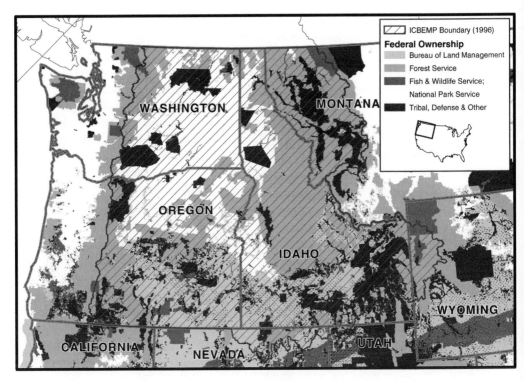

Map 4. Interior Columbia Basin Ecosystem Management Project. Roughly 63 million acres of Forest Service and BLM public lands are covered by the joint ICBEMP initiative, basically the entire Columbia River drainage east of the Cascade range crest and south of the Canadian border. The crosshatched area represents the ICBEMP project as originally conceived; lands situated in Nevada, Utah, and Wyoming were eventually deleted from the project as a result of congressional political pressures.

rigid management standards, and various timber harvest limitations. Environmentalists endorsed the ecological restoration emphasis but condemned the active management theme, projected increases in logging levels, and absence of protected nature reserves. Most important, the regulatory agencies—namely, the U.S. Fish and Wildlife Service, NMFS, and Environmental Protection Agency, with jurisdiction over key species, habitat conditions, and water quality—responded that the project's emphasis on active restoration, particularly the forecasted increase in timber harvesting, could imperil terrestrial and aquatic habitat as well as water quality. Amid such widespread dissatisfaction, Congress again entered the fray, threatening to defund the project. To avoid a congressional confrontation, the secretaries of interior and agriculture agreed to modify the project by establishing only general ecosystem restoration goals for the region, leaving specific implementation decisions to local public land managers.[78]

In response, ICBEMP managers prepared a supplemental draft EIS (SDEIS), which borrowed several features from the Northwest Forest Plan and again endorsed an active restoration management strategy. According to the agencies, the proposed alternative represented the "best strategy for: restoring the health of the forests, rangelands, and aquatic-riparian ecosystems in the project area; recovering plant and animal (including fish) species; avoiding future species listings; and providing a predictable level of goods and services from the lands administered by the BLM and the Forest Service."[79] The SDEIS eschewed detailed regional management standards and instead devised general strategies for restoring landscape-scale ecological processes (including prescribed burning), improving terrestrial habitat for diverse indigenous species, and reinvigorating aquatic habitat and riparian conditions. To enhance regional biodiversity, the SDEIS created designated nature reserve areas, denominated as Terrestrial (T) Habitat, Aquatic (A1/A2) Subwatersheds, and Riparian Conservation Areas (RCAs). Covering a substantial portion of the region's public lands, these reserve areas would be managed conservatively to promote species recovery objectives by disallowing intensive industrial activities. To address scale problems, the SDEIS adopted a stepdown analysis process that land managers would incorporate into their planning and project decisions. This process would help them to determine whether resource proposals— when assessed at increasingly smaller geographic scales (subregional, subbasin, watershed, and site-specific)—were consistent with the ICBEMP management goals. The SDEIS also adopted an adaptive management strategy to enable managers to adjust resource decisions based on an extensive monitoring and evaluation program. Under the proposed restoration alternative, timber harvesting was projected to increase by 5 percent (though smaller logs may deter some sales); livestock grazing was expected to decline by 10 percent; and the acreage subjected to prescribed burning would increase by more than 700 percent. The SDEIS forecast only negligible effects on overall employment in the basin, anticipating approximately 4,000 new public land–related jobs, mostly in stewardship timber harvesting and prescribed fire activities. With its emphasis on active restoration initiatives, protected nature reserves, large-scale analysis processes, and adaptive management strategies, the ICBEMP proposal represented a dramatic shift away from the traditional commodity-based and output-driven resource management agenda that has long characterized public land policy.[80]

In its brief history, the ICBEMP initiative has been whipsawed between its commitment to science-based ecological policy reforms and the hard realities of congressional politics. Although the ICBEMP was originally conceived in 1993 when the White House and Congress were under Democratic control, that political alignment changed dramatically one year later when the Republicans captured Congress, bringing with them an avowedly anti-environmental agenda. As the ICBEMP grew in size, complexity, and cost, the new Republican Congress's initial skepticism was transformed into outright hostility, forcing substantial changes in its scope and detail. First, when the original ICBEMP data suggested that the region's economy overall was robust and diverse, the local congressional delegations used an appropriations rider to force the agencies to prepare a more detailed study of local economic and social conditions. When the new data revealed that the region's rural communities—representing 10 percent of the basin's population—depended heavily on public lands, it spotlighted the effect policy modifications would have on this populace and stiffened local resistance to the ICBEMP's ecological restoration agenda. Second, the project's sheer size made it an easy target for local politicians concerned about protecting their home turf. As originally conceived, ICBEMP boundaries corresponded perfectly to the interior Columbia River Basin watershed, making it truly an ecosystem-based planning project. But with mounting congressional opposition, the agencies were forced to reconfigure the project boundaries, eliminating Nevada, Utah, and Wyoming from its coverage. Third, to avoid a major congressional defunding rider fight, the secretaries of interior and agriculture agreed to further scale back the project by refocusing it on key regional issues while leaving other, more detailed management decisions for local resolution. In short, the project's boundaries were reshaped, its policy reform aspirations reduced, and its political vulnerability exposed.

Nevertheless, despite sustained congressional scrutiny, the ICBEMP's fundamental commitment to ecological restoration on the region's public lands remained intact through the 2000 elections. The reasons were painfully obvious. If ecological conditions on public lands continued to decline, the region faced the very real prospect of protracted ESA litigation, intervention by the regulatory agencies, and destructive catastrophic fire events. No one relished the prospect of another spotted owl standoff or a devastating fire season. Unable to change the governing environmental laws or to mandate a major logging campaign, Congress had few viable options left. But problems still plagued the ICBEMP initiative. It had already cost over $40 million to gather the necessary scientific information, develop restoration management alternatives, and prepare environmental documents, far exceeding original estimates and raising serious questions about the fiscal viability of similar large-scale proposed ecosystem management projects. Estimates ran as high as $67 million in annual funding to implement the proposed restoration agenda. In addition, as the project evolved through its various iterations, the interested constituencies remained either hostile or ambivalent about the announced changes, raising the specter of litigation over any final decision the agencies might make. And no one was certain exactly how the proposed ecological restoration strategies would work on the ground.[81]

Perhaps inevitably, following the 2000 presidential election, the ICBEMP project has been dramatically reshaped to reduce its legal import yet retain its scientific insights. With the EIS process complete, the Clinton administration chose not to finalize the ICBEMP record of decision, leaving the project in limbo as the Bush administration took office. Confronted with an assortment of unhappy constituencies and the prospect of legal challenges that could delay agency planning throughout the basin, the Bush administration adroitly converted the ICBEMP initiative from a multiagency action document into a less meaningful strategy document. This new document—endorsed through a memorandum of understanding and signed by the affected land management and regulatory agencies—incorporates the final EIS's fundamental commitment to ecosystem management and restoration without mandating specific planning or species recovery standards. Drawing upon the wealth of scientific and other information developed during the extended six-year project, it articulates general principles for the land management agencies to utilize in their planning and project decision processes. It calls upon them to consider terrestrial and aquatic habitat protection and restoration opportunities, to utilize multiscale environmental analyses, to coordinate among themselves and with neighbors, and to employ adaptive management techniques. Where the original ICBEMP proposal would have set clear resource management priorities and standards, the strategy document is purposefully more flexible, giving individual national forests and BLM districts greater leeway in how they implement its principles to accommodate local circumstances. Unlike the counterpart Northwest Forest Plan, it is ecosystem management and restoration without legally enforceable teeth.[82]

It is probably too soon to write the final chapter for the ICBEMP ecological regionalism initiative, though some observations are in order. As a legal matter, by eschewing mandatory standards, the strategy document does not create obvious litigation opportunities, therefore leaving local Forest Service and BLM managers largely unaccountable for planning or project decisions that depart from its key ecosystem management or restoration principles. This outcome dovetails neatly with the Bush administration's professed concern that legal processes and appeals have hindered agency decision making and thus unnecessarily delayed timber sales, forest thinning, and other resource projects. The strategy document approach also can be understood in partisan political terms: a conservative Republican administration interested in promoting extractive activities on the public lands has little interest in constraining itself with the rigid ecosystem plan developed by its ideological opponents and heavily weighted toward an environmental protection agenda. At the agency level, faced with implementing an extraordinarily large, costly, and untested regional ecosystem restoration program, Forest Service and BLM officials have understandably sought to maintain flexibility and managerial discretion. Yet unable to overlook the accumulated scientific data on ecological conditions across the region or the very real ESA consultation and NEPA cumulative effects analysis requirements confronting them, the agencies made an important if ambiguous commitment to species recovery and habitat protection, multilayered environmental analysis, and regional coordination. Indeed, the strategy document's flexible approach reflects the uneasy tension that persists between the land management agencies (whose primary concern is overseeing their own jurisdictional domain) and the regulatory agencies (whose statutorily mandated focus is on regionwide species and environmental quality concerns). This tension will almost certainly continue to manifest itself as the land management agencies proceed to amend their plans and to render individual project decisions—most of which will be reviewed by the regulatory agencies under ESA consultation requirements and the like. Reminiscent of the earlier multiagency Greater Yellowstone vision document experience, which also culminated in vague commitments to regionalism, the expansive and expensive ICBEMP experiment may have pushed untested ecosystem management and restoration concepts to their geographic and political limits.

Elsewhere, the federal government is actively engaged in other large-scale ecosystem restoration projects, which now dot the public landscape. Congress has authorized many of the projects, which variously aim to restore degraded public lands, free flowing rivers, and native biodiversity. Most projects employ an integrated, ecosystem-based planning process, using regional assessments to marry ecological restoration goals with local economic and social concerns. In California's Sierra Nevada mountains, the Forest Service has overseen the Sierra Nevada Ecosystem Project (SNEP), which has completed a regional assessment and a regional EIS that revised forest management policies in response to forest health and biodiversity concerns. Similar regional restoration proposals have surfaced in the Greater Yellowstone Ecosystem, southern Rockies, Colorado Plateau, and Great Basin. Other major ecological restoration initiatives include the massive south Florida Everglades project designed to restore free flowing water to the Everglades National Park environment, the San Francisco Bay Delta project (nicknamed CALFED) designed to enhance fresh water flows into San Francisco Bay to protect its fragile ecology and safeguard endangered salmon runs, and the southern Appalachian Man and the Biosphere program designed to promote sustainable development and ecosystem management in the region.[83] Relatedly, several federal projects are designed to restore degraded river ecosystems. In one highly publicized experiment, the Bureau of Reclamation modified its Glen Canyon dam water releases in an effort to mimic the high-flow, spring runoff conditions that historically characterized the Colorado River ecosystem. The Park Service is moving forward with the Elwha River dam removal proposal, which would unclog historic salmon migration routes into the Olympic National Park backcountry. And the chorus supporting the still controversial proposal to breach the lower Snake River dams to restore salmon migration patterns on the Columbia River system is growing ever louder. As these initiatives proliferate, astute observers submit that ecological restoration—undoing the environmental mistakes and miscalculations of the past—will soon define the next era on the western public lands.[84]

Ecosystem management and restoration concepts have plainly developed a momentum of their own, becoming major considerations in contemporary public land policy. Designed to overcome our past mistakes, these interrelated concepts are driven by the need to ensure our biological heritage and to sustain functioning ecosystems. Standing alone, however, these concepts cannot fully achieve such lofty goals. We must also integrate our nature reserves—the national parks, national wildlife refuges, and wilderness areas—into a comprehensive biodiversity conservation strategy built around ecosystem maintenance and restoration goals. But as the spotted owl and wolf controversies illustrate, even our most expansive reserves are not sufficient to protect imperiled species from endangerment. The answer, according to most biologists, must include enlarging our nature sanctuaries, which means expanding both the size and scope of our protected land systems. Yet setting aside public lands strictly for preservation purposes has proven deeply unpopular in nearby communities. As controversial as our nascent ecological restoration policies are proving, an expanded federal preservation agenda evokes even more strident political resistance. Nonetheless, an integrated preservation strategy must be part of any meaningful ecological management policy that aims to maintain and repair our public land ecosystems.

SIX

Shaping a New Heritage

Preservation in the Age of Ecology

The first principle is to point out the need for complete Nature
Sanctuaries. They are essential if any of the original nature . . .
in North America is to be saved for future generations.

VICTOR E. SHELFORD,

ECOLOGICAL SOCIETY OF AMERICA (1933)

We are certain to lose the mountain lion, or cougar, and the grizzly bear
unless the remnants of our wilderness areas are preserved.

WILLIAM O. DOUGLAS (1965)

Although preservation has deep roots in public land policy, the same cannot be said for the role of ecology in preservation policy. Aesthetics, scenic grandeur, and outdoor recreation—not ecosystems, biodiversity, or scientific inquiry—have been the primary forces behind our national parks, wilderness areas, and other protected land designations. To be sure, grizzly bears, bison, and other wild creatures have availed themselves of the millions of acres of public land placed off-limits to industrial development. Rarely was their welfare, however, the principal reason for preserving these wildlands. But we now understand that undeveloped wilderness serves as the last refuge for many of our most imperiled species, making it both the real and symbolic heart of any ecosystem management strategy. We also now understand that our traditional enclave strategy of nature conservation does not meet contemporary biodiversity concerns and ecological needs. Only large and interconnected wilderness preserves will do that.

Public land preservation decisions are inherently political decisions. Congress has reserved for itself the power to create new national parks and wilderness areas. And it has routinely deferred to local political preferences in shaping new wilderness legislation, effectively devolving this power to local congressional delegations. Large-scale wilderness preservation proposals must therefore surmount major political obstacles, which often leaves them languishing in legislative limbo. This is true for Montana's national forestlands as well as Utah's Bureau of Land Management (BLM) lands, and in several other cases too. Nonetheless, a plethora of administrative initiatives—ranging from national monuments to new roadless area policies—has significantly advanced the preservation agenda, creating a multitude of new protective designations that incorporate ecological concerns into the design. But the local resistance is fierce, and even these sizable designations fall short of the margin of safety necessary to ensure our biological heritage.

As expansive as they are, public lands simply cannot carry the full weight of a landscape-scale preservation strategy grounded in science. Too much ecologically significant habitat falls outside the boundaries of the public domain. This means that private lands also are an important component of any ecosystem-driven biodiversity preservation strategy. And that raises the specter of private property rights and the accompanying expectations of ownership. Whatever the political opposition is to more preservation on public lands, it can be doubled or tripled when private lands are added to the preservation mix. Put simply, the science and politics of preservation are not proving easy to reconcile.

To Save the Wild: An Unfinished Agenda

Ever since Congress passed the Wilderness Act of 1964 and enshrined the wilderness concept in law, the protection of wilderness landscapes has assumed a prominent position on the public land policy agenda. More often than not the wilderness designation question is framed in rather stark choices: Is the land more valuable for its development potential and the related economic opportunities, or should it be left undisturbed as a nature sanctuary to benefit wildlife and intrepid adventurers? For the natural resource industries and many rural communities, placing large areas off-limits to any resource development activities or to motorized access is an unacceptable alternative, portending the loss of local control and related economic benefits. Sentiments run equally high among environmentalists and their allies, who see wilderness protection as the last best opportunity to save fragments of the primitive American landscape for aesthetic, recreational, and increasingly ecological purposes. Given the high stakes, it is not surprising that the conflict over wilderness continues unabated, both on the national forests and BLM public lands. This has become a battle for the soul of the public lands. And there are few places where the discourse and politics have been so passionate as over the fate of Montana's roadless national forestlands and southern Utah's expansive BLM lands.

THE MONTANA WILDERNESS DEBATE

In the popular imagination, wilderness and Montana are virtually synonymous, reflecting the state's unique geography and heritage. Montana—the Big Sky State—is the fourth largest of the 50 states, boasting vast open spaces, seemingly endless mountain ranges (hence the state's name), and a population of fewer than a million people. The state is really two different places: the more heavily populated western portion of Montana contains rugged mountains, expansive coniferous forests, and bucolic river valleys; the rather sparsely populated east is flat and mostly dry high plains country. The economy in western Montana is somewhat diversified, though logging has long been a major regional industry. In eastern Montana, cattle ranching and wheat farming are key local industries. Mining has a long and colorful history in the state's economic and political life, though the energy fuels—coal, oil, and gas—have assumed much greater importance during the past several decades. Ever since the Lewis and Clark Expedition traversed the state in its epic journey of discovery, Montana has stood as a symbol of the western frontier and its wilderness attributes. That symbolism was affirmed early in the twentieth century when the nation's forest reserves were established, giving Montana one of the largest federally administered landscapes among the lower 48 states. Although the state's national forests have not escaped industrial-scale logging and the accompanying road corridors, they are still surprisingly intact, with expansive undeveloped tracts suitable for wilderness designation.

Nature preservation has a lengthy history in Montana. Both Glacier and Yellowstone national parks, two enduring icons in the national park system, are connected with the state. By the early twentieth century, American presidents had set aside over 15 million acres as forest reserves across western Montana—lands now encompassed within the state's nine national forests. Montana is also home to more than 20 national wildlife refuges, along with several federally designated wild and scenic rivers. Beginning in 1912, the state made a commitment to wildlife preservation, establishing the first of several state game refuges to protect vital habitat for its valuable big game populations. During the past half century, Montana has been one of the last strongholds for the grizzly bear, perhaps the ultimate symbol of wilderness. Nor was it a surprise when the first extirpated Rocky Mountain wolf reappeared in Montana, signaling its return to the American West.

The Wilderness Act of 1964 initiated what has become a decades-long campaign to preserve Montana's remaining wildlands as federally protected wilderness areas. Yet before then, the Forest Service had designated over 1.5 million acres of Montana's undeveloped national forestlands as "wilderness" or "wild" areas, an administrative classification that placed them off-limits to timber harvesting and other industrial activities. According to its terms, the Wilderness Act automatically converted these administrative designations—namely the Bob Marshall, Selway-Bitterroot, Anaconda-Pintler, Cabinet Mountains, and Gates of the Mountains areas—into "instant" wilderness, reinforcing the state's wildlands heritage. The Wilderness Act also obligated the Forest Service to study its preexisting "primitive" areas—another 417,000 acres in Montana—for possible later wilderness designation. With these initial designations, Montana boasted one of the largest national forest wilderness assemblages among the 50 states.[1]

The Montana congressional delegation—notably, Senator Lee Metcalf—was deeply involved in shepherding the original wilderness legislation through Congress. Several other western politicians, including senators Frank Church of Idaho and Henry Jackson of Washington, also played lead roles in the long drama surrounding passage of the Wilderness Act. (That Montana and Idaho politicians played such a key role in the original wilderness legislation makes it doubly ironic that Montana and Idaho are the only two western states without post-1964 statewide wilderness bills.) Not everyone in the West, however, was happy with the Wilderness Act or its instant wilderness designations. The mining, logging, and ranching industries initially opposed the wilderness bill, then organized against any further protective designations. They regarded the statute as an ill-advised federal effort to limit access to potentially valuable forestland and to undermine the economic lifeblood of rural, resource-dependent communities. Since then, motorized recreation groups have emerged as another principal wilderness opponent, objecting to prohibitions on any mechanized access. But in Montana, the conservation community has long supported wilderness. The Montana Wilderness Association (MWA) was organized in 1958 to garner grassroots support for protecting the state's undeveloped—or roadless—national forestlands. National environmental groups, including the Wilderness Society and National Wildlife Federation, have also long sought to protect Montana's pristine national forest landscapes. Once the Wilderness Act was adopted, the battle over Montana wilderness was joined, and it soon devolved into a long series of bitter struggles over specific roadless areas.

Under the Wilderness Act, the Forest Service was directed to review its existing "primitive" areas and to make recommendations on their wilderness suitability. Seeking to expand this process, wilderness proponents urged the Forest Service to include all its roadless lands—more than 55 million acres nationally—in this initial inventory, arguing that these de facto wilderness lands also merited special protection. Early in this review process, Montana activists mounted a lengthy grassroots campaign to gain wilderness protection for the unroaded Lincoln backcountry area, which extended into three national forests at the southern terminus of the newly created Bob Marshall Wilderness Area. Congress finally acceded in 1972, creating the 240,000-acre Scapegoat Wilderness. The Scapegoat legislation is significant in several respects: it represented the first time that Congress had legislatively protected de facto wilderness lands; it demonstrated the power of a grassroots, citizen-driven wilderness campaign as a political strategy; it initiated what has become a pattern of individual wilderness bills to protect Montana's roadless forestlands; and it substantially augmented the earlier Bob Marshall Wilderness Area, creating an expansive interlocked network of protected wildlands extending over 1.5 million acres along the Rocky Mountain spine from Glacier National Park to Lincoln in mid-state Montana. Eventually, seeking a comprehensive national resolution of the wilderness issue, the Forest Service prepared two Roadless Area Review Evaluations (dubbed RARE I and II) that assessed the wilderness qualities of all its unroaded lands. Neither of the RARE processes, however, came to fruition in legislation, the victims of lingering controversy over too many areas that were either included in or excluded from the agency's final recommendations. Nevertheless, by the mid-1980s, most of the western states had reached agreement over the wilderness issue, and Congress had memorialized these understandings in individual statewide wilderness bills. That did not occur, however, in Montana, where the various factions were unable to achieve any consensus.[2]

Thus began the long and sometimes torturous process of addressing the Montana wilderness question one area at a time, punctuated by periodic efforts to cultivate an elusive consensus over statewide wilderness proposals. When the Forest Service released its initial RARE I inventory in 1972, it contained miserly Montana wilderness recommendations, omitting such well-known roadless areas as the West Pioneers and Big Snowies. At the behest of the MWA and other groups, Senator Metcalf and the Montana delegation persuaded Congress to pass the Montana Wilderness Study Act of 1977, which granted wilderness study area status to nine roadless areas covering almost 1 million acres. These areas (also known as S. 393 lands, after the Senate bill number) are effectively protected from development until Congress acts to either protect or release them. Other potential Montana wilderness study areas—the Elkhorn and Great Bear areas—were granted similar protection under the Forest and Refuge Omnibus Act of 1976. Since then, responding to local political sentiments, Congress has gradually added several Montana sites to the national wilderness system, literally one area at a time. The list of additions is impressive: in 1975, the Mission Mountains Wilderness Area; in 1976, three separate National Wildlife Refuge wilderness areas; in 1978, the Welcome Creek, Absaroka-Beartooth, and Great Bear wilderness areas; in 1980, the Rattlesnake Wilderness just outside Missoula; in 1983, the 254,000-acre Lee Metcalf Wilderness north of Yellowstone National Park, named after the late Senator Metcalf, who ardently championed Montana wilderness during his congressional career.

Truly fascinating grassroots mobilization, coalition building, and political negotiation stories lurk behind these individual designations, each illuminating—with local variations—the recurrent struggle between the state's environmental community and its extractive industries over the wilderness question. The net result has been the establishment of over 3.5 million acres of wilderness in Montana's national forests, though key areas—the Great Burn and Crazy Mountains, for example—remain unprotected. Wilderness proponents have been unable to muster sufficient local consensus to pass a statewide wilderness bill, as has occurred in most other western states, though not from a lack of effort.[3]

Since the early 1980s, several statewide Montana wilderness bills have surfaced, only to die in the throes of political infighting. From then until now, political power has gradually shifted in Montana from the Democrats to the Republicans, and the state's Republican political leaders have steadfastly opposed any

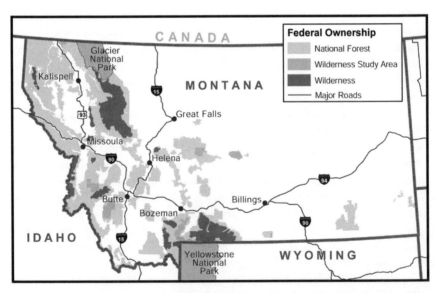

Map 5. Montana National Forest Lands. Most of Montana's national forestlands are located in the western portion of the state. Shown here are congressionally designated wilderness areas and S. 393 lands or wilderness study areas, which the Forest Service is charged with managing to protect their wilderness character. Conservation groups have also sought wilderness protection for several million acres of other unroaded Montana forestlands.

significant additions to the national forest wilderness system. The proposed Montana Wilderness Act of 1984, born of a tenuous consensus within the state's congressional delegation, failed badly, opposed by environmentalists because it protected only 750,000 acres and would have released over 5 million acres to potential logging and other multiple-use activities. Four years later, the various factions actually reached a compromise, and the Montana delegation convinced Congress to pass the Montana Natural Resources Utilization Act of 1988, which would have designated 1.4 million acres of new wilderness. But on the eve of the 1988 elections, President Reagan pocket vetoed the bill, the first time ever that a president vetoed wilderness legislation. The reason was crass political opportunism: wilderness had become a critical issue in the bitter Montana Senate campaign, and it was ultimately credited with defeating incumbent Democratic senator John Melcher, who had sponsored the ill-fated legislation. Two years later, seeking an alternate approach to the wilderness issue, several environmental groups entered into negotiations with western Montana's millworker unions over wilderness designation on the Lolo and Kootenai national forests. They ultimately reached an agreement known locally as the Lolo-Kootenai accords, which then became the basis for a local wilderness bill. However, opposition from other environmentalists and anti-wilderness forces, both of whom feared the precedent that might be established, doomed this local consensus approach to the wilderness issue. The subsequent Montana National Forest Management Act of 1991, which incorporated parts of the Lolo accord, met a similar fate the following year. During the next session of Congress, the House of Representatives did pass the Montana Wilderness Act of 1994, which contained more than 40 new wilderness or wilderness study area designations and created a unique Northern Rockies Ecosystems Study Panel. Opposition from recalcitrant environmental groups and anti-wilderness forces, however, doomed the bill in the Senate. Since then, with the state badly divided over the wilderness issue, no serious effort has been made to advance statewide wilderness legislation.[4]

The political stalemate, however, has not removed wilderness protection from the public agenda in Montana. Instead, the debate has moved in other directions, as wilderness proponents have refashioned their arguments and sought other forums to protect the state's unroaded forestlands. In several Montana locations, the environmental community has used ecological concepts to link unprotected roadless lands with adjacent national parks and existing wilderness areas, arguing that these public land complexes should be managed as a "greater ecosystem." This has been the strategy in southwest Montana, where Yellowstone National Park has been conjoined with the Gallatin, Beaverhead, and Custer national forests into the northern segment of the Greater Yellowstone ecosystem. The same strategy has been employed in northern Montana to delineate the Crown of the Continent ecosystem, which encompasses Glacier National Park, the Bob Marshall wilderness complex, and adjacent national forestlands. In northwest Montana's Cabinet Mountains, similar ecosystem preservation arguments have been advanced using the grizzly bear as the ecologically unifying force to explain the need for additional wilderness protection. On a much grander scale, several groups have developed regional wilderness proposals, using conservation biology principles to link Montana's existing protected areas with more distant protected areas through wilderness corridors. Examples of these broader-scale approaches to wilderness protection are the proposed Northern Rockies Ecosystem Protection Act (NREPA) and the Yellowstone to Yukon (Y2Y) initiative, both of which are described in more detail later. Some wilderness proponents, however, fear that these transcendent regionalism proposals cannot command a sufficient consensus to pass political muster in Congress, where individual state concerns are the stock in trade. Nonetheless, undaunted by the lack of a statewide wilderness bill, Montana environmental activists have effectively raised the stakes in the debate, using scientific principles to tie existing wilderness proposals to larger ecoregion preservation concerns.

Meanwhile, Montana wilderness proponents have fought a lengthy rearguard action to protect the wild character of the state's roadless forestlands. Over the years, the threats confronting these lands have been many and varied: proposed timber sales, oil and gas projects, road construction, and motorized recreational activities. The principal strategy has been to challenge the legality of such proposals, in either court or agency administrative proceedings. During the late 1970s, for example, the Montana Wilderness Association sued to enjoin the Forest Service from granting Burlington Northern road access to its private checkerboard timberlands across the Gallatin National Forest to safeguard the wilderness character of the surrounding roadless forestlands. Although the MWA's challenge ultimately failed, it highlighted the problem and set the stage for congressional approval of two major land exchanges that eliminated the checkerboard inholdings and thus protected the forest's remaining wilderness values. During the mid-1980s, with federal officials poised to issue oil and gas leases in the Lewis and Clark National Forest's remote Rocky Mountain Front area, local environmentalists won a major court victory blocking the leases until comprehensive National Environmental Policy Act analyses and endangered species consultations were completed. Since then, the Forest Service has administratively withdrawn the Rocky Mountain Front region from any energy leasing or hardrock mining entry. In 2001, the Montana Wilderness Association successfully prosecuted a lawsuit to protect the S. 393 wilderness study areas from increased motorized recreational activities. A Montana federal district court found that the Forest Service, by allowing all-terrain vehicles (ATVs) onto backcountry trails, had violated the 1977 Montana Wilderness Study Act, which required "wilderness study areas . . . be administered . . . to maintain their presently wilderness character." Besides enjoining the Forest Service to control the ATVs, the court ordered the agency to restore the wilderness character of those areas that had been damaged. Elsewhere, Montana's wilderness proponents have employed similar legal and political tactics in order to preserve the wilderness option until Congress can be persuaded to act.[5]

During the waning days of the Clinton administration, the Montana wilderness controversy took yet another turn with potential far-reaching implications. Acting under the president's orders, the Forest Service released a nationwide Roadless Area Conservation EIS that administratively protected over 56 million acres of unroaded national forestlands from further timber harvesting, road construction, and other development activities. As a practical matter, the Forest Service's Roadless Area EIS was RARE III, but with legal teeth because it was translated into a binding federal regulation. For Montana, the so-called roadless rule protected over 2 million acres, including the S. 393 wilderness study areas and other de facto wilderness lands. Although the roadless rule was unpopular with most western congressional delegations, who view wilderness protection as their prerogative rather than the president's, Congress was not prepared to reverse this expansive administrative initiative. But angry western states, led by Idaho, secured a federal court injunction that temporarily barred the rule from being implemented. Once the George W. Bush administration took charge, as we have seen, the Forest Service indicated it would revise the roadless rule. The uncertainty surrounding the roadless rule once again leaves the fate of Montana's unroaded forestlands in limbo, just as they have been for the past 35 years. And there is no indication, given the anti-wilderness sentiment that prevails within the state's congressional delegation, that the stalemate will end soon.[6]

It is worth noting, however, other related developments that could affect the future direction of the Montana wilderness debate. As is true in other western states, Montana has been beset by an influx of newcomers—both new permanent residents and second home owners—most of whom have chosen the Big Sky State for its quality of life attributes, environmental amenities, and open space. Their mere presence and influence could eventually help to move the state's wilderness debate off dead center, perhaps toward additional protective designations. Even without a formal wilderness designation, the presence in Montana's national forests of the wide-ranging grizzly bear, which is protected under the Endangered Species Act, imposes significant limitations on the Forest Service's management options and thus adds a layer of legal protection to those unroaded forestlands that encompass grizzly habitat. In addition, with environmental and budget watchdog groups united in their opposition to unwarranted federal subsidies, the Forest Service's road building program—historically fueled by generous congressional appropriations—has ground to a halt. A Clinton administration regulation designed to curtail new national forest road construction should further reduce the pressure on Montana's roadless forestlands.[7] And finally, with the Forest Service poised to begin the second round of NFMA-mandated planning on the state's national forests, that planning process—with the agency's new emphasis on ecosystem management—should provide both the agency and the public with an opportunity to reassess the wilderness option for the yet unprotected roadless lands. In sum, with no statewide consensus on the horizon, the skirmishing over wilderness in Montana's national forests seems destined to continue apace, with new ecological arguments pressing for recognition and expanding the stakes in the state's fractious wilderness politics.

SAVING UTAH'S DESERT LANDSCAPES

Utah's redrock desert landscapes, although a sharp contrast to Montana's verdant national forestlands, are also a flash point in the high-stakes wilderness protection controversy on BLM public lands. The state of Utah sits astride the juncture of three major geographical regions: the Colorado Plateau, the Great Basin, and the Rocky Mountains, which give the state one of the most diverse topographies in the country. The Utah Colorado Plateau—or redrock—country, much of which falls under the aegis of the BLM, is characterized by high plateaus, deep riverine canyons, jumbled rock formations, and spectacular vistas, which have attracted an ever growing number of visitors seeking solitude, recreation, and other outdoor pursuits. The plateau's unique geologic structures contain oil, gas, coal, uranium, and other valuable mineral resources, and its parched landscape has long supported a hardscrabble local ranching industry. Utah's Basin and Range lands, also largely in the BLM's portfolio, compose a vast, dry, sagebrush-studded landscape, home to a small but hardy number of ranchers, mining firms, and the military, which has used the region's empty spaces for target practice. In contrast, the state's Rocky Mountain region is characterized by rugged mountain ranges and their alpine ecosystems; much of this land is managed by the Forest Service, for logging, grazing, and mining as well as manifold recreational benefits that include a flourishing ski industry. In total, the federal government owns nearly two thirds of Utah, with BLM public lands accounting for roughly 22 million acres and the national forests another 8.2 million acres. With this ownership pattern and the sheer size of the acreage involved, wilderness designation has proven controversial. Although Congress passed national forest wilderness legislation in 1984 and established twelve new wilderness areas covering mostly alpine terrain, the BLM wilderness debate has festered unresolved, complete with dueling wilderness bills, acrimonious legislative showdowns, and regular legal confrontations.[8]

The notion of preserving public land from development is not new to Utah, though such initiatives have always stirred strong emotions and local resistance. The state is home to 5 national parks (including Canyonlands, Zion, and Bryce Canyon), 7 national monuments (including the Grand Staircase–Escalante and Dinosaur), 2 national recreation areas, 4 national wildlife refuges, and 750,000 acres of national forest wilderness lands. All the national parks except Canyonlands, initially lacking local political support, began as presidentially designated national monuments only to be subsequently converted to national parks. During the 1930s, Secretary of the Interior Harold Ickes proposed establishing an Escalante National Monument that would have embraced most of southeast and south central Utah, but intense local opposition ultimately killed the proposal. Over the years, the state's national parks and monuments, most of which are situated in southern Utah's scenic redrock country, have become magnets for tourism, which has become one of the state's largest private sector employers.[9]

Outside Utah's national parks and monuments, however, the fragile desert landscape lies largely unprotected and, until recently, little appreciated for its scenic, biological, or recreational qualities. Writers like Wallace Stegner and Edward Abbey helped change that, just as an upsurge in mountain biking, river rafting, and other outdoor pursuits exposed more and more people to the desert's aesthetic charms, recreational possibilities, and environmental fragility. Once the BLM started its congressionally mandated wilderness review process, the southern Utah slickrock country had an instant constituency that was set to push for a large wilderness commitment. Its preservation fervor was further fueled by the sense that the state's wild places had been shortchanged in the miserly 1984 Utah national forest wilderness bill. But the state's rural communities, which depended primarily on the mining, energy, and livestock industries, were equally steadfast over maintaining the region's working landscapes and their traditional access privileges to public lands. A perhaps predictable major confrontation ensued—one that has, in fact, become a lengthy and often frustrating stalemate.[10]

The BLM wilderness designation process was set in motion over a quarter century ago. Under the Federal Land Policy and Management Act of 1976 (FLPMA), Section 603 directed the secretary of the interior to inventory all roadless public lands exceeding 5,000 acres, to determine their suitability for wilderness designation, and then to prepare specific wilderness recommendations. The secretary's recommendations were to go to the president, who would then make his own recommendation to Congress, which retained the final designation authority. Those lands determined to be suitable for wilderness designation during the inventory process were designated "wilderness study areas" (WSAs), and the BLM was obligated to manage them "so as not to impair the suitability of such areas for preservation as wilderness." As a practical matter, once the BLM completed its inventories and recommendations, the WSAs became de facto wilderness pending final congressional action on the proposal—a view confirmed by the federal courts when they ordered the BLM to protect the wilderness integrity of these areas against threatening development. Faced with this new wilderness designation responsibility, the BLM developed a *Wilderness Inventory Handbook* that outlined standards for determining an area's potential suitability for wilderness protection: it must contain 5,000 acres of contiguous roadless public land; it must be in a natural condition, with humanity's imprint substantially unnoticeable; and it must offer an outstanding opportunity for either primitive recreation or solitude. With these criteria, the BLM plainly had room for subjective judgments in making its wilderness recommendations.[11]

Perhaps inevitably, given its traditional mining and ranching constituency, the BLM found few of its lands qualified for potential wilderness designation. Nationally, the BLM reviewed over 180 million acres and eventually recommended only 24 million acres as suitable for wilderness designation. Confronted with these paltry recommendations, the environmental community concluded that the BLM had not fairly evaluated many areas that actually qualified for wilderness protection, a fact supported by Interior Secretary James Watt's efforts to dilute the agency's eligibility criteria.[12] By the mid-1990s, however, Congress had approved a statewide BLM wilderness bill for Arizona and resolved the contentious California desert controversy. But the story was quite different elsewhere. None of the other western states were near agreement on BLM wilderness designations, neither on the acreage recommendations nor on water rights, release language, or other key terms. The stalemate was perhaps most pronounced in Utah, which boasted a spectacular array of potentially eligible BLM lands but found itself locked in mortal combat over the very notion of wilderness on these lands.

The Utah BLM wilderness debate has raged for more than two decades with little obvious progress toward a final solution. The controversy erupted in 1980 immediately after the BLM completed its initial Section 603 inventory of the state's potentially suitable wilderness lands. When the BLM announced that it believed only 2.5 million acres qualified for further wilderness study, the state's conservation community protested vehemently, asserting that the BLM had improperly overlooked many locations that merited WSA status in order to protect mineral development opportunities or in deference to phantom roads that did not really exist. Following a prolonged series of administrative appeals and Interior Board of Land Appeals decisions, wilderness proponents forced the BLM to add over 600,000 acres to its initial WSA inventory, which meant that roughly 3.2 million acres were finally given WSA status. In the process, the Utah Wilderness Coalition (UWC) was created as an umbrella organization to coordinate the lobbying efforts of the myriad local and national environmental organizations that supported an expansive state wilderness bill. The BLM, meanwhile, initially recommended that a mere 1.89 million acres be formally designated wilderness, a figure that increased to 1.97 million acres by the time the final EIS was released. In response, the UWC initiated the first of two major citizen field surveys, eventually determining that 5.7 million acres actually qualified for wilderness designation. By then, though, Utah Republican Representative Jim Hansen had seized upon the BLM's initial fieldwork and introduced a bill into Congress that called for 1.4 million acres of wilderness. His Democratic counterpart, Representative Wayne Owens from Salt Lake City, countered with a competing bill that drew upon the coalition's fieldwork and sought to protect 5.4 million acres. The legislative battle had been joined, but neither side had the clout to advance its agenda in a politically divided Congress.[13]

In 1994, however, the political landscape changed dramatically when the Republicans wrested control of Congress from the Democrats for the first time in 40 years. The time appeared right for Utah's solidly Republican congressional delegation to push through a small-acreage BLM wilderness bill and to settle finally this festering dispute. Joined by the state's popular Republican governor, Mike Leavitt, the Utah delegation introduced legislation that would protect only 1.8 million

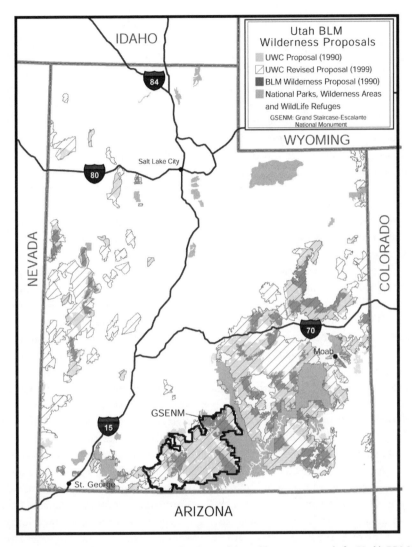

Map 6. Utah BLM Wilderness Proposals. Three of the wilderness proposals for Utah's BLM public lands are illustrated here. The BLM's 1990 wilderness proposal covered approximately 2 million acres; the Utah Wilderness Coalition's (UWC) original wilderness proposals encompassed 5.7 million acres; and the 1999 UWC proposal encompasses over 9 million acres. Other federally protected lands in Utah include 5 national parks, 7 national monuments (including the Grand Staircase–Escalante), and nearly 750,000 acres of national forest wilderness.

acres as wilderness; their bill also contained "hard release" language that would have opened all undesignated public lands for multiple-use management without any opportunity for further wilderness consideration, and it would have permitted pipelines, reservoirs, transmission corridors, and other development activities inside designated wilderness areas, while denying any federal reserved water rights. The UWC vigorously contested the bill, rallying several thousand supporters to voice their opposition at a series of hearings held around the state. Outside Utah, once the environmental community succeeded in focusing national attention on the issue, the public opinion polls and editorial commentary strongly indicated that a majority of Americans opposed the Utah delegation's wilderness bill too. With the issue joined, Senator Bill Bradley, from New Jersey, led a successful Senate filibuster against the bill, while the Utah House delegation was forced to pull its bill without a decisive vote. Nonetheless, the Utah delegation's heavy-handed political power play did not go unnoticed among Democratic officials within the executive branch.[14]

Less than six months after the Utah wilderness bill failed, President Clinton entered the fray, using his executive power under the Antiquities Act to establish the 1.7 million–acre Grand Staircase–Escalante National Monument in southern Utah. The president's monument designation added a new layer of protection to more than 880,000 acres of potentially vulnerable WSA lands. Predictably, the president's action provoked angry responses from the state's Republican political leaders as well as its rural communities, where both the president and his secretary of the interior were hung in effigy on the day of the announcement. Opponents of the new monument objected that the president had overstepped his authority, by neither consulting with state and local officials nor limiting the size of the new monument to "the smallest area compatible" with its protective purposes. They immediately filed suit over the monument designation, convened a congressional investigation ostensibly to uncover the president's true political agenda, and introduced a spate of congressional bills to rewrite the Antiquities Act. Their strategy was evident: to challenge the president's authority and actions in any forum other than the executive branch, which was, of course, firmly under Democratic control. Moreover, during a subsequent congressional hearing, Utah Representative Hansen challenged Secretary of the Interior Babbitt to substantiate his assertion that the initial BLM inventory was flawed: "All I am saying is to re-inventory it. . . . I keep hearing these comments about all this additional acreage [that qualifies for WSA status] but I have yet to see the criteria; I have yet to see the first acre of ground . . . why don't we just come up with some work and find out where it is?"[15] It was a challenge he would immediately regret.[16]

Within days, the secretary announced that the BLM would reinventory the state's public lands to determine exactly how much acreage above the original 3.2 million acres of WSAs were eligible for wilderness status. Although the state of Utah obtained a court-ordered injunction from a sympathetic local federal judge blocking the reinventory, the ruling was eventually overturned on appeal because the lawsuit was premature. When the BLM finally released its reinventory in 1999, the new study showed that, indeed, more than 5.8 million acres qualified for wilderness study status. Faced with this new reality, the BLM began reassessing its management obligations, specifically whether this newly discovered potential wilderness acreage should be granted WSA status and thus protected from further development. Meanwhile, in the aftermath of the 1995 legislative debacle, the UWC launched another citizen-conducted inventory to reassess which BLM public lands still qualified for wilderness designation and to identify previously over-looked wilderness lands. By now, the UWC and other wilderness supporters were also intent on broadening the wilderness debate beyond the traditional criteria of solitude and recreational potential to include the value of wilderness as an important repository for the state's biodiversity resources and unique ecological systems. Ultimately, the UWC concluded that more than 9.1 million acres of BLM public lands qualified as wilderness, with much of the new acreage located in the state's remote west desert region. As the stakes kept going up, the possibilities for political compromise kept diminishing.[17]

But even if both sides were intransigent on the overall acreage question, might they nevertheless compromise over a smaller piece of the Utah wilderness puzzle? Was it possible to negotiate a local wilderness bill, something less than a statewide resolution? At least two of the main players thought so. Despite the re-criminations surrounding the Grand Staircase–Escalante monument designation, Utah's Governor Leavitt and Secretary of the Interior Babbitt managed to negoti-ate an exchange of the monument's state school trust sections for federal lands elsewhere. The exchange rid the monument of potentially troublesome inholdings and enabled the state to "block up" its school trust lands into larger, more man-ageable parcels with mineral development potential and thus realize some real rev-enue for the state's chronically underfunded public schools. With this problem behind them, Leavitt and Babbitt sought to build upon their success by jointly endorsing a piecemeal strategy for addressing the bitter BLM wilderness contro-versy. They began by proposing a West Desert wilderness bill, believing that it might be possible to reach a compromise between the large-acre wilderness advo-cates and their small-acre rural antagonists in the remote western reaches of the state, where passions had not historically run as high as in southern Utah. The idea was to resolve the wilderness issue there and then bring that momentum to bear on the issue in another quadrant of the state.[18]

But they were too late and perhaps too optimistic about bridging the gap be-tween the opposing wilderness forces. The UWC, having recently identified over 2.5 million acres of potential wilderness acreage in the West Desert area during its second citizen survey, was not about to compromise away these new lands. The state's rural counties, still smarting over the new monument and suspicious of the recent school trust land exchange, were likewise unprepared to accept any sizable new wilderness areas anywhere. Besides, the environmental community was wor-ried that it could lose control over the wilderness issue if it agreed to local negotia-

tions. Emery County in eastern Utah had already come forward with its own locally conceived legislative proposal for resolving the wilderness impasse over the
scenic San Rafael Swell. But the county was not interested in establishing any new
wilderness areas in the swell; rather, it wanted to create separate legacy and conservation districts that would eliminate further mineral development but still permit
motorized access throughout the region. The state's environmental community,
noting that it had not been included in the county's deliberations, was vehemently
opposed to anything less than real wilderness. In fact, environmental leaders began to fear that the entire wilderness issue could devolve into a series of locally negotiated compromises that would discount their national constituency and undermine their national legislative strategy. They also feared the more significant
precedent that such locally negotiated legislation might establish in the BLM wilderness context, especially when so many other western states still did not have
BLM wilderness bills. As a result, they opposed both the West Desert proposal and
the San Rafael Swell legislation, and succeeded in killing both bills, though the
state's delegation managed to salvage the West Desert school trust land exchange.[19]

Final resolution of the Utah BLM wilderness controversy is not on the immediate horizon. Given the national prominence now attached to the issue within
the environmental community, any further effort to devolve the issue into a series
of local compromises is sure to meet sustained resistance, particularly since both
the UWC and BLM reinventories have dramatically expanded the acreage debate.
After the 2002 elections, with the Republicans in control of Congress and the executive branch, there is little reason to believe that the state's political establishment or its rural counties will soon embrace any big-wilderness proposal. In fact,
the 2002 elections saw the Emery County electorate defeat a locally conceived San
Rafael Swell National Monument proposal that had been endorsed tentatively by
both Utah's governor and the Bush administration. The wilderness designation
stalemate is therefore likely to continue indefinitely until one side or another
achieves sufficient political leverage to force home its proposal. Until then, the
skirmishing will continue, in Congress, the executive branch, and the courts, with
each side seeking an elusive decisive advantage.

Rethinking the Preservation Paradigm

Historically, nature conservation in the United States has been about
boundaries. We have created various well-defined enclaves—national parks, national wildlife refuges, wilderness areas, and the like—to preserve our natural heritage. Within these enclaves, the management goal has been to protect nature in its
various manifestations, including unique natural features, scenic vistas, resident
wildlife, native vegetation, and other such objects. Outside the enclaves, quite
different ownership expectations and management objectives have prevailed, often in direct conflict with those governing the protected areas. As a result, even our
largest protected areas have suffered, experiencing lost species, disrupted ecologi-

cal processes, and habitat fragmentation. Two facts have become painfully evident. First, our synthetic enclave boundaries are too small to ensure species preservation over the long term. And second, these same boundaries are just too permeable, placing the very objects we sought to protect at risk. If meaningful and enduring ecological preservation is the goal, a new and more expansive strategy of nature protection is required.

THE ECOLOGY OF PRESERVATION

A quick review of the traditional American approach to preservation on public lands reveals its shortcomings. The national park system, which traces its lineage to the creation of Yellowstone in 1872, mainly consists of rectilinear preserves set aside for their aesthetic qualities and scenic splendor with little regard for biological realities. Even the largest national parks have proven too small to protect the many species that call these enclaves home; one oft-cited study catalogues major species loss at all the major western national parks. Besides, during much of their history, the national parks completely disregarded ecological realities and eliminated predators and other undesirable species to improve the visitor experience. National wildlife refuges suffer from similar problems: they tend to be relatively small and scattered enclaves set aside to protect specific species without consideration of the larger ecological setting. Like the national parks, they too were historically managed for preferred species, usually those that would maximize hunting opportunities. The national wilderness preservation system, which had its congressional genesis in 1964, is composed largely of national forestlands, typically high-altitude and mountainous terrain. Although sometimes quite large, most wilderness areas were selected for their scenic or recreational rather than ecological values; they are rarely biologically rich or diverse landscapes. Various wild and scenic river designations have further diversified the preservation system, but these protected river corridors are only thin ribbons in the larger landscapes they drain. To be sure, in settings like the Yellowstone region, an assortment of protected lands might be cobbled together to enhance the ecological value of the individual designations, but even then the area's wide-ranging species still remain at risk. In fact, even when we have used ecosystem criteria to design national parks, we have sometimes subsequently discovered that the original boundaries were inadequate to protect the reserve. The expansive Everglades National Park is a case in point, having sustained enormous ecological damage from upstream water diversions far beyond its outermost boundaries.[20]

The scientific community, alarmed by worsening extinction problems and the obvious shortcomings in our protected lands system, has begun reexamining our nature preservation strategies. To set the agenda, scientists have identified interrelated conservation goals that should inform any biodiversity protection strategy: represent all native ecosystem types and seral stages across their natural range of variation; maintain viable populations of all native species in natural patterns of abundance and distribution; maintain ecological and evolutionary processes, such as disturbance regimes, hydrological processes, nutrient cycles, and biotic interactions; and manage landscapes and communities to respond to environmental change and to maintain the biota's evolutionary potential. Given these conservation goals, one study succinctly captures the prevailing scientific view: "Nature reserves should be as large as possible, and there should be many of them." Why the focus on largeness and multiplicity? Because an effective biodiversity conservation strategy requires broad-scale ecosystem protection over evolutionary time frames.[21]

The basic concepts governing nature reserve design are now widely accepted within the scientific community and can be used to redesign existing preservation strategies. During the spotted owl controversy, the Interagency Scientific Committee identified key principles underlying an effective nature reserve system, which were eventually incorporated into the Northwest Forest Plan. Put succinctly, the committee believed that species can be sustained only over the long term if they are well distributed on large, contiguous, and interconnected blocks of habitat across their native ranges. The basic idea is to ensure secure habitat and avoid fragmentation in order to facilitate interchange and dispersal. To that end, most biologists now agree, given the fragmented nature of modern landscapes, roadless blocks of habitat inaccessible to humans provide more secure habitat than do accessible and roaded areas. And noting that some species occur only in isolated or distinctive locations, most biologists concur that these biodiversity hot spots merit special protection, preferably through complementary "coarse" and "fine filter" conservation strategies that provide an extinction safety net.[22]

What do these nature reserve design principles mean on the ground? Put simply, they envision an expansive regional network of interconnected nature reserves configured to accomplish landscape-scale conservation objectives. The basic model encompasses fully protected core areas surrounded by buffer zones and connected by corridors to other similarly inviolate reserves. Ideally, core areas should contain representative ecosystem types, biodiversity hot spots, and adequate habitat to sustain self-perpetuating populations of the native species, some of which would serve as focal or umbrella species for reserve design purposes. The core areas would be strictly regulated and give priority to native species protection. Some scientists advocate active restoration strategies in these cores, while others endorse a "hands off" management approach. The buffer zones would accommodate increasingly more intense human uses away from the core; they would be adaptively managed to meet species preservation goals. The corridors would permit species migration to seasonal ranges outside the core area as well as dispersal to facilitate genetic interchange with other population segments. The corridors should be large enough to promote species movement and ecosystem integrity; the intensity of management might vary depending on the corridor's purpose and location. If so configured, a regional reserve network could not only accommodate the needs of large, wide-ranging species, such as elk, bears, wolves, and cougars, but it could also absorb natural disturbance regimes, such as fires, floods, disease, and avalanches, that are integral parts of the ecosystem processes that sustain the native biodiversity.[23]

The size or acreage implications of such an expansive reserve system can be quite significant, depending upon the ecosystem and species being protected. For large carnivores, like the grizzly bear or wolf, or for habitat specialists, like the northern spotted owl, the potentially affected acreage can be very large. Many biologists believe that 50,000 square miles of habitat are required just to sustain a viable population of 2,000 grizzly bears; the Northwest Forest Plan placed over 18 million acres in protected reserves for the spotted owl. Top predators are key ecological components that exert an important cascading influence on the surrounding ecosystem: at the pinnacle of the food chain, they effectively regulate abundance among other species further down the chain, which in turn can have major ecological repercussions on the surrounding landscape. A classic example of this cascading effect is seen in the relationship between wolves, deer, and vegetative browse. Wide-ranging, top-of-the-chain species can also serve as an umbrella for other species within the ecosystem; if they are secure then others further down the food chain are also secure, thus minimizing extinctions and promoting ecological integrity. Conservation biologists therefore suggest that from 25–75 percent of regional landscapes should be protected in core areas and regulated buffer zone designations. Dr. Reed Noss, for example, believes that "at least half of the land area of the forty eight coterminous states should be encompassed in core reserves and inner corridor zones (essentially extensions of core reserves) within the next few decades," which he asserts can "be done without great economic hardship." By comparison, roughly 10 percent of the nation's land base is currently protected under national park, wilderness, and similar designations, with much of that acreage being located in Alaska.[24]

Plainly, public lands must play a key role in designing any such expansive nature reserve system. Public lands cover large landscapes with a diverse topography, representing an array of ecosystems. They are relatively less developed than most private or state lands; they are not used for residential or agricultural purposes, though logging, mining, grazing, and other activities occur on the multiple-use lands. And they already serve as the last refuge for several dwindling species, including grizzly bears, wolves, and other predators. Indeed, most scientists view our preserved public lands—the national parks, national wildlife refuges, and wilderness areas—as the presumptive cores for any nature reserve system. They also view the adjacent multiple-use lands as potential buffer areas, where extractive development activities would be regulated in a concentric manner to protect core regions, with more intensive activities permitted further from the core. And they envision the core preserved areas linked together with designated corridors to enable species safe passage across an already fragmented landscape. To meet these objectives, some locations will require active restoration programs to repair over logged and otherwise tattered lands; in other locations, adjacent private lands may have to be enlisted in the conservation effort, which raises an entirely different set of political, legal, and moral issues. But public lands and existing reserves are obvious starting points for revisiting our preservation strategies.[25]

TOWARD A NEW AGENDA: VISIONS FOR THE FUTURE

What strategies, therefore, might be employed to augment the existing nature reserve system? At the conventional level, the traditional nature conservation tools include expanding the number of designated nature reserves, enlarging existing boundaries, and acquiring new lands. The usual strategy is to increase the number of preserved areas by creating new national parks, wildlife refuges, and wilderness areas, with a view toward diversifying the ecosystems that are legally protected. Several studies have identified underprotected ecosystems, including the Great Plains grasslands, the Southwestern and Great Basin desert landscapes, and the California coastal region. Another strategy is to extend the size of existing protected areas, and thus enhance their effectiveness as nature reserves. In recent years, however, Congress has resisted most expansive preservation proposals, as reflected in its steadfast opposition to the large-scale wilderness proposals for Montana's national forests and Utah's BLM lands. But Congress is not the only institution with preservation powers. As we shall see, the Clinton administration aggressively used the Antiquities Act to enlarge the federal protected land base, which then prompted Congress to create yet other landscape-scale conservation areas. A related strategy is for the government either to purchase ecologically valuable private land or to engage in public–private land exchanges to strengthen federal biodiversity conservation efforts. Both Congress and the public land agencies have consummated ecologically strategic land exchanges, such as the Gallatin National Forest exchange in southwestern Montana to enhance wildlife habitat protection and the Utah school trust land exchange to eliminate federal inholding problems. But these strategies are only piecemeal approaches to the larger biodiversity conservation problem; they are very site-specific, generally require congressional approval, and lack the overall coherence necessary to create a true system of nature reserves.[26]

At another level, the ecosystem management concept represents an alternative means to strengthen the federal commitment to preservation on public lands. To mesh an ecosystem management strategy with broader landscape preservation objectives, biodiversity conservation proponents advocate the "greater ecosystem" concept, which is designed to conjoin diverse federal lands with existing protected areas to promote coordinated regional management across the landscape. One obvious example is the Greater Yellowstone Ecosystem; it has effectively linked the region's national parks, national wildlife refuges, wilderness areas, and multiple-use forestlands around its wide-ranging, charismatic wildlife populations—grizzly bears, wolves, elk, and bison—in an effort to better protect these species and their habitat needs. Other examples include the Greater Glacier–Bob Marshall

Ecosystem complex, the Greater Grand Canyon Ecosystem, and the North Cascades Ecosystem, each of which links an important national park with surrounding protected and multiple-use public lands. Under ecosystem management policy, as noted earlier, these regions would be managed as integrated entities, giving priority to biodiversity conservation goals, promoting interagency coordination at a regional scale, employing adaptive management techniques, and utilizing scientific data to develop conservation strategies. And that, of course, is precisely what the opponents of ecosystem management fear—a de facto wilderness preservation policy that "locks up" public lands, eliminating resource development and other economic opportunities. But that is not really ecosystem management; it is better understood as an effort to zone—not eliminate—commercial activity on the public lands in order to protect critical habitat and ecological processes.[27]

At the visionary level, the Northern Rockies Ecosystem Protection Act (NREPA) has already attracted political support. It seeks to establish a 16-million-acre interconnected network of public land wilderness areas stretching across five states, Montana, Idaho, Wyoming, Washington, and Oregon. The NREPA proponents, noting that the region still harbors a full complement of native biodiversity as well as large, nationally recognized wilderness areas, fear these key resources are imperiled by the Forest Service's commitment to commercial logging, road construction, and other development activities. First introduced as a congressional bill in 1993, the NREPA would create five new ecosystem-level wilderness areas (including the Greater Yellowstone Ecosystem and the Greater Glacier Ecosystem) and designate biological connecting corridors, some of which would be managed as wilderness. It would also establish a National Wildland Recovery Corps to restore the region's damaged landscapes and provide new local employment opportunities. And it would empower a new interagency public–private management team to oversee the region, while requiring the agencies to follow adaptive management protocols. In short, the act would create an interconnected regional wilderness network and introduce ecosystem management across these public lands, similar to what the Northwest Forest Plan has done in that region. Although the bill's proponents argue that it will enhance rather than harm the area economy, neither the public land agencies nor the region's politicians have rallied behind the legislation. Nonetheless, the NREPA bill offers a concrete legislative model for how new preservation concepts might be used to reshape the regional landscape.[28]

The Yellowstone to Yukon Conservation Initiative (Y2Y) envisions a linked network of protected public lands extending the length of the Rocky Mountain chain from northern Canada to the Yellowstone country in the United States. According to conservationists, the region is badly degraded and fragmented, the result of transportation corridors, logging and other industrial development, community expansion, and local population growth. Political and legal fragmentation also prevails: two countries along with various state, provincial, and local governments have jurisdiction over the area. As a result, the region's large carnivores—its grizzly bears, wolves, cougars, wolverines, lynx, and the like—are at risk of increased isolation and thus local extinctions. An array of ten large national and provincial parks and other protected areas, such as Yellowstone, Glacier, Waterton, Banff, Jasper, and the undeveloped Yukon wilderness landscape, dot the region, providing critical but inadequate core sanctuaries for the region's wildlife populations. The conservation value of these preserved lands would be greatly enhanced if they were stitched together into a "gigantic linear ecosystem." The fact that the United States and Canada have historically cooperated over mutual conservation concerns, as manifested by the Glacier-Waterton International Peace Park and the North American Waterfowl Management Plan, offers hope that a similar cooperative approach might be developed here. To promote international-level coordination, the Y2Y initiative proponents have begun identifying critical linkage corridors between the existing core park-wilderness areas, and examining the region's shifting economic trends to secure necessary local support for its overall ecosystem protection strategy. In short, it is moving forward on both scientific and political fronts to protect what has been described as "the wild heart of North America."[29]

The Wildlands Project also takes preservation to a continental level, envisioning eco-regional networks of nature reserves designed to "protect and restore the ecological richness and native biodiversity of North America." Marrying conservation biology theory, GIS mapping technology, and grassroots environmental advocacy, the project would create an expansive wilderness recovery network consisting of interlinked core reserves and buffer areas connected across the landscape. The concept is known as "rewilding." The project's principal founders are former Earth First! activist Dave Foreman, whose passion for wilderness has long stirred hard-core environmentalists, and retired biology professor Dr. Michael Soule, a founder of the Society for Conservation Biology. Convinced that the conservation movement must "think much bigger and on a longer time scale," they envision cadres of wilderness advocates first using conservation biology principles to identify and map the critical habitat needed to maintain and restore large predators and other native species, and then employing grassroots political organizing tactics to secure wilderness, national park, and similar protective designations for entire regions. Already the project has completed the Sky Islands Wilderness Conservation Network Plan that offers a science-based road map for restoring over 10 million acres of mainly public lands in southern New Mexico, Arizona, and northern Mexico. Similar ecoregion plans are slated for the Southern Rockies Ecosystem, Adirondacks, southern Appalachians, and Oregon's Klamath-Siskiyou region. But regardless of the project's scientific merit, the sheer number and size of the potentially protected landscapes portend formidable political challenges ahead.[30]

GETTING THE DEBATE STARTED

Indeed, these expansive preservation proposals have provoked substantial opposition from local political leaders and communities. The typical response to

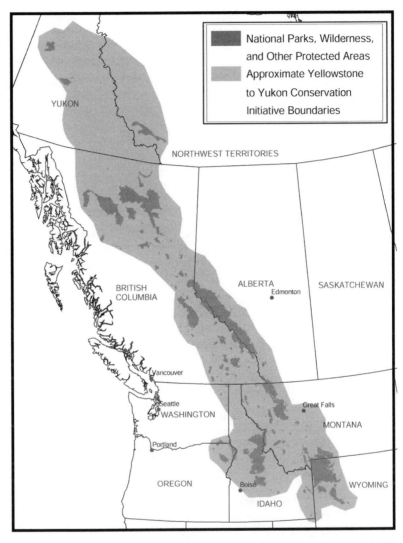

Map 7. Yellowstone to Yukon Conservation Initiative. The Yellowstone to Yukon (Y2Y) initiative envisions a continent-long network of protected areas that are linked together to protect wide-ranging species along the spine of the Rocky Mountains, namely the grizzly bear, wolf, elk, cougar, lynx, and wolverine. Y2Y proponents have labeled the area "the wild heart of North America."

any new wilderness-like proposal is: "How much is enough?" Preservation opponents note that more than 40 percent of the nation's public lands are already either "locked up" in protective legislative designations or otherwise off-limits to resource development activities through administrative withdrawals, road closures, or other restrictions. Lamenting the loss of the resource extraction industries as well as local employment opportunities, they explain that these industries usually support relatively high-paying jobs in rural areas and that alternative tourism-based employment opportunities are often low-paying, seasonal jobs. Recreational groups and their industry counterparts are frequently divided over new preservation proposals. Advocates of motorized recreation—off-road vehicles, snowmobiles, jet skis, and the like—generally resist wilderness-type protective designations, fearing that this means the end of any motorized access. But back-packers, hikers, climbers, and other recreational users tend to support new protective designations, perceiving they will still enjoy access for their own pursuits without the intrusion of motors. The opposition tends to be strongest in communities located near proposed preserved lands, while support for additional protective designations is usually urban-based and strongest on both the east and west coasts.[31]

The fundamental question, therefore, is whether these new large-scale preservation proposals represent politically viable options. Is it possible to address the needs of communities dependent on public lands and motorized recreation groups while preserving an ecologically intact landscape? According to biodiversity conservation proponents, the answer is yes, and they are prepared to make strategic concessions from strict wilderness protection in order to achieve paramount species protection goals. Briefly stated, the principal ideas for altering preservation policy include: encourage the growth of local recreational and tourism industries that are perceived as more benign than the traditional extractive industries to offset job losses and expand the preservation constituency; promote an active ecological restoration agenda (perhaps even in wilderness areas) to not only repair the landscape but also provide local employment opportunities and address local fire management concerns; create new protective designations, such as multiple-use national monuments and diversity maintenance areas, that allow some economic uses to continue inside the reserve; permit alternative economic activities, such as bioprospecting, to occur inside reserves and thus generate revenues for scientific research and other management purposes; and sanction direct local involvement in managing protected areas. Put simply, these new ideas would alter the strict standards governing wilderness and other preserved lands in order to accommodate economic uses and more local control as the trade-off for expanding the system. Whether these ideas will be acceptable to traditional wilderness opponents is an open question, though the ideas certainly address key objections to expansion of the preserved land system. They have set off alarm bells in other circles, however.[32]

Among traditional wilderness advocates, the notion of opening protected public lands to more economic uses or local management is anathema. It has ignited a "great new wilderness debate." Defenders of the current system fear the precedent that would be set if the strict legal protections governing wilderness and other protected areas were diminished, and they fear losing the political momentum required to complete the nation's still-incomplete national park, refuge, and wilderness systems. Having regularly battled obstructionist local communities to create new wilderness areas and other protected designations, they believe that relinquishing control of these lands to such communities would undermine the very preservation purposes for which they were created. They also are convinced that an active restoration program could compromise the essential naturalness of wilderness, encouraging the public land agencies to begin remaking nature in their own image. And they see any proposal aimed toward promoting local recreation and tourism industries as placing the surrounding landscape at risk from further development and an increased human imprint on the land. Although the debate has thus far been joined mostly in academic circles, certain policy changes—the new multiple-use national monuments and bioprospecting in national parks, for example—have already occurred, which suggests the issue has real world consequences. And in that world, the debate is fundamentally political, yet deeply rooted in the federal government's long-standing commitment to preservation on public lands.[33]

The Politics of Preservation

From the beginning, nature preservation has been as much about politics as anything else. The nation's first national park—Yellowstone—was a deliberate political creation, the result of Congress taking specific legislative action to protect this unique region from private exploitation. The same holds true for all our principal nature reserves—national parks, national wildlife refuges, and wilderness areas—all of which have been accorded legal protection through myriad national political judgments. Over the past century, Congress has been the major player in this process, periodically employing its legislative powers to conserve remnants of the public landscape in a natural state. But the president, the public land agencies, and courts have also occupied key roles in the various political dramas surrounding the creation and management of our nature reserves. Throughout this process, the rationale underlying our nature conservation efforts has gradually evolved, shifting from aesthetic concerns to primitive recreation and then to ecological preservation. As a result, both the origins and the justifications for our preserved public lands are as diverse as the landscapes themselves. Yet there remains one enduring reality: nature preservation is inherently a political matter.[34]

PARKS AND REFUGES: QUINTESSENTIAL RESERVES

National parks and national wildlife refuges are the quintessential nature reserves. In the case of the national parks, Congress has been primarily responsible for the establishment and evolution of the national park system. Beginning with Yellowstone in 1872, Congress has created our parks, passing individual enabling statutes to set up each new one. In 1916, Congress established the national park system as well as the National Park Service to administer it under a new preservationist charter. The Park Service soon became a champion of new national parks, but Congress very deliberately retained the authority to designate new parks, which it still jealously guards. From these modest beginnings, the national park system has evolved into a truly diverse array of more than 380 units that cover nearly 80 million acres, including national recreation areas, national historical parks, national seashores, national rivers, and other such designations. Moreover, Congress has periodically expanded the boundaries of existing parks to better protect sensitive resources and to more accurately reflect ecological realities. In some instances, nearby communities have promoted new park designations for economic and other purposes; in other cases, a national preservationist constituency has forced the issue on a reluctant local populace. But even then, local popular support can make or break a park proposal, even when nationally significant features or resources are at stake.[35]

The national park system, however, does not owe its existence solely to Congress. Individual presidents have also shaped the system, primarily through their power under the Antiquities Act to designate new national monuments. Several prominent national parks, including Death Valley, Grand Canyon, Olympic, and Zion, were first protected as national monuments and then eventually transformed by Congress into national parks as local opposition receded. The president, according to the courts, has broad discretion in making national monument designations and in determining their size. Beginning with the Grand Canyon National Monument in 1908, every president—except Nixon, Reagan, and George H. W. Bush— has routinely used the Antiquities Act to safeguard sensitive scientific and historic resources, occasionally even setting aside large areas for protection. In 1978, for example, President Carter used the Antiquities Act to protect over 50 million acres in Alaska by designating 17 new national monuments. During the late 1990s, President Clinton protected large swathes of public land with multiple national monument designations, several of which expanded existing park boundaries. But the unilateral nature of the presidential Antiquities Act power remains quite controversial in many quarters, despite the fact that history has consistently vindicated individual monument decisions. Which probably explains why Congress has not changed it, thus leaving the president with an important political tool that he can use to pursue a national preservationist agenda.[36]

The political origins of our national wildlife refuge system are even more diverse than the national park system. The nation's first national wildlife refuge was created in 1903 by fiat when President Teddy Roosevelt unilaterally used his executive withdrawal powers to designate Pelican Island as a federal bird reservation. Shortly thereafter, Congress endorsed these nascent federal wildlife preservation efforts when it authorized a game reserve in Oklahoma's Wichita National Forest Reserve in 1905 and then purchased the National Bison Range in Montana one year later. Since then, new refuges have been established through an array of devices: by presidential directive, by congressional legislation, and by congressionally approved agency purchases from private landowners. Most of the refuge acreage (approximately 97 percent) has been carved from existing federal lands; the remainder represents acquired private property. Despite periodic congressional efforts to give some coherence to the refuge system, it was 1997 before Congress finally adopted organic legislation that established uniform management and planning policies. Under this legislation, biodiversity conservation has now become a specific refuge purpose, effectively validating the U.S. Fish and Wildlife Service's earlier decision to employ an ecosystem-based strategy to manage its refuges.[37]

The national park and wildlife refuge systems, while representing an impressive political commitment to nature preservation, are not adequate to meet the nation's biodiversity conservation needs. Most of the early parks were set aside for their scenic grandeur and aesthetic features without regard for ecological considerations or representativeness. Over the past century, species have been lost from even the largest national parks. Although the more recently created parks—Everglades, Alaska, and the California Desert, for example—were designed with boundaries roughly mirroring regional ecosystems, they have still proven too small to meet biodiversity protection needs. And several ecosystem types are still not represented in the national park system. The national wildlife refuges, given their diverse origins and scattered locations, also lack much ecological coherence and do not encompass a full array of ecosystem types. In the case of both systems, Congress has been forced to acknowledge that perimeter developments on adjacent lands—often referred to as external threats—can imperil their resources and undermine their ecological integrity. But sensitive to local prerogatives, Congress has been unwilling to give the responsible agencies any explicit authority to control threatening external activities. As the creator of these renowned preservation systems, Congress serves as the final arbiter of any ecological or other controversies.[38]

As important as the national park and national wildlife refuge systems are in the nation's biodiversity conservation efforts, they are not likely to experience significant new growth in the near future. Although Congress has periodically created new parks, monuments, and refuges, it has been increasingly reluctant to set aside large areas for biological or other preservation purposes. During the mid-1990s, given the opportunity to establish a large national park at the Baca Ranch in central New Mexico, Congress balked at dedicating the ranch exclusively to nature preservation and instead created a new multiple-use trust arrangement. And Congress has been reluctant to enlarge existing parks or refuge boundaries, absent a strong local or statewide consensus, as was the case with the California Desert lands. Interagency relations have also played a key role in the preservation political debates. Neither the Forest Service nor the BLM has been keen to relinquish their valuable landholdings to a rival agency. The fact that both agencies now have important wilderness and national monument lands in their portfolios has stiffened their resistance to new park or refuge proposals and has essentially validated their ability to manage ecologically sensitive lands. Given these political realities and interagency dynamics, biodiversity protection proponents candidly acknowledge that the national parks and refuges represent only one element in securing ecosystem-level protection on public lands. Another key element is expanding the wilderness system, which introduces other contested political considerations into the nature preservation debate.[39]

WILDERNESS: WHERE CONGRESS REIGNS

Wilderness preservation has proven to be a high-stakes political drama played out in all our governmental institutions. The reason is perhaps obvious: wilderness represents the most restrictive designation that can be placed on public lands. By law, it is "an area where the earth and its community of life are untrammeled by man . . . an area of Federal land retaining its primeval character and influence . . . managed so as to preserve its natural conditions."[40] Development activities, such as logging and mining, are forbidden in wilderness areas, as are roads, lodges, and motors. Faced with these restrictions, the extractive industries, in concert with natural resource–dependent communities, have regularly opposed new wilderness proposals. Their opposition has focused on Congress, which has retained the power to designate new wilderness. As this political drama has unfolded, Congress has converted the national wilderness debate into a predominantly local political question, with each state's congressional delegation holding vetolike power over unacceptable wilderness proposals. But localism has not always triumphed. The executive branch and the courts have also employed their respective powers to preserve future wilderness options, often to the consternation of wilderness opponents. The interplay among these governmental institutions over the wilderness question is a fascinating study in grassroots politics, executive prerogative, and the rhetorical power of the wilderness concept.

Although the Wilderness Act dates from only 1964, the origins of wilderness as a land protection concept are much older, harking back to the Forest Service's early years. In 1924, two farsighted Forest Service employees, Aldo Leopold and Arthur Carhart, convinced the agency to designate a "primitive area" in the Gila National Forest and to adopt regulations that limited uses in designated roadless areas. Within five years, the Forest Service promulgated much stricter regulations to better protect these primitive areas against development, partly to avoid losing more of its scenic landscapes to the rival National Park Service. But these administrative initiatives were not enough. By the mid-1950s, conservationists were actively promoting wilderness legislation to ensure permanent legal protection for undeveloped public lands. A hard-fought battle over damming the Green River at Dinosaur National Monument, along with the Forest Service's penchant for administratively altering primitive area boundaries to accommodate timber sales, convinced wilderness proponents that only a national wilderness system protected by law is truly secure. Thus began a decade-long campaign to persuade Congress to create another major preservation system on public lands.[41]

In 1964, Congress finally passed the Wilderness Act and gave its blessing to what has become an expansive national wilderness preservation system designed to leave the landscape "untrammeled" by humanity. Besides creating 9.1 million acres of "instant" wilderness from preexisting national forest primitive areas, the Wilderness Act directed the Forest Service, National Park Service, and U.S. Fish and Wildlife Service to each prepare a presidential report identifying additional lands for wilderness protection. The president was then charged with making wilderness recommendations to Congress, which retained the final designation power. Early drafts of the wilderness bill contemplated that the president would be the final decision maker, but Congress quite intentionally reserved the power to designate new wilderness areas. This decision shifted wilderness designation authority from an administrative to a congressional forum, transforming it into an overtly political process. With Congress having the final word, wilderness proponents quickly realized that grassroots political activism was the key to expanding the fledgling wilderness preservation system.[42]

Opposition to the wilderness bill came primarily from western congressional representatives who feared the legislation would limit or deter local economic activities. They succeeded in extracting numerous concessions, including several grandfather provisions that allowed preexisting mining and grazing activities to continue within designated wilderness areas. Mining proponents even negotiated a 20-year window during which new projects were allowed in designated wilderness areas; ranchers eventually secured a permanent limitation on any grazing reductions. Ironically, the Forest Service opposed the original wilderness legislation, fearing that it would lose control over its own lands. And the National Park Service also initially opposed the legislation, reluctant to have the rival Forest Service intruding into its preservation domain. To forestall further interagency jealousies, Congress decided that wilderness designations would simply overlay existing land management boundaries, leaving each agency responsible for wilderness management on its own lands.[43]

Once the Wilderness Act was law, controversy erupted over wilderness designation on other national forestlands. Expressly charged with making further wilderness recommendations, the Forest Service made few, much to the consternation of wilderness proponents. Committed to its own "purist" interpretation of wilderness, the Forest Service questioned whether any of its roadless lands were pristine enough for inclusion in the system. And the agency further angered wilderness proponents when it scheduled timber harvests in roadless areas potentially eligible for wilderness designation. In the case of the Gore Range Primitive Area in Colorado, only the intervention of a federal court stopped proposed timber sales and thus saved the area for eventual wilderness designation. In 1972, seeking to recapture a role in the wilderness debate, the Forest Service launched the first of two Roadless Area Review Evaluations designed to settle definitively which national forestlands were suitable for wilderness designation. To its credit, the Forest Service brought ecological considerations into the wilderness debate, even at this early date. The RARE I criteria for determining whether an area merited wilderness status required the agency to include representative ecosystems in the national wilderness system. But when the Forest Service's RARE I survey produced only paltry wilderness acreage recommendations, environmental groups again turned to the courts and blocked the agency from finalizing its recommendations. Even during these early years, the courts had assumed a key role in preserving the wilderness option for eventual congressional consideration.[44]

Impatient with the Forest Service's lack of progress, Congress seized the initiative. In 1975, Congress passed the Eastern Wilderness Areas Act and created over 200,000 acres of wilderness in 13 states, ignoring the Forest Service's purist view that these previously disturbed eastern forestlands did not qualify as wilderness. But before the House Public Lands Subcommittee would approve any of the proposed wilderness areas, each congress member with a proposed area in his or her district was required to sign off on it—an early indication that local political support is vital in any wilderness protection campaign. Three years later, Congress adopted the Endangered American Wilderness Act of 1978, which not only extended the liberal congressional wilderness definition to the western states but also added another 1.3 million acres to the system. And during its 1980 session, Congress used some of the Forest Service's RARE recommendations to adopt statewide wilderness bills for Colorado, New Mexico, and three other states. Having finally embraced the wilderness concept, Congress seemed intent on fashioning a truly national wilderness preservation system, regardless of the Forest Service's reluctance toward the concept it had pioneered. But Congress was also giving state congressional delegations an increasingly larger role in wilderness designation determinations.[45]

The national wilderness lobby clashed outwardly with local interests over the wilderness issue in Alaska. In 1980, after more than two decades of acrimonious controversy, Congress finally passed the Alaska National Interest Lands Conservation Act (ANILCA) and dramatically expanded the nation's conservation systems. On its face, the ANILCA legislation added 56 million acres to the national wilderness preservation system, and established 13 new national parks, 16 new wildlife refuges, and 13 new national wild and scenic rivers. The expansive designations were drawn to protect entire ecosystems; the statute's proclaimed purpose was "to preserve in their natural state extensive unaltered arctic tundra, boreal forests, and coastal rainforest ecosystems" and "to maintain opportunity for scientific research and undisturbed ecosystems." Although proponents of the act lamented its numerous boundary compromises, it still tripled the acreage devoted to nature preservation on the nation's public lands. Acknowledging that many Alaskans rely on the land for their living, however, the legislation did allow subsistence hunting and fishing practices to continue even within the new national parks. Significantly, Congress passed ANILCA over the state of Alaska's strenuous objections, one of the rare instances where a state's congressional delegation has not succeeded in preventing passage of a locally objectionable preservation initiative. By casting the Alaska wilderness campaign in "national interest" terms, supporters of the act effectively cobbled together a vocal national constituency behind their cause, which enabled them to override the Alaska delegation's opposition. But even then, significant concessions were made to local concerns, notably boundary adjustments and subsistence provisions.[46]

Problems loomed, however, for the fledgling wilderness system. In 1980, Ronald Reagan assumed the presidency, bringing with him a cadre of officials like Secretary of the Interior James Watt who were dedicated to promoting intensive development activity on public lands. With two Arab oil embargos fresh on the nation's mind, Secretary Watt sought to open existing wilderness areas, including the expansive Bob Marshall Wilderness Area in Montana, to oil and gas leasing. Watt's proposal generated an avalanche of criticism, as well as an eventual congressional rebuke, when Congress placed a moratorium on wilderness oil and gas leasing. But faced with a recalcitrant administration and deepening local opposition to wilderness, Congress was itself stalemated over wilderness designation proposals, unable to complete the process envisioned in the original wilderness bill. Meanwhile, under the 1976 National Forest Management Act, the Forest Service was poised to begin a comprehensive new planning process, which would include further wilderness recommendations. So in another effort to resolve the wilderness issue comprehensively, the agency undertook a second roadless area inventory, dubbed RARE II. Once again, however, the agency's wilderness acreage recommendations proved disappointingly small, prompting the state of California to challenge its underlying environmental analysis. And once again, a federal court invoked the National Environmental Policy Act (NEPA) and enjoined the agency from finalizing its wilderness recommendations. Another stalemate loomed, leaving a cloud of uncertainty hanging over the roadless lands of the national forests.[47]

The logjam finally broke in 1984, when Congress passed 20 separate state wilderness bills, creating 6.8 million acres of wilderness in the western states, though not in Montana and Idaho. Although the timber industry initially favored a comprehensive national wilderness bill, Congress acceded to the wishes of other constituencies—the states, grassroots environmental activists, and local communities—by addressing the issue on a state-by-state basis. No one got all of what they sought, but over 4 million acres were set aside as wilderness in California, Oregon, Utah, Washington, and Wyoming. The roadless Forest Service lands not included in this acreage were "released" for multiple-use management, though technically they remained eligible for wilderness designation during the next round of forest planning. But an important precedent was set: Congress was no longer making wilderness designation decisions at a national or regional level; rather, these decisions had devolved and were being handled through a state-by-state negotiation process. To be sure, the timber industry, environmental organizations, and other groups were still active participants in the wilderness debates, but individual state congressional delegations had claimed a paramount role in negotiating future wilderness decisions involving local public lands. Without the support or at least acquiescence of the state congressional delegation, it would prove virtually impossible to secure new wilderness designations.[48]

Outside the national forests, the wilderness designation process languished following passage of the Wilderness Act. In the case of national parks and national wildlife refuges, environmental organizations consciously deferred pursuing wilderness legislation, deciding that these lands were already protected and that their limited resources could be better spent focusing on the more vulnerable multiple-use national forestlands. Nevertheless, Congress eventually added a wilderness overlay onto several national parks, including Mount Rainier, Olympic, Sequoia–Kings Canyon, and Yosemite, and scattered national wildlife refuges as part of larger statewide wilderness bills.[49] In the case of the BLM, the problem was quite different. The Wilderness Act made no provision for BLM lands, which meant there was no legal basis for pursuing wilderness protection on these little-known landscapes. During the original Wilderness Act debates, there was no vocal constituency for BLM wilderness, and Congress was still undecided over how to manage the unreserved public lands.[50] In 1976, though, everything changed when Congress passed the Federal Land Policy and Management Act. The FLPMA established a federal retention policy for public lands, charged the BLM with actively managing them, and included a wilderness provision that instructed the BLM to review its roadless lands for their wilderness potential. The act also provided special protection for the California Desert Conservation Area and legally protected any BLM-designated wilderness study area from impairment until Congress acted on the agency's recommendations. All the principal federal land management agencies were now charged with wilderness review and management responsibilities.[51]

By the mid-1980s, the national preservation agenda was focused on the BLM wilderness designation process. Although the FLPMA had singled out the California Desert Conservation Area for special treatment, efforts to reconfigure the BLM's southern California landholdings were stalemated primarily due to opposition from the state's Republican political leaders. In 1990, with the California issue still unresolved, the Arizona congressional delegation shepherded a consensus statewide wilderness bill through Congress, creating more than 1.1 million acres of BLM wilderness and another 1.34 million acres of new national wildlife refuge wilderness. Four years later, with the Democrats firmly in control of the California congressional delegation and also ensconced in the White House, Congress finally passed the long-delayed California Desert Protection Act. The bill significantly expanded the region's national parks and added over 3.5 million acres of BLM land to the national wilderness preservation system, with many of the units defined primarily by ecologically determined boundaries. The pattern for BLM wilderness designation was following the same political path as the national forest process: preservation decisions were being negotiated on a state-by-state basis using the FLPMA-mandated wilderness study area inventories as the baseline.[52]

For the most part, wilderness designation proposals made little headway in Congress during the 1990s. On the national forest front, Congress was notably quiet, adding isolated areas to the wilderness system only in those rare instances where there was clear local consensus. Neither the Montana nor Idaho wilderness impasses were resolved, despite valiant efforts to devise local solutions that might move the process forward. On the BLM front, other than California and Arizona, there was little evidence of any common ground over wilderness legislation. Indeed, the acrimonious Utah BLM wilderness debate demonstrated just how difficult it was to achieve any local consensus over these contested lands. Beyond Utah, few other western states had produced even a draft BLM wilderness bill. With a largely anti-wilderness Republican majority holding the balance of power in Congress, wilderness proponents had little hope of overcoming local resistance to wilderness protection, absent a strong national constituency. And having embraced new ecological arguments for large protective designations, wilderness advocates were not disposed to compromise for smaller acreage designations. Clearly, any further legal protection for vulnerable public lands would have to come from outside Congress, where recalcitrant local interests did not hold such sway. With the executive branch under the control of a receptive and environmentally friendly Clinton administration, preservationists turned their attention in that direction. Their strategy, as it turned out, was well rewarded.

PURSUING THE PRESERVATION AGENDA IN OTHER FORUMS

The primary tool of choice was the Antiquities Act. With the wilderness designation process stalled in Congress, President Clinton stepped into the void and aggressively employed his Antiquities Act powers to create an array of new multiple-use national monuments on the public lands. In fact, during his tenure, President Clinton designated 22 new national monuments in 10 different states, covering over 6 million acres—a preservation legacy that rivals Teddy Roosevelt's a century earlier. In a major break with tradition, most of the new monuments are administered by the BLM or Forest Service rather than the preservation-oriented Park Service, which has typically assumed responsibility for new national monuments. The process of creating BLM-administered national monuments began in 1996, when President Clinton decreed the Grand Staircase–Escalante National Monument in southern Utah, which at 1.7 million acres was designed in ecosystem terms to protect the region's fragile biological and other resources. Although the subsequent Clinton monument designations were not as large, the individual proclamations indicate that many of them were also designed to safeguard threatened ecological resources. Drawing upon the Grand Staircase example, the new monuments also took local concerns into account, generally allowing livestock grazing, hunting, and other nonextractive activities to continue. With these new designations, the BLM is now responsible for more than 43 million acres of protected lands, and it has acquired a new preservation-focused constituency. The bureau has responded by establishing a new National Landscape Conservation System to consolidate its national monuments, national conservation areas, and wilderness landholdings for management purposes. More such lands will undoubtedly be added to its portfolio once the congressional wilderness designation logjam is broken. In the meantime, preservation has become an important management responsibility for all the public land agencies.[53]

Beyond the White House, the public land agencies have undertaken diverse preservation initiatives. These agencies each have very real and legally enforceable biological conservation responsibilities with direct land preservation overtones. Under the Endangered Species Act, all federal agencies are responsible for conserving listed species and ensuring that their actions do not jeopardize or take any protected species. Both the Forest Service and the U.S. Fish and Wildlife Service also have explicit statutory biodiversity conservation obligations that extend across their respective lands. In tandem, these legal obligations have significantly affected public land planning and management decisions, essentially compelling the agencies to ensure adequate habitat to sustain native wildlife populations. In the case of the northern spotted owl and other old growth species, for example, the Northwest Forest Plan created an extensive network of late-successional and riparian nature reserves where timber harvesting and other development activities were prohibited. The lesson is clear: when species protection is at issue, the public land agencies have the administrative authority to establish nature reserves to meet their biological conservation legal obligations. Relatedly, under the FLPMA, the BLM has the authority to designate areas of critical environmental concern (ACECs) "to protect . . . fish and wildlife resources or other natural systems or processes." Although the BLM has been reluctant to designate large-scale ACECs for ecological purposes, it nonetheless possesses adequate authority to create a network of nature reserves to supplement its National Landscape Conservation System. It has used this authority at least once in the Las Vegas area, placing over 743,000 acres in ACEC status to protect desert tortoise habitat. In short, a diverse array of statutory provisions addressing biodiversity conservation on public lands empowers—and in some instances actually compels—the public land agencies to implement land preservation strategies. The result is a nascent but discernible patchwork of administratively created nature reserves that complement the existing system of legislatively designated reserves.[54]

In several controversial moves, the Clinton administration Forest Service invoked its administrative authority to protect remaining national forest roadless lands, creating nearly 60 million acres of de facto wilderness. Pursuant to a 1999 presidential directive, the Forest Service adopted a new roadless area regulation eliminating commercial timber harvesting and new road construction from its undeveloped lands. Noting that national forest roadless lands have been the focus of unrelenting contention for 20 years, the Roadless Area Conservation EIS highlighted the ecological importance of these undeveloped forestlands: they cover 31 percent of the national forest system acreage; they serve as "biological strongholds for terrestrial and aquatic plants and wildlife and as sources of high-quality water"; and more than half the ESA-listed species found on national forestlands depend on habitat situated in these roadless areas. Without adequate protection, the agency feared further forest fragmentation and habitat loss. The EIS also asserted that the national forests were more valuable left undisturbed: "A growing number of people value Federal lands as a repository of biodiversity and conservation. Many people appreciate national forest system lands more for their inherent naturalness than for the commodities, such as timber, minerals, and grazing, that they can provide." Because less than 7 percent of national forest timber comes from these roadless lands, the agency felt it could be replaced by other sources and that a proposed financial assistance program would adequately address rural job losses. Relatedly, the Forest Service imposed an 18-month road construction moratorium across the national forests and then adopted another rule revising its road management policies. The new policy links road construction decisions to local resource management plans; it requires a scientific analysis for any new construction and establishes an active road decommissioning program. At a smaller scale, the Forest Service withdrew nearly 500,000 acres on the Rocky Mountain Front in northern Montana from hardrock mining as well as oil and gas leasing to protect its natural qualities. These interrelated administrative initiatives not only protected large expanses of undeveloped national forestlands and enhanced their value as ecological reserves but also paralleled other developments signaling the agency's growing commitment to an ecological conservation agenda.[55]

Not surprisingly, the extractive industries and their allies have vigorously opposed these preservation initiatives, turning to Congress and the courts for relief. In their view, the president and the public land agencies seriously overstepped their statutory authority. Yet despite several efforts to curtail the president's unilateral national monument designation powers, Congress has steadfastly refused either to rescind the Antiquities Act or to give itself a larger role in the monument creation process. Two court challenges to President Clinton's individual national monument designations have been rejected because the president historically has been given wide latitude under the Antiquities Act, both in making individual monument decisions and in defining the necessary boundaries. The Forest Service's opponents also adopted a litigation strategy to challenge its new protective policies. They initially won an injunction prohibiting the agency from implementing its new roadless area regulations, but that ruling was overturned on appeal. An earlier industry court challenge to the Forest Service's Rocky Mountain Front mineral leasing withdrawal decision failed when a Montana federal court affirmed the agency's authority to preclude new development on its roadless lands. As we have seen, however, the George W. Bush administration has begun revising these policies, provoking a corresponding political outcry as well as the likelihood of further litigation from the environmental community.[56]

Meanwhile, it is clear that the Clinton administration effectively employed its administrative authority and withdrawal powers to expand the nation's protected lands inventory, which can only enhance federal ecological conservation efforts over the long term. If the recent initiatives withstand the intense political and judicial scrutiny they have received, then important new executive branch venues will have been identified and deployed in the ongoing struggle over preservation on public lands. But even if the Bush administration reverses or amends some of these initiatives, the basic administrative preservation strategies have now been exposed, and they can be revitalized and deployed again at a later date. In other words, barring a congressional change in the governing law, these same tools remain available to advance future preservation objectives.

PRESERVATION AND THE POLITICS OF DEVOLUTION

Nonetheless, Congress is still the major institutional player in the preservation arena. And in Congress, public land wilderness debates have increasingly devolved toward locally negotiated compromises. Although Congress has not passed any major new wilderness preservation legislation since the 1994 California Desert bill, it has adopted preservation proposals addressing local controversies. During the 106th Congress, western state delegations, confronted with President Clinton's aggressive national monument designation agenda, seized the opportunity to resolve previously deadlocked public land controversies. The Oregon delegation, for example, sponsored legislation creating a 425,000-acre Steens Mountain Cooperative Management and Protective Area on BLM lands in southeast Oregon, not only protecting this ecologically unique landscape from further development but also adding more than 150,000 acres to the BLM's wilderness inventory. Similar compromises were negotiated to create the Santa Rosa and San Jacinto Mountains National Monument in southern California on adjacent Forest Service and BLM lands, the 112,500-acre Colorado Canyons National Conservation Area on BLM lands in western Colorado (including 75,000 acres of new wilderness), the Las Cienegas Conservation Area in southeast Arizona, the Black Rock Desert National Conservation Area in Nevada, and the Great Sand Dunes National Park in southwest Colorado. In the contentious Utah BLM wilderness debate, however, a locally conceived bill to create a San Rafael Swell Western Legacy District and National Conservation Area failed after environmental opponents successfully argued that they had been excluded from the negotiations. Thus, while Congress may be stalemated over national wilderness policy and ill-disposed toward statewide wilderness proposals, several smaller-scale, locally negotiated preservation proposals have gained legislative approval. Devolution is plainly alive and well in the preservation policy debates, at least in the legislative arena.[57]

But devolution and ecology do not always mix well. From an ecological perspective, Congress's repeated failure to take a comprehensive approach to preservation on public lands has left major deficiencies in the nation's nature conservation efforts. The preserved lands simply do not encompass the full array of ecosystems that merit protection to meet biodiversity conservation needs. Most of our large protected landscapes—namely national parks and national forest wilderness areas—consist of high-elevation alpine terrain that is neither biologically diverse nor rich. Congress has protected rock and ice landscapes while leaving more ecologically diverse and commercially valuable, lower-elevation lands open to exploitation. In fact, the Pacific Northwest forest controversy manifested precisely this problem, as the federal courts discovered. The absence of an ecolog-ically sensitive approach to preservation on the region's public lands left the courts with little choice but to force the agencies into adopting a landscape-scale bio-diversity conservation strategy. From a political perspective, the devolutionary trend toward locally negotiated compromises over wilderness and other preserva-tion issues can also be problematic. Such negotiations are not comprehensive by definition; they often overlook larger ecological issues inherent in the preservation debate, and they ordinarily do not address cross-jurisdictional preservation prob-lems. By any measure, they are inherently parochial, raising the specter that preser-vation will take a back seat to local economic and other concerns. Hence a para-dox: as ecology has expanded our nature preservation needs, the nation's political dynamics have shrunk the scale of our legislative possibilities.

The resulting vacuum, as we have seen, has been addressed by default in two different forums, namely the executive branch and the judiciary. But neither of these forums is without inherent problems. Within the executive branch, President Clinton and the public land agencies pursued diverse administrative initiatives—national monument designations, roadless area protection policies, and other re-lated strategies—in an effort to protect substantial chunks of the public domain. The initiatives were motivated in large part by ecological preservation concerns, as reflected in the large acreage protected and the rationale advanced to support these decisions. But from the perspective of the affected states and communities, the ini-tiatives were driven by a national preservation agenda, and both the input and con-cerns of these locales were regularly disregarded. As a result, there is at best mixed local support for the newly protected lands, and much local resentment, which the Bush administration has used as a justification for its revisionist proposals. The courts, too, have helped promote a preservation agenda on public lands. Applying an array of planning and environmental laws, the courts have essentially compelled the public land agencies to interject biodiversity conservation objectives into their resource management agendas, which has forced them to designate administrative nature reserves to meet these obligations, adopt new ecosystem management poli-cies, and otherwise curtail extractive activities on the public domain. Although these judicially inspired initiatives have enlarged the overall acreage subject to preservation-oriented management, they are nonetheless subject to congressional revision. Indeed, Congress can always change the governing laws, just as it can over-turn presidentially created monuments and other administrative designations.

That has not occurred, however. Congress has not undone any protective designations or otherwise undermined future preservation opportunities. Con-gress has not decommissioned any national parks, wildlife refuges, or wilderness areas; nor has it decommissioned any of the national monument designations. Instead, Congress historically has converted national monuments into national parks, and it has regularly expanded national park boundaries to better protect park resources and more accurately reflect ecological realities. Congress even continues to create new parks, monuments, wilderness areas, and other protected areas, though the number and size of these new designations has shrunk in recent years. And, as important, Congress has neither repealed nor significantly amended any key preservation laws. Despite the Clinton administration's aggressive use of the Antiquities Act, Congress has not reduced the president's unilateral authority to designate new national monuments. And despite much political rhetoric, the Endangered Species Act and other key environmental laws are still intact, enabling the agencies to implement ecosystem-based management policies that include administratively designated nature reserves. Congress has also given the U.S. Fish and Wildlife Service new biodiversity conservation and comprehensive planning responsibilities for managing national wildlife refuges, and it has given the National Park Service a new science mandate to complement its statutory preservation responsibilities. In short, fundamental preservation tools remain firmly in place and available to pursue a more systematic, ecosystem-oriented preservation agenda.[58]

The challenge is to build the political consensus necessary to create a biologically comprehensive system of protected nature reserves. Somehow the ecology and politics of preservation must be reconnected, much as our fragmented reserves must be joined to form a coherent system. The political strategy must be both national and local in scope. A national focus is necessary to ensure that transboundary ecological factors are part of the preservation equation, to garner the political support needed to pass the requisite legislation, and to counteract any lingering parochial opposition. But a local focus is important too: to ensure that reserve boundaries match on-the-ground realities, to assuage the economic and social concerns of nearby communities and residents, and to neutralize recalcitrant local opposition to such conservation efforts. That is a major challenge, and it may call for new alternatives to wilderness—that is, a nature reserve design that does not curtail all economic activity but leaves alternatives open to local citizens to earn a living and to participate in management decisions. As we have seen, such preservation alternatives are available, but they are still quite controversial and only a few have been converted into concrete legislative proposals. It may be time for an open and robust debate over how to integrate biodiversity conservation needs with our traditional preservation models to advance a politically viable preservation agenda. Until that debate occurs, however, the various multiple-use monuments, administrative nature reserves, and ecosystem management initiatives will help demonstrate whether such an approach to preservation is feasible. If so, then we may finally witness the marriage of politics and ecology in the name of nature preservation.

The Private Lands Dilemma

Even the most comprehensive nature reserve system cannot meet the nation's biodiversity conservation needs. Despite their size and diversity, public lands do not embrace a full array of ecologically important landscapes. Many of our most sensitive ecosystems are not federally owned; they remain in private hands. To fully ensure our biological heritage, these private lands must be included in any national ecosystem protection policy. How can privately owned lands be integrated with public lands into a viable nature reserve system? This presents a much more difficult ecological and legal challenge than exists on public lands.

THE INTERMIXED OWNERSHIP PROBLEM

Even a cursory examination of the map reveals the problem: federal, state, and private lands are in interspersed ownership, creating a veritable patchwork quilt on the western landscape. To be sure, the federal government has a vast 670 million–acre estate, most of which is situated in the American West, with more than 270 million acres managed primarily for conservation purposes. But that still leaves much of the landscape in state and private ownership, where the management objective is rarely conservation. Often these lands are intermixed with surrounding federal lands, the checkerboard-like heritage of the railroad construction era, state school trust land grants, and early homesteading laws. When this is the case, ecological management problems are endemic, including divergent ownership goals and related spillover problems. In addition, much of the region's low-elevation riparian land is privately held, reflecting early settlement patterns and the importance of water. These lands encompass sensitive riparian ecosystems that also provide critically important wildlife habitat, particularly during the harsh winter months. Outside the West, most land is in private ownership, and much of it has already been converted to agricultural purposes. This pattern does not bode well for establishing an expansive, ecologically driven system of interconnected nature reserves.[59]

Indeed, the prevailing mixed ownership system has badly fragmented the landscape, posing major challenges for protecting the nation's biodiversity resources. Many sensitive ecosystems are privately owned. Recent studies indicate that one quarter of the major terrestrial and wetland ecosystems are not within the federal estate. Outside Alaska, roughly 75 percent of wetlands are located on private property. Nearly 40 percent of the species listed under the Endangered Species Act are not represented on federal lands, and over 90 percent of the listed species have some or all of their habitat on nonfederal lands. Moreover, human population pressures and rapid growth are putting ever greater pressures on our undeveloped lands, whether publicly or privately owned. As the West's cities have expanded, their growth boundaries have increasingly absorbed adjacent agricultural lands and bumped up against nearby federally owned lands. Even in predominantly rural areas, agricultural and range lands are now regularly being subdivided into ranchettes and second home sites. More than a few ranchers, faced with marginal economic returns and growing regulatory constraints, see few realistic options other than to sell their lands for development. The net effect is obvious: more people, houses, and roads now dot previously uncluttered landscapes, a potentially deadly scenario for human-sensitive species and habitat specialists. Put simply, private land use decisions are exacerbating fragmentation problems on the region's already intermixed public and private lands.[60]

The legal system governing nonfederal lands is as fragmented as the landscape itself. Whereas a relatively uniform set of federal laws and policies govern the nation's public lands, a diverse set of state and local laws govern private land use decisions. Most states have adopted zoning and local land use planning laws, but these laws are often discretionary rather than mandatory, and they usually contain broad waiver provisions. Although some progressive states, such as California, Oregon, and Vermont, have included ecological protection provisions in their planning laws, most states do not require consideration of ecological concerns. Some states also have adopted their own endangered species legislation or other biodiversity conservation laws, but these laws are notoriously weak at the state level, and several of the western states have turned a blind eye toward such legislation. In fact, deeply resentful of new environmental regulations, several states have adopted takings compensation legislation in an effort to curb any further land use restrictions. To the extent that federal law reaches state or private lands, its impact is limited and often quite controversial, as in the case of federal endangered species and wetlands protection provisions. Even then, these laws do not protect the full array of ecosystems or species potentially at risk, and they, too, can trigger constitutional takings challenges.[61]

The simple fact is that private landowners have come to assume legal rights that are basically not compatible with nature conservation goals. State law has traditionally recognized that property owners possess an autonomous right to decide how to use or develop their land, subject only to minimal zoning and judicially created nuisance constraints. Indeed, property law has evolved primarily as a rights-based system built on private ownership and autonomy; it not only promotes transferability and divisibility but also focuses on present uses as well as future gains. The Constitution explicitly protects private landowners from government expropriation without just compensation. And according to the Supreme Court, it also protects owners from excessive governmental regulation, requiring compensation whenever a regulatory provision goes too far. The modern Court has essentially adopted a view of property that shortchanges nature; a majority of the justices seem to see no value in land left in its undeveloped state. In short, the law has long viewed land as a fungible commodity that individual landowners may sell, use, or develop as they see fit, with few if any obligations to the public at large, existing ecosystems, or future generations. To the extent that conservation merits any consideration, it is a voluntary exercise on the part of individual landowners.[62]

But this prevailing view of property is slowly but inexorably eroding, owing largely to the compelling logic behind Aldo Leopold's notion of a "land ethic." Although Leopold despaired of our ability to see land in other than economic terms, he nonetheless called upon property owners to assume "individual responsibility for the health of the land" and to "preserve the integrity, stability, and beauty of the biotic community." Evidence is mounting that his call is being heeded. Replacing the view that land is simply another fungible commodity is the view that land performs critical and unique ecological functions that merit legal recognition and protection. Property is increasingly being seen as a vital ecological resource that confers important communal benefits. As part of an interconnected ecological web, individual pieces of property play an important role within the larger landscape. From this perspective, property owners may no longer be entitled to absolute autonomy; they may be required to take account of whether the parcel performs critical riparian, habitat, or other ecological functions. When this is the case, owners may owe an ecologically based obligation to the community to protect these functions, which may require modification or relinquishment of development plans. Put simply, property ownership is coming to be seen as entailing public responsibilities as well as individual rights.

To be sure, Leopold's view of property is still quite controversial and not yet widely shared. But property law has always been evolutionary; legal principles governing property ownership rights and responsibilities have been adapted—time and again—to meet changing social conditions. With some state legislatures and courts now beginning to integrate ecological concerns into the concept of ownership, that evolutionary process appears to be taking a new turn and finding its way into the law. Given the public's steadfast commitment to environmental protection, this trend can be expected to continue, though on a piecemeal basis.[63]

RESHAPING THE FEDERAL ESTATE: ACQUISITION AND EXCHANGE

Throughout American history, the federal government has been in the business of acquiring (and reacquiring) lands, many of which have become valuable parts of our natural heritage. But federal ownership of public lands has always been controversial in the West, so any effort to reconfigure existing ownership patterns is likely to generate controversy. Several western states, anxious to protect the local tax base and development opportunities, have endorsed the proposition of no net increase in federal ownership. Bills seeking to establish this principle have even been introduced in Congress, but thus far to no avail. Indeed, both Congress and the public land agencies continue to acquire state and private lands to augment the federal estate, though not without considerable hand wringing and local opposition. The basic federal acquisition tools are straightforward: the outright purchase of private lands, or the exchange of federal lands for state or private lands. Increasingly, these transactions are being driven not just by management efficiency concerns but also by ecological considerations. A prestigious National Academy of Sciences panel strongly recommended that federal acquisition policy utilize landscape-scale selection criteria and concentrate on protecting a full array of representative natural areas. The same report encouraged the public land agencies to deploy an array of land protection strategies, including fee simple purchases, less than fee simple acquisitions, exchanges, land use regulations, and collaborative partnerships.[64]

The federal government can either purchase ecologically sensitive land through consensual transactions or invoke its eminent domain power against recalcitrant landowners. Because the money for these purchases comes from the public treasury, Congress has usually played a critical role in the acquisitions process, both appropriating the necessary funds and passing legislation authorizing the purchases. Congress also has given the public land agencies limited authority to consummate their own purchases. Most federal acquisition dollars come through the Land and Water Conservation Fund, which has financed the purchase of almost 5 million acres for wildlife and outdoor recreation purposes. Although the fund is financed directly from offshore oil and gas revenues, Congress must still appropriate the money for individual purchases, thus retaining a powerful oversight role and injecting politics into the process. In New Mexico, as we have seen, Congress approved acquisition of the ecologically important Baca Ranch while also creating a unique private citizen trust arrangement to oversee federal management of these lands. In Nevada, Congress has authorized the BLM to sell its increasingly valuable lands located near fast growing Las Vegas and to use the sale proceeds to acquire environmentally sensitive lands throughout the state, with an eye toward enhancing biodiversity and developing a multispecies habitat conservation plan for the Las Vegas area. During the late 1990s, a bipartisan alliance of conservation organizations sought unsuccessfully to establish permanent funding—through proposed legislation known as the Conservation and Reinvestment Act (CARA)—for land purchases in order to avoid recurrent budget battles. Taken together, the Baca Ranch and Las Vegas acquisition legislation suggests that Congress is actively exploring alternative models for augmenting the federal estate. If these new models prove economically and politically workable, they should help reduce local resistance to new federal acquisitions and thus enhance national ecological conservation efforts.[65]

Over the past couple of decades, the public land agencies have pursued an increasingly aggressive land exchange agenda, often motivated by ecological and biodiversity concerns. Faced with diminished acquisition funds during the 1980s budget deficit era, the agencies turned to land exchanges as an alternative means to consolidate federal land ownership and to address related environmental concerns. The Clinton administration, in particular, regularly used land exchanges to resolve endangered species problems and other contentious issues. The Federal Land Policy and Management Act establishes the basic legal standards governing federal land exchanges: the exchange must be for equal value and serve the public interest. In addition, the exchanged lands must be from within the same state, the exchange must conform with existing land management plans, and agencies can use cash to equalize appraised values. In 1988, to expedite the sometimes cumbersome exchange process, Congress passed the Federal Land Exchange Facilitation

Act, which has brought greater uniformity to the appraisal process and otherwise streamlined these transactions. Although land exchanges can be proposed by anyone, many large-scale exchanges have corporate origins—a fact that has prompted critics to denounce some transactions as merely sweetheart deals between the public land agencies and their corporate clients. Unlike outright purchases, most exchange transactions do not require congressional authorization, which means the public land agencies are usually in charge. That does not eliminate accountability, however. Any significant exchange proposal will require NEPA review, opening the transaction to public scrutiny and to subsequent legal challenge.[66]

As the federal land exchange program has grown in size and complexity, so too have the criticism and litigation. In 2000, at Congress's behest, the General Accounting Office reviewed the program's recent history and recommended a moratorium on further exchange transactions. According to a report issued by GAO auditors, neither the Forest Service nor the BLM were following the governing statutory directives: they had not received equal value in numerous exchanges; they were not employing standard appraisal procedures; and they had regularly failed to meet the public interest standard. The report also noted the valuation problems inherent in public land exchanges, whether from the absence of a functional market for particular lands or from the unique qualities associated with specific parcels. Other vocal critics have chastised the agencies for too often exchanging ecologically valuable lands for lesser-value corporate lands, for undervaluing public lands with real development potential on the periphery of growing cities, and for frequently bowing to corporate pressure in order to complete dubious exchanges. In one oft-cited example known as the Huckelberry Land Exchange, the Forest Service relinquished 4,300 acres of prime federal timberland and received 30,000 acres of high-elevation, cut-over private timberland of considerably less value.[67]

The courts have also begun examining specific land exchange proposals, with mixed results thus far. Legal challenges based on alleged agency failure to consider the pubic interest have met with little success, as the courts have routinely deferred to agency discretion in making such determinations. Valuation-based challenges have fared better. In the *Desert Citizens* case, the court of appeals ruled that the BLM had wrongly undervalued property as mere open space or wildlife habitat rather than valuing it as a potential landfill site, and it enjoined an already consummated exchange. In other cases, the courts have held the public land agencies to full NEPA compliance as part of the exchange process. In the *Muckleshoot Tribe* case, for example, the court of appeals invalidated a Forest Service–timber company exchange because the agency's EIS did not address the potential cumulative effects of past and foreseeable land exchanges involving nearby lands. In short, the federal land exchange program has come under intensifying scrutiny, with the courts beginning to actively review both the merits and procedures involved in individual exchanges.[68]

Not surprisingly, when particularly large or contentious exchanges are at issue, Congress has inserted itself into the process, adding an overtly political dimension to these transactions. In two well-publicized cases—one involving Utah's extensive school trust landholdings and the other involving checkerboarded Gallatin National Forest lands in Montana—Congress invoked its legislative authority to facilitate these transactions. The 1998 Utah Schools and Lands Exchange Act resolved a two-decade impasse, during which time the federal government faced recurrent inholding controversies and the state was unable to develop its scattered school trust lands. Precipitated by the Grand Staircase monument designation, the legislation removed all state inholdings from within the new monument as well as Utah's national parks, national forests, and Indian reservations, while the state received title to comparable, nonsensitive federal lands along with mineral interests and a $50 million payment. Two years later, Congress approved the Utah West Desert Land Exchange Act, enabling the BLM to acquire 106,000 acres of sensitive state school trust lands in or near wilderness study areas in the state's remote western reaches, including nearly 500 acres of state-owned desert tortoise habitat outside rapidly growing St. George, Utah. In exchange, the state received equivalent federal acreage with no evident environmental value, though some of the land is situated in the scenic road corridor approaching Zion National Park and has thus proven controversial. The Gallatin National Forest land exchanges in Montana ultimately required two separate congressional bills for the Forest Service to add nearly 100,000 acres with strategically located wilderness or wildlife habitat value to the federal land base. The new lands were obtained through a complex series of exchanges and purchases negotiated with private timber companies, though Congress required the Forest Service to use local timber receipts to finance part of the transaction. In each instance, the congressionally authorized exchanges resolved long-standing environmental problems and enabled the public land agencies to consolidate their ownership, which should bear long-term ecological and other benefits. The legislative exchange process also offered a degree of flexibility not available under the existing FLPMA exchange provisions and related environmental laws, as illustrated by the subsidiary desert tortoise habitat provisions included in the Utah west desert exchange.[69]

The congressional exchange process, however, has generated intense criticism of its own. Some environmental organizations, budget watchdog groups, and others object that legislative exchanges can be used to negate important appraisal requirements, to override key environmental laws, and to eliminate public involvement opportunities. Put simply, they fear that political deals will be struck without adequate public scrutiny. And that can certainly happen, as it did with the controversial Utah Snowbasin Land Exchange Act of 1995. Using the pretense of the 2002 Winter Olympics, the Utah congressional delegation pushed through legislation authorizing the Forest Service to exchange 1,320 acres of its land at the base of the Snowbasin ski area in the Wasatch-Cache National Forest with the Sun Valley Company to provide it with valuable real estate for development. In return, the Forest Service received several parcels totaling 4,100 acres scattered throughout the forest, while Congress allowed the Sun Valley Company to arrange for the necessary appraisers and shielded its elaborate ski area development plans from the environmental laws, including any further NEPA review. It was, from all appearances, a corporately driven sweetheart deal, made all the more troubling when the Forest Service's lead negotiator subsequently went to work for Sun Valley after taking early retirement. Thus, even though the legislative exchange approach is subject to scrutiny through the political process, it can still be abused to achieve local political or economic benefits. And once Congress has endorsed an exchange, there will be little opportunity for any meaningful judicial oversight.[70]

ECOLOGY AND PRIVATE OWNERSHIP:

REGULATIONS, INCENTIVES, AND COLLABORATION

Yet the federal government cannot acquire all the ecologically important land currently in private ownership. If the conservation goal is an integrated, ecosystem-level management regime, then what strategies might be employed to engage private landowners in such a common enterprise? Given the legal rights that attach to land ownership and given the predominant state role in administering the property system, the federal government has traditionally occupied a much more limited role on private lands than on its own public lands. The Congress, however, does have broad constitutional authority that it could employ to regulate private land use and wildlife. In fact, reflecting the growing national commitment to ecological conservation, the federal government now regulates endangered species and wetlands on private lands, though its involvement in these matters is often deeply unpopular. An alternative to regulation, of course, is to employ marketplace and financial inducements in order to pursue ecological conservation goals, which is what an array of nonprofit organizations are now doing. New voluntary collaborative processes are also being employed to address ecosystemwide concerns, especially in areas of intermixed public and private lands.

To promote a regional ecological conservation agenda, the federal government could potentially assert its formidable array of constitutional powers to regulate or otherwise constrain private land use practices. The courts have consistently interpreted the Article IV property clause as giving Congress broad authority to regulate activities on nearby lands that threaten the purposes for which federal lands are being managed. To protect wilderness values, for example, Congress can prohibit motorboats or snowmobiles on nonfederal lands situated within a designated wilderness area. If Congress has delegated regulatory power to the agencies, then they too can proscribe potentially harmful activities on adjacent lands to protect common resources. To illustrate the point, the courts rejected a broad constitutional attack on the Columbia River Gorge National Scenic Area Act, which gives an interstate regulatory commission the authority to control both public and private land uses within the scenic area. Also, Congress can invoke other constitutional provisions to regulate or control adjacent land uses or activi-

ties that have an adverse impact on public lands or resources, including the commerce clause, treaty power, and spending authority. In the case of the ESA-sanc-
tioned red wolf reintroduction program, for example, the courts ruled that
Congress has sufficient authority under the commerce clause to prohibit private
landowners from removing wolves from their property when they stray outside
the wildlife refuge. To be sure, any expansive federal regulatory initiative that limits a private land owner's prerogatives is subject to a potential takings challenge,
but the courts have regularly rejected such claims, including those based on
wildlife-related damages. And given the state sovereignty concerns that any federal
land use initiative provokes, both Congress and the agencies face serious political
constraints whenever the subject of extending federal regulatory control over private lands surfaces.[71]

Nonetheless, the powerful and controversial Endangered Species Act extends federal regulatory authority to private land use matters. Under the ESA,
the Section 9 "no take" provision prohibits anyone from harming a protected
species, whether on public or private land; it even forbids landowners from altering or destroying habitat necessary for the species' survival. Property owners can
avoid these constraints, however, by preparing a habitat conservation plan (HCP)
and simultaneously securing an incidental take permit. The Clinton administration, faced with vocal political opposition from property rights advocates, significantly modified the act's HCP program in order to mitigate its impact on individual landowners. Two of the new HCP policies are particularly noteworthy: first,
the program was expanded to encourage planning at an ecosystem level, seeking to
integrate multiple species as well as adjacent public and private lands into a comprehensive ecological network; second, a "no surprises" policy sought to assure private landowners who entered an HCP agreement that the government would not
impose any additional land use limitations but instead would itself bear the burden of any further conservation needs on public lands. The result has been an
enormous increase in both the number and size of HCP agreements. The Plum
Creek Native Fish HCP in western Montana, for example, covers more than
1.6 million acres of privately owned timberlands and addresses 17 different fish
species. In short, to forestall political opposition to an enhanced federal involvement in ecological conservation, the Clinton administration cleverly expanded
the scope of ESA coverage on private lands by offering landowners nonmonetary
incentives for their compliance.[72]

Recognizing the limitations inherent in any federal regulatory strategy, environmental groups have begun pursuing alternative nonregulatory, market-oriented approaches to ecological conservation on private lands. Groups like The Nature Conservancy and Trust for Public Lands are actively involved in purchasing
sensitive lands, acquiring either full ownership of the property or conservation
easements to preclude development. The Nature Conservancy regularly employs
biodiversity and other ecologically derived criteria to set its land acquisition priorities, utilizing the extensive natural heritage inventories it has developed for each

state. Sometimes the strategy is to purchase sensitive land and hold it for subsequent sale to government agencies; other times the strategy is to acquire property and then actively manage it for ecological purposes while still deriving an income from it. Conservation easements have proven especially effective in preserving ranch- and farmlands from subdivision development, as is proving true in the Malpai Borderlands region. But this strategy is available only when landowners are willing to encumber their property with easement limitations and to accept the corresponding loss in real estate value. Changes in federal and state tax laws have helped to make conservation easements an economically viable alternative to outright sale or subdivision, particularly when estate tax concerns are part of the financial equation.[73]

Some environmental organizations have also offered direct compensation or subsidies to private landowners to encourage ecologically responsible stewardship. One prime example is the Defenders of Wildlife's wolf compensation fund, which reimburses ranchers for livestock lost to depredating wolves. To further encourage acceptance of wolves, the Defenders also offer a subsidy (or "bounty") payment if wolves successfully den and breed on private lands. In both instances, landowners are provided financial incentives by a nongovernmental organization to promote nationally important biodiversity objectives, thus relieving the localized economic costs associated with restoring these controversial predators. While helpful—and perhaps even essential—to the wolf recovery effort, these entirely voluntary compensation programs do not begin to address the myriad biodiversity and ecosystem restoration concerns that extend across the broader landscape. Several observers believe landowners should be encouraged to undertake active ecological restoration initiatives, such as prescribed burning on their property. One ambitious proposal calls for a federally funded stewardship program that would compensate private landowners for actively managing their lands to accomplish ecological objectives, with payments disbursed through regional organizations overseeing large-scale planning efforts. Whether or not such an expansive stewardship program is feasible, public and private policies subsidizing ecologically sensitive landowner practices can only help mitigate the objections to environmental regulation and make landscape-scale conservation an economically attractive option.[74]

Alternatively, the intermixed public and private lands dilemma is being addressed through new collaborative initiatives that are bringing the federal agencies together with private citizens to fashion more ecologically sensitive management policies. These cooperative processes provide an opportunity for federal, state, and local government officials, along with private landowners, environmental groups, and others, to fashion a common plan for managing shared public–private landscapes and to devise appropriate implementation strategies. When everyone can agree upon mutual goals and strategies, the instinctive resistance to land use planning and concomitant regulatory limitations often disappears. It would be blinking at reality, however, not to acknowledge that the coercive legal mandates found in the ESA and other federal laws have proven essential to convincing private landowners and local governments to acknowledge paramount federal biodiversity goals and other such ecological concerns. But within these constraints, the law is often flexible enough to accommodate innovative place-specific solutions to local resource controversies. Indeed, an astonishing number of watershed groups, public–private partnerships, and similar collaborative initiatives have sprung up across the western landscape, offering more opportunities to inject ecology into local land use practices.[75]

Manifold challenges remain to integrate the West's public and private lands into any comprehensive nature reserve or ecological management program. The above-described initiatives—whether regulatory, incentive-oriented, or collaborative—cover only a small portion of the landscape and its ecosystem components. Neither the federal government nor the nonprofit community has the financial resources required to purchase the ownership interests necessary to construct a truly landscape-scale system of interconnected nature reserves. Existing federal and state endangered species protections may forestall some damaging development proposals, but these laws are of limited scope and generate such intense resistance that government officials are often reluctant to employ them aggressively on private lands. Moreover, there are few incentives for private landowners to undertake active ecological restoration activities, such as reintroducing fire or altering livestock grazing patterns. Yet the West is changing, and the evidence is mounting that ecological concerns are entering the land ownership equation. Resistance certainly persists in some quarters, but the rapid growth of new collaborative public–private initiatives bespeaks a more environmentally sensitive approach to the landscape. How that is occurring and what it may mean merit further examination.

SEVEN

Collaborative Conservation

Building Sustainable Communities

We believe the appropriate range of representation includes not just the
obvious direct interests, such as grazing, recreation, mining, fish and
wildlife, and wilderness, but the professor, the laborer, the townsman,
the environmentalist and the poet as well.

PUBLIC LAND LAW REVIEW COMMISSION (1970)

The love of the land that brought so many people to the West and keeps
them there is common ground on which westerners can articulate and
enact a commitment to a shared agenda of living well in a well-loved place.

DANIEL KEMMIS (2001)

Over the past several decades, the West has seemed at war with itself. Western communities, faced with jarring economic dislocations and rapid social changes, have struggled to reconcile the evolving imperatives of a "new West" with the time-honored traditions of the "old West." Much local anger has been directed toward the region's omnipresent federal landlord, reflecting a widely shared concern that people remain able to avail themselves of public lands for economic and other purposes. The Sagebrush Rebellion of the late 1970s was a prelude to the Wise Use movement of the 1990s, each venting frustration over tightening federal regulatory policies and seeking to wrest control of the federal estate from the agencies that oversee it. Although local in nature, both movements employed an extensive array of political tactics designed to safeguard traditional prerogatives. The basic strategy—whether pursued in Congress, the federal courts, or the county commission meeting—has been to contest new federal policy reforms, strengthen private property rights, and assert local control over public lands. While the details vary by location, the net effect has been an outpouring of local hostility to the public land agencies and their new ecosystem management initiatives.

There is, however, another western story, one that is being written in a new spirit of collaboration. In many communities, local officials, citizens, and agency employees have chosen a different path, beginning to cooperate with one another to address thorny environmental issues. Whether born from frustration over the gridlock that has paralyzed public land decision processes, from a renewed commitment to community spiritedness and civic dialogue, or from an abiding fear of the Endangered Species Act and its regulatory strictures, such cooperation has become increasingly common across the western landscape. Initiatives like the Canyon Country Partnership in the Moab area, the Malpai Borderlands Group in the Southwest, and the Quincy Library Group in the northern Sierras are emblematic of this new spirit of cooperation, prompting some observers to proclaim a new era of collaborative stewardship on the public lands. Indeed, it is fair to say that these new community-based conservation initiatives represent devolution in action. The best of these new collaborative ventures have also embraced the logic of ecological management, creating an important nexus between science and politics while further legitimizing ecosystem management as a new policy.

But there are storm clouds on the horizon, and the largest one may be the shibboleth that recreation and tourism are the economic salvation for the West's rural communities. With the decline of logging, mining, and grazing across the region, many communities have turned to recreation and tourism for their daily sustenance. They are, in effect, selling their scenery and lifestyle rather than the raw commodities of the land. Environmental organizations have both promoted and endorsed this shift in priorities, and the public land agencies have joined the chorus. Industrial-scale recreation, however, is proving every bit as problematic as the extractive industries of yore; both are hard on the landscape and its natural communities. Second homes, ranchettes, and other symbols of sprawling development are the inevitable by-products of any new amenity-based economy, creating open-space and urban–wildland interface problems as well as wildlife habitat fragmentation concerns. And the growth in outdoor recreation, particularly the accelerating popularity of off-road vehicles, snowmobiles, and personal watercraft, has deepened the schism between those who view public lands as a place for solitude and those who see them as a vast playground. The fact that many rural westerners have not found a niche in the tourism and recreation industries has created a cadre of disillusioned sideline spectators to these changing economic dynamics. How to integrate the new amenity-driven economy with new ecological realities to sustain both the landscape and existing communities poses a new challenge every bit as vexing as the old ones.

Confronting the New West: The Moab and Catron Experiences

Is there such a thing as the "new West"? For some time now, various think tanks, media prognosticators, and others have extolled the arrival of a "new West" that has irrevocably altered everyday life beyond the 100th meridian. Continued growth of the region's urban areas—as reflected in cities like Boise, Denver, Las Vegas, and Salt Lake City—is one dimension of a changing western landscape and culture. But the rural West has also changed, portending even greater shifts in the region's economic and social structures. The towns riven by such changes are legion—Telluride, Steamboat Springs, Santa Fe, Taos, Park City, Moab, Ketchum, Sandpoint, Jackson, Bend, Bozeman, and Flagstaff, to name some of the most prominent examples. Though the details vary, the rough outlines of what such a community transformation entails are clear: the diminished presence of the historically dominant mining, ranching, and logging industries, along with a greater reliance on tourism, recreation, and service sector industries to support rural economies; a stronger commitment to environmental values and to protecting rather than developing public lands; and the growing presence of newcomers who embrace these values, often to the chagrin of longtime residents. In some locations, these demographic, economic, and attitudinal changes have generated corresponding political changes; elsewhere, there is little evidence of any political impact except a hardening resolve to resist further changes. Although it would be unrealistic to suggest the "new West" has arrived everywhere, few western communities have escaped the cascading effect of these larger market forces and demographic changes. The point is perhaps best made by examining two different communities—Moab, Utah, and Catron County, New Mexico. Both have come face to face with the hard realities of change and reacted very differently.

MOAB: FROM MINING TO MOUNTAIN BIKES

For the first half of the twentieth century, Moab could appropriately be described as a quiet town on the northern fringes of the Colorado Plateau. Perched alongside the upper reaches of the Colorado River, Moab is also nestled against soaring redrock cliffs that give southern Utah its distinctive scenic aura. The area was settled during the late 1870s, its economy dominated by farming and ranching for decades. Mining figured into the regional economy too, and it was destined to play a larger role in the town's development. During the twentieth century the community realized that it might capitalize on the area's stunning scenery. After Hollywood started filming on-location western movies in the Southwest during the 1930s, Moab began touting itself as a prime film location, eventually landing several popular western movies, television commercials, and the like. Moreover, the town's leaders embraced President Herbert Hoover's decision to designate the nearby Arches area as a national monument in 1929, just as they supported its expansion nine years later. Although the outbreak of World War Two brought any local aspirations of a new tourism economy to a halt, the seeds of preservation had been planted. They would bear more fruit for Moab during the latter part of the century.[1]

After the war, Moab entered a second distinctive phase in its evolution, proudly proclaiming itself the "uranium capital of the world." With the advent of atomic power, the federal government needed uranium to fuel the nation's nuclear arsenal as well as its incipient nuclear power industry. The Colorado Plateau was where this strategically precious mineral could be found. In 1952, prospector Charlie Steen discovered the nation's largest high-grade uranium deposit outside Moab, which transformed the town virtually overnight. Overrun with fortune seekers, the town's small population nearly doubled to 2,775 people in less than a year, as thousands of prospectors fanned out across the plateau in search of the new yellow gold. Five years later, Steen constructed a $12 million uranium milling plant on the north bank of the Colorado River, giving the town its first major production facility and valuable industrial jobs. At the same time, the Texas Gulf Sulfur company started building a $40 million potash mining and processing plant north of town, further linking the town to the mining industry. By the early 1960s, however, Moab's mining boom had run its course. In short order, the Atomic Energy Commission stopped taking new uranium orders, the Texas Gulf Sulfur mine was closed following a deadly explosion, and the miners departed in droves. The local economy spiraled downward in the classic western boom and bust tradition, leaving longtime residents to wonder how the town would survive.[2]

The surrounding desert landscape offered more than mineral wealth, however; the scenery also might be sold to invigorate the local economy. In fact, a nascent tourism industry was already evident in Moab, spurred not only by the availability of wartime surplus jeeps, which opened up the desert, but also by Hollywood's depictions of the scenic southwestern country. By the early 1960s, local officials were actively promoting additional national park designations. Congress

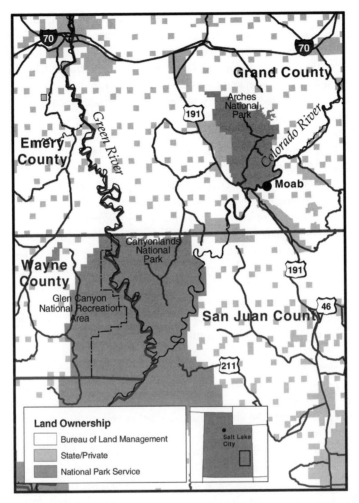

Map 8. Moab and Surrounding Lands. The Moab region in southeastern Utah with its extensive federal lands, including Canyonlands and Arches national parks, Glen Canyon National Recreation Area, and BLM public lands. The scattered dark squares are primarily state school trust lands, which Utah has elsewhere exchanged with the federal government to improve management efficiency. The Sand Flats area is situated east of Moab on adjacent BLM public lands.

obliged in 1964 by establishing Canyonlands National Park, adding over 257,000 acres to the national park system and giving Moab another local scenic attraction. In 1968, Congress doubled Arches in size and then redesignated it a national park three years later. Yet economic stability continued to elude the town. An upswing in the uranium market during the late 1960s briefly rekindled the local mining industry, and the reopened Texas Gulf Sulfur mine provided 150 residents with regular employment. Visitation was steadily increasing at the adjacent national parks, while the number of local outfitters and guides was expanding to meet growing public interest in the area. By the late 1970s, author Edward Abbey had published two major works, *Desert Solitaire* and *The Monkey Wrench Gang*, which helped popularize the desert Southwest and particularly the Moab environs. But as newcomers arrived and as the federal agency presence grew, notable fissures began to appear within the community.[3]

During the 1980s, open conflict erupted over environmental issues. First, after the Federal Land Policy and Management Act (FLPMA) of 1976 gave the Bureau of Land Management (BLM) new wilderness inventory responsibilities, the Sagebrush Rebellion spilled over into Moab. On July 4, 1980, incensed over the BLM's local wilderness study area (WSA) decisions, a local road crew supervisor used a county bulldozer to scrape a road into nearby Negro Bill Canyon, one of the BLM's recently designated WSAs. The widely publicized incident highlighted the county's opposition to wilderness and to any federal regulation of local roads crossing public lands. Second, after the Atlas Corporation closed the processing mill in 1984, the county commission endorsed a new proposal to establish a nuclear waste storage facility adjacent to Canyonlands National Park in an ill-advised effort to bolster the local economy. An acrimonious debate over the project's environmental and public health implications followed, with the community rejecting it in a decisive referendum vote. It was no longer acceptable to seize blindly any economic development proposal that surfaced; instead, Moab residents were beginning to embrace the broader environmental values that typified the nation, cognizant of the area's scenic beauty, tourism potential, and natural fragility. Before the decade ended, the county passed another referendum opposing a hazardous waste incinerator proposal. And local opposition was gathering to a long-standing proposal to build a road across the remote Book Cliffs country to connect Moab with Vernal in Uintah County. But the community's new commitment to environmental values was not complete. When the BLM finally released its wilderness inventory, the Grand County Commission promptly announced its opposition to any new wilderness in the county, drawing strong local support for its position.

The seeds of change had taken root, however, as the events of the next decade revealed. By 1989, Grand County was officially deemed a "tourism-dependent" county.[4] The allure of Moab's slickrock was spreading, largely attributable to two seemingly unrelated events. During the late 1960s, Moab started sponsoring an annual Easter Jeep Safari, designed to attract off-road vehicle enthusiasts to the town as visitors. The safari steadily grew in popularity, soon becoming a key event in the town's tourism economy. In 1983, after being laid off from the uranium mines, the Groff brothers opened Rimrock Cyclery, hoping to capitalize on the burgeoning popularity of the new fat-tire mountain bikes that enable riders to travel off-road into the backcountry. Once the national press extolled the Moab area's biking possibilities, an ever increasing number of mountain bike enthusiasts began arriving, eager to test their skills and gear against the challenging slickrock trails. Soon the town was billing itself as the "mountain biking capital of the world"—a remarkable turnaround from its earlier identity. Meanwhile, visitation at the nearby national parks shot up by over 20 percent, representing a further testament to the area's tourism potential.

With the town recast as a tourism and recreation destination, Moab underwent a dramatic transformation. The number of mountain bikers using the popular Sand Flats Slickrock Trail had grown from 140 in 1983 to nearly 100,000 annually in the early 1990s. Recreational use on the surrounding public lands increased by over 300 percent in just five years. One knowledgeable observer described the impact on the community in graphic terms: "Grand County's population swells from approximately 7,000 to about 16,000 during the regular tourist season . . . [making it] almost impossible to get a motel room. . . . Real estate prices have soared. Traffic snarls the streets. The infrastructure is inadequate to deal with the crowds. The town is quickly changing face with espresso bars, fast food restaurants, pricey gift shops, and exclusive art galleries."[5] After the real estate developers appeared, new subdivisions began to sprout on the desert landscape, creating new sprawl, open space, and infrastructure concerns. In the same entrepreneurial spirit, two chairlifts were built across the nearby sandstone walls, effectively commercializing the town's scenic backdrop. Soon, the grim reality of Moab's new industrial-strength tourism economy reared its ugly head. The decisive event occurred during the 1993 Easter weekend, when Moab was overrun by mountain bikers, youthful spring break revelers, and the annual Jeep Safari participants. As the weekend crowds swelled in the Sand Flats recreation area, neither the BLM nor the local police were able to control the unruly partygoers. Vandals tore up the fragile desert terrain, driving their jeeps everywhere, ripping up plants, and leaving human excrement strewn about. The event, which became known as the Easter riot, left nearly everyone in Moab reconsidering the town's future as a tourism and recreation haven.

The stage was plainly set for a change in direction, but any such change would require an unprecedented degree of federal and local cooperation. The fact is that Moab is virtually surrounded by federal lands, including the BLM-managed Sand Flats bike trail and the adjacent national parks. A local BLM-initiated ecosystem management symposium created a public forum to begin discussing the Sand Flats dilemma, which presented a classic interjurisdictional resource management problem. More meetings followed, culminating in a new agreement between the BLM and Grand County to share governance responsibilities for the renamed Sand Flats Recreation Management Area. The highlights included: a new $3 daily fee requirement, with the fee being collected by the county and then used by the BLM to maintain and restore the area; a local Citizens Stewardship Committee responsible for allocating the funds; and a Community Sand Flats Team (initially composed of AmeriCorps volunteers) to maintain and oversee the bike trail, subject to the BLM's ultimate authority over the land. Despite some initial misgivings, the arrangement has worked well. The new fee has generated over $100,000 annually, which is used to maintain the bike trail, nearby campsites, and the surrounding landscape. The chaos of the 1993 spring break has not recurred.

The Sand Flats imbroglio also spawned the Canyon Country Partnership, which has thus far produced mixed results as an experiment in collaborative stewardship. Confronted with an array of common natural resource management problems that spanned political boundaries, the Sand Flats principals convened the partnership in an effort to begin jointly addressing issues on a regional (or ecosystem) scale. The partnership, which covers over 5 million acres across southeastern Utah, boasts a diverse membership from the federal agencies as well as state and local government. According to its charter, the partnership was designed to "maintain the basic health and sustainability of ecosystems, *and* serve the needs of people depending upon such ecosystems for commodity or non-commodity values."[6] To promote its agenda, the partnership created a governing board, a separate scientific committee, and an array of ad hoc committees to address specific issues. Flush with the Sand Flats success, the members readily agreed to begin developing a coordinated regional recreation strategy, to address air tour overflight problems, and to prepare an ecological and socioeconomic regional profile.

Without a galvanizing event (like the Sand Flats Easter riot) and secure funding, however, the partnership was soon adrift without a clear direction or common philosophy. The overflights issue proved too complex to resolve locally. Not only did the partnership lack jurisdiction over the region's airspace, but a local environmental organization threatened costly litigation if the partnership did not adhere strictly to cumbersome federal advisory committee requirements. Shortly thereafter, San Juan County terminated its membership, frustrated over the partnership's unwillingness to tackle the contentious wilderness and local road ownership issues. Next, Emery County chose to ignore the partnership and develop its own wilderness proposal for the nearby San Rafael Swell region. Recognizing its inherent limitations as a voluntary association, the partnership has now retrenched, focusing primarily on public education and information sharing. It continues to meet regularly, and the members uniformly report that it has helped establish trust and strengthen local relationships. But it does not currently aspire to any larger role in regional governance or coordination, which leaves everyone free to follow their own instincts and mandates.[7]

Meanwhile, with Moab's shifting social and cultural climate, major local political changes were afoot. Grand County, like most western counties, had long been governed by the traditional three-person county commission form of government. As elsewhere, the Grand County Commission was dominated by ranchers and other longtime residents with deep roots in the local business community. Still inclined to support every new economic development opportunity that presented itself, the commission endorsed the Book Cliffs road proposal when it resurfaced in the early 1990s. The decision, however, incensed many of Moab's new residents, who put a premium on open space and environmental concerns.

They began mobilizing opposition to the road. But faced with an intransigent county commission, they found themselves forced to organize an initiative to revamp the commission's structure by expanding it from three members to seven and by requiring geographic representation. This was an obvious and unprecedented effort to alter the local balance of power. When the initiative triumphed at the polls in November 1992, a local historian deemed the election "a watershed development in county history, symbolizing the triumph of the new Moabites over the old guard, since the initiative was prompted by the desire of community members for more liberal representation on the hitherto arch-conservative county council."[8] The restructured council not only heralds a further evolution in Moab's political history but also reflects much greater local sensitivity to environmental concerns and to the town's new economic priorities.

Indeed, Moab is a much different place today than it was just 20 short years ago, reflecting both the good and the bad of the "new West." The town is still struggling to reconcile its new tourism identity with the small-town and open-space characteristics that have made it such an attractive place to live, work, and play. The question is no longer whether the community will survive economically, but rather how to control the growth that is largely attributable to the town's intangible environmental amenities. Zoning, sprawl, and infrastructure concerns tend to dominate the local debate. As a testament to its past, the town has also been forced to confront the question of how to clean up the mammoth Atlas tailings pile, fearful that the leftover uranium may leak into the Colorado River and contaminate the local aquifer. Regarding public lands, the Canyon Country Partnership provides a forum for federal, state, and local officials to exchange ideas and to anticipate potential resource management conflicts. Cooperation rather than hostility seems to characterize these relationships, and environmental concerns are a key part of most discussions. Even wilderness is no longer quite the same hot button issue it once was. When the Utah wilderness debate heated up during the mid-1990s, the revamped county council supported a substantial BLM wilderness designation in the county—a far cry from the 1980 Negro Bill Canyon bulldozing episode. In sum, confronted with economic and social forces beyond its control, Moab has made a difficult transition, not only capitalizing on its natural assets but also reshaping the local political landscape. It is a prime example of how the West is changing.

CATRON COUNTY: PRESERVING LOCAL CUSTOM AND CULTURE

Far to the south of Moab, in rural Catron County, New Mexico, a very different story has emerged. The county, like Moab, has experienced its own economic rollercoaster ride over the past 20 years, along with an often hostile relationship with the federal agencies that oversee the surrounding public lands. Unlike Moab, however, the county has experienced few of the "new West" trappings that can reshape a traditional rural community. Catron County residents have long been wedded to logging and ranching for their economic sustenance, though the local sawmill closed over a decade ago with a devastating impact on the local economy. Since then, the county has fought a running battle with federal agencies and the environmental community over public land policies and related endangered species protection requirements. The controversy first focused on national forest logging practices, and more recently has involved public land livestock grazing policies. Along the way, the county has aggressively asserted itself through a series of confrontational ordinances and its own litigation agenda. Struggling to retain its traditional rural lifestyle, Catron County has laid the blame for its problems squarely at the feet of the federal government and its environmental policies.[9]

Situated in a remote portion of southwestern New Mexico, Catron County sprawls across 7,800 square miles of land, covering an area larger than the state of Connecticut. The county's population, however, numbers fewer than 3,000, scattered in small towns and ranches across the vast landscape. The county seat is located in the 600-person town of Reserve, which takes its name from the late-nineteenth-century federal forest "reserve" policy. The area's original native inhabitants were first displaced by eighteenth-century Mexican land grants and then by European settlers who arrived during the 1850s. The early settlers relied upon the land for their living, particularly its minerals, timber, and grasses. Indeed, the region's history is inextricably linked to the surrounding landscape: Geronimo and his Apache followers traversed the deep canyons and rugged mountains during the Southwest's Indian wars; prospector James Cooney discovered silver in the Mogollon mountains during the early twentieth century and established what soon became the state's most productive mining district; and the legendary conservationist Aldo Leopold served as the local forest ranger during the 1920s and helped establish the first-ever wilderness area in nearby Gila National Forest. Although the Native Americans are gone and the last operating mine closed during the 1980s, the wilderness concept has taken hold, and two congressionally designated wilderness areas now cover 750,000 acres of national forestland. In fact, more than 70 percent of the county's land base is publicly owned, including the 3.3-million-acre Gila National Forest, along with smaller amounts of other national forest, BLM, and state-owned lands. The Gila's changing management policies have served as the focal point in the ongoing local controversies.

The timber industry occupied a key role in the county's economy during most of the twentieth century. By mid-century, more than 25 sawmills were scattered across the county, dependent upon the nearby national forest for logs. Logging provided seemingly reliable and well-paying jobs for local residents. Over time, though, the smaller mills closed, leaving the large Stone Forest Industries mill in Reserve as the county's principal employer, with a workforce of several hundred people. Trouble was afoot, however, arriving in the form of the Mexican spotted owl. Distantly related to the northern spotted owl, the Mexican spotted owl depends on the region's old growth forests, much of which was being lost to chainsaws. But the Gila National Forest's 1986 forest plan contemplated an annual timber harvest of 30 million board feet; it barely even mentioned the then little-known owl. In late 1989, a regional environmental activist petitioned the U.S. Fish and Wildlife Service (FWS) to protect the dwindling Mexican spotted owl population; three years later, the owl was listed as a threatened species. Meanwhile, the Forest Service had implemented regional Mexican spotted owl guidelines to govern local timber management practices, which reduced access to valuable old growth timber. Eventually, just as in the Pacific Northwest, environmentalists secured a federal court injunction barring any commercial timber harvesting on the Southwest's 11 national forests—a prohibition that extended for 18 months while the Forest Service rewrote its local forest plans to protect Mexican spotted owl habitat and thus gain the FWS's approval to resume logging. Since then, the Forest Service has designated the forest-dwelling northern goshawk as an indicator species and adopted regional management guidelines to safeguard the bird's habitat. The net effect is evident: with both the Mexican spotted owl and northern goshawk now legally protected, any new timber harvest proposals will face major obstacles before they can be approved.[10]

From Catron County's vantage point, the Mexican spotted owl–logging controversy was catastrophic. Prevented from accessing the Gila's old growth timber, residents saw the forest's harvest levels plummet to near zero. In 1990, the Stone Forest mill announced it was closing, eliminating virtually all the local timber-related jobs. Although the mill was already on shaky financial footing and the owl's actual listing was still two years away, county officials nevertheless placed the blame squarely on the Mexican spotted owl and federal mismanagement. They chastised the Forest Service for not consulting regularly with county officials over its management policies.

To highlight local frustrations, the Catron County commissioners adopted a "custom and culture" county ordinance, thus giving birth to the short-lived county supremacy movement. The 1991 ordinance mandated that federal land managers must comply with the county's interim land use plan, and it sought to maintain current timber production and grazing levels while prohibiting any further wilderness designations. Related ordinances required local gun ownership and obligated environmentalists to register with the county. As might be expected, an environmental group soon challenged the Catron County ordinances, but a local federal court dismissed the suit, finding that no one had yet been injured by the county's actions. Although the ruling fell short of validating the ordinance, the county's efforts to preserve its local "custom and culture" struck a responsive chord across the rural West. Several dozen other counties soon adopted similar ordinances. The courts, however, eventually struck down these ordinances, but not before the Catron County initiative had garnered national headlines and focused renewed political attention on the public land agencies. Meanwhile, sensitized to the power of the Endangered Species Act (ESA), Catron County launched its own litigation campaign. Facing several new endangered species listings, the county sued the FWS to force the agency into employing NEPA processes—namely its

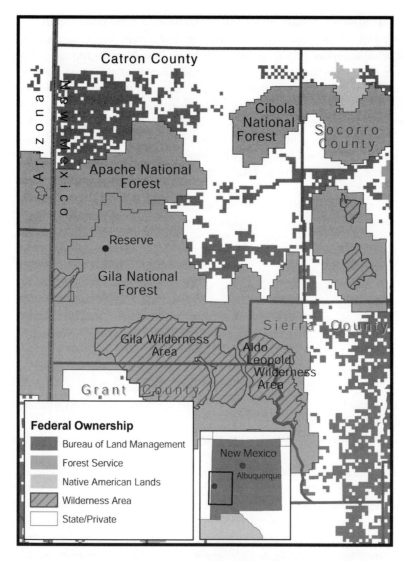

Map 9. Catron County Region. Nearly 70 percent of Catron County, New Mexico, is composed of federally owned lands, primarily national forestlands that include the Gila and Aldo Leopold wilderness areas. With dramatic cutbacks in logging to protect the Mexican spotted owl, more than 25 local sawmills closed over the past 20 years, including the large Stone Containers mill in Reserve, the county seat. During the early 1990s, county commissioners adopted a "custom and culture" ordinance designed to override federal authority on the county's public lands.

public involvement requirements—before making any critical habitat determinations. When it prevailed, the county won the right to have its local economic and social concerns addressed in that process.[11]

In an effort to reduce local friction, the Forest Service and county officials began reexamining the county's role in forest planning and management processes. After a year's protracted negotiations, they reached consensus and formalized their relationship in a 1994 Memorandum of Understanding (MOU). Freighted with legalese, the lengthy MOU acknowledges that each party retains final decision authority within its own jurisdiction, obligates both parties to notify each other of pending decisions, creates joint planning opportunities, establishes dispute resolution procedures, and elaborates specific coordination obligations designed to ensure that the county's economic and social concerns are considered. The Catron County MOU, widely circulated in other rural venues, has become a model for similar agreements elsewhere. To further improve community relationships, the Forest Service also joined the locally conceived Catron County Citizens Group, which was designed both to reduce growing tensions and to address public land issues. Although local environmentalists originally participated in the Citizens Group, most have since left, lamenting its tendency toward lowest common denominator solutions that shortchange environmental concerns. Yet the group has persisted in its efforts to address local timber and range issues, experiencing a modest success when the Forest Service approved one of its forest thinning timber sale proposals. There is little indication, however, that the Citizens Group has significantly altered federal management policies; it seems primarily to serve an information sharing and relationship strengthening role.[12]

Following their timber harvest victory, environmentalists turned their attention to livestock grazing practices on the region's national forests and scored another major victory. The tool of choice was again the Endangered Species Act: the "listed" southwest willow flycatcher, Gila trout, loach minnow, and spikedace all served as surrogates to safeguard riparian habitat. Although hardly household names, these species have achieved real notoriety in the desert Southwest. In the case of Gila National Forest, range conditions have long been classified as sub par, owing to the general lack of water and forage, as well as historical overgrazing practices. Livestock levels are a third of what they were in the early twentieth century, more than half the allotments are considered in poor condition; and most ranchers are not grazing their full allotments. As is true throughout the arid West, livestock grazing takes its heaviest toll in riparian areas, since cattle tend to congregate near water, where the grass is richest. But on the Gila, that is also vital habitat for several of the forest's protected species, which effectively provides layered legal protection for these riparian areas. After local environmentalists sued the Forest Service to reform its grazing practices, the federal government agreed to fence local river bottoms to exclude cattle and to monitor the affected riparian areas. Despite protests from ranchers, the fences have been erected, with Congress appropriating $400,000 to cover the construction costs and lessen the sting a bit. It remains to be seen whether these new riparian protections will force more cows into the forest's dry uplands, creating other ecological problems.[13]

The grazing controversy was not finished, however. Catron County successfully sued the FWS for not integrating local economic concerns into its critical habitat designation process. Although the county prevailed in court, the litigation simply forestalled the inevitable, as the agency eventually expanded its original critical habitat determination. Moreover, the Forest Service had already reduced grazing allotment levels in an effort to improve forage conditions. In one well-publicized incident, a young ranching couple—Kit and Sherri Laney—who held the expansive Diamond Bar allotment refused to accept the Forest Service's proposed reduction (which cut their herd size from 1,200 to less than half that number), asserting that they had the right to graze without federal oversight. The Laneys ultimately lost the argument along with their entire grazing lease, much of which was situated in the Gila Wilderness Area. Some local ranchers fear the case may presage an effort to eliminate ranching in designated wilderness areas. Regardless, like the earlier timber controversy, the ESA has forced the Forest Service to revise its local livestock grazing program, though not without litigation and recrimination.[14]

The Catron County wildlife conflicts do not stop here: wolves and elk have also fomented local controversy. During the mid-1990s, the U.S. Fish and Wildlife Service initiated a Mexican wolf reintroduction program in the Southwest over vehement protests from the region's ranching community. The New Mexico livestock industry, however, lost its suit to block the program, which treats the reintroduced wolves as a nonessential experimental population (like their Yellowstone counterparts). Even though the actual releases are occurring elsewhere, the remote Gila Wilderness Area is serving as a relocation site for problem wolves to the dismay of local ranchers. Given that endangered species restrictions have already curtailed local logging and grazing activities, Catron County residents fear additional limitations from the wolves' mere presence, as well as the inevitable depredation incidents. In addition, after the New Mexico Game and Fish Commission reintroduced elk to Gila National Forest in the 1950s, the elk population mushroomed, creating another conflict with livestock. With elk numbers at an all-time high, ranchers have pushed to reduce the local herd size, but hunters have objected that too much forage is being allotted to cattle. Although local ranchers derive some income from selling state-issued elk permits for their private land (a trophy bull elk permit can fetch $10,000), the principal beneficiaries are the larger landowners, and many ranchers fear further reductions in their federal grazing allotments. A creative proposal that would allow ranchers to sell more trophy elk permits contingent on reducing their livestock numbers has surfaced, but knowledgeable observers are skeptical about whether the idea can command the necessary political support. It appears that the county's beleaguered ranchers can expect further livestock management changes owing to the government's renewed commitment to maintaining and restoring the region's native wildlife species.[15]

With environmental protection taking center stage on the Catron County public lands, one might expect the "new West" to be arriving too. But that is not the case. When the Reserve sawmill closed in 1990, the local population dipped precipitously: the school district went from 350 to 150 students as younger families left town in droves. Those remaining faced a bleak economic outlook; several were forced into bankruptcy. Although parts of the county are quite scenic with development potential and the nearby wilderness areas serve as an outdoor recreation magnet, the county is isolated from major population centers. Nonetheless, the county experienced a 38 percent population increase during the 1990s, and local merchants reported an upturn in their business levels. Today, subdivision notices and "for sale" signs are evident on ranchlands, and some new homes are beginning to appear. But it is too soon to herald an economic revival, and the county's various towns—Reserve, Glenwood, Aragon, Quemado, and Datil—are not each evenly affected by new development activity. The local political scene has not changed either. New officials may occupy county commission seats, but the basic political philosophy remains quite conservative, and residents are still very suspicious of the federal agencies and local environmentalists. There is no enthusiasm for additional wilderness designations, nor for the translocated wolves or any other endangered species. Local guiding and outfitting businesses, which have focused mostly on big game hunting, have grown in number, and several outfitters now offer summer recreational trips. Ideas about luring high-tech firms or other such businesses have gone nowhere. In fact, still clinging to its past, the county has purchased the closed Reserve sawmill, and it is seeking a new operator, hopeful that timber may once again flow from the national forest. Catron County thus stands as a stark symbol of a rural West that is being forced to conform to new environmental realities but seems unable to embrace the corresponding economic and social adaptations that might enable it to capitalize on its remaining assets.

Indeed, Catron County's relationship with the Forest Service seems to be moving in new and yet uncertain directions. In 1996, county officials advanced a pilot project proposal that would have had the Forest Service relinquish responsibility for managing the national forest to the county, exempt from any endangered species oversight. The proposal, given the long history of antagonism, went nowhere politically, though a similar pilot project proposal surfaced in California's northern Sierras from the Quincy Library Group and gained congressional approval. The Forest Service and the county continue to abide by the MOU terms, while the Citizens Group continues to provide a local forum for discussion and small-scale projects. With no commercial timber harvesting in the forest, local citizens have become increasingly apprehensive over the increased forest fire risk, following decades of government fire suppression. The Forest Service is allowing fires to burn in the Gila Wilderness Area backcountry, and it has initiated a major watershed restoration and tree thinning project on Negritos Creek in the forest's northern reaches. County officials believe the project could provide small-diameter trees to the retooled local sawmill. But fearful that the Negritos project represents a thinly disguised logging operation, local environmentalists have thus far blocked it through administrative appeals. And the county has not forsaken its confrontational tactics either. Embracing a controversial New Mexico Senate bill enacted in the aftermath of the disastrous Los Alamos forest fire, Catron County has adopted another county ordinance admonishing the Forest Service to thin the local fire-prone forests or the county will do it. The ordinance may be merely another symbolic gesture or the first volley in an effort to reopen the national forest to logging. In either event, lawsuits and controversial ordinances continue to mark relationships on the region's public lands, reminding us that the transition from an "old West" is not proceeding smoothly everywhere.[16]

Counteroffensive: The Old West Fights Back

Despite the recurrent headlines announcing the demise of the mythic "old West," large parts of the rural West are still strongly linked to the traditional mining, logging, and ranching industries. With notable exceptions, few rural communities have seen many tangible benefits from the new economy, be it high-tech industry, recreation, or tourism. Indeed, the old West—symbolized by the region's ranchers, loggers, and miners—has actively fought changes in federal policy, launching various counteroffensives designed to curtail the growing federal regulatory presence and to preserve the region's traditional economies and lifestyles. With industry support, the old West has flexed its political muscles through such grassroots vehicles as the Sagebrush Rebellion, the Wise Use movement, and the Jarbidge Shovel Brigade. It has used local political venues as well as Congress and the courts to challenge new federal environmental initiatives and policy reforms. Whatever the venue, the basic rallying cry has been the same: war is being waged against the West that could cost the region its traditional culture and economy. In short, the old West has fought to ensure that the new age of ecology does not leave either people or the economy out of the ecosystem equation, and that local residents are heard in setting priorities on public lands.[17]

WISE USE, COUNTY SUPREMACY, AND LOCAL ANGST

Perhaps the clearest expression of the old West's frustration with new federal public land policies is the Wise Use movement, a rebirth of the earlier Sagebrush Rebellion. But the Wise Use movement of the 1990s differs markedly from its 1970s predecessor: it does not seek to divest the federal government of its landholdings in favor of state ownership; it boasts a much broader constituency than the rancher-dominated Sagebrush Rebellion; it has been well funded by the extractive industries; and it has used a broader arsenal of weapons, including county supremacy ordinances, strategic litigation, old-fashioned civil resistance, and grassroots legislative lobbying. Its primary targets have been the federal public land agencies, their new environmental policies, and the environmental community (the so-called Green advocacy groups). These make inviting targets for the rural West, which is being buffeted by distant market forces, globalization trends, and other influences that defy local control. The frustration is both palpable and understandable, just as the region's counteroffensive tactics fit within the American tradition of local resistance and political dissent. But the results, as we shall see, have too often fallen short of the mark, opening the door for an alternative, more cooperative approach to addressing the inevitable changes that the region can only forestall but not halt.[18]

In an early effort to reassert the principle of local control over western public lands, the Wise Use movement endorsed the county supremacy ordinance concept. Flush from its success in derailing the Greater Yellowstone Vision Document in the early 1990s, Wise Use proponents perceived that these ordinances might force federal land managers into bowing to local priorities on the public domain. The prototype was the 1991 Catron County ordinance, which mandated that federal land managers "shall comply with" the county land use plan or face local prosecution. The county's plan sought "to protect the custom and culture of County citizens through protection of private property rights." Ostensibly linked to federal planning statutes that provide for coordination with state and local governments, the ordinance actually went much further: it tried to limit federal acquisition of land within the county, regulate federal wild and scenic rivers, maintain current livestock grazing and timber production levels, and prohibit any further wilderness designations. Attracted by these notions, other rural counties soon pushed the county supremacy idea even further, promoting local ordinances that declared the western states and not the federal government owned the public lands. Reminiscent of Sagebrush Rebellion era legal theories, these latter ordinances were spawned in Nye County, Nevada, following a widely reported confrontation between a local county commissioner and resident Forest Service officials. When federal officials tried to stop Commissioner Dick Carver from bulldozing a road into a protected wilderness study area, he responded that the federal government had no authority over either the national forestlands or the roadbed. Ultimately, several dozen counties either proposed or adopted similar ordinances, reflecting the rural West's mounting frustration with public land management policies that were seen as crippling the traditional ranching, logging, and mining industries.[19]

Widely perceived as an attempt to intimidate local public land managers, the county supremacy ordinances enjoyed a brief but notorious life. In 1996, the federal government challenged the Nye County ordinance and readily convinced a Nevada federal district court to invalidate it. The court rejected the county's equal footing doctrine argument that the state rather than the federal government owned public lands, ruling that the doctrine applies only to submerged, not dry, lands. The court recognized that the United States had broad legislative power over public lands, and that the county's assertion of jurisdiction over roads traversing these lands violated constitutional supremacy principles. In a related dispute, the federal courts rejected similar claims advanced by a Nevada ranching couple whose grazing permit was revoked after they ignored federal livestock grazing requirements. Agreeing with the earlier *Nye County* decision, the Court of Appeals for the Ninth Circuit curtly ruled that neither the equal footing doctrine nor the Tenth Amendment required the United States to divest itself of public lands acquired through an earlier treaty with Mexico. In addition, in a case brought by local environmentalists, the Idaho Supreme Court found that a Catron County–style "custom and culture" ordinance likewise violated constitutional supremacy principles to the extent that it tried to displace federal law to accommodate local preferences. In short, the courts completely rejected the tenuous legal theories underlying these county supremacy ordinances, reaffirming the federal government's basic ownership rights and management authority.[20]

But the county supremacy ordinances were not the only local challenges to federal public land policies. During the early 1990s, the news media regularly reported on direct confrontation incidents involving federal agency employees and irate local citizens. Surely the most widely reported incident was the bulldozing confrontation in Nye County, but other illegal road openings, livestock trespasses, and similar incidents created a hostile atmosphere on public lands. More than one federal employee reported receiving physical threats, and the agencies adopted buddy-system travel policies to protect against violence. In the worst cases, a Forest Service employee's house was bombed in Nevada, and another bomb exploded at the state BLM office in Reno. Because neither of these crimes was solved, it remains uncertain who was responsible. But the state's national forest supervisor, Gloria Flora, eventually resigned her position, submitting her resignation in a widely circulated letter that decried the atmosphere of fear and intimidation confronting federal employees throughout the state's rural communities.[21]

Yet tensions never really subsided in rural Nevada. Another unseemly confrontation surfaced in the tiny town of Jarbidge, situated in the remote northeast corner of Elko County. After the Jarbidge River flooded in 1995 and washed out an adjacent dirt road that accessed a nearby wilderness area, the Forest Service decided not to rebuild the road. Its decision was triggered by the U.S. Fish and Wildlife Service's determination that reconstruction would jeopardize the endangered bull trout, a federally protected species. Claiming ownership of the road, the local community took matters into its own hands and started bulldozing a new roadbed through the river itself. Federal officials promptly intervened, securing a federal court injunction halting the reconstruction. The community then converted the dispute into a regionwide issue; sympathetic loggers, miners, and ranchers rallied to the town's cause. In a show of solidarity, they sent hundreds of symbolic shovels for use in reopening the road, thus creating the Jarbidge Shovel Brigade. A local Fourth of July demonstration shrouded in patriotic rhetoric highlighted the town's grievances and launched a perfunctory effort by the Shovel Brigade to realign the road. Ultimately, following the 2000 presidential election, the controversy was settled with the county gaining ownership of the road subject to federal oversight, which meant it cannot be reconstructed without the necessary federal authorizations and permits. This episode reconfirms that tensions still run high within the West's rural communities, fostered by a steadily mounting federal regulatory presence, as well as noticeable decreases in logging, mining, and other extractive activities.[22]

TURNING TO CONGRESS AND THE COURTS

Well aware that Congress and the courts wield extensive powers over public land policy, the Wise Use movement also sought both legislative and judicial relief from new ecologically oriented laws and policies. In Congress, the Wise Use agenda has been both offensive and defensive, drawing strong support from the West's largely Republican congressional delegations. Offensively, the movement's congressional allies have sought to amend the Endangered Species Act, National Forest Management Act, and other environmental laws that are perceived as putting biological concerns ahead of human ones. They also have sponsored federal takings legislation designed to curb environmental regulations that limit the use of private property. Although none of these statutory reform proposals have advanced very far, their proponents have also used an alternative appropriations rider strategy that has achieved related short-term objectives. Most notably, the Republican-controlled 104th Congress passed the 1995 Timber Salvage rider, which effectively opened the West's national forests, including the Pacific Northwest's spotted owl forests, to unregulated timber harvesting. Defensively, the West's congressional delegations have fended off major legislative reform proposals designed to revise federal grazing and mining laws, and they have blocked any large-scale wilderness legislation since 1994. Congress also regularly used legislative riders to delay various Clinton administration policy reforms, including new mining reclamation regulations, new R.S. 2477 road maintenance regulations, and the Bitterroot grizzly bear reintroduction proposal. Thus, unable to advance an affirmative legislative agenda, the Wise Use movement's congressional allies have stymied reform proposals aimed toward greater environmental protection on public lands.

Outside the congressional arena, the federal courts have provided the "old West" with another forum to air its grievances against the advancing federal environmental agenda. By the early 1990s, when the Clinton administration took office, the extractive industries and their local allies had already developed a sizable public lands–related litigation docket. Several lawsuits had been filed (and frequently lost) over such issues as: opening the Overthrust Belt national forests to oil and gas leasing; maintaining national forest timber sale levels in order to keep a local mill afloat; compensating ranchers for forage consumed by wild horses; and protecting livestock from marauding grizzly bears. But faced with the new Clinton administration's ecologically driven reform agenda, the Wise Use movement and its allies ramped up their litigation efforts. In some cases, the litigation was brought by individual western states, several of which have created litigation funds to bankroll challenges to federal public land policies or decisions. In other cases, the litigation was brought on behalf of an industry or individual public land users, often with support from the Mountain States Legal Foundation or other such conservative advocacy organizations. While the docket is varied, several themes are apparent: constitutional takings doctrine continues to be a principal line of attack; the bedrock logging, mining, and ranching industries merit protection from any change in the status quo; and the Endangered Species Act's prohibitive mandates must be blunted. The fact that the Wise Use movement has registered few lasting judicial victories should not obscure the fact that the threat of litigation and an occasional injunction have perceptibly slowed the pace of federal reform efforts, particularly the ecological management agenda.[23]

The constitutional takings argument—that federal environmental regulation constitutes an illegal taking of private property—has been bolstered by a spate of U.S. Supreme Court decisions revitalizing the Fifth Amendment takings clause as it applies to privately owned land. These decisions, however, have only limited application in the public land setting, where most takings claims involve public—not privately owned—property and where there is a long history of extensive government regulation. But because the Constitution trumps other legal rights, the courts have seen a steady flow of takings claims. In several cases, ranchers have asserted a legally protected property right in their federal grazing permits, but the courts have regularly rejected these claims, finding that the permits are contingent upon compliance with federal grazing regulations and can therefore be revoked by the agencies without creating a taking. In a related case, Nevada rancher Wayne Hage argued that federal grazing restrictions precluding him from utilizing established water rights on public lands constituted a taking of his property interest in the water and appurtenant grazing rights. The Court of Federal Claims has accepted a limited version of this argument as it pertains to ditch rights of way across national forestlands. Takings arguments have also been invoked to challenge ESA limitations on timber harvesting and other activities on public lands, but they have usually been rejected. In sum, absent an unconditional permit, lease, or other clearly defined property interest in public resources, the takings provision does not wield much power on public lands.[24]

The Wise Use movement's litigation docket has otherwise sought to blunt the Clinton administration's public land reform initiatives. Although the cases were numerous, the successes were few, as reflected in the following illustrative decisions. The ranching community mounted a frontal assault, carried all the way to the Supreme Court, on Secretary of the Interior Bruce Babbitt's rangeland reform regulations, ultimately losing on virtually every claim. When the groundbreaking Yellowstone wolf reintroduction plan was finalized, the Wyoming Farm Bureau tried to halt it in a Wyoming federal court but ultimately lost when a federal appeals court found that the program fully complied with the ESA's experimental population requirements. Litigation seeking to block the Babbitt-initiated BLM wilderness reinventory of its Utah lands failed when the appellate court determined that the challenge was premature. In a lawsuit over a Forest Service decision reducing timber sale levels in a northern Montana forest to protect grizzly bear habitat, the D.C. Circuit Court of Appeals ruled that the agency's decision fell within its multiple-use discretion and complied with relevant ESA requirements. Repeated efforts by western rural counties to use the nineteenth-century R.S. 2477 statute to establish unfettered local control over roads crossing public lands have also failed, with the courts concluding that the federal public land agencies retain considerable authority over these roads. The federal courts also have rejected challenges to President Clinton's national monument designations. And the Court of Appeals for the Ninth Circuit has reversed an Idaho lower court ruling that enjoined implementation of the Forest Service's new roadless area conservation policy. In short, the courts have accepted few of the legal arguments advanced against these new ecologically based policies or initiatives, thus effectively legitimizing them as viable public land management strategies.[25]

GOING AFTER THE ENDANGERED SPECIES ACT

To appreciate fully the Wise Use movement's counteroffensive strategy, it is instructive to examine how it has tried to defang the powerful Endangered Species Act. At least since the northern spotted owl controversy, it has been apparent that the ESA could significantly constrain extractive uses on public lands. In response, the timber industry and its allies turned to Congress and the courts for protection, scoring temporary victories in both venues. They were unable, however, to secure a long-term legislative solution to the old growth controversy, and their litigation strategy likewise failed to limit the ESA's prohibitory strictures. Since then, Wise Use proponents have continued to make the case that the ESA is fundamentally unfair because it gives ecological concerns priority over economic interests. Their arguments continue to meet with little success in the congressional arena, but the courts have softened some of the act's provisions. Equally important, their litigation efforts have kept the statute in the spotlight, providing additional political fodder for those advocating a major overhaul of this powerful law.

As the Pacific Northwest's spotted owl controversy unfolded, local political leaders and timber industry executives shuddered when the FWS announced it was listing the owl as a threatened species. They understood the ESA would now become a major weapon in the battle over the region's old growth forests. They immediately turned to Congress for relief, seeking either to amend the ESA outright or to blunt its impact with an appropriations rider. With the ESA scheduled for reauthorization in 1992, the timing seemed propitious for a statutory overhaul. During previous reauthorizations, Congress had tinkered with the law, notably adding the requirement in 1982 that economic considerations must be factored into critical habitat determinations. But Congress had also steadfastly refused to alter either the act's basic structure or its principal science-driven mandates. Nevertheless, the ESA's congressional critics introduced several reform bills during the early 1990s, seeking to narrow the statutory "taking" definition and to preclude critical habitat designations when the economic costs outweighed the benefits. None of these proposals ever made it to the floor for debate, as Congress was unable to agree on any of the bills. Instead, it began extending the ESA on an annual basis, indefinitely deferring any reauthorization debate. But Congress also enacted annual appropriations riders that established a high-volume regional timber harvest quota and temporarily suspended the ESA and other environmental laws, thus limiting judicial review of agency logging plans. By the early 1990s, however, the Northwest's congressional delegations were unable to muster much support for these annual riders. If the ESA's opponents were going to alter the statute, they would have to do so through the courts, because they could not openly amend the law.[26]

Once the northern spotted owl was listed as a threatened species, timber industry allies filed two major legal challenges designed to limit the ESA's reach. In one suit, county commissioners from timber-rich Douglas County, Oregon, sued to enjoin the designation of nearly 7 million acres of public land as critical habitat for the spotted owl, alleging that the FWS ignored the National Environmental Policy Act (NEPA) by not preparing an environmental impact statement (EIS) for its critical habitat determinations. The Court of Appeals for the Ninth Circuit, however, found that Congress did not intend NEPA to apply to the ESA critical habitat designation process, concluding that the related Administrative Procedures Act (APA) rule making process provided adequate opportunity for public comment on the economic consequences associated with such designations.[27] In a second suit, the timber industry and its allies sought to narrow the scope of the ESA's most powerful provision, the Section 9 prohibition on taking any protected species. The taking prohibition applies to any person (not just federal agencies) and extends to endangered and threatened species. Relying on the ESA's definition of "take" as encompassing "harm" to a protected species, the secretary of the interior adopted regulations that prohibit any "significant habitat modification or degradation," which meant the spotted owl was fully protected on both public and private lands. The plaintiffs argued that the secretary had wrongly interpreted the statute and unlawfully extended the "take" prohibition to habitat modification activities. The Supreme Court, however, thought otherwise. The Court upheld the ESA taking regulation, finding that the secretary's broad reading of the "take" provision was consistent with the ESA's central purposes, which included providing "a means whereby the ecosystems upon which endangered species and threatened species depend may be conserved." By reaffirming the legal linkage between species protection and ecosystem preservation, the Court's ruling ensured the ESA would remain a potent legal factor in future natural resource controversies involving a protected species. And it signaled a major judicial defeat for the ESA's opponents.[28]

After the 1994 election, when Congress changed hands, the Wise Use movement and its allies refocused their attention on a legislative strategy, sensing that the new Republican-led Congress might be receptive to revamping the ESA. Relying on unsubstantiated anecdotes, the statute's opponents boldly asserted that the ESA was singlehandedly responsible for blocking critically important economic activity on both public and private lands.[29] Several ESA reform proposals were introduced during the 104th Congress, including proposals that would have effectively eviscerated the statute's key protective provisions. The more radical bills sought to eliminate the mandatory species recovery provision, to make critical habitat designations more difficult, to allow waiver of the consultation process, and to require financial compensation whenever the presence of an endangered species diminished property values. In fact, the Young-Pombo bill, co-sponsored by two of the ESA's most vocal critics, would have required the FWS to seek outside peer review before finalizing any listing, delisting, critical habitat, or jeopardy review decision. Although Congress again turned a blind eye to these bills, the ESA's congressional critics still attached a rider to a defense spending bill that temporarily placed a moratorium on new species listings and reduced the FWS's ESA compliance funding. Beyond the public lands, these same ESA critics also pursued crippling amendments to the statute's habitat conservation planning provisions, effectively forcing the Interior Department to promulgate regulations that established "no surprises" and "safe harbor" policies in an effort to placate irate private landowners. Since then, the Wise Use movement has continued to push its anti-ESA legislative agenda. Badly split over the issue, however, Congress has rebuffed every reform proposal, leaving the agencies to make the statute work, subject to the watchful scrutiny of the federal courts.[30]

Stymied in Congress, the Wise Use movement also has pursued litigation in a further effort to undermine the statute's biological preservation objectives. In an important victory, Oregon ranchers convinced the Supreme Court that they had legal standing to challenge the FWS's ESA jeopardy determinations, even though their interests were primarily economic rather than environmental. In the interior West, the Wise Use movement has succeeded in inserting economic concerns into the ESA critical habitat designation process. The Court of Appeals for the Tenth Circuit, in a case from Catron County, ruled that the FWS must prepare an EIS when making a critical habitat determination because economic considerations must be taken into account. In another Catron County case, the same court overturned an FWS critical habitat designation decision because the agency's policy effectively eliminated any economic calculus from its review process. Relatedly, the Court of Appeals for the Ninth Circuit ruled that the FWS cannot require an "incidental take permit" for livestock grazing on public lands under the ESA's Section 7 consultation provision unless it establishes that actual harm will befall the protected species. These decisions collectively are part of the ranching community's counterattack against the environmentalists' ESA-based campaign to curtail cattle grazing in southwestern national forests. Moreover, an Oregon district court overturned as arbitrary the National Marine Fisheries Service's "ecologically significant unit" listing decision policy distinguishing natural from hatchery salmon—a potentially significant ruling that calls into question other species listing determinations. Although none of the cases undermines the ESA's species preservation priorities, the Wise Use movement's litigation strategy has opened the courthouse doors to the statute's opponents, forced the FWS to address economic impacts in its critical habitat determinations, and limited its Section 7 consultation authority. In short, the courts have judicially softened but not fundamentally altered the statute.[31]

MOVING BEYOND CONFRONTATION

Whatever its chosen counteroffensive strategy, the Wise Use movement has focused on derailing new ecological management reform policies. As such, the principal targets have been the public land agencies and the environmental movement, which one battle-hardened Wise Use attorney refers to as extremists "seek[ing] to turn everything from the 100th meridian to the Cascade Mountains into a vast park."[32] The primary message is a mistrust of government and the need for local control over public lands, generally by claiming private property rights to curtail regulation. This strategy, however, masks a curious irony. The changes in public land and natural resource policy decried by rural westerners are as much the result of transcendent market forces and broad social changes as they are of any conscious government policies or environmentalist conspiracy. It is ironic that the agenda of the old West, a region that regularly espouses its strong commitment to free market principles and related property rights, is designed to forestall the very market pressures and social forces that created existing property interests. The federal public land agencies have little if any control over these forces, nor do environmental organizations. As we have seen, in the face of dynamic economic and social changes, even protected property rights will eventually give way to a new order, reflecting the evolutionary quality of our legal system and the private rights it chooses to bestow. And when it comes to public lands, property rights arguments have generally proved unavailing, absent an explicit governmentally created property interest.[33]

The Wise Use movement strategy is not based solely on property rights claims, however; it also employs the endangered community concept to generate public support for its efforts to preserve a disappearing way of life. The endangered community argument has several strands. From an economic perspective, rural westerners are deeply troubled by the loss of well-paying extractive industry jobs, a corresponding decrease in the local tax base, and the absence of any comparable alternatives to replace this lost income and revenue. As a social matter, many rural western communities feel increasingly isolated from the region's urban centers, prompting a sense of geographic inequity and marginalization. In fact, the rapid spread of urban archipelagoes across the West is viewed as further evidence of the growing isolation and frustration that the region's rural areas are experiencing. As a cultural matter, the growth of tourism, recreation, and preservation and corresponding shifts in federal policy priorities on public lands has prompted many rural communities to worry about losing their traditional identity and culture. That sentiment is often expressed in terms of not wanting to become another Moab or Telluride. Class tensions—pitting urban-bred newcomers attracted to the region's scenic and other natural amenities against longtime rural residents still committed to making a living off the land—may simmer just below the surface in communities undergoing change.

To address these concerns, the rural West has portrayed itself as the beleaguered underdog, or—rather curiously—as a new, nearly disenfranchised rural white minority. At the extreme, the contemporary plight of the rural western community has been likened to that of Native Americans, who have a long, sad history of oppression at the hands of the dominant white culture. Whether this comparison with the Native American experience merits the extension of additional legal protection for rural communities is debatable at best and disingenuous at worst. But the imperiled community image has resonated in political venues, at least enough to forestall some important policy changes.[34]

The endangered community concept may not be powerful enough to stave off the broader economic, social, and political changes confronting the rural West, but it has given the Wise Use movement's strategy a decidedly preservationist tenor. That preservationist strategy may, in turn, provide an opportunity to mesh the Wise Use agenda with the environmental agenda. Both are committed to preservation—community and cultural preservation in one case, biodiversity and ecosystem preservation in the other. As we have seen, ranchland conservation and other open-space initiatives, which have clear cultural preservation overtones, are also critical to any large-scale biodiversity conservation initiative in the rural West. Furthermore, although the county supremacy ordinance movement fell flat, it did acknowledge—albeit while embracing yesterday's industries—the need for land and resource planning to address the changes that are afoot. This is a significant concession for the rural West, where planning has historically been viewed with suspicion or worse. And while traditional individually held property rights may not be sufficient to forestall extant market and social forces, the notion of group or community-held property rights may hold some promise to protect traditional lifestyles and economies. Though there is little reason to expect a complete reconciliation among the contending factions, underlying similarities in their respective agendas may point the way toward some common ground.[35]

Indeed, with gridlock prevailing in Congress and the courts, there are indications that the rural West is moving away from the strident rhetoric of the Wise Use movement toward establishing a dialogue with its environmental adversaries over public land policies. Both sides in these pitched controversies have managed to hold their own in Congress, achieving an uneasy stalemate. Since the mid-1970s, there have been few comprehensive legislative changes involving the multiple-use public lands, save occasional state-based wilderness designations. The legislative record has been basically the status quo. In the courts, the environmental community probably holds the upper hand, particularly with its triumph in the northern spotted owl litigation and its regular ESA successes. The Wise Use movement has scored some transient judicial victories, though it has failed to undermine the environmental laws, including endangered species safeguards, that are driving contemporary public land policy. In truth, however, the courts alone are unlikely to remake federal public land policy in any significant way. Rather, the Wise Use movement's real power base lies in the West's rural communities and among the interior western states, though that does not typically translate into enough votes to redirect national policy. It does suggest that local venues might be an available forum for addressing many of the controversies facing public lands, but not if the strategy is purely a local one. Instead, an inclusive and broad-based dialogue where the various interests are fully represented holds real promise for breaking the prevailing gridlock and for ensuring that local issues are addressed constructively while keeping the overall public interest in sight. And that is precisely what has happened with the emergence of myriad place-based, collaborative initiatives across the western landscape.

Community-Based Conservation: Seeking a More Collaborative Model

That local communities would reclaim a major role in the public land policy agenda just as ecosystem management concepts are taking hold should come as no surprise. As evidenced by the Moab and Catron County developments, both communities have acknowledged the need to coordinate planning efforts with the federal land management agencies. Indeed, public land policy has never been divorced from the western states or the region's rural communities; local connections to these lands are too strong and the consequences of new policies are too immediate to be ignored. Thus, while new ecological concepts are forcing the agencies to shift their priorities and broaden their visions, a plethora of locally conceived collaborative initiatives are pressing community concerns into the new federal ecosystem management equation. And though local involvement with the public land agencies is not new in the West, these initiatives appear different in kind and focus from earlier interactions, which typically found federal managers hewing the community line. They are more extensive in terms of the issues being addressed; they are more inclusive in terms of the interests represented; and they are more expansive in terms of their geographical reach. As we shall see, these collaborative initiatives hold great promise and great peril.

COLLABORATION AND COMMUNITY: A NEW SYNTHESIS

The concept of community-based conservation, as it is emerging in public land policy, can be traced to multiple sources. At one level, the concept is linked to the national distrust of centralized government, as reflected in the growing devolution movement in federal policy, the emergence of civic republicanism theory, and the judiciary's reinvigoration of long dormant federalism principles. The movement toward greater local control captures the sense that important public policy decisions are best made by those who will be affected by them and who understand their full ramifications. At another level, the concept embraces the well-established trend toward encouraging public involvement in governmental decision processes, predicated on the realization that most policy decisions involve choices among competing values. Technical experts may help us to understand the choices to be made, but the ultimate decision will involve a value judgment, and who better to make such judgments than those who will live with the consequences? At yet another level, the concept is a further manifestation of the now widely accepted role of citizen participation in public land planning and project decisions, which Congress has ordained through NEPA and other laws to promote better-informed decisions and greater intergovernmental coordination.

The notion of community-based conservation also has deep roots in international nature conservation policy, particularly recent efforts to preserve parks and wildlife in the underdeveloped world where local residents are directly dependent on the land for their subsistence and where central regulatory strategies have often proven ineffective. Although the analogy between third world conservation and the modern American West's environmental controversies can be carried too far, local involvement in government planning and management decisions can help build critical popular support for new environmental policies.[36]

On public lands, the common experience with public participation in planning and management decisions has not been an entirely happy one. Traditionally, the responsible agencies sought citizen input through public hearings that routinely degenerate into political posturing rituals, each party adamantly staking out its own resolute position. With little real engagement over the issues, this basically adversarial model of public involvement left the agencies with little to do but announce their decisions and then defend them from the inevitable appeals that followed. We had a form of devolution in theory, but tightfisted federal control in practice. After the spotted owl controversy erupted into a legal donnybrook, nearly everyone seemed to agree that there must be another way to make decisions. Protracted litigation and endless legislative battles were draining to all, and they often yielded little concrete improvement on the ground. Seeking an alternate approach, former adversaries started meeting together, tentatively exploring potential points of agreement over locally contested terrain. A thoroughly pragmatic crisis response, these initial conversations soon solidified into concrete community-based initiatives, each somewhat different in purpose but sharing a common commitment to collaborative dialogue.[37]

The result has been truly phenomenal. In less than a decade, collaborative stewardship, consensus processes, watershed councils, community dialogues, public–private partnerships, and other such grassroots initiatives have sprung up everywhere, each committed to seeking common ground on public lands and elsewhere. The immediate origins of these initiatives are as diverse as their purposes. Some like the Canyon Country Partnership were the brainchild of agency officials seeking to promote a new ecosystem management agenda; others like the Trout Creek Mountains Working Group were catalyzed by an impending crisis, often an endangered species listing; still others like the Catron County Citizens Group were convened to reduce gridlock and to ease local tensions; and yet others like the Malpai Borderlands Group were locally designed to engage the community collectively in shaping its future. Secretary of the Interior Bruce Babbitt, after encountering one of these initiatives during the early days of his tenure, incorporated collaborative local stewardship principles into his range reform proposals, putting an official stamp of approval on the emerging collaborative conservation movement and simultaneously linking it to the Clinton administration's new ecological management agenda. Regardless of their origin or pedigree, these new collaborative initiatives are remaking traditional relationships between the public land agencies and their constituencies, thus reshaping the way these agencies do business.[38]

The collaborative process movement is closely related to the new federal ecosystem management agenda. Collaborative processes provide a forum where new resource management policies and proposals can be tested and shaped to fit local circumstances. They provide the agencies with a sense of community priorities and concerns, while also helping to identify areas of uncertainty or contention. They offer the community an opportunity to participate in solving local resource controversies, giving participants a stake in the process and outcome. Local knowledge about the landscape and its resources can be injected into agency deliberations, thereby improving final decisions. Partnerships can also strengthen community relationships, help reduce acrimony, and redirect resources from litigation to more productive purposes. For private landowners and others suspicious of the federal government, they can offer an informal and less adversarial setting for exploring alternative solutions and strategies. Partnerships can also help coordinate resource management decisions across an ecosystem and ensure that proposals take full account of potential environmental and other impacts. And because ecological management proposals are inherently entangled in human concerns, these new collaborative initiatives provide a forum for identifying and integrating local economic, social, and other concerns into the proposal. In short, partnerships and other collaborative approaches can be used to marry ecosystem-based management goals with community interests to produce more durable policies and decisions. They are, at their best, a form of devolved ecological management.[39]

Given the predominantly local nature of these initiatives, how should the relevant community be defined for collaboration purposes? In public land policy, the concept of community has traditionally been defined in geographical terms. Western communities have long depended on adjacent public lands for their economic sustenance, namely the minerals, timber, grass, or recreational opportunities associated with these lands. Within individual communities, citizens generally shared a common interest in the local sawmill, mine, or ski area, which often employed large numbers of local residents, afforded spinoff benefits to local businesses, and generated vital tax revenues to support government services. Cognizant of this reality, the Forest Service early on adopted a community stability policy designed to stabilize the fortunes of timber-dependent communities by ensuring a continuous flow of logs to local mills. But the agency's policy proved short-lived, owing to the changing dynamics of the national timber market and the competing demands that were increasingly being placed on national forestlands. Elsewhere, the notion of community involvement in public land policy was also defined in narrow, primarily geographic terms: for rangelands, the Taylor Grazing Act of 1934 provided for grazing advisory boards composed solely of local livestock permit holders; for water, state-created water district boards were limited to irrigators within the relevant watershed; and for wildlife, state game and fish commissions were composed primarily of hunters, ranchers, and other consumptive wildlife users. In an era when the West was a relatively homogeneous place, when the demands on public lands were relatively light, and when formal public involvement in government decision processes was rare, this narrow definition of community may have made sense. At least it did not ignite much controversy, though these arrangements are now cited as classic examples of "agency capture"—instances where a preferred user group is able, by virtue of its political power, to dictate agency policy.[40]

It is now apparent, however, that the concept of community must be broadened to embrace the manifold interests vitally concerned with public land policy. In a time when pluralism is the dominant characteristic of our political life, an array of diverse constituencies have laid claim to public lands, demanding a voice in how they are managed. Local political boundaries rarely correspond very well to the ecological concerns underlying contemporary resource management problems, and few communities reflect the myriad constituencies that claim an interest in the resources at stake. Rather than defining the community in purely geographic terms, therefore, it is appropriate to define it in broader terms—as a community of interests. Multiple factors account for this more expansive concept of community: evolving social and demographic patterns; greatly expanded markets and changing economic opportunities; public involvement legal requirements; major improvements in information gathering and sharing capabilities; the growing concern over governmental accountability; and the sheer geographical scale of new ecosystem planning processes. For public land management purposes, the concept of community has thus taken on new meaning; it now embraces not only loggers, miners, ranchers, and hunters, but also hikers, mountain bikers, birdwatchers, environmentalists, and others. This does not obviate geography as a factor in setting public land policy, but it changes its role: rather than being the sole criterion in defining community boundaries, geography has become the common interest that binds these diverse constituencies together. It is a shared "sense of place" that brings everyone to the table.[41]

The collaborative model of environmental decision making has been widely though not fully accepted across the western landscape. In 1997, seeking an alternative to the West's polarized environmental debates, the Western Governors' Association adopted eight overarching principles to guide resolution of such controversies. The resulting program—christened Enlibra—endorses local, ecosystem-based, collaborative problem solving processes, governed by national standards and fact-based science yet incorporating cost-benefit concerns as well as market-based solutions. The George W. Bush administration has also signaled its commitment to local involvement in public land decisions. In one high-profile instance, after criticizing the Clinton administration for ignoring local input during its national monument designation process, Secretary of the Interior Gale Norton requested the affected states and communities to advise her whether boundaries should be redrawn or management standards revised. Several environmental organizations, however, have rejected the Enlibra model and criticized the Bush administration's emphasis on local consultation, fearing these approaches are intentionally designed to disenfranchise national environmental constituencies and undermine national environmental standards. More broadly, they wonder whether such devolved decision making is really just a clever political strategy devised by the Wise Use movement to thwart the environmental community's public land reform agenda. Their position has a certain *real politik* logic to it: If the new ecological management agenda is prevailing at the national level, then why cede any authority to a locally conceived group? It is, pure and simple, a strategic matter of selecting the most hospitable forum.[42]

Nonetheless, the new community-based conservation movement is gathering momentum. Much like ecosystem management, these collaborative processes are not expressly sanctioned by law, and the relevant concepts are being stitched together, a mixture of political pragmatism, homespun experimentation, and sheer determination. Often touted as models for promoting sustainability, many of these new initiatives are consciously being designed to address the needs both of the ecosystem and the community, thereby becoming a vital part of ecosystem management lore. In the best cases, the initiatives provide a useful forum for identifying and resolving the scientific, economic, and social issues that lie at the heart of most natural resource controversies. Or, put differently, a well-designed collaborative process is the means for integrating human concerns into the ecological management equation.

THE COLLABORATIVE EXPERIMENT:
BUILDING A PRINCIPLED PROCESS

Having emerged organically in the shadow of the law, the nascent collaborative process movement is governed by few clearcut standards. This raises a fundamental question: What are the essential requisites for a viable and legitimate community-based collaborative initiative? As with any alternative dispute resolution process, the governing principles are easier to identify than to apply: fairness, equity, accountability, sustainability, and scientific integrity. The first three principles are basically procedural in nature, while the final two add substantive dimensions to the process. Community-based conservation is thus best understood as a collaborative process employed to achieve broad ecological, economic, and social goals in a truly sustainable and scientifically credible manner.

Collaborative processes must be fair and equitable. The procedures governing any collaborative effort should afford all participants a full opportunity to engage in the dialogue, with each voice given equal respect. The rules of participation should be clearly established in advance, known by everyone, and applied uniformly. The process also must be inclusive. This means that everyone with an interest in the issues—whether a local resident or a concerned "outsider"—should have the opportunity to participate and to be treated similarly. Defined in community of interest—rather than geographic—terms, community-based collaborations are therefore diverse groups; they involve the myriad constituencies interested in public lands—ranchers, miners, loggers, hikers, hunters, four-wheel-drive enthusiasts, environmentalists, and others. Each of these constituencies will bring different factual knowledge and expectations to the table, which should broaden and enrich the ensuing dialogue. Moreover, the collaborative process must be open and transparent, permitting others to join the dialogue or to simply observe its progress. A fully open and transparent process is likely to foster a sense of trust among the parties, enabling them to move beyond initial "us–them" conceptions and related posturing to begin building durable relationships. When everyone has an opportunity to share ideas and build relationships, new and innovative solutions may well emerge, often converting a seemingly zero sum game into a win-win proposition.[43]

Community-based conservation processes also must be accountable, in both a political and legal sense. Indeed, accountability is crucial to the legitimacy of any collaborative initiative; it forces both the process and the outcome from the shadows into the full glare of public scrutiny. Given the informal and representative nature of most collaborative processes, the participants must be prepared to answer for their decisions to their separate constituencies. Unless the interested constituent groups are satisfied that their views have been represented in the process, they may question both its legitimacy and outcome. When federally owned lands or resources are at issue, the final outcome of any collaborative group also must be assessed in terms of the broader public interest. This means that national as well as local views must be factored into the process. Beyond these essentially political concerns, collaborative processes also must comply with the relevant federal laws that govern any such ad hoc initiative. Dissatisfied parties are entitled to question both the process itself and any final decisions through administrative appeals and judicial review. Thus, whether the collaborative conservation movement is understood in terms of public-spirited citizenship or merely private gain, accountability is crucial to ensuring the legitimacy of the process.[44]

As a substantive matter, community-based collaborative processes should produce tangible and sustainable results that can be verified. Meaningful collaboration cannot be based solely on process; it must be held to a standard of achievement. When the goal is community-based conservation, that standard will generally be measured in sustainability terms, that is, whether the initiative produces enduring environmental benefits while also meeting overall community needs. According to the scientists who helped revise Forest Service planning regulations, sustainability means "meeting the needs of the present generation without compromising the ability of future generations to meet their needs. . . . it calls for integrating the management of biological and ecological systems with their social and economic context, while acknowledging that management should not compromise the basic functioning of these systems."[45] As so defined, sustainability differs from the traditional approach to natural resource management; it takes a long-range (rather than short-term) view, and it treats the landscape as an integrated and holistic entity (rather than as a collection of individual resources). Because sustainability predictions are often uncertain and contingent, sustainability is best measured over time. Such verification will ordinarily require regular monitoring to ensure that the established goals are being met and that revisions are made as necessary. In other words, the same adaptive management protocols that are part of ecosystem management should be incorporated into any community-based conservation initiative.

If sustainability is the overarching goal, then the collaborative process must integrate credible science into the community dialogue. Although science cannot dictate final outcomes when human value judgments are at issue, it can provide communities with key information regarding risk, trends, and uncertainty that must be factored into any final decision. In some instances, science may actually set the bottom-line requirements, as when legally mandated endangered species standards are at issue. In other instances, scientific data will serve to define risk thresholds, which may counsel against proceeding with irreversible environmental modifications when the ramifications of doing so are uncertain and potentially serious. Outside the ecological sciences, technical information concerning economic and social trends will also help participants understand the potential opportunities and trade-offs involved in selecting one option over another. And the scientific method—trial, error, and readjustment—basically mirrors the adaptive management principles that are essential to measuring ecological sustainability and other conditions. In short, science informs what would otherwise be merely an ad hoc process, giving collaborative conservation initiatives both integrity and credibility.[46]

Community-based conservation, when pursued according to these basic principles, adds a new and important dimension to the ecosystem management equation. It ensures that human and scientific concerns are both part of the agenda by offering a meaningful process for identifying and resolving community controversies. Put simply, it provides a democratic forum for shaping future natural resource management policies, thus confirming the marriage between politics, science, and posterity.

THE LIMITS OF COLLABORATION: LAW AND DEVOLUTION

The rapidly evolving collaborative conservation movement is not without limitations, however. Because public lands and resources are at issue, the responsible federal agencies face very real legal constraints on the degree to which they can devolve their authority to any collaborative group. Because the Constitution expressly vests primary responsibility for public lands in Congress, federal law and policy will clearly prevail under basic supremacy principles over any contrary state or local pronouncements—a lesson reiterated by judicial decisions invalidating the various county supremacy ordinances. In other words, the law generally preserves federal authority and seeks to ensure basic fairness. In the few cases that have involved collaborative-type process initiatives, the courts have scrutinized the arrangements skeptically, insisting on agency accountability and procedural regularity. But the law does not preclude such arrangements. In fact, the public land agencies have specific legal obligations to engage others in their decision making processes.

In the new world of collaborative conservation, a key question is whether the agencies can devolve—or delegate—their congressionally derived management authority to a citizen group. When the National Park Service (NPS) sought to relinquish its statutory planning authority for the Niobrara National Scenic River to an advisory committee composed primarily of local government officials and private landowners, a federal court promptly invalidated the arrangement, citing the unlawful delegation doctrine. According to the court, although the Park Service retained the power to terminate its Interlocal Agreement with the Niobrara Council, the management arrangement was an unlawful delegation "because NPS retains no oversight over the Council, no final reviewing authority over the Council's actions or inaction, and the Council's dominant private local interests are likely to conflict with the national environmental interests that NPS is statutorily mandated to represent."[47] The court was quite troubled that the Park Service was no longer accountable for managing the Niobrara River area, that it could not ensure compliance with the Administrative Procedures Act, NEPA, and other important federal laws, and that its only control option was the draconian—and thus unlikely to be used—power of terminating the agreement. In another instance, when the BLM essentially transferred its livestock grazing management responsibility to local ranchers, a federal court struck down the arrangement, finding that the agency could not "tie [its] own hands with respect to [its] authority to modify, adjust, suspend, or cancel [the grazing] permits."[48] These same delegation principles would seem to limit the U.S. Fish and Wildlife Service's ability to delegate its endangered species management authority to a local Citizens Management Committee, as the Bitterroot grizzly bear reintroduction proposal contemplated. Plainly, whether framed as an unlawful delegation or as an accountability problem, the law requires the public land agencies to retain final management authority over public lands, absent a clear expression of contrary congressional intent.

By their ad hoc and experimental nature, collaborative conservation processes raise rudimentary fairness concerns. What procedural obligations—if any—must a collaborative group meet before it can legitimately participate in public land management decisions? The once obscure Federal Advisory Committee Act (FACA) provides legal guidance here, requiring federal agencies to follow rigorous procedural requirements before receiving or soliciting advice from any group of persons who are not full-time federal employees. Under FACA, an agency seeking outside advice must organize and charter a committee, ensure that its membership is balanced, publish notice of its meetings, and permit public participation in those meetings. During the mid-1990s, as the collaborative process movement was gaining momentum, several lawsuits raised FACA-based objections to these new entities. One of the most prominent cases was the timber industry's FACA challenge to the Northwest Forest Plan and the Clinton administration's budding ecosystem management policy, which sought to derail both initiatives. Although the federal courts agreed with the industry plaintiffs that the Federal Ecosystem Management Assessment Team (FEMAT)—which included five state university professors among its core membership and which conducted its work in secret—constituted an illegal advisory committee that violated FACA's basic openness requirements, they refused to overturn the forest plan or to block the agencies from using the FEMAT report. In other cases, the courts have occasionally chastised the agencies for various technical FACA violations, but they have usually refrained from prohibiting these same agencies from using an advisory committee's work product. Nonetheless, these initial FACA court cases reverberated far beyond their immediate contexts, casting a cloud of uncertainty over many of the new multiparty collaborative initiatives. The federal land management agencies reacted by disassociating themselves from many of these processes, fearing that their involvement with nonfederal constituents could trigger a FACA challenge. Citizen-initiated collaborative groups were also confounded, both by the federal agency retrenchment and by the unexpected impact of these abstruse legal requirements. Thus, despite its ambiguous results, the FACA litigation slowed many collaborative processes and recast the federal role in local, ecosystem-based management initiatives.[49]

Confronted with this unexpected threat to the nascent collaborative conservation movement, Congress responded by exempting certain intergovernmental relationships from the FACA requirements. The vehicle was the 1995 Unfunded Mandates Reform Act, a key part of the Republican-controlled 104th Congress's devolution agenda. Congress amended FACA to exempt any meetings held exclusively between federal officials and their state, local, or tribal counterparts when the purpose was to share information or advice related to their joint responsibilities. The strict FACA procedural requirements, therefore, no longer apply to most intergovernmental groups or communications, which play a key role in promoting interjurisdictional coordination for ecosystem management purposes. State, local, and tribal governments have effectively attained a privileged status in their government-to-government dealings with the public land agencies. However, the FACA procedural requirements still apply to collaborative processes involving private parties or nongovernmental organizations, assuming the group fits the definition of an advisory committee. When this is the case, the choices are twofold: the group can either comply with FACA and have the public land agencies involved in the collaborative process, which few local groups have yet chosen to do, or continue the collaboration without direct federal agency involvement, the path chosen by the Malpai Borderlands Group, Quincy Library Group, Applegate Partnership, and others. So long as the group complies with FACA's strict procedural requirements, its recommendations and proposals are entitled to a presumption of legitimacy, which is forfeited when the group chooses to ignore the FACA strictures.[50]

Because the restrictive Endangered Species Act is widely regarded as the ultimate expression of federal power on public lands, are there opportunities to soften this law's sometimes harsh impact through collaborative processes? Can the federal regulatory agencies—the USFWS and the NMFS—bargain with the states, local communities, and others over endangered species listings, jeopardy opinions, takings limitations, and other ESA-dictated restraints? In one highly publicized instance, when federal officials agreed not to list an Oregon coho salmon run in deference to a locally negotiated state recovery plan, a federal court struck down the arrangement, finding there was no assurance the state plan would adequately protect either the species or its habitat. Oregon's Governor John Kitzhaber, concerned that an ESA listing would bring harsh federal restraints to bear on the state's timber companies, ranchers, and other landowners, conceived an alternative ecosystem-based Oregon Plan to restore salmon habitat and to regulate logging and ranching practices. Produced through a self-initiated local collaborative process, the Oregon Plan sought to enlist private landowners in salmon habitat restoration efforts, which the governor argued were more critical to the species' recovery than its listing on the federal ESA registry. Although the plan was formally endorsed by the NMFS, which agreed to hold its previously announced listing decision in abeyance, local environmental groups balked. Asserting that the timber companies and others could not be trusted voluntarily to protect salmon habitat, they sued on the theory that the NMFS was legally obligated under the ESA to "list" and protect the salmon. A federal court agreed, finding the ESA listing criteria required "existing regulatory mechanisms" to protect salmon habitat, not vague or untested assurances about future protection plans. Thus, although the ESA regulatory agencies have explored alternative collaborative approaches to species recovery, they have limited ability to devolve their explicit protection authority to others.[51]

Of course, the ultimate devolutionary action under the Endangered Species Act is to delist species, which returns management responsibility from the federal government to the states. On several occasions, state officials have decried the FWS's alleged reluctance to declare species recovered, while the federal government has made headlines in the few instances when it has upgraded the status of individual species, such as the peregrine falcon, American alligator, and bald eagle. Several mostly industry-backed lawsuits have been litigated to compel the FWS to delist species, but with little success thus far.[52] On public lands, a major delisting debate is brewing over the grizzly bear in the Greater Yellowstone Ecosystem. The surrounding states—Idaho, Montana, and Wyoming—are intent on recapturing management responsibility for the bear, and the FWS has indicated that bear population numbers are approaching recovery targets. A key issue in the debate will be whether the states can develop adequate regulatory mechanisms to ensure the bear's well-being, especially on federal lands where the state's authority is weakest. Among the three states, only Montana has a state endangered species act, and it is notoriously weak. A similar debate is also brewing over delisting the recently reintroduced wolves, because recovery population targets have nearly been reached. But as we have seen, Wyoming still treats the wolf as an unprotected predator, while Idaho has steadfastly refused to help manage reintroduced wolves, raising serious questions whether these states are truly prepared to meet post-delisting management responsibilities. Put simply, the states must navigate a rigorous legal obstacle course before they can reclaim management authority for ESA-listed species.[53]

CONFRONTING THE HARD QUESTIONS: OF POLITICS AND POWER

Beyond legal constraints, the critics of collaborative processes have raised serious political and policy concerns that counsel caution before wholeheartedly embracing such a devolutionary approach to public land planning and management. The criticisms range from the paucity of proven accomplishments to the potential disempowerment of urban, environmentally friendly constituencies, to the fear that new bureaucratic burdens will impede efficient decision making. Stripped to the essentials, the primary concern is really over who controls the policy agenda and thus holds the reins of power. The proponents of collaboration meet these concerns with a very pragmatic response, arguing that there are few other effective alternatives for implementing public land policy in the present overheated atmosphere. Given the diversity of issues, settings, and interests involved in any public land dispute today, they may have the better argument, at least when the issue is not one of obvious national significance.

Perhaps the most accurate measure of any natural resource decision process is the result that it produces on the ground. What, therefore, have these new collaborative conservation initiatives actually accomplished: Are species and their habitats more secure? Is the ecosystem healthier? Are local economies more sustainable? Critics of the collaborative conservation movement contend that very little has been accomplished despite the proliferation of local initiatives, asserting that collaborative groups tend toward lowest common denominator solutions. While generalizations are difficult (given the diversity of the initiatives, the disparate local conditions, and limited time most have existed), some tentative observations can be made. Several projects have plainly achieved noteworthy results: examples are the Canyon Country Partnership's Sand Flats bike trail project outside Moab; the Malpai Borderlands Group's fire restoration and grass banking projects; and the Flagstaff fire restoration initiative. But these projects have addressed primarily local matters. The story is quite different when tough, national-level issues, such as endangered species listings and wilderness designation, are at stake. The Canyon Country Partnership, for example, was unable to resolve air tour overflights, wilderness, roads, or other such issues, and some members quit the partnership because of its perceived ineffectiveness. Elsewhere, no species have been delisted as a result of any collaborative initiatives, while several species—like the jaguar and various salmon runs—have landed on the endangered species registry over local objections. In fact, Governor Kitzhaber's ill-fated Oregon salmon restoration plan vividly illustrates the lowest common denominator problem: rather than seeking agreement over concrete regulatory measures, the locally negotiated plan was built on voluntary and thus unenforceable agreements, that while popular with the timber and ranching industries, proved unacceptable to the environmental community. Nevertheless, it is wrong to discount entirely the local results that have been achieved through innovative collaborative projects, and it would be equally wrong to expect these projects to resolve national-level debates that will plainly have to be addressed in a larger, and inevitably more adversarial, federal forum.

Because of the local and self-appointed nature of most collaborative conservation initiatives, critics have repeatedly questioned whether these groups are truly balanced and representative. At the conceptual level, critics note that place-based initiatives, which by definition include local residents and focus on specific local concerns, tend to exclude or overlook broader national interests from consideration. One prominent critic makes the point bluntly: "The allocation and conservation of federal lands and resources necessarily and properly are national, not local, questions."[54] But others disagree, arguing that few public land issues are of truly national significance and that most public land communities are sufficiently diverse today that local representatives effectively serve as surrogates for more distant urban interests.[55] As a practical matter, whenever government agencies, industry representatives, and environmental organizations are involved in a collaborative process, there is often an inherent imbalance between the parties that can undermine effective participation. Government and industry generally can bring more substantial resources to the table than can private citizen participants. Not only are their representatives being paid while involved in the process, but they usually have extensive financial resources, scientific information, and technical support readily available to them. That is not the case for most environmental organizations and other voluntary private participants, who can find themselves overwhelmed and forced into concessions. Also, because most collaborative processes involve place-based initiatives and local negotiations, environmental critics believe this seriously disadvantages their largely urban membership, who are often unable to participate in distant meetings. When federal lands or resources are at issue, therefore, a collaborative group must minimally comply with the basic FACA membership balance and procedural fairness requirements before it can enjoy privileged status. Or better yet, collaborative initiatives should be defined in community-of-interest rather than community-of-place terms, and they should be consciously designed to facilitate broad-based participation throughout the process.

The critics also question whether many of the alleged benefits of collaborative processes are actually illusory. Its proponents have touted collaboration as a more efficient and less costly alternative to litigation or legislation for resolving natural resource controversies. They note that by seeking consensus, participants build goodwill and strengthen community relationships, thus enhancing local problem solving capabilities. And they argue that these qualities should enable collaborative groups to resolve controversies fairly and efficiently. To the critics, however, any collaborative process requires a heavy investment of time and money from multiple parties; the tougher the issue is to resolve, the heavier the investment must be. Because most collaborative efforts are voluntary and lack meaningful coercive enforcement powers, critics fear they will only add another bureaucratic layer and expense to the decision making process. If the issue is difficult enough, the likelihood of achieving consensus is remote, which means the traditional adversary tools—litigation and legislative lobbying—will probably end up being deployed anyway. In addition, consensus usually depends on achieving win-win solutions, but many of today's resource issues—given prevailing forest, rangeland, and watershed conditions—can produce only win-lose outcomes. So why not, the critics ask, simply rely on our traditional decision processes, which have built-in procedural protections and can often produce more efficient results?[56]

Another troublesome issue is the matter of scale, as both a theoretical and practical matter. Many collaborative conservation proponents see local groups as the logical precursor to a successful ecosystem management initiative, conceiving them as the catalyst that is required to build trusting relationships before larger ecosystem-level issues can be addressed. At its core, this view rejects a top-down, federally driven ecosystem management agenda in favor of an organic, locally based strategy for resolving natural resource controversies. But for ecosystem management proponents, this view of collaborative conservation puts the very idea of ecosystem management at risk; it basically ignores the need to plan at an ecosystem scale in deference to local political sensitivities. Rather than setting the scale by scientific standards, collaborative conservation would use political standards, even though political boundary lines are responsible for many of our current ecological problems. And even when the collaborative process involves a localized, place-based initiative, the assumption that local communities or residents will unite around a common commitment to the landscape can prove unfounded. Rural communities sharing the same public lands do not always have the same priorities. Witness San Juan County's decision to withdraw from the Canyon Country Partnership, because it did not share neighboring Grand County's views on recreational development or wilderness designation. Similarly disparate views have sometimes surfaced between a state and its communities, as was the case when Idaho resisted the U.S. Fish and Wildlife Service's Bitterroot grizzly bear reintroduction proposal even though it represented a locally negotiated agreement. The practical lesson is clear: as the scale of the initiative grows in size or as the issue grows in precedential importance, the number of people involved must increase and the likelihood of finding ready consensus decreases.

Critics, citing the immature nature of the collaborative conservation movement, also wonder whether the many mostly voluntary community-based initiatives can be sustained over the long term. Indeed, even proponents of collaborative processes do not deny that the presence of a crisis or "cause" is often critical to sustaining the process. Absent a proposed ESA listing or threatened litigation or land use plan revisions, there may be little incentive for a community or various interest groups to invest the time and energy necessary to solve what are invariably difficult natural resource allocation problems. The Canyon Country Partnership, for example, has continued in the wake of the Sand Flats controversy, but it has had difficulty sustaining local involvement, as witnessed by San Juan County's defection. Equally important, federal involvement has been a key ingredient in many of these new collaborative initiatives. Most observers agree that the federal agencies must participate if the process is to work; only they have the decision making power, the scientific data, the technical know-how, and the money necessary to sustain the process. When federal agencies have not been active participants in such processes, the initiatives have invariably been subjected to harsh criticism and experienced serious implementation problems, as we shall see in the case of the Quincy Library Group. Only time will tell whether these new initiatives will prove self-perpetuating, but federal involvement is plainly emerging as a critical element of success.[57]

These problems should not be discounted. They not only help put the collaborative conservation movement in proper context but also offer important provisos for it. Collaborative processes are not an all-purpose substitute for the traditional dispute resolution processes of legislation, lobbying, and litigation; instead, they should be viewed as a supplemental mechanism to these conventional forums for addressing natural resource controversies. Collaborative initiatives are an appropriate strategy for addressing some natural resource controversies, so long as they operate within the existing institutional and legal framework. A properly designed and legitimately constituted collaborative group can help the public land agencies formulate workable solutions to many natural resource controversies, particularly localized, place-based problems that do not raise nationally significant issues. Clearly, a local collaborative group can bring much to the table—its collective knowledge, its passion for the place, and its resolve to find an acceptable solution. And the benefits of such problem solving cannot be gainsaid: a renewed sense of community spirit, a reduction in federal–local tensions, and a commitment to sustaining the shared landscape. But even then, the ultimate decision authority must reside in the managing agency, which also must ensure the fairness and integrity of the collaborative process. And the process itself, as well as the agency's final decision, must be subject to public scrutiny, including technical review and administrative appeal where available. With these limitations and safeguards, accountability concerns should be resolved and community-based conservation should merit a place beside ecosystem management as a legitimate and mutually reinforcing approach to contemporary public land management.[58]

Storm Clouds on the Horizon: Recreation as a Panacea?

Ecosystem management and collaborative conservation are not the only major changes affecting public lands. The past several decades have witnessed a veritable explosion in outdoor recreational activities, and a corresponding growth in tourism throughout the western states. Beginning with the earliest national parks, the scenic and historical attractions on the West's public lands have served as a magnet for visitors, creating an important link with the region's tourism industry. Over the years, the nation's growing population has continued to turn its sights westward, drawn by opportunity and open space. Many of the region's rural communities, once rocked by a local mine closure or decommissioned sawmill, have embraced recreation and tourism as an economic salvation. Moab certainly swallowed this bromide, only dimly aware of the changes lying ahead, both for itself and for adjacent public lands. Rather than putting environmental conflicts to rest, recreation and tourism have brought a new set of controversies to the fore, including the wisdom of an industrial-scale recreation economy and an acrimonious debate over motorized versus nonmotorized activities. The basic question is not the propriety of recreation or tourism on public lands but how to mitigate their impacts in order to protect the ecosystems that nourish these alluring landscapes.

Opening Pandora's Box: Recreation and Tourism Unleashed

To say that recreation is a growth industry on the public lands would be a gross understatement. From Moab to Missoula and numerous points beyond, recreation has effectively displaced mining, logging, and ranching as the major local industry, generating revenue and job opportunities that far outstrip conventional extraction industries. The figures for recreational use are truly impressive. On the national forests, visitation has skyrocketed, growing from 27 million visitors in 1950 to 172 million in 1970, and to 345 million in 1995. Recent estimates have these figures approaching a billion annual visitors within the next decade. The national parks, national wildlife refuges, and BLM lands have experienced similarly dramatic growth patterns. The national parks, for example, saw tourism surge from 33 million visitors in 1950 to 280 million in 2001. In specific locations, the numbers are equally sobering. Visitation at the Colorado Plateau's 27 national parks soared from 8.6 million in 1981 to 16.7 million in 1994. At Yellowstone National Park, visitation jumped from 2 million in 1980 to over 3 million in 2000, including a major boost in winter visitors during this same period. Similarly, the ski industry has experienced remarkable growth during the past half century; expansive all-season resorts have sprung up across the national forests, often transforming sleepy mountain valleys into small cities. By any measure, recreation has become an independent force on the public lands, supported by diverse new constituencies every bit as intent on securing access rights as their counterparts in the extractive industries.[59]

Historically, recreation played an important but often overlooked role on public lands, treated quite differently by the individual agencies. The Park Service has long regarded itself as the nation's primary recreation agency, and outdoor recreation has traditionally been an important part of the park visitor experience. On the national wildlife refuges, early preservationist impulses gave way to a strong hunter lobby, which has sanctified recreational hunting and fishing as a legally endorsed refuge activity. On multiple-use lands, however, recreation often seemed as much an afterthought as a primary management objective. Following World War Two, however, the nation took a fresh look at how the public lands might better accommodate society's resurgent interest in outdoor recreation pursuits. The Park Service responded by launching its ambitious Mission 66 campaign, designed to attract new park visitors by expanding roads and visitor facilities. A congressionally created Outdoor Recreation Resources Review Commission endorsed a plethora of new proposals aimed toward enhancing recreational opportunities; its recommendations were soon translated into the Wilderness Act of 1964, the Land and Water Conservation Fund Act of 1965, the Wild and Scenic Rivers Act of 1968, and the National Trails System Act of 1968. At the same time, Congress began creating specially designated national recreational areas, starting with Lake Mead in 1964, designed to accommodate a wider array of motorized and other activities than were ordinarily allowed in the national parks. The principal strategy was evident: develop new specialized designations to expand, enhance, and diversify recreational opportunities on public lands.[60]

The boom in outdoor recreation extended beyond the public lands, reaching into the nearby towns and creating new entrepreneurial opportunities and job possibilities. With it, another important link was forged between the public lands and adjacent communities—one founded on recreation and amenity values rather than resource extraction and development. Of course, the gateway community phenomenon has long been connected to the national parks, exemplified by such towns as Jackson Hole and Estes Park. Once the Forest Service started permitting ski areas in the national forests, downhill skiing quickly grew in popularity, soon giving rise to new destination resorts hooked inextricably to the surrounding public lands and nearby communities, such as Aspen, Telluride, Park

City, and Ketchum. Local dude ranches, river rafting companies, and outfitting businesses created yet other connections to nearby public lands. With these myriad linkages, the public land agencies could no longer ignore the local economic implications of recreation and tourism in their planning and management decisions. As more and more recreation-oriented businesses emerged to accommodate the increasing number of visitors, the agencies were soon regularly negotiating special use permit arrangements that gave these commercial enterprises access to public lands, generally at modest fees. Increasingly viewed as attractive places to live and play, many of these communities experienced tremendous growth themselves, spreading right up against nearby federal boundaries and creating a new set of urban–wildland interface problems, including fire control and public access concerns.[61]

Until recently, the public domain has served primarily as an unregulated commons for recreation. Access for recreational purposes has been virtually free, and the courts have viewed recreationists as licensees who enjoy unfettered access so long as they do not bother others.[62] The exploding interest in outdoor recreation, however, has brought a corresponding explosion in recreational activities and yet more pressure on public lands. Both the number and variety of activities are difficult to comprehend. Hunting, fishing, hiking, and sightseeing, once our primary outdoor pursuits, are today only part of an ever-growing assortment of muscle-powered and motorized sports that include: skiing, snowboarding, snowmobiling, cross country skiing, hut-to-hut ski touring, snowshoeing, backpacking, rock climbing, caving, horseback riding, off-road driving, motorcycling, dirt biking, mountain biking, hang gliding, orienteering, river rafting, kayaking, sail boarding, marathon running, and assorted competitions and jamborees associated with each of these activities. Not surprisingly, as public interest in outdoor recreation has grown, conflicts between recreationists and the extractive industries have proliferated; few people relish hiking or camping in a field of clear-cut stumps or beside an abandoned mine site. And as off-road vehicles have become more popular, similar conflicts have arisen between motorized and nonmotorized users, many of whom object to the sound of roaring motors and the smell of exhaust fumes. The mere presence of more people recreating on the public lands has generated concomitant pollution, erosion, and habitat incursion problems. Vast as they are, these lands are no longer expansive or durable enough to accommodate every type of recreational activity in every location.

To address these conflicts, the federal agencies have begun imposing direct and indirect limitations on recreational users and activities. Direct regulatory constraints include visitation ceilings, permit requirements, activity prohibitions, and zoning limitations; indirect constraints include such devices as user fees, facility location and design decisions, availability of services, and the presence or absence of roads, trails, and signs. Various user groups have mobilized in opposition, mounting court challenges and lobbying campaigns against these new limitations. Sensitive to the tradition of open access, the agencies have treaded lightly, reluctant to prohibit specific activities or even to curtail motorized uses except when absolutely necessary. The courts, for their part, have ruled that the agencies have broad authority to regulate recreation on public lands, sustaining regulations that outlawed motorized vehicles in backcountry locations, limited river rafting numbers, and prohibited mountain biking in national park areas. For its part, Congress has played an equivocal role in these matters: it has created new wilderness areas and other special designations; provided federal funds for new roads, trails, transportation systems, and visitor facilities; and inaugurated a new recreation fee demonstration program to provide additional maintenance funds. But Congress has not endorsed any comprehensive legislation that would either alter agency resource management priorities or shift the balance of power among the various user groups. As a result, the agencies enjoy considerable managerial discretion over recreation on the public lands, which they might—or might not—employ to protect environmental values or visitor experiences.[63]

Across the West, the tourism industry has also experienced enormous growth, with major implications for the region's public lands and communities. Western tourism has long been a de facto public–private partnership: national parks and adjacent public lands have offered unparalleled scenic vistas, historic sites, and recreational opportunities that attract visitors, while adjacent towns and communities have provided accommodations, guides, and other necessary services. A similar symbiotic relationship has evolved with the region's ski towns, where the ski area is typically situated on leased national forestland, while lodging and services are available on privately owned base lands and in nearby towns. Not surprisingly, the economic implications of tourism are substantial: in 1995, tourists spent over $26 billion in the nine interior western states, while the travel and tourism industries accounted for nearly 500,000 jobs in these same states.[64] Because a tourism-based economy offers an array of economic opportunities, ranging from real estate sales and development to new retail and service industry niches, tourism is now the largest private employer in several western states.[65] In addition, many vacation visitors have returned to the region as new residents, often bringing outside incomes with them that are injected into the local economy. Their arrival may also bring new local amenities, which can range from restaurants and bookstores to a seasonal symphony orchestra or an airport expansion. Not all communities, of course, can or even want to embrace tourism and recreation, but those that do have usually enhanced their own economic and social diversity.[66]

But tourism and recreation have a dark side too. Economically, the associated employment opportunities are often seasonal and relatively low-paying, particularly when compared to wage scales in the extractive industries. An amenity-based economy may afford new entrepreneurial opportunities, but much of the investment capital for infrastructure development may come from outside sources, fostering the same sense of distant ownership and exploitation that historically characterized the West's extractive industries. The corporate concessionaires in the region's largest national parks—Yellowstone, Grand Teton, Grand Canyon, and

Glacier—have few ties to the surrounding communities; and few of the region's ski resorts are locally owned or controlled anymore. Many resort towns have spawned a gaping divide between wealthy new arrivals and longtime residents, as well as a new stratum of maids, dishwashers, and other service employees, most of whom are poor and minorities. As Moab's evolution illustrates, basic social, cultural, and political differences often separate new and old residents, reflected in attitudes toward the environment, nearby public lands, and the community itself. Too often the new residents are seasonal; they absorb and may even change a place without participating directly in the local society. In short, the advent of a tourism- and recreation-based economy tends to benefit some more than others while generating new schisms and classes, often leaving a transformed community in its wake.

The new tourism-recreation economy can also have a transformative effect on the landscape, as Moab has also discovered. Rather than clearcuts and open pit mines, its legacy is suburban-like sprawl, new ranchettes, and mega-resorts that chop once pastoral landscapes into smaller and smaller fragments. As new homes and secondary roads spread across vacant agricultural lands, open space begins to disappear, winter wildlife habitat is lost, seasonal migration routes are disrupted, and erosion problems are exacerbated. Although perhaps more subtle, the impacts can be equally troublesome on adjacent public lands, particularly when ski resorts and other large-scale recreation complexes are involved. But it is the sheer number of people attracted to destination resort areas and nearby public lands that may pose the greatest problems. Unlike the site-specific impacts associated with a mine or timber sale, recreationists are ubiquitous; the mere presence of more people will generate more human waste, create more unauthorized travel routes, and disturb more wildlife. Motorized recreational users often compound these problems, particularly as they demand more roads and trails into undisturbed areas. The cumulative impact of myriad recreational users and related sprawling development places the very environmental qualities that lured them in the first place at risk.

NEITHER PEACE NOR QUIET: THE MOTORIZED RECREATION DEBATE

The overriding recreational issue on public lands may well be the intensifying conflict over motorized activities. The past 30 years have witnessed phenomenal growth in motorized recreation sports, specifically the use of off-road vehicles (ORVs), all-terrain vehicles (ATVs), dirt bikes, snowmobiles, and jet skis. The Forest Service estimates that by 2020 it will log 118 million ORV visitor days annually to the national forests. In the decade between 1988 and 1998, Utah recorded a threefold increase in ORV registrations, which jumped from 22,000 to over 68,000. Yellowstone National Park reports that snowmobile traffic grew from fewer than 10,000 riders in 1968–69 to over 38,000 during the 1982–83 winter, and then to more than 85,000 by the 2000–2001 season. Since 1976, the California Desert Conservation Area has witnessed a running battle over motorized recreation; its BLM managers have faced repeated lawsuits challenging their ORV management decisions. Similar confrontations are under way in Colorado's national forests and southern Utah's redrock country. Put simply, the conflict pits environmental organizations, ORV enthusiasts, industry groups, and nonmotorized recreation interests against one another, each intent on gaining a greater slice of the public domain. Though the public land agencies appear caught in the middle, they have not helped the situation, having long ignored the most egregious abuses and mounting tensions. And with more and more communities hitching their economic future to the recreation-tourism economy, their job is no easier today.[67]

The White River National Forest in north central Colorado is a case in point. The forest faces interrelated recreation, ORV, and resort development issues that were foremost in a new forest planning process. The 2.3 million–acre White River forest lies less than an hour's drive west of the rapidly growing Denver metropolitan area. It is one of the nation's most heavily visited national forests, with visitation at 8.4 million annually and expected to almost triple over the next 20 years. The surrounding economy is dominated by tourism; 12 ski areas dot the forest, and more than 80 percent of the local jobs are in the trade, service, and construction sectors. Environmental problems and user conflicts are endemic, including habitat encroachment, pollution and erosion concerns, motorized versus nonmotorized confrontations, and ski area expansion proposals. The White River's 2000 draft forest plan was geared toward protecting biodiversity and sensitive wildlife habitat, minimizing new developments, and reducing escalating recreation conflicts. It proposed significant road closures, ATV and snowmobile access limitations, mountain bike regulations, and ski area expansion curbs. The 2002 revised forest plan, however, takes a more restrained approach to recreation management. Plainly influenced by local political sentiment, the Forest Service backed away from several earlier limitations. The new plan still forbids off-road motorized travel and closes specified backcountry roads, but it also provides for new roads and defers most road management decisions for a further planning process. Whereas the draft plan severely limited ski area expansion, the final plan approves resort expansions in areas situated closest to Denver. It also portends increased timber harvesting in roadless areas and decreases in biodiversity levels over the draft proposal. Overall, the White River Forest Plan has disappointed knowledgeable observers who viewed the planning process as an important test of the Forest Service's willingness to confront the full ecological effects of recreation and tourism on the national forests.[68]

In southern Utah's redrock country, similar recreational controversies are being played out against the backdrop of a major wilderness protection campaign. In Utah, the BLM oversees 23.5 million acres of public lands, and it has traditionally allowed ORV use unless an area is specifically closed. To the chagrin of wilderness advocates (who believe that 9.1 million acres of Utah's BLM land qualify for wilderness designation), the BLM has designated only 3.2 million acres as wilderness study areas, and it has closed only 1.5 million acres to ORV use, covering less than 500,000 WSA acres. Excessive ORV use has created rutted trails in several locations and thus compromised wilderness quality, forcing the BLM to drop some WSA acreage from its inventory figures. In this sensitive desert environment, ORV use can also damage ecologically important cryptobiotic soils, erode fragile streambanks, displace wildlife, and harm nearby archaeological sites. Conflicts between ORV users and solitude-seeking hikers are legion. Although the BLM has the clear legal authority to minimize environmental degradation and recreation conflicts, it has not aggressively regulated ORV use on its southern Utah lands. Even the Park Service has been reluctant to curtail ORV access in southern Utah's expansive backcountry, prompting a court-ordered closure of one popular Canyonlands route. Yet even were the BLM to restrict ORV use, the accompanying enforcement issues cannot be ignored; the agency has but two law enforcement officers for the entire 1.8 million–acre San Juan Resource Area. Convinced that the desert's ecological well-being is in jeopardy, environmental groups have sued the BLM in an effort to force the agency to confront its growing ORV dilemma and to limit motorized access.[69]

In the national parks, the signature motorized recreation issue is whether snowmobiles should continue to be permitted in Yellowstone National Park. During the 1960s, Yellowstone decided to allow snowmobiles to traverse its remote snow-covered winter roads—a decision that seemed innocuous at the time, when few people ventured into the park during the harsh winter months. But that is no longer true: during the peak winter of 1993–94, for example, more than 143,000 people visited the park, including 87,600 snowmobilers. Over the intervening years, the town of West Yellowstone, Montana, rejuvenated its moribund winter economy by proclaiming itself the snowmobile capital of the world. More than one-third of the town's annual revenue is now derived from the winter tourism season, and most local businesses strenuously resist any effort to curtail park snowmobile use. The scientific evidence, however, has steadily mounted that winter snowmobile use has imperiled important park resources: air pollution often exceeds legal limits wherever snowmobiles congregate; collisions with park wildlife occur regularly; the groomed road surface invites winter-stressed bison to migrate out of the park where they may be shot for fear that they might transmit brucellosis to domestic livestock; and the ear-piercing whine of two-cycle engines penetrates far into the park's backcountry solitude. Under the Clinton administration, the Park Service evaluated this evidence, concluded that snowmobile use violated its Organic Act nonimpairment obligations, and announced it would ban them from the park. The George W. Bush administration, however, decided to reverse that decision, allowing recreational snowmobiling to continue at higher than previous levels subject to tighter phased-in emission controls and guide requirements. The decision, given Yellowstone's iconic status, represents a major triumph for motorized recreation groups, establishing an unwelcome precedent for similar issues elsewhere on public lands.[70]

To be sure, the principal antagonists in the motorized recreation controversy see the landscape quite differently. Environmentalists tend to focus on the damage that is occurring to sensitive resources, lamenting increased levels of water and air pollution, erosion problems, exotic introductions (often from seeds transported by ORV tires), habitat incursion, wildlife displacement, and aesthetic consequences. With ORV use on the rise, they see the cumulative impact of dispersed motorized recreation activities as a major threat to ecological integrity, every bit as harmful as logging, mining, and grazing. They particularly fear the pressure to open more remote backcountry to motorized recreation, where hunters and other users regularly object to the noise of internal combustion engines shattering the natural quiet. One question is being asked with growing frequency: Should solitude or silence be treated as a distinct resource on public lands—one that merits legal protection? The Wilderness Act expressly includes solitude as an attribute of wilderness, and Park Service management policies place a premium on protecting the natural sound scape. In today's more populated West, though, it is becoming difficult to escape the sounds of civilization, particularly the gas-powered engine. If public lands cannot provide a respite for the average citizen, there may be few alternative locations left to commune with nature. All of which suggests the sound of silence is an ever more scarce resource with value in its own right.[71]

For their part, ORV users and their industry supporters have framed their access arguments primarily in equitable rather than environmental terms. They see the environmental opposition as elitists who are unwilling to share the vast public lands with those who enjoy motorized transport as part of their outdoor experience. They wonder how the infirm or aged can experience wild nature without vehicular access. Having long enjoyed nearly unlimited access to nonwilderness public lands, ORV proponents view any effort to regulate their activities as infringing on established legal access rights and, more ominously, as a precursor to eliminating all human uses from public lands. They also dispute claims of environmental degradation, pointing to improved technology and a conservation ethic among ORV users. In fact, ORV groups frequently assist agencies with trail maintenance and related projects, earning both the respect and support of public land managers for their efforts. Not surprisingly, ORV user groups have joined ranks with the extractive industries to contest new wilderness proposals and other environmental restraints on access to public lands.[72]

The law governing motorized recreation on public lands provides little specific guidance to the agencies. The Forest Service and BLM are governed by similar multiple-use mandates that explicitly recognize recreation as a permitted use, and both agencies have long allowed motorized recreation on their lands with few limitations. Two presidential executive orders also address ORV use on public lands: in 1972 President Nixon effectively enjoined the agencies to zone their lands for ORV use based on environmental criteria and public input, and in 1977 President Carter directed that ORV use should be banned in areas where continuing use "will cause or is causing adverse effects." Faced with court decisions treating these executive orders as legally enforceable obligations, the agencies have relied primarily on their own resource planning laws, NEPA requirements, and ESA strictures in making ORV management decisions. Significantly, the courts have consistently sustained agency decisions closing or limiting motorized recreation activities. The Court of Appeals for the Ninth Circuit, for example, upheld a Forest Service decision closing a trail to ORV use in order to avoid user conflicts and provide semiprimitive recreational opportunities. When agency officials have ignored environmental objections and chosen not to limit motorized activities, the courts initially tended to sustain these decisions, but they have more recently begun forcing agencies to justify their positions. A Washington federal district court, for instance, reversed a Forest Service ORV trail reconstruction decision that was made without any detailed environmental analysis, citing documented wildlife disturbance problems and potential user conflicts. When the controversy has involved motorized access to protected lands, the courts have more rigorously held the agencies to their preservation obligations, as evidenced by court decisions in Utah, Minnesota, and Montana curtailing motorized activities in national park, wilderness, and wilderness study areas. In short, public land agencies have abundant legal authority to regulate motorized access either to redress environmental problems or to minimize user conflicts.[73]

The motorized recreation controversy is also a political matter, however, which has brought Congress and partisan politics into the fray. Once Congress passed the Symms Act in 1991 (now known as the National Recreation Trails Program), public land agencies began receiving trail construction and maintenance funds from the federal gasoline tax, most of which has been spent on building and upgrading ORV trails, with little money going to hiking or horse trails. For their role in shepherding the Symms Act through Congress, ORV groups endeared themselves to agency officials grateful for the additional funds. Their willingness to help the cash-strapped agencies maintain trails has only strengthened these connections. Individual congressional delegations have regularly pressured the public land agencies on behalf of motorized recreation groups, as reflected in the Wyoming delegation's well-publicized efforts to maintain snowmobile access into Yellowstone Park. The Bush administration's decision to revise the Park Service's Yellowstone snowmobile closure rule and to review the Forest Service's roadless area rule has clearly signaled where its sympathies lie. To counter these political overtures, the environmental community has sought its own congressional allies, just as it used its influence with the Clinton administration in an effort to curb abuses from motorized vehicles. Environmental groups, however, have sometimes found themselves at loggerheads with potential allies over recreation restrictions, perhaps most notably with the growing mountain bike constituency, which has balked at efforts to restrict bikes from public lands. And there are deep disagreements over Congress's experimental fee demonstration program, with opponents convinced that the new fees create a perverse incentive favoring more intensive recreational uses. Given the strident nature of the debate, conventional coalition building strategies as well as creative incentives will play an important role in shaping the future of motorized recreation on public lands.[74]

The wilderness designation debate is intrinsically linked to the controversy over motorized recreation on public lands. The presence of roads precludes official wilderness designation, while an official wilderness designation prohibits any motorized access. The ORV constituencies have joined with the extractive industries and various resource-dependent communities to form a powerful anti-wilderness coalition. They view restrictive wilderness designations as anathema to ORV opportunities, asserting that wilderness precludes a large segment of the public from ever experiencing the backcountry and thus benefits the select few who have the time and stamina to hike into wilderness venues. In southern Utah and elsewhere, ORV users have openly blazed rutted trails into wilderness study areas, then asserted that the area no longer qualifies for wilderness consideration because it is neither roadless nor untrammeled. They have also invoked the little-known (and now repealed) R.S. 2477 law, which grants counties a right-of-way for pre-1976 roads across public lands. Their position is that preexisting ORV routes and similar trails are legally protected R.S. 2477 roads, another reason for rejecting the wilderness option. They even helped defeat a locally conceived proposal to designate the popular San Rafael Swell region as a multiple-use national monument, which would have served as an alternative to wilderness and still allowed ORV access. Environmental groups, however, see the growing ORV threat as a principal justification for more wilderness protection, arguing that R.S. 2477 should not be read broadly. In the case of the San Rafael Swell, they applaud the idea of additional protection for these sensitive lands but feared the monument proposal would simply create an unregulated ORV playground. The real question may well be whether there are politically acceptable alternatives to wilderness that can accommodate limited motorized uses.[75]

The motorized recreation debate also cannot—and should not—be divorced from new ecosystem management concepts and collaborative problem solving processes. Part of the answer to peace and quiet on public lands is entwined with the wilderness debate, and it will require a congressional resolution of the various wilderness designation controversies. In some locations, it may be possible to design compromise designations short of outright wilderness that would allow limited motorized access while still protecting ecological values. Part of the answer will also require a conclusive determination of whether gasoline-powered recreation is appropriate in our national park sanctuaries. And part of it will require a definitive resolution of the R.S. 2477 controversy, most likely through a series of federal court rulings. Beyond these patently national-level issues, new ecosystem-based planning and assessment processes should enable public land agencies to identify critical ecological concerns associated with motorized recreation and to design appropriate, science-based limits on roads and ORV access. Operating at this larger scale, agencies and affected constituencies may be able to reach zoning compromises that divide the landscape among the competing users.

Specific conflicts may require a more localized collaborative process focused on the potentially affected site, including related environmental implications and user conflicts. Adaptive management strategies should also be integrated into the process, enabling adjustments when necessary to resolve unanticipated problems. It will be a give-and-take process, befitting the political nature of public lands yet also sensitive to new ecological concerns and the value of true silence.

AVOIDING PAST MISTAKES: OF RIGHTS, FEES, AND DISCRETION

With the surge in outdoor recreation and the concomitant growth in nearby tourism-oriented communities, public land agencies face daunting new resource management challenges. As rural economies continue to shift toward recreation and tourism, the pressure mounts on these agencies to provide more commercial opportunities on their lands and to offer such businesses greater security of tenure to facilitate capital investments. Through an experimental recreation fee program, agencies are now realizing direct economic returns from recreational users—a new arrangement that could tempt them to expand these activities in order to enhance their own revenues. Faced with escalating visitation pressures and mounting evidence of related environmental degradation, agency officials must begin to incorporate recreation concerns into their new ecosystem management processes and to assert their inherent regulatory authority to safeguard sensitive resources. Beyond the boundary line, the emerging recreation-tourism economy raises oddly familiar concerns about the economic dependency of adjacent communities on public lands. It also raises related urban–wildland interface issues that are forcing agencies and communities into new cooperative planning relationships. These new concerns are further testing the role of ecological management principles and collaboration strategies in shaping a sustainable future.

Recreation and tourism on public lands have an overt commercial dimension that is more about economics than the environment, and therein lies the potential problem. With ever-increasing numbers of people visiting public lands, commercial outfitting and guide services have grown in popularity. Many visitors are unprepared to face the wilderness without the assistance of a guide, and fewer still are ready to navigate the Grand Canyon or Salmon River on their own. With their financial welfare obviously tied to the public domain, outfitters and other recreation concessionaires are anxious to have secure access rights to necessary resources, just like their counterparts in the mining, timber, and ranching industries. A private party's commercial rights on public lands are typically defined in contract terms, through either a lease, license, or special use permit. Outfitters in the national forests, for example, must obtain a special use permit, which details their rights and obligations in using forestlands. In the case of national park concessionaires and national forest ski area operators, though, Congress has statutorily granted them long-term tenure rights. A recent congressional bill—the Outfitter Policy Act of 2001—similarly seeks to expand the rights enjoyed by guides, outfitters, and other recreational service providers on public lands, granting them more secure tenure and transfer rights than they currently enjoy. From a resource management perspective, however, any new set of statutory rights creating long-term property interests could tie the agencies' hands in addressing the growing impacts of recreation and tourism, including potential ecological problems and user conflicts. Indeed, one of the Forest Service's most difficult problems in its transition to ecosystem management has been resolving the legal claims arising from ecologically driven timber sale modifications. Neither Congress nor the agencies should be eager to extend a new set of property or contract rights to another industry on public lands, at least not before the full ramifications of the emerging recreation and tourism economy are better understood.[76]

A new congressionally authorized recreation fee program could presage a significant shift in management philosophy on public lands. Under the Land and Water Conservation Fund Act of 1965, Congress prohibited public land agencies from charging recreation user fees except for developed facilities, such as campgrounds, marinas, and the like. In 1996, however, with the agencies facing recurrent budget shortfalls and a growing maintenance backlog, Congress passed a little-noticed budget rider authorizing the four public land agencies experimentally to charge recreation user fees and to retain the funds "to demonstrate the feasibility of user-generated cost recovery for the operation and maintenance of recreation areas or sites and habitat enhancement on Federal Lands." Proponents of the new fee program believe it will infuse badly needed market principles and economic incentives into the agencies, and also promote more balanced management decisions by placing recreation on a par with other revenue generating uses, such as timber and grazing. They also see the new charges as leveling the competitive playing field between the private and public sectors in providing recreation services. Opponents, however, view the fee as inequitable double taxation that violates the long tradition of free recreational access to public lands, as well as another perverse incentive that seriously threatens environment values. More ominously, citing the recreation industry's strong role in the legislation, they fear the fee program is the opening salvo in a larger plan to commercialize recreation on public lands, signaling the triumph of an industrial-scale recreation agenda that could also trigger a privatization movement. Despite gradually mounting opposition, Congress has thrice reauthorized the program that now runs through fiscal year 2005, and the agencies have uniformly embraced it as a vital source of much needed revenue. If the recreation fees serve to counterbalance the role of commodity uses on the public domain and if they are utilized to maintain trails and protect habitat, then industrial recreation fears may be ill-founded. But if the fees are used as an indirect marketing device or to expand infrastructure and pave new roads, then the perils of commercialization cannot be discounted. Regardless, the recreation fee concept represents an important enough shift in management philosophy to merit a full and open congressional debate directly addressing these competing concerns before it is incorporated permanently into public land policy.[77]

To cope with mounting commercial recreation pressures and related equitable access claims, public land agencies should focus their attention on environmental rather than equity concerns, utilizing ecosystem management concepts to safeguard important ecological values. This is precisely what the White River National Forest proposed to do in its controversial draft forest plan, but then failed to accomplish in its final plan. As we have seen, each of the four public land agencies has comprehensive planning responsibilities as well as broad regulatory authority over its lands, and the courts have consistently sustained reasonable use, location, and timing limitations on recreational activities. The power, in other words, is available to control recreational excesses in the name of environmental protection. For the Park Service, this means taking seriously its preservation mandate and nonimpairment obligations to prohibit snowmobiling and other environmentally damaging activities inside the national parks. For the Forest Service and BLM, it means employing rigorous environmental analysis standards to determine whether ecological resources are at risk, and then regulating access when necessary to protect the land base or to minimize user conflicts. With the advent of ecosystem management, agency planning responsibilities also extend beyond the boundary line, which entails coordination with adjacent land managers as well as local governments and private landowners. In fact, recreation and tourism planning can best be addressed on this larger scale, where the agencies can take account of broader legal, ecological, economic, and social factors. Already the Park Service and Forest Service have collaborated on a joint winter recreation assessment in the Greater Yellowstone region, while the Forest Service and BLM have prepared a joint EIS on ORV use in Montana, North Dakota, and South Dakota. Planning on this larger scale presents the agencies with the opportunity to zone these recreational activities across a broader landscape, and thus to better match uses and locations. But whether this ecologically based approach to recreation and tourism will ultimately prevail depends in large measure on curbing the evident political pressures and on redirecting the debate toward environmental concerns.[78]

As local economic priorities shift toward an amenity-based view of public lands, the federal agencies will face new demands from nearby communities for recreational and visitation access. Whereas these communities once depended on public timber or minerals, a new recreation-based dependence may well emerge. Two classic cases in point are Moab and West Yellowstone; businesses in both communities now depend heavily on nearby public lands for recreational access. A similar relationship has evolved in Colorado's White River National Forest, where its 12 ski areas and the surrounding communities are directly dependent on federal ski area management and related recreation decisions. Just as public lands have proven unable to sustain a high-volume timber industry, however, the same may also prove true for the growing recreation and tourism industries, particularly as they expand in size and scope. Despite their rugged appearance, public lands cannot endure an unending barrage of commercial recreation demands or an infinite number of tourists and other users without compromising the very environmental and aesthetic qualities that attract visitors in the first place. To address this problem, both the agencies and neighboring communities must begin to think in true sustainability terms, which implies both ecological and economic diversification. At a minimum, sustainability contemplates a high-quality environment on public lands that can help attract new businesses and residents to further diversify the local economy. A related strategy entails incorporating ecological restoration into the public land agenda, which can help maintain diverse job opportunities while enhancing environmental values. In fact, these two strategies can be employed in a complementary fashion to further sustainability objectives and to avoid single-industry dependency. Also, collaborative planning processes utilizing ecosystem management principles can provide a local forum for mutual education on the environmental and economic implications of the new amenity-based industries.[79]

Beyond these recreation implications, the agencies and nearby communities cannot ignore the growth and related urban–wildland interface issues that accompany the new amenity-based economy. Because most of the development is occurring on private lands, state laws and local zoning ordinances are the primary legal tools for addressing sprawl and other environmental problems. But these private land use decisions inevitably affect public resources too, creating new fire hazards, wildlife habitat encroachments, access impediments, aesthetic concerns, and the like. This gives federal public land agencies a direct stake in local land use decisions. Conversely, public land management decisions, such as ski area siting or prescribed fire policies, can affect adjacent development patterns, making the federal relationship with local communities a reciprocal matter. Coordination is plainly the key here, as well as the mutual willingness to view the landscape as an integrated or ecological whole. Collaborative planning processes based on ecosystem management principles sensitive to private property rights are one means to address boundary problems and other shared concerns. Federal land exchanges can also be used strategically to rearrange the landscape in an environmentally sensitive yet economically viable manner. Regardless of the chosen strategy, the public land agencies have little choice but to become actively engaged in local growth and development issues if they are to fulfill their own environmental stewardship and good neighbor responsibilities. This reality only further confirms that the new landscape of recreation and tourism is not an ecological panacea for public lands.[80]

The elusive quest for sustainability, long an aspiration in public land policy, is now being written in terms of ecosystem management, collaborative conservation, and a new amenity-based economy. Ecosystem management concepts have broadened agency planning efforts and elevated biodiversity conservation on the policy agenda. Collaborative conservation reconfirms that human communities must be part of the planning equation, consistent with our long-standing commitment to democratic principles and the notion that everyone is entitled to a voice in the decision making process. The growth of recreation and tourism offers an alternative vision for the future of public lands while reflecting the dynamic nature of the marketplace. How to harness these diverse—and sometimes seemingly contradictory—forces to create new management strategies is a monumental challenge that is now playing out in myriad locations across the western landscape. Nowhere has the experiment gained as much momentum—or notoriety—as in a small Sierra timber town that has, de facto, become a test case for linking new collaborative conservation processes with ecosystem management concepts to plot a new sustainability course.

EIGHT

Toward a New Order

Ecosystems and Democracy

Our conservation must be not just the classic conservation of
protection and development, but a creative conservation of restoration
and innovation. Its concern is not with nature alone, but with the total
relation between man and the world around him. Its object is not just
man's welfare but the dignity of man's spirit.

LYNDON BAINES JOHNSON (1965)

We will always have disputes over land, water, minerals, and animals. . . .
The overarching concern therefore is not to deny that conflict will occur
but rather to acknowledge an ethic that sets standards for resolution
and, as importantly, provides a method for dealing with disputes.

CHARLES WILKINSON (1992)

The town of Quincy, California, is not the first place that comes to mind when the topic is public land policy. But in a few short years, Quincy came to represent all that is either right or else wrong with evolving new ecological management policies and related collaborative processes. Meeting together, a diverse array of local citizens and former antagonists—dubbed the Quincy Library Group—cobbled together a 2.5 million–acre experiment in national forest management predicated upon ambitious ecological restoration goals that were also designed to maintain the local timber industry and to minimize fire danger. When the Forest Service balked at implementing the proposal, the Quincy group convinced Congress to legislate it, though not without a major confrontation with the national environmental community. Yet even with Congress's blessing, the group found its proposal enmeshed in dismaying legal complexities as well as a broader regional national forest planning initiative. With its future in doubt, the Quincy experiment raises major substantive and procedural questions that cannot be ignored in wedding ecosystem management concepts with local community concerns through collaborative processes. It is an object lesson in the politics of devolution and the science of ecology.

Nonetheless, these two promising new ideas—ecosystem management and collaborative conservation—are reshaping public land policy. There are powerful reasons, as reflected in the Quincy experiment, to seek a merger between the two concepts. And there are equally troublesome problems that can arise with such a merger. Place-based initiatives may present a promising opportunity to consummate such a merger; small-scale ecosystem management experiments make intuitive sense as a transitional strategy, while local communities have an obvious interest in the surrounding public domain. That said, however, the emerging national interest in biodiversity conservation cannot be subordinated to local interests, nor can management authority be blithely transferred to a local collaborative group. With the current level of controversy unlikely to subside soon, more place-based initiatives will undoubtedly surface and press for recognition. Their experiences should further improve our understanding of how to meld these two important concepts together to meet existing ecological and political challenges on public lands. As that process unfolds, national ecosystem management legislation will also merit our attention. Its passage would put a final stamp of legitimacy on this new direction in public land policy.

The Quincy Library Group Experience: Evolution or Devolution?

TIMBER AND THE CALIFORNIA SPOTTED OWL

The town of Quincy is located in rural Plumas County, California, amid the remote, timber-rich northern Sierra Nevada mountains. Here the moist west side of the mountains sustains a commercially valuable mixed conifer forest that includes Douglas fir and Ponderosa pine. On the more arid east side, the forest consists mostly of less commercially attractive pine and fir species, much of which was heavily logged during the early twentieth century. Over 70 percent of Plumas County's land base has national forest status, most of it in the 1.14 million–acre Plumas National Forest. At least since the 1950s, Quincy and surrounding towns have been heavily dependent on the timber industry for their economic sustenance; traditionally, nearly 10 percent of Quincy's 5,000 residents have worked in the timber industry. Indeed, the nearby Plumas forest, along with the adjacent Lassen and Tahoe national forests, have been the Sierra Nevada region's major timber production forests over the past several decades. Quincy itself has long supported a large Sierra Pacific Industries sawmill, while other nearby towns have their own smaller mills. But as timber production fell dramatically throughout the Sierras during the early 1990s, half the Quincy area mills were closed with concomitant job losses.[1] Yet though Plumas County is a decidedly rural setting, its southern reaches are near enough to Sacramento that urban refugees are beginning to arrive, spurring new resort developments and nascent shifts in local demographics.[2]

Controversy first erupted over logging on the Plumas forest during the 1980s. Early in the decade, Congress began debating a California national forest wilderness bill that Quincy and other logging communities vigorously opposed, fearing the loss of access to commercially valuable timber. But in 1984, wilderness advocates prevailed when Congress designated several thousand acres of wilderness on the Quincy area's national forests. Soon thereafter, during the first round of forest planning mandated by the National Forest Management Act (NFMA), local Quincy environmental activists (known as the Friends of Plumas Wilderness) joined with national environmental organizations to advance a protective management alternative that would have modestly reduced timber harvest levels while also protecting the forest's remaining roadless areas. At the local timber industry's behest, however, the Forest Service rejected this proposal, instead maintaining the forest's harvest levels at the prevailing 265-mbf annual sale quantity. Under the new forest plan, timber's predominant position in the forest seemed relatively secure.[3]

But that was before the spotted owl made its presence known. In the nearby Pacific Northwest, the northern spotted owl had already surfaced by the late 1980s as a potential obstacle to that region's high-volume historical timber harvest levels. By the early 1990s, the distantly related and old growth–dependent California spotted owl was also becoming an ecological cause celebre in the Sierra Nevada forests. Once the owl made it onto the Forest Service's radar screen, logging cutbacks became a fact of life throughout the region, triggering protests from timber interests. The town of Quincy served as the focal point for the logging industry's national Yellow Ribbon Coalition campaign, designed to protest new federal timber harvesting restrictions. And Quincy had reason for concern over the local industry's long-term prospects. Once timber stopped flowing in the Pacific North-west, some Oregon companies began bidding on northern California national for-est timber sales, threatening to divert the logs from nearby mills. More important, local environmentalists—after successfully challenging several timber sales in the southern Sierra forests and forcing the Forest Service to reexamine its local herbi-cide policies—were appealing the Plumas forest plan decision. They contended that the plan did not adequately protect roadless areas, old growth forests, riparian zones, or at-risk species. And they had indicated that they were contemplating fur-ther litigation over the California spotted owl. By then, the newly elected and en-vironmentally friendly Clinton administration was poised to take the reins of power. With many of the same forces that had brought logging to a standstill fur-ther north at play in the Plumas area, the future suddenly looked quite uncertain for the local timber industry.[4]

COLLABORATION COMES TO QUINCY

In late 1992, representatives from Sierra Pacific Industries and the Plumas County supervisors approached the Quincy environmental activists with a star-tling proposal: to avoid a local timber war, we are prepared to accept your earlier forest plan proposal. In short order, the parties were in face-to-face negotiations over the details of a much different timber management policy for the Plumas Na-tional Forest, one that would emphasize group selection harvesting and thinning, and also protect roadless areas and riparian zones. Recognizing the momentous-ness of the occasion, the initial group quickly expanded its membership: represen-tatives from Collins Pine, another local timber firm, were added, as were represen-tatives from the adjacent Lassen County government, which also perceived an opportunity to avert a local timber donnybrook. (These early additions to the group were critical in shaping the boundaries for the ultimate proposal, which were designed primarily for economic purposes to ensure an adequate supply of logs to keep the local mills afloat—a persistent point of criticism.) Even though invited, representatives from national environmental organizations declined to participate, citing travel difficulties and other more pressing matters. The local ranching community initially attended some early meetings, but they soon opted out when it became apparent that the group was focused predominantly on forestry issues.

If one factor united the group members early on, it was their distrust of the Forest Service. As a result, the group did not invite the Forest Service to participate formally in its deliberations, though agency representatives regularly observed the group's meetings. Once additional members were aboard, it was apparent that the group needed new meeting quarters. They selected the local Quincy Public Li-brary because it was deemed neutral turf and had enough space for public meet-ings—thus giving the group its unlikely name. Before long, the group's members had agreed on the rudiments of a new local forest management policy, which was then endorsed at a public meeting attended by over 150 residents, with only 4 dissenting votes. The speed of the agreement was startling, even to the original participants. But the expedited process would also generate later criticism, with opponents asserting that the group failed to include all interested parties at the negotiating table.[5]

In 1993, the Quincy Library Group embraced a community stability proposal that served as the basis for its subsequent negotiations with the Forest Service and for its later legislative initiative. Covering nearly 2.4 million acres in the Plumas, Lassen, and Tahoe national forests, the proposal's ultimate goal was to restore "an all-age, multi-story, fire-resistant forest approximating pre-settlement conditions." To achieve these objectives, the proposal endorsed four integrated ecosystem management strategies: the use of group or individual selection timber harvesting techniques rather than clear-cutting; an active fire and fuels management program that included proposed fuel breaks, later denominated "defensible fuel profile zones" (DFPZs); the elimination of logging in roadless areas (roughly 500,000 acres) or California spotted owl protected activity centers (PACs); and the establishment of an active riparian area and watershed restoration program. These strategies, according to the group, would not only expand the actual land base open to timber harvesting, but also make some old growth timber available for logging. To achieve its economic goals, the group sought to limit nearby national forest logging opportunities only to local timber operators, using such devices as small business set-asides and the antiquated working circle concept. Acknowledging the experimental nature of its proposals, the group envisioned a five-year pilot project to test the feasibility of its recommended management strategies. The proposal was endorsed by 39 diverse local participants, including representatives from the local environmental community, timber industry, county government, and labor unions. But the terms of the agreement were problematic for the Forest Service, which saw it as a threat to its own managerial authority.[6]

The Forest Service, however, had more than the Quincy proposal occupying its attention in California. A 1991 Pulitzer Prize–winning series in the *Sacramento Bee* newspaper entitled "The Sierra in Peril" had highlighted badly deteriorated environmental conditions across the California national forests. During the early 1990s, faced with mounting concern over the California spotted owl, the Forest Service undertook a technical biological assessment of the owl's overall population trends and habitat conditions. In 1992, the scientists reported that although owl population trends were uncertain, the old growth habitat that owls primarily used in the Sierra Nevada was rapidly declining. To meet its NFMA-based species viability legal responsibilities and to preempt the U.S. Fish and Wildlife Service (FWS) from listing the owl as a species protected by the Endangered Species Act (ESA), the Forest Service responded by adopting interim regional management guidelines designed to protect the owl and its old growth forest habitat. The so-called CASPO (California spotted owl) Interim Guidelines imposed dramatic new constraints on timber harvesting in the Sierra Nevada forests, establishing

300-acre no-cut zones around known owl sites, limiting the size of each harvested tree to under 30 inches in diameter in owl habitat, and imposing canopy closure requirements. Although the CASPO guidelines dissuaded the FWS from listing the owl and thus protected the Forest Service's managerial prerogatives, they did not sit well with the regional timber industry or its allies. The industry sued to overturn the guidelines, but was rebuffed in federal court. Across the 11 Sierra Nevada national forests, harvesting levels dropped sharply (from approximately 1,500 mbf to 800 mbf), and several smaller mills closed their doors. Confronted with the CASPO Interim Guidelines, Quincy and other logging-dependent communities faced major local economic changes.[7]

The Quincy Library Group incorporated the new CASPO guidelines into its proposal and then began pursuing its agenda aggressively. Its strategy of choice, endorsed by the Clinton administration, was an administrative one: it would enlist the Forest Service to adopt its proposal by amending the existing forest plans for the three affected national forests. But it met a bureaucratic stone wall. The Forest Service was obviously troubled by the prospect of effectively relinquishing its managerial authority to an ad hoc citizens group that was driven by local concerns and lacked the agency's technical expertise. When presented with the Quincy plan, then Forest Service chief Jack Ward Thomas quietly resolved to kill it, later reportedly exclaiming, "Who the hell turned over my forests to [Sierra Pacific Industries]?"[8] In addition, the Forest Service was concerned about the potential multimillion-dollar costs associated with the Quincy group's ambitious forest thinning and fuel reduction plans. The broader environmental community was also growing increasingly uneasy with the Quincy proposal, particularly its hard annual timber targets. They worried that the Forest Service would use these timber targets to reestablish a high-volume local logging program. They viewed the DFPZ concept as another timber harvest subterfuge, as well as raising serious forest fragmentation concerns. For them, the Quincy proposal simply involved too much active restoration to suit their tastes. Delicate national-level negotiations, overseen by the undersecretary of agriculture, eventually broke down, with the various environmental factions blaming one another for the impasse. A major sticking point was the Quincy Library Group's insistence on pursuing congressional legislation if it was unable to secure an adequate administrative commitment to implement its pilot project proposal.

A CONGRESSIONAL ENDORSEMENT

When the debate over the Quincy proposal moved to Congress, the criticisms sharpened and the schisms widened. After its experience with the Pacific Northwest spotted owl imbroglio, Congress was clearly intrigued with the fact that loggers and environmentalists had come together and negotiated a local compromise to a budding timber controversy. The California congressional delegation quickly endorsed the Quincy proposal and introduced the necessary legislation.

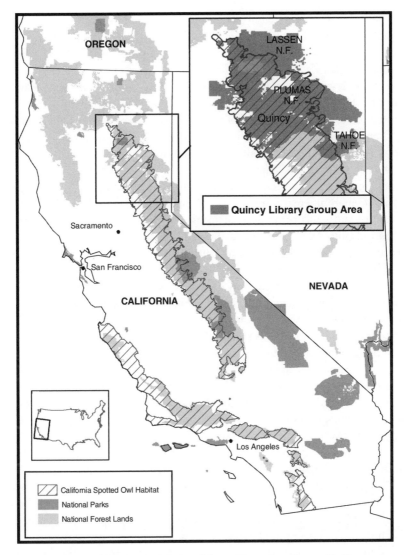

Map 10. Sierra Nevada Region and Quincy Library Group Area. Located in the northern Sierra Nevada mountain range, the Quincy Library Group project covers the Plumas and Lassen national forests, as well as part of the Tahoe forest. The Sierra Nevada Forest Plan Amendment initiative embraces 11 separate national forests, reorienting Forest Service management policy toward protecting and restoring forest ecosystems and controlling wildfire in urban–wildland interface areas. Note that the California spotted owl range extends nearly the entire length of the Sierra Nevada range, including the Quincy area forests.

But the national environmental community objected, now greatly troubled by the precedent that such congressional legislation could establish for managing the national forest system. Reiterating their concern over the proposal's hard timber targets and the DFPZs, an environmental coalition sought to derail the Quincy legislation, but it was too late. In 1998, the House of Representatives overwhelmingly passed the Herger-Feinstein Quincy Library Group Forest Recovery Act by a 429–1 vote, and the Senate eventually followed suit, though only after California Senator Diane Feinstein quietly attached the Quincy bill as a rider to the 1998 Interior appropriations bill. The final bill, however, was redrafted to require the Forest Service to follow applicable federal laws as well as the CASPO guidelines, including preparation of an environmental impact statement (EIS) assessing the proposal's potential environmental implications. Otherwise, the bill mirrored the Quincy Library Group's 1993 community stability proposal: it placed roadless areas and California spotted owl habitat areas off-limits for timber harvesting purposes, gave special protection to riparian areas, authorized mechanical thinning to create DFPZs on 40,000–60,000 acres per year, and sanctioned group selection timber harvesting. Because the project was a five-year experiment, the Forest Service was required to prepare a final status report, evaluating its resource management and economic consequences. After having defaulted in the earlier northern spotted owl controversy, Congress was back in the business of legislating regional national forest policy, albeit at the behest of a local collaborative group.[9]

The real test for the Quincy proposal came one year later when the Forest Service released its EIS decision, which detailed how the agency would implement the legislation and amend the existing forest plans. The EIS adopted most of the Quincy group's proposal, though it added a controversial California spotted owl mitigation measure. To meet the Forest Service's NFMA-based biodiversity and species viability obligations, the EIS contained a measure that forbade any timber cutting within "suitable California spotted owl habitat, including nesting habitat and forage habitat." It was, according to the Quincy Library Group, a "poison pill." The mitigation measure not only eliminated nearly one-third of the proposed DFPZs, but also called into question whether the proposal could sustain the local mill, a key concern in the group's original proposal. The measure effectively shifted most group selection cutting from the timber-rich west side of the Sierras to the relatively timber-poor east side, prompting the Plumas National Forest to rescind several lucrative, pre-1993 west-side timber sales and further undermining the proposal's economic viability. After the final EIS was released, nearly everyone filed an administrative appeal, but the Forest Service rejected them all. Despite a tangential court action, the Forest Service has moved ahead with the proposal, simultaneously responding to recurrent administrative appeals of project decisions and persistent criticism from the Quincy group that projects are moving too slowly.[10]

In the meantime, the Forest Service has completed a multiyear, regional EIS that amends forest plans throughout the 11 Sierra Nevada national forests, including those within the Quincy proposal boundaries. The 2001 Sierra Nevada Forest Plan Amendment EIS grew out of an earlier regionwide assessment known as the Sierra Nevada Ecosystem Project (SNEP), a congressionally mandated, independent scientific study designed to assess ecological conditions across the region's forests. According to the 1996 final SNEP report, the region's forests had been significantly altered by historic logging, fire suppression, and grazing practices, which put species relying on old growth habitat at increased risk. To address these problems, the SNEP report outlined alternative management strategies designed to increase the region's old growth forest cover and restore fire to the ecosystem. Many of the SNEP ideas were subsequently incorporated into the regional EIS (or framework document), which closely parallels other regional forest plan amendment processes, including the Pacific Northwest's Forest Ecosystem Management Assessment Team (FEMAT) report and the Interior Columbia Basin Ecosystem Management Plan (ICBEMP) initiative. Reflecting a federal commitment to managing on an ecosystem scale and restoring ecological sustainability, the framework document seeks to protect old growth forest ecosystems and riparian areas, to sustain viable California spotted owl populations, and to establish a fuels treatment program emphasizing protection of urban–wildland interface areas. It effectively eliminates commercial timber harvesting on the Sierra Nevada forests, while channeling nearly two-thirds of the allowed cutting into the three Quincy proposal forests. Moreover, it overrides the earlier Quincy EIS, imposing new and more onerous California spotted owl restraints on any timber harvesting in the Sierra national forests. By limiting the size of trees that may be cut and requiring denser canopy closures, the regional EIS further reduces group selection harvesting opportunities within the Quincy project area and forces additional modifications in the DFPZs. In short, to better protect California spotted owl habitat and to once again avoid an ESA listing of the owl, the regional EIS has further shrunk the original Quincy group proposal, further diminishing the project's economic feasibility.[11]

THE CONTINUING QUEST FOR COMMON GROUND

The Quincy Library Group has remained firmly united behind its original community stability proposal, having first negotiated its terms among themselves, then defended it in high-level administrative negotiations, and then shepherded it through Congress as a legislative mandate. The group appealed the regional EIS decision, hoping to modify the final standards to better accommodate its own forest management proposal. It asserted that the Forest Service, by precluding commercial timber harvesting and aggressive thinning strategies, had increased the risk of catastrophic fire, endangered the California spotted owl, and proposed an economically unsuitable management strategy. Indeed, as the Forest Service has chipped away at the original proposal, first in the 1999 Quincy EIS and then in the 2001 regional EIS, the group has solidified in its collective suspicion of the agency and its motives. Trust among the group's diverse membership has also deepened, a development that can be traced to the Forest Service's 1994 salvage timber sale in the roadless Barkley fire area, which the timber industry chose not to bid on as a gesture of good faith because the area was off-limits under the group's original proposal.[12] The multiyear Quincy Library Group process has strengthened relationships within the community too; former adversaries are now on more friendly terms, the general level of civility is much improved, and a spirit of relative harmony seems to prevail. Reflecting its commitment to the project, Sierra Pacific Industries retooled its Quincy mill to accommodate the smaller-diameter logs that are available under the proposal. A new wood-based ethanol plant is on the drawing board, which would provide another outlet for smaller logs. To keep the project moving forward, the Quincy group has lobbied Congress for the necessary funding and also cajoled local Forest Service officials to implement as many DFPZs and other projects as possible.

For its part, the Forest Service has found its options limited, particularly by the need to protect the California spotted owl and to meet its other biodiversity conservation obligations. From the beginning, the Forest Service basically took a hands-off position toward the Quincy Library Group, skeptical of both its fuel treatment proposals and the likelihood that Congress would enact its proposed legislation. But once the Herger-Feinstein bill was adopted into law, the agency turned its attention to implementing the proposal through the EIS process and to engaging the Quincy group in an ongoing dialogue. With the Forest Service simultaneously engaged in the Sierra Nevada regionwide EIS project, the lingering question was how to fit the Quincy and regional initiatives together. Not surprisingly, the Sierra-wide framework document and its additional California spotted owl protections has taken precedence, though the Forest Service also committed to implementing the Quincy project within the limits of these constraints. Because the Herger-Feinstein Act and the Quincy EIS both establish hard acreage timber targets, local Forest Service officials have been under constant pressure to advance project proposals in order to meet these targets. Clearly, one of the Forest Service's driving motivations in both the Quincy EIS and the framework EIS, each with its own stringent spotted owl protections, has been its desire to forestall the U.S. Fish and Wildlife Service from listing the owl and thus investing itself with review authority over national forest management decisions. Throughout both EIS processes and since then, the Forest Service has continued to attend regular Quincy Library Group meetings as an observer and to keep the group apprised of its plans and problems.

The broader environmental community, in contrast, has pursued a strategy designed to further limit the Forest Service's discretionary management authority as well as any timber cutting that might be planned under the Quincy group's proposal. Its efforts thus far have been rewarded: in Congress, the Herger-Feinstein bill was amended to require full compliance with all relevant environmental laws; during the Quincy EIS process, the Forest Service inserted the California spotted owl mitigation measure provision, thus tightening timber harvest opportunities beyond those required under the interim CASPO guidelines; and during the regional Sierra Nevada EIS process, the Forest Service again tightened timber harvest restrictions on behalf of the owl. But even with these constraints, many environmentalists fear another full-scale timber harvesting program across the Sierra range, mindful of their experiences with the 1995 timber salvage rider. They note that nearly two-thirds of the scheduled timber cutting in the regional framework document will occur in the Quincy project area forests, deepening their suspicion that the project's fuel treatment cutting will really amount to commercial harvesting in disguise. And they note that the Bush administration's proposed changes in national forest policy could ultimately permit even more local logging without adequate environmental review.

To curtail further timber harvesting, several groups have petitioned the U.S. Fish and Wildlife Service to list the California spotted owl, Pacific fisher, and various amphibians as threatened or endangered species. If successful, they will have vested the FWS with direct oversight of the Forest Service's timber program and also assured themselves direct access to the courts to enforce ESA protections against the Forest Service's future cutting plans. They also have administratively appealed individual Quincy project decisions, delaying group selection logging proposals as well as several DFPZ projects. In addition, the environmental community has initiated a new California wilderness protection campaign, which would legally protect several roadless areas within the Quincy project boundaries. Although a statewide wilderness bill covering over 2.4 million acres has been introduced in Congress, it remains unclear whether Congress will act on this expansive proposal. In sum, the broader environmental community not only has significantly limited the sweep of the Quincy group's community stability proposal, but also is intent on further constraining the Forest Service's ability to harvest timber across the Sierra range.[13]

The Sierra Nevada framework document and the Quincy Library Group processes are dynamic works in progress for which the final chapter has yet to be written. The framework document was adopted during the final days of the Clinton administration as part of its ecosystem management legacy. Despite its close ties to the extractive industries, the George W. Bush administration seemingly endorsed the framework document by rejecting wholesale the 234 administrative appeals filed against it. The Forest Service, however, is reexamining the framework's fire policy assumptions with an eye toward more aggressive fuel treatments, which could presage more extensive logging to reduce fire risks. It also has initiated an administrative study for the Quincy area forests designed to test how different group selection, DFPZ, and other harvesting strategies will affect the forest structure and California spotted owl habitat. Meanwhile, a frustrated Quincy Library Group has threatened to sue the Forest Service to implement fully its experimental timber cutting programs, arguing that the Herger-Feinstein legislation should take precedence over the framework guidelines in local forest management decisions. As a further sign of its frustration, the Quincy group briefly suspended its regular public meetings and closed other committee meetings to nonmembers. Exactly how the regionwide framework review and the Quincy area administrative study will be resolved remains to be seen, though the specter of litigation hangs over each process. The debate over Sierra Nevada forest management is plainly moving into new and uncharted political-legal arenas, which is perhaps where it was destined from the beginning.[14]

Getting Ecosystem Management Right: Lessons from Quincy and Beyond

The Quincy Library Group experience provides a window into the new age of ecology and democracy on public lands, illustrating just what a marriage between ecosystem management and collaborative processes might mean on the ground. Indeed, the Quincy experience reveals the multiple complexities involved in translating ecological principles into concrete planning goals through a community-based collaborative process. Manifold questions are evident: Has adequate attention been given to biodiversity conservation and ecological sustainability? What are the relative merits of active versus passive management strategies in achieving ecosystem maintenance and restoration goals? How much weight should local economic and social concerns receive? What is the appropriate geographical scale and time frame for an ecosystem management initiative? How should adaptive management strategies be employed? What role should local collaborative groups play in formulating ecosystemwide policies? How can such groups be structured to ensure fairness and equity? What role should Congress and legislation play in this transition to ecosystem-based management? How much weight should be accorded national versus local concerns in establishing public land policies and priorities? What legal constraints do the public land agencies face in integrating their regional, local, and project-level planning and management responsibilities? Some answers are beginning to emerge, while others are still obscure, leaving the marriage between ecosystem management and collaborative processes still very much a work in progress. And leaving the jury still out on whether the Quincy experiment represents a step forward or a step backward.

BIODIVERSITY PROTECTION AND ECOSYSTEM RESTORATION

If ecosystem management has a core meaning, it is the maintenance of sustainable ecosystems, including the protection and restoration of biodiversity. Taking a holistic approach to natural resource policy, ecosystem management differs from traditional approaches that have emphasized single resources and their utilitarian value. As we have come to understand ecological complexity, our focus has shifted to the ecosystem as a whole, using the species that inhabit it as a proxy for ecological health and resilience. Laws like the ESA, NFMA, and National

Environmental Policy Act (NEPA) have legitimized biodiversity protection and ecosystem integrity as a public land management goal, as reflected in the Pacific Northwest spotted owl court rulings. In fact, the 24 million–acre Northwest Forest Plan was constructed around a multispecies protection strategy, with timber harvesting and other resource management goals subordinated to first-order biodiversity conservation goals. Elsewhere, too, biodiversity conservation has attained an ascendant status. Multiple ESA species listings, critical habitat designations, and indicator species determinations now overlie much of the public domain; spotted owls, salmon, red cockaded woodpeckers, lynx, loach minnows, bull trout, and others have forced modifications in logging, mining, and grazing activities that before had the public lands pretty much to themselves. Freshly sensitized to this new hierarchy in values, the public land agencies have undertaken the Sierra Nevada Framework Plan, the ICBEMP project, and other regional planning initiatives that give priority to biodiversity conservation and ecosystem restoration. While each of these initiatives has its unique history and nuance, they share a common commitment to maintaining ecological integrity on the public lands. It is a commitment that can be traced directly to the prevailing federal law and a growing societal concern over extinction problems.

In the case of the Quincy Library Group proposal, critics have complained that it does not adequately protect the California spotted owl or other sensitive forest-dwelling species. Plainly alert to biodiversity concerns, the Quincy Library Group's 1993 community stability proposal pointedly incorporated the CASPO Interim Guidelines, placing spotted owl habitat areas and activity centers off-limits to timber harvesting. Although the 1998 Herger-Feinstein legislation reaffirmed the CASPO guidelines, it also mandated that the forthcoming EIS process must comply with all relevant environmental laws, including the NFMA biodiversity provision. Over the Quincy group's vehement objections, the final Herger-Feinstein EIS included additional California spotted owl protections that forbid any logging within suitable owl habitat, thus effectively eliminating nearly one-third of the Quincy project's DFPZs and otherwise reducing local timber sale opportunities. Then in the 2001 Sierra Nevada Framework EIS, faced with a potential ESA listing and its own NFMA biodiversity obligations, the Forest Service established even more onerous California spotted owl protections, further reducing timber harvesting opportunities in the Quincy project forests. In both instances, the Quincy Library Group appealed these increasingly more protective owl standards, arguing that they were scientifically unjustified and undermined critical local fire protection needs. Meanwhile, seeking even greater California spotted owl protections, national environmental organizations petitioned the U.S. Fish and Wildlife Service under the ESA to list the California spotted owl as a threatened species.[15] Regardless of who prevails in this biological debate, it is noteworthy that the Quincy group has consistently taken a less protective position toward the owl, instead giving more weight to local economic concerns and other forest health objectives.

A similar schism has occurred over the Quincy Library Group's commitment to restoring forest ecosystem health and absent ecological processes. The group has endorsed an active management regime to achieve forest restoration objectives, while its environmental critics would prefer to leave the forest largely untouched. The Quincy proposal is focused mainly on forest health and restoration; its premise is that a restored forest ecosystem will sustain the California spotted owl and other such species by re-creating necessary old growth habitat and minimizing destructive fires. To achieve its restoration goals, the Quincy proposal relies heavily on timber harvesting and prescribed fire: logging companies will remove overgrown, smaller-diameter trees to reopen the forest understory to its historical condition; strategically located DFPZs will be constructed to minimize fire size and to protect local communities and property; and prescribed fires will be used to clear the remaining understory. Revenues from these timber harvesting activities could be used to finance the costly forest restoration effort. The broader environmental community, however, objects on each score. It views the Quincy timber program as an invitation to resume large-scale commercial logging; it sees the DFPZs as an unnecessary insult to the forests, characterizing them as a series of "lineal clear-cuts" that will further fragment the forest, with adverse consequences for local species as well as the ecosystem itself; and it would rather allow nature to take its course by permitting natural fires to burn, employing prescribed burning where appropriate, and thinning only where essential to protect lives or property. Even then, it would prefer that the federal agencies directly subsidize the needed restoration work rather than relying upon revenues from logging companies intent on extracting a commercial profit from the forest.

If the Quincy experience is typical, then the lessons for biodiversity protection and ecosystem restoration on public lands are evident. Because biodiversity and ecosystem integrity are broad, amorphous, nationally shared goals, local collaborative groups may not always be committed to protecting them when the cost of protection will be borne locally. In the case of the California spotted owl, the Quincy Library Group was unwilling or unable to sacrifice local economic well-being to pursue more diffuse biodiversity protection goals. In the case of forest restoration, faced with a choice between an active or passive ecosystem restoration agenda, the local community will tend to favor the former, particularly if it means more jobs and greater security from fire or other natural hazards. Although the Quincy restoration debate is framed in ecological, forest health, and fire management terms, it also includes distinct acreage harvest targets as the starting point or bottom line.[16] Without a clear and unambiguous national biodiversity conservation standard—reflected in the overlapping ESA and NFMA legal mandates—local economic and related concerns may well have taken precedence on the area forests. In short, when it comes to protecting biodiversity and ecosystem integrity, local collaborative groups are more likely to serve the local rather than national interest, which could compromise important ecological objectives on public lands.

WHITHER THE HUMAN ELEMENT?

Any ecosystem management strategy developed through public involvement is not just about ecology; it is also about integrating human concerns into natural resource policy decisions. Most knowledgeable observers agree that the ultimate goal must be sustainability, defined broadly in ecological, economic, and social terms. The Forest Service's revised NFMA regulations are very clear that the goal of forest planning is "sustainability, composed of interdependent ecological, social, and economic elements . . . meeting the needs of the present generation without compromising the ability of future generations to meet their needs."[17] Put simply, sustainability contemplates a healthy environment that supports the human communities dependent on it over the long term. With public involvement playing a key role in planning and management decisions, adequate opportunity exists to inject economic and social concerns into the decision process. The question is not whether human-centered economic and social concerns are part of the policy equation, but how they are to be reconciled with competing ecological considerations. Does—or should—one trump the other?

Reflecting these basic sustainability principles, the Quincy Library Group proposal endorsed a vision of the community's future that incorporates ecological, economic, and social concerns. A major tension, however, is evident in the relationship between the proposal's ecological and economic goals. Ecologically, the Quincy proposal sought to protect roadless areas, California spotted owl habitat, and riparian ecosystems while also reintegrating fire into the forest ecosystem. Economically, the proposal was designed to save the local mills and ensure a continuous supply of timber, to maintain job opportunities and local tax revenues. The proposal actually increased the volume of timber coming off the forests by establishing annual acreage targets and allowing some old growth cutting, though not clear-cut logging. A pragmatic response to contentious timber-related issues, it was plainly designed to address the local timber industry's concerns as a means of stabilizing the local economy. Not surprisingly, the Quincy critics believe the group's proposal—with its logging provisions and annual acreage targets—emphasizes economic goals to the detriment of ecological concerns. Long committed to curtailing (or even eliminating) commercial logging in the Sierra Nevada forests, the environmental community simply does not trust the Forest Service not to convert salvage or thinning programs into full-scale logging operations. These same critics also do not believe the Quincy proposal will strengthen the local economy over the long term, arguing that its focus on preserving the town's major Sierra Pacific lumber mill will deter necessary diversification efforts. Some critics have further suggested that by focusing on saving the large Sierra Pacific mill, the Quincy proposal actually jeopardizes the smaller local mills or independent timber workers, who are being forced into an ever more dependent relationship with the larger mill.[18]

Similar tensions pervade the relationship between the Quincy proposal's ecological and social goals. There is considerable evidence that the Quincy process has helped resolve long-standing community social problems. According to Quincy group members and outside observers, a sharply divided community has begun a healing process. A sense of camaraderie and dogged perseverance is plainly evident among the Quincy Library Group's diverse membership, emblematic of how long-standing local divisions have fallen by the wayside in pursuit of a common goal. Even the Quincy group's most vocal critics acknowledge these positive developments in community relationships. But though an improvement in local social relationships is important within Quincy, it is much less important to outside environmental groups who are focused on their own individual agendas, namely ecological protection. The local community may benefit socially from the Quincy collaborative process, but that does not mean that forest conditions and the California spotted owl are similarly benefited.

The hard question, of course, is whether local economic and social values have triumphed at the expense of equally important ecological values. Inasmuch as ecosystem management is about promoting long-term sustainability, these diverse values should be seamlessly integrated into any collaboratively developed agenda. But that has not historically been the case on the public lands, where the extractive industries have pretty much had their own way. And it may not have been true with the original Quincy proposal, given its overt commitment to maintaining the local timber industry. Even with the rhetoric of ecosystem management, it may well belie reality in a place-based collaborative process to expect any community long dependent on public lands for its economic sustenance to ignore this fact and endorse resource policies primarily for ecological reasons. Because these are publicly owned lands, the community should not have the final word on setting the relative management priorities. With so much at stake, sustainability demands a longer-term perspective framed in less parochial terms. And in the Quincy case, of course, Congress legislatively added the requirement that the project must comply with federal environmental laws, thus effectively striking the final balance between environmental, economic, and social interests on these national forests. In a democracy, where there is no perfect answer to where the public interest lies, this congressional resolution may be as good an answer as can be expected.

SCALE, COMPLEXITY, AND ADAPTIVE MANAGEMENT

Sensitive to biological realities, ecosystem management seeks to accommodate nature on the landscape. It acknowledges not only that ecosystems transcend conventional political boundaries, but also that they are complex, dynamic, and ever-changing entities. As a geographical matter, ecosystem management initiatives ordinarily should conform to ecological rather than political boundaries in order to address the full array of species, processes, and ecosystems on the landscape. As a temporal matter, ecosystem management proposals should be conceived and framed with a view toward evolutionary rather than conventional time frames. To accommodate complexity and unpredictable change, ecosystem management initiatives usually employ adaptive management strategies, namely monitoring, reevaluation, and adjustment protocols to address changes in scientific knowledge and human values. The goal is to ensure that management strategies are calibrated to ecological realities, including biodiversity conservation and ecosystem integrity concerns. However, when an ecosystem management project is conceived in a community-based collaborative process, these scientific concerns and strategies must also be reconciled with competing political and economic concerns.

The Quincy Library Group proposal illustrates the difficult geographical scale issues embedded in any ecosystem management initiative. The proposal covers two and a quarter national forests, or 2.5 million acres, an area larger than Delaware and Rhode Island combined. Nearly everyone involved agrees that the Quincy proposal boundaries were designed primarily for economic purposes, to ensure an adequate supply of logs to keep the local mills afloat. The boundaries also reflect calculated political judgments: they mirror preexisting national forest (or ranger district) boundaries; they were summarily expanded to include the Lassen National Forest at the request of the neighboring county commission; and they omit Sierra Pacific Industries' private lands. The boundaries have an ecological dimension too: public lands embraced within the proposal nearly match the Feather River watershed and cover other smaller adjacent watersheds, but they do not fully encompass the California spotted owl's habitat requirements. Without question, the Quincy proposal has effectively expanded the Forest Service's vision beyond the confines of individual forests or ranger districts, forcing the agency to coordinate its management priorities and strategies on a larger than normal scale. And because the Quincy project boundaries are linked to economic and political considerations, the project has garnered community support and can draw upon local knowledge to sustain the initiative. Quite simply, the boundaries are based on a pragmatic mix of economic, political, and ecological factors.

The Quincy proposal, however, may be criticized for encompassing either too little or too much land within its boundaries. As an ecological matter, the Quincy boundaries are much too small to completely address important biodiversity goals, namely ensuring the California spotted owl's survival and its habitat needs. For the owl, the appropriate ecological scale would be nearly the entire Sierra Nevada range, which is the landscape covered by the framework document that tightened the Quincy proposal's California spotted owl obligations. Relatedly, critics have noted that Sierra Pacific Industries, a key Quincy Library Group participant and the state's largest landowner, deliberately excluded its own private lands from the group's discussions. Given the obvious ecological interconnections between these adjoining public and private lands, the Quincy proposal does not fully address even local biodiversity, forest health, and related issues. In each instance, the Quincy Library Group obviously made a calculated political judgment when deciding on the project's boundaries. It had no legitimate basis to assert any authority over Sierra Nevada forests far removed from its own environs, nor did it have any knowledge about these distant forests or the needs of nearby communities. And it had no legal power or claim over Sierra Pacific's private lands without the property owner's consent.

Paradoxically, Quincy's environmental critics also believe the project encompasses too much land, because it embraces 2.5 million acres of public land extending far beyond the immediate environs of Quincy and other nearby timber-dependent towns. This criticism is framed in political not ecological terms; it contends that the Quincy group has illegitimately asserted itself beyond its appropriate sphere of interest or influence. It is driven by the environmental community's concern over the Quincy group's hard target timber harvest acreage figures. Quincy supporters respond that while the proposal covers 2.5 million acres, the initiative's timber-related projects will directly affect something less than 300,000 acres, or only 10 percent of the overall acreage. They assert that the proposal's landscape scale is based on sound science as outlined by the SNEP scientists. And they note that nearby communities as well as Congress have endorsed the proposal, which gives it political legitimacy and has helped generate the federal funds necessary for the project. Regardless, these scale criticisms of the project boundaries reflect the difficulties inherent in matching ecological and political realities on the landscape. A local collaborative ecosystem management project ignores these realities at its own risk.[19]

Ecosystem management is designed to bridge these jurisdictional problems through coordinated, landscape-scale planning and management. The bedrock principle is interagency and interjurisdictional coordination to address transboundary ecological concerns. In the case of the Quincy Library Group proposal, its geographic and biodiversity shortcomings have been addressed through the Forest Service's regional Sierra Nevada framework document, which superimposes ecosystemwide California spotted owl restraints on the Quincy project and other forest management decisions. The relationship between the framework document and the Quincy proposal illustrates how local collaborative initiatives can be nested hierarchically within larger ecosystem plans to facilitate coordinated planning at an appropriate scale. Under this nesting approach, planning can be scaled down from the landscape as a whole to discrete ecosystems or ecological components, and then to specific locations, like the Quincy area. Or for watershed planning purposes, subwatersheds can be nested in watersheds, which can then be nested in entire river basins. Of course, the key to making such a nesting approach work is coordination among the various agencies and jurisdictions with authority over the affected landscape. At the federal level, the public land agencies each have legally mandated coordination responsibilities in their planning and NEPA analysis processes. And when ESA-listed species are at issue, the U.S. Fish and Wildlife Service has regulatory authority that extends across jurisdictional boundaries. A hierarchical nesting strategy would seem to require that coordination efforts be explicitly noted and explained in the relevant planning documents. Even better, a separate interagency coordination statement requirement would strengthen this ecosystem management obligation.[20]

A more difficult question is how private landowners within an ecosystem might be brought into a collaborative process to achieve landscape-scale planning objectives. The Quincy proposal, for example, would be strengthened greatly as an ecosystem management plan if Sierra Pacific Industries' private landholdings were included within the areawide plan. When such private lands are at issue, the basic strategic choices are limited either to coercive legal mandates or to voluntary inducements. Private landowners typically resist coercive regulatory approaches to ecosystem management, just as they are reluctant to encumber their private landholdings with ecological obligations. Voluntary inducements therefore represent a more attractive strategy for achieving ecological coordination objectives on private lands. Such inducements can include land exchanges for less encumbered federal property or the obvious tax benefits associated with conservation easements. When these voluntary approaches fail, however, the only recourse may be to adopt a more coercive strategy. In the Quincy case, if the California spotted owl is listed under the ESA, then federal regulatory authority under the Section 9 no-take provision will extend onto private lands throughout the owl's range. At that point, Sierra Pacific Industries could either devise its own California spotted owl conservation strategy or perhaps join the Quincy ecosystem management proposal. In either case, the ESA habitat conservation plan no-surprises and safe-harbor policies offer private landowners like Sierra Pacific a powerful incentive to cooperate in ecosystemwide planning efforts.[21]

Scale can also be understood in temporal terms for ecosystem management purposes. In the Quincy case, this raises the question whether the project's goals and strategies are designed from an ecological perspective, taking full account of the evolutionary nature of ecosystem processes. Put differently, is the Quincy approach likely to succeed in restoring the ecosystem to its natural range of variability? The project, with its emphasis on restoring historical fire regimes and forest conditions, is plainly sensitive to long-term ecological health and restoration goals. By relying on selective logging and the DFPZs, however, the project interjects a short-term active management strategy into this evolutionary equation. Sensitive to local political realities, these strategies are a very pragmatic response to immediate economic and human safety concerns. They also reflect the Quincy group's belief that an aggressive cutting strategy is necessary to reverse ecological decline in the area forests owing to past fire suppression and timber harvesting policies. While the scope of the proposed cutting is troubling to the project's environmental critics, the active management strategy has a plausible scientific foundation that takes account of temporal ecological concerns. In short, the Quincy group has married long-term ecological objectives with a short-term implementation strategy that ensures local political support for the project. Whether the strategy will succeed or not remains to be seen, which highlights both the contingent nature of the initiative and our limited knowledge about ecosystem processes.[22]

Because ecosystems are complex and dynamic entities, ecosystem management employs adaptive management techniques to address change and uncertainty. Conceding that our scientific knowledge is limited, adaptive management involves establishing baseline conditions, monitoring the ensuing changes, reevaluating the situation, and then adjusting management strategies to incorporate new ecological information as well as related changes in human values. From the outset, the Quincy proposal has been viewed as an experimental five-year project that must be reassessed to determine whether its forest restoration and other goals are being met. The 1998 Herger-Feinstein legislation instructed the Forest Service to prepare an initial EIS documenting the prevailing environmental conditions, and it imposed explicit monitoring and reporting requirements in order to evaluate the project's progress. In addition, the Forest Service has initiated its own administrative study to determine how the California spotted owl and the area forests respond to different fuel treatment strategies, including those provided for in the Quincy legislation. The net result should be extensive baseline data on forest conditions and species trends, as well as updated information that can be used to reassess management strategies. The framework for adaptive management is in place.[23]

The law, however, may limit experimental opportunities in the Quincy area forests, revealing a tension between law and science for adaptive management purposes. The Quincy critics, fearing the group's annual hard acreage timber cutting targets, pressed Congress into requiring the Forest Service to abide by federal environmental laws when implementing the project. In their view, it was essential that the agency be obligated legally to adjust its management strategies as necessary to protect the California spotted owl. The Forest Service, as we have seen, responded to this mandate, first in the Quincy Library Group EIS and then in the 2001 Sierra Nevada framework decision, both of which curtailed the original timber cutting proposals in an effort to safeguard the owl. In both instances, the Quincy Library Group appealed these decisions, asserting that they compromised the experimental value of its agreed-upon forest management strategy. The group was concerned that the Forest Service would not implement enough timber harvest and DFPZ projects to make a worthwhile assessment of how the forest is responding. The Forest Service's framework document review may or may not alter the situation. Nonetheless, the Quincy group appeals highlight an evident quandary: when strict legal standards are imposed on such projects, the law may limit the flexibility necessary to test alternative management strategies. Although this potential conflict between law and science could hinder experimental initiatives, few would advocate giving agency officials unfettered discretion for adaptive management purposes. The challenge is to design meaningful yet flexible standards that promote scientific learning while protecting important ecological and other values.

A MATTER OF PROCESS

The new collaborative conservation movement has made significant inroads on the public lands, influencing management policies and decisions in numerous locations. At the same time, collaborative processes have received intense scrutiny, from both academic commentators and the broader environmental community. The key question is what constitutes a principled and thus legitimate collaborative decision process. Given the ad hoc and experimental nature of the collaborative conservation movement, few firm legal standards govern these arrangements. Nonetheless, certain fundamental collaboration principles have been identified: the process must be inclusive, fair, and equitable. National as well as local interests should typically be represented and considered. The process must be open and transparent. Basic accountability standards must be met, from political, legal, and scientific perspectives. When public lands are at issue, an important consideration is what role the responsible land management agency has played in the process. Although some degree of latitude must be granted those willing to enter the uncharted territory of collaborative conservation, that flexibility necessarily shrinks when publicly owned lands and resources are at stake.[24]

In the case of the Quincy negotiations, critics have questioned whether the process was inclusive and fair. They do not believe the Quincy Library Group is sufficiently representative of the multifarious national and local interests that have a stake in the Sierra Nevada national forests. Not only was the Forest Service excluded from direct involvement in the group's original negotiations, but the broader, nonlocal environmental community was also not represented. The ranching community and related national forest grazing issues were also not part of the Quincy process, further illustrating the limited scope of the community stability initiative. Put simply, its critics believe the Quincy group represents a narrow community of place, not an extended community of interests. Recalling the Forest Service's history of capture by the timber industry, one critic even suggests that this has happened again in Quincy, though this time the industry has captured the Quincy Library Group rather than the agency. Its critics, therefore, view the Quincy group as an unrepresentative and unaccountable alliance of specific local interests that, by virtue of its political clout and skill, has gained a privileged position in setting policy for the local forests.[25]

The reality of the Quincy Library Group process is more complex, however. The actual negotiations leading to the Quincy community stability proposal were quite brief. To nearly everyone's surprise, once the initial overture was made, the negotiations quickly came to a head. Sensing that they held a losing hand, the timber industry representatives readily agreed to several long-standing environmental demands, namely protection for roadless areas and riparian zones, no further clear-cutting, and a new focus on restoring forest health. The original Quincy participants, representing a limited but important segment of the local community, were intent to capitalize on this unexpected opportunity to resolve long-standing forest management issues. Recognizing the need to broaden its base of support, the Quincy group early on sought to expand its membership and even contemplated expanding its agenda to include nontimber issues. The group's environmental activists tried to engage out-of-area environmentalists in the initial negotiations, but they declined, citing travel logistics and similar concerns. Before finalizing its original proposal, the group convened a community meeting to endorse the initiative, thus effectively legitimizing itself within the local community. That the group's environmental objectives were clothed in community stability terms and involved a modified timber harvesting program confirms that local economic concerns were a major factor in the negotiations, but that should be expected in any meaningful give-and-take negotiation where local jobs, tax revenues, and the like are at stake. Given the various environmental protections included in the Quincy proposal, there is little hard evidence that the timber industry actually captured the process or that ecological concerns were ignored. One might wonder whether the Quincy environmentalists could have cut a better deal, but they plainly obtained several long-standing environmental objectives. Yet unable to broaden its original base, the Quincy group soon found itself operating as a place-based initiative driven primarily by its local members and their principal concerns. And that, rightly or wrongly, has cast a parochial pall over its proposal and subsequent activities.

For the most part, the Quincy Library Group process has been open and transparent. Soon after convening, the participants shifted their meetings to the town library to accommodate all interested parties in a neutral setting. The group's meetings have been regularly scheduled, noticed in advance, and open to outside observers or new participants. Everyone may participate in the meetings, and information is widely shared. As the process has evolved, however, the Quincy group has sometimes proven reluctant to reexamine or modify its basic approach. In early 1999, when faced with new requests from outside environmentalists, the group voted to exclude nonmembers from its Pilot Project Consultation Committee deliberations, citing the need to avoid disruptions from opponents while monitoring the Forest Service's implementation efforts.[26] In effect, the group has closed ranks around its original agreement, which is clearly a fragile political deal among competing factions who have strongly divergent views on timber management policy and ecosystem protection. Having reached a hard-earned consensus among themselves, and then having shepherded a modified version of their proposal through Congress, the Quincy participants have been reluctant to reopen the agreement just to appease disaffected observers who did not participate in the original negotiations. While understandable, this reluctance to reexamine its handiwork calls into question the group's commitment to openness, in terms of both new participants and new information. Given the tenuous legitimacy of such collaborative processes, an unqualified commitment to openness can help counteract charges of procedural unfairness.

The Forest Service's role—both in the initial Quincy deliberations and in the subsequent EIS implementation phase—raises additional process concerns. Astute observers of collaborative processes agree that federal agency involvement is often vitally important to a group's ultimate success; agencies can bring technical expertise, financial resources, legal authority, and general credibility to the process. But a federal agency faces legal constraints on its interactions with an ad hoc collaborative group: the agency cannot relinquish its ultimate decision making powers, and it must adhere to Federal Advisory Committee Act (FACA) procedural limitations. In the Quincy case, the Forest Service was excluded from the initial negotiation process; the group's members were collectively suspicious of the agency, and they used this distrust as a rallying point to help hold their disparate membership together. But this meant that the Quincy proposal was eventually presented to the Forest Service as an all-or-nothing proposition, which disregarded basic inclusiveness and fairness principles. The Forest Service, nevertheless, was a regular observer at the Quincy gatherings, sharing its concerns over the potential costs involved in the ambitious forest restoration program but not (to its discredit) the agency's unstated opposition to the proposal itself. Had the Quincy group brought the Forest Service into a closer alliance during its original negotiations, perhaps it could have strengthened the relationship and secured better cooperation during the implementation phase. Of course, Congress ultimately interceded and created a new and more clearly defined working relationship between the Forest Service and the Quincy group. And under the Herger-Feinstein legislation, the Forest Service retained its overall management authority, as reflected in its subsequent decisions altering the Quincy proposal to address overarching spotted owl concerns. All of which confirms that the federal agency role cannot be gainsaid on public lands.

Assessing whether the Quincy Library Group initiative meets basic political, legal, and scientific accountability standards is not an easy task. From a political perspective, the Quincy group was plainly responsive to local concerns and constituencies: the community ratified its proposal in a near unanimous vote; local political leaders were active group participants; and most surrounding communities have also endorsed the proposal. Nonlocal environmental groups, however, contend that the proposal did not adequately reflect national interests. Although the real problem may have been the national environmental community's refusal to engage in the process, the Quincy group was eventually forced to accommodate broader biodiversity concerns. By placing its stamp of approval on the Quincy group's proposal, Congress gave the Quincy process political legitimacy, but only after superimposing national environmental standards over it. From a legal perspective, the Quincy group originally operated without any clear legal guidelines, leaving it to invent its own process absent any direct federal involvement. As the Quincy proposal has evolved, though, various federal legal obligations—NFMA biodiversity standards, NEPA procedures, FACA limitations, and agency decision making requirements—have been introduced into the process. Although the Quincy group has resisted some of these legal constraints, both its proposal and process have now been brought into legal compliance, primarily because of the inevitable federal involvement in the initiative. From a scientific perspective, the Quincy group has emphasized the forest restoration dimensions of its proposal, leading some critics to question its commitment to biodiversity conservation. While only time will tell if the Quincy proposal was well-conceived, the Forest Service not only redesigned the proposal to better protect California spotted owl habitat, but it has also initiated the administrative study to assess whether the DFPZs and selective harvesting techniques will achieve the group's forest restoration goals. Thus, what began as a local public land initiative has been legitimized as a national scientific experiment, albeit in a significantly modified version that takes account of overriding environmental concerns.

POLITICAL AND LEGAL RAMIFICATIONS

The Quincy initiative illustrates the political and legal difficulties involved in advancing an ecosystem management agenda through local collaborative processes. Politically, the Quincy critics were most dismayed over the precedent of Congress adopting national legislation to establish resource management policy on individual national forests at the behest of a local group. National environmental groups strenuously resisted the Quincy Library Group's efforts to secure congressional endorsement for its community stability proposal, eventually losing this battle when Congress passed the 1998 Herger-Feinstein legislation. According to them, Congress has traditionally treated the national forests as a system, opting for a uniform set of systemwide organic legislative mandates rather than a series of forest-specific laws. By endorsing the Quincy Library Group proposal and establishing site-specific forest management standards, the Herger-Feinstein legislation seemed to ignore that congressional tradition. As a result, the Quincy experience could encourage other local collaborative groups to pursue their own place-specific legislation and thus effectively subvert the uniform policies governing the national forests. Contesting such piecemeal public land legislation would place an enormous strain on environmental groups and their limited resources.[27]

But is the Quincy site-specific legislation so unusual? As we have seen in the Pacific Northwest and elsewhere, Congress has regularly used the appropriations process to mandate short-term, local forest management policies. Congress also uses site-specific legislation to accomplish preservation goals, when establishing national parks, national conservation areas, wilderness areas, and similar designations. The trend in wilderness legislation is likewise toward more localized, site-by-site wilderness bills rather than broader regional or statewide bills. But the Quincy legislation only partially follows these models. By revising the Herger-Feinstein bill to require the Forest Service to follow national environmental laws, Congress fundamentally recast the Quincy legislation into an amalgam of national legal standards overlaying locally negotiated resource management priorities. Whether Congress intended to establish a new precedent with this hybrid

Quincy legislation is a more difficult question. Nearly anything that Congress does can be seen in precedential terms, given its ultimate responsibility for public lands. But Congress usually has many motives when it acts, as was evident in the Quincy case: it resolved a festering local forest management controversy; gave a qualified endorsement to local collaborative process experiments on public lands; endorsed a new resource management strategy; and reinserted itself into the national debate over ecology's role in forest policy. In short, the Herger-Feinstein legislation is not unprecedented; it did not relieve the Forest Service of its well-established environmental responsibilities, and any precedent it sets is still bounded by existing federal law.

The Quincy Library Group experience also highlights Congress as a forum where national and local interests have traditionally been reconciled over public land policy issues. By assembling a diverse group of local citizens, industry representatives, and political leaders, the Quincy group initially bridged partisan politics and gained widespread support for its proposal, regardless of party affiliation. The early support of nearly the entire California congressional delegation was critical to moving the proposal forward in this legislative arena. The fact that the proposal included important environmental provisions—roadless area, California spotted owl, and riparian protections—was attractive to Democratic representatives, while the timber harvesting provisions and the local nature of the Quincy group was attractive to their Republican counterparts. Indeed, the 429–1 House of Representatives vote in support of the Herger-Feinstein legislation demonstrated the potential political power that local bipartisan coalitions might wield over place-based public land matters. The bipartisan coalition, however, came unwound in the Senate after national environmental groups voiced strong opposition to the proposal. When they convinced the Senate to insert provisions in the bill requiring adherence to national environmental laws, the legislation was transformed from a purely local initiative into one that incorporated ecosystemwide concerns. Not only does the Quincy legislative experience validate Congress as a forum for resolving national–local debates over public land policy, but it demonstrates how national biodiversity concerns might be integrated into a locally negotiated management proposal. And it demonstrates why the environmental community, having succeeded over the past quarter century in nudging Congress toward more environmentally prescriptive and preservation-oriented legislation, may be reluctant to embrace a public land policy agenda based on devolution and local control.

The Quincy proposal has given rise to vexing legal complexities. In fact, one of the greatest Quincy project challenges has been how to reconcile the various laws that govern the initiative. At one level, the Forest Service is faced with integrating its NFMA, NEPA, and ESA responsibilities into a truly multitiered ecosystem management program that operates on several scales: from the Sierrawide framework decision to the multiforest Herger-Feinstein Quincy initiative to each individual forest and then to separate project-level decisions. At another level, the agency must reconcile its congressionally sanctioned Sierra Nevada Forest Plan Amendment process with the Herger-Feinstein legislation, determining whether one trumps another and how to mesh the two laws together. Lurking in the background, of course, is the prospect that the U.S. Fish and Wildlife Service may decide to list the California spotted owl or other local species, which would severely diminish the Forest Service's managerial discretion across the region's national forests. It is no secret that the environmental community and other interested constituencies are poised to challenge any perceived missteps in court; one lawsuit has already been litigated over the Herger-Feinstein EIS, while other lawsuits and administrative appeals have been filed on Quincy project proposals and the Sierra Nevada framework decision. And the Quincy group has threatened to sue too, asserting that its site-specific legislation takes precedence over more generic forest management statutes. Thus far, sensitive to the regional implications of its decisions and its related legal obligations, the Forest Service has insisted on adjusting the Quincy project to fulfill its broader biodiversity and ecosystemwide responsibilities. The evolving legal precedent, therefore, is one of adapting locally negotiated priorities to meet national standards while also adjusting planning protocols to incorporate local initiatives into the process.

Beyond the Quincy experience, legal complexity has become a fact of life in managing the public domain. Not only are the public land agencies constrained by multiple legal mandates, but they are also subject under the ESA to regulatory oversight by the U.S. Fish and Wildlife Service, which can veto or revise resource management proposals. The welter of overlapping organic planning strictures and NEPA environmental documentation requirements can sometimes seem an inefficient and wasteful exercise in redundant paperwork. In fact, Forest Service chief Dale Bosworth has mounted a campaign against the current legal structure, complaining loudly that his agency is hamstrung by "analysis paralysis" and unable to meet its resource objectives. His apparent solution would be less law and fewer mandates—a solution, as we have seen, that the Bush administration has actively pursued through its forest health initiative. But these complaints ignore one fundamental fact: it is the law—with its collective environmental protections, planning obligations, public involvement, and judicial review provisions—that has inculcated a new ecological management imperative on public lands. The law has injected biodiversity into the management equation, forced agency officials to do cumulative environmental analyses, opened agency planning processes to public scrutiny, and rendered the agencies accountable for their management decisions. Rather than jettisoning the entire existing legal structure, much of the current frustration could be obviated by one simple legislative revision—drop the facade of multiple-use management and establish clear resource priorities for public lands. And those new priorities, in this age of ecology, should explicitly include biodiversity conservation and ecosystem protection.[28]

Although the foregoing might be read as an unduly harsh critique of the Quincy Library Group and its community stability proposal, it is primarily intended to highlight the problems involved in translating ecosystem management principles into a concrete plan through a community-based collaborative process. Devolution and ecology are uneasy bedfellows on public lands. As the Quincy experience illustrates, any local group contemplating an ecosystem planning experiment on public lands faces difficult challenges, both in defining ecologically sound resource management goals and in devising workable implementation strategies. Any shortcomings in the proposal or process will open the door for legal challenges, legislative revisions, scientific criticisms, or bureaucratic recalcitrance. That said, the Quincy strategy has proven remarkably resilient and adaptable, and therefore should not deter the search for alternative models to strengthen the connection between ecosystem management and collaborative conservation. Opportunities exist to build on the Quincy experience and to construct a new approach to public land policy. In fact, we are beginning to see new models and proposals designed to accomplish just this transformation.

A Merger of Sorts: Ecosystem Management and Collaborative Processes

Ecosystem management and collaborative conservation have come to the fore in public land policy. At every level and in every forum, these ideas not only resonate but have been translated into concrete proposals, several of which have been sanctioned by law. The federal commitment to ecosystem management—whether expressed in terms of new institutions, legislative mandates, regulatory reforms, agency policies, or court decisions—is bringing a new set of relatively uniform policies for integrating ecological science into public land decisions. But there is a nonfederal side to the ecosystem management experiment, and it is decidedly local in character and origins. The emerging community conservation movement, as symbolized by the Quincy Library Group experiment, has claimed a role in public land policy debates. Unlike the new federal policies that apply uniformly across the public domain, these collaborative process initiatives are generally local in nature and scope, focused on a specific place or environmental controversy. The participants usually seek to integrate local ecological, economic, and social concerns into a new set of priorities and strategies for managing the nearby public domain. The important question, therefore, is whether the nationally driven ecosystem management agenda can—or should—be merged with the locally driven community conservation movement. And if so, then how should this merger be consummated?

THE ALLURE OF PLACE-BASED INITIATIVES

The ecosystem management and community conservation movements share similar organic origins. Like many of our original public land policies, they can each be viewed as pragmatic responses to pressing national priorities or local concerns. The obvious connection between these two movements is best seen in the plethora of grassroots, place-based initiatives that have surfaced in recent years, many of which have already been described in some detail. These groups share a commitment to addressing local natural resource problems with a new sensitivity to ecological realities and to meaningful public participation. According to one study, "because the success of ecosystem management is largely dependent on who takes responsibility for it and carries it out, linking it to local communities that benefit from it can build powerful local incentives and improve the likelihood that it is done well." Indeed, the Western Governors' Association's Enlibra principles strongly endorse the community-based conservation movement, promoting local solutions as the key to resolving federal natural resource controversies while acknowledging the need for national environmental standards. The logic behind these place-based initiatives is rather compelling: because ecological, economic, and social conditions vary from place to place, locally conceived strategies are necessary to respond to these diverse conditions. Or as one astute observer puts it: "nature tends to decentralize." Devolution in governance thus should, at least in theory, correspond to ecological diversity on the landscape.[29]

It is important, however, to distinguish between the collaborative process itself and the resulting management proposals. For the most part, these collaborative groups have contented themselves with developing and then advocating consensus-based proposals to the responsible agencies as a means of solving difficult problems and reducing local controversy. They have sought greater local influence over management decisions, but not local control over the lands themselves. Within the legal constraints governing the public land agencies, federal officials have been receptive to these suggestions but unwilling to relinquish management authority to these groups. It is, in essence, a soft form of devolution. But there is another side to the collaborative process movement, one that seeks to displace the public land agencies and to assume control over federal lands. It too is place-based, but it no longer trusts the agencies to manage the resources within their portfolios. These groups seek the full relinquishment of authority from the federal to the local level, a much harder and more radical form of devolution. Absent explicit statutory delegation authority, new congressional legislation is ordinarily required before a nonfederal entity can be vested with such management authority. Both models of community conservation are evident in the modern collaborative process movement, and they raise quite different issues for integrating ecosystem management principles with collaborative conservation processes.

The ecosystem management and collaborative conservation movements are widely identified with different constituencies and levels of government, which presents potential political obstacles as well as opportunities. Ecosystem management policy is ordinarily traced to the environmental movement, and often associated with its preservationist agenda. After the Clinton administration endorsed ecosystem management to resolve the spotted owl litigation, each public land agency proceeded to develop its own ecosystem management policies, while the pre-1994 Democratic Congress seemed poised to embrace this new idea too. Later Clinton administration initiatives, including the national monument designations and the roadless area rule, cast its commitment to ecosystem management in strongly preservationist terms. As a result, ecosystem management has been viewed as a top-down federal policy initiative linked to the national environmental community and directed toward expanding the preservation agenda on public lands. In contrast, the collaborative conservation movement has a distinctly local grassroots quality to it, reflected in the numerous citizen-driven groups that have made their presence known across the western landscape and elsewhere. Lacking a clear foundation in federal law, the groups have a distinct extralegal character to them, which has often caused the federal agencies to distance themselves from these initiatives. Most collaborative groups are focused on particular landscapes or watersheds, and they are quite cognizant of the economic and other local implications of their proposals. The post-1994 Republican Congresses have generally encouraged these initiatives, seeing them as a potential antidote to the Clinton administration's large-scale ecosystem management experiments. So too has the George W. Bush administration, which has strongly endorsed local involvement on public lands. There is, in short, an obvious political mismatch between ecosystem management and collaborative conservation that could stand as a wedge between these two promising new ideas.

That has not yet occurred, however, and therein lies the opportunity to consummate a functional merger between these seemingly disparate public land management strategies. As a legal and scientific matter, ecosystem management has a strong claim to legitimacy; the courts have endorsed key ecosystem management principles, while its basic scientific underpinnings have not been seriously challenged. Its political status is more tenuous, however, given the current congressional mood and presidential administration, as well as the public's evident interest in decentralized governance. On the other hand, community-based conservation has no clear legal mandate on public lands. It is driven less by science than by local concerns and the related desire to ratchet down the level of controversy that swirls around public lands. Because it can make peace in divided communities, the collaborative process idea is attractive to politicians regardless of their partisan affiliations. In other words, these two rather different approaches to public land management can be seen in complementary terms. In effect, they need each other as neither can carry the day alone. Because our governing institutions cannot—and will not—overlook either the law or science in managing public lands, local control advocates cannot ignore national environmental goals as reflected in endangered species and biodiversity legislation. And because the same institutions likewise cannot—and will not—ignore the human communities or political realities that shape public land policy, environmental activists must consider how to participate effectively in these local forums. (As a collateral matter, these collaborative processes offer the environmental community an opportunity to broaden their constituent base among rural working-class westerners with economic ties to public lands.) One obvious way to merge the two, then, is through place-based initiatives responsive to national ecological goals as well as local concerns.

In fact, Congress has a long history of crafting place-based legislation to resolve thorny public land and natural resource controversies. The numerous enabling acts creating individual national parks, monuments, and recreation areas can be seen in these terms; to defuse controversy, Congress has frequently established preserved areas where intensive development is precluded or limited. Wilderness and wild and scenic river legislation can be understood in similar terms. Other examples, however, do not involve strict preservation. One of the earliest was the federal compact legislation creating the Lake Tahoe Regional Planning Agency, which serves as a bistate planning and regulatory body now responsible for maintaining and restoring environmental quality across the watershed. The Columbia River Gorge National Scenic Area is another such example, where Congress has established an innovative federal–local planning scheme to safeguard the area's unique resources. Perhaps the most obvious example is the Quincy Library Group legislation, which places a congressional imprimatur on a locally negotiated national forest management experiment conducted under the Forest Service's auspices subject to the full panoply of national environmental laws. The Valles Caldera trust arrangement can be seen in similar terms, though there Congress chose to forgo federal management over these newly acquired lands for oversight by a citizen board of trustees. Thus, though Congress typically deals with public land issues on a systemic basis, it also has used place-based legislation to meld national and local interests into an acceptable and often locally brokered political compromise. Federal place-based legislation is therefore very much within the public land tradition, adding both legitimacy and credibility to the community conservation movement.[30]

Advocates of community-based conservation have developed yet other proposals to devolve decision authority for public lands to local citizen groups. One of the most vocal proponents of this approach is Daniel Kemmis, who makes a powerful argument for vesting western communities with management responsibility for nearby public lands on a regional basis. Kemmis envisions congressionally approved regional compacts that would create a functional regional governance structure to oversee the lands while ensuring that they remain public and that they are managed under a sustainable ecological health standard. Richard Behan, another longtime student of public land policy, essentially endorses Kemmis's devolutionary approach, arguing that local citizens concerned with nearby public lands can serve as effective proxies for more distant citizens who might be interested in these same lands. Similarly, the so-called Lubrecht group proposed experimentally transferring some national forests from federal control to a local collaborative group that would manage them free from the legal mandates and traditions binding the Forest Service bureaucracy, thus testing the thesis that local community groups can better and more efficiently manage the national forests than can the existing agency. Not surprisingly, the George W. Bush administration has seized the idea and unveiled its own charter forests proposal, which would empower "a local trust entity" to administer individual national forests "outside the Forest Service structure." The model presumably would be the New Mexico Valles Caldera trust arrangement, where Congress has forgone federal management of newly acquired public lands in deference to a citizen board of trustees. Or a related model might be the U.S. Fish and Wildlife Service's proposal vesting a local citizen management committee with oversight responsibility for an experimental population of reintroduced grizzly bears in the Selway-Bitterroot region of central Idaho and western Montana. The bottom line: a local citizen group would actually displace a federal agency and assume management responsibility for publicly owned lands or resources. This would go well beyond the typical consultative collaborative process, raising additional concerns over the fit between devolution and ecosystem management.[31]

DEVOLUTION REVISITED:

THE NATIONAL VERSUS LOCAL INTERESTS DEBATE

Clearly, there is reason to pause before endorsing wholesale devolution of management authority to local communities for public lands. Does the new rhetoric of devolved decision making match the reality? Will devolution, as a public land policy, really merge local interests with broader ecosystem management principles? When the stakes are small and mostly local, a local citizen body might well be entrusted with a management role under clearly defined standards. But when the stakes are large and the ramifications widespread, a local body may not be well equipped to understand or reconcile the diverse resources, interests, and values at play. Where public lands and resources are at issue, there are often fundamental differences between local and national interests. These long-standing federalism tensions are particularly acute during times of change.

A recurrent theme in the debate over both ecosystem management and community conservation is how to reconcile divergent national and local interests in establishing resource priorities for public lands. At one level, the debate is over what level of government should set management policy—distant and often ill-informed federal officials, or parochial and often self-interested local citizens, to put the matter starkly. At another level, the debate is over which interests—national or local—should ordinarily prevail in determining on-the-ground priorities, a controversy often portrayed as pitting national environmental interests against local economic ones. Or put differently, the question is whether directly affected local interests should have a larger voice in resource management decisions on public lands than more distant and thus less directly affected national interests. Given the growing interest in devolving management authority to local collaborative groups, the debate is not purely an academic one; it is having powerful real-world consequences. The fundamental issue, then, is whether the states or local communities can—or should—be entrusted with responsibility for managing public lands.

Federalism theory can help us better understand the benefits and costs of devolution for the public lands. Under the Constitution, the states are sovereign entities that deserve respect from the federal government, including the federal agencies that oversee public lands within their borders. The states often serve as laboratories for democracy; local communities can fulfill a similar role by devising experimental new resource management policies and strategies. The federal agencies, confronted with an array of diverse conditions on their landholdings, can benefit from drawing upon local knowledge about unique or different conditions. Because governance at the state and local levels is closer to the people, local management would promote civic participation and engagement in decision processes. Some commentators, asserting that the states have significantly improved their school trust land management capabilities, believe they deserve an opportunity to extend their new skills to public lands. But there are also powerful arguments against devolution on public lands. The transboundary dimensions of most modern environmental problems can be addressed only at the federal level; state and local governments lack both the authority and knowledge to effectively address these issues. The federal government is uniquely situated to prevent or ameliorate the distributional inequities that arise where one state benefits from resource development while another bears its costs. Only the federal government can ensure a level playing field among the states and thus prevent interstate discrimination that could see states hoarding scarce public resources for themselves. Compared to the states, the federal government has far more fiscal and technical resources at its disposal for management purposes. Given the powerful role that big business has historically played on the public lands, the federal government is probably better positioned than the states to deflect corporate influence and power. And according to some observers, the western states have not improved their resource management capabilities, particularly with respect to new biodiversity conservation and other ecological concerns. These concerns are magnified, of course, when management responsibility passes from the state level to local governments and then to ad hoc citizen groups.[32]

Indeed, the potential problems with devolving actual management authority over public lands to a local citizen council are manifest, particularly given the important ecological resources at stake. Were a public land agency to relinquish control over its lands, the federal laws governing them would presumably be replaced by counterpart state laws. Even a cursory examination, however, reveals that the western states do not have adequate environmental laws in place to manage public land resources. Most western states lack an endangered species act or their endangered species laws are quite weak. Even fewer states have any biodiversity protection laws. Several western states still lack an environmental policy act, which would at least require an environmental impact analysis before development proposals are approved. Even though Montana has both an endangered species act and a state environmental policy act, the state legislature has been as-siduously trying to dismantle these laws, as well as the state's clean water laws. Faced with recurrent budgetary problems, few western states have the financial re-sources or technical capabilities to undertake the sophisticated planning processes or adaptive management strategies necessary to manage public lands. After stren-uously resisting the federal wolf reintroduction effort, Montana, Idaho, and Wyoming have indicated they will need substantial federal financial assistance be-fore they can assume responsibility for managing any ESA-delisted wolves or griz-zly bears. And there is no reason to believe nonresidents would have any meaning-ful role in framing state management policies. Bluntly put, most western states have demonstrated little legal, political, or financial commitment to an ecological management or restoration agenda, which argues against blithely entrusting them with federally owned lands or resources.[33]

The Bitterroot grizzly bear reintroduction controversy provides a telling glimpse into the hard realities of devolution or local control of public resources. Here, Montana and Idaho, the timber industry, nearby communities, local resi-dents, and environmental groups were all involved in fashioning a grizzly bear reintroduction strategy. The proposal empowered a 15-member local citizen man-agement committee to oversee the reintroduced bears, subject only to the require-ment that their decisions promote recovery of the bear population. The proposal not only contemplated local collaborative management, but also was a locally ne-gotiated arrangement. Not everyone agreed with the proposal, however, fearing that reintroduced grizzly bears would bring more federal regulation and pose a threat to human safety. Ignoring the devolutionary aspects of the reintroduction plan and his own bedrock conservative political principles, Idaho's governor sued in federal court to block the reintroduction proposal and sought political redress within the friendly confines of the George W. Bush administration. Although Sec-retary of the Interior Gale Norton had unreservedly endorsed local involvement in federal resource management decisions during her confirmation process, she swiftly acquiesced in Idaho's arguments, withdrew the reintroduction proposal, and jettisoned the citizen management experiment.[34]

As is so often true in natural resource federalism debates, the Bitterroot grizzly bear controversy revealed once again that the issue is of paramount importance, not the level of government responsible for making the decision or managing the re-sources. The real concern was not with the decision process but with the substance of the decision itself. The lesson is simple: devolution is not a first-order principle when it comes to managing public lands or resources. This reality cuts strongly against relinquishing federal control over the public domain, particularly without any assurance of what priorities or standards local officials will attach to these lands.

Some public land issues are simply too important or contentious to resolve other than at the national level. Judging from the controversies engulfing the pub-lic land agencies and key concepts attached to new ecosystem management ideas, endangered species and wilderness are two such issues. The fact that Congress has retained sole authority to designate new wilderness areas and has vested all federal agencies with explicit endangered species protection responsibilities is powerful evidence of the overriding national character of these issues. Both the northern spotted owl controversy and the Clinton administration's national forest roadless area rule activated broad national constituencies, effectively transcending the local communities potentially affected by these issues. Conversely, the locally conceived San Rafael Swell conservation area legislation went nowhere in Congress, primarily because its proponents were insensitive to the broader wilderness implications of their proposal. But national-level public land issues are not solely environmental in character; the Bush administration's energy policy proposals raise important national defense and economic concerns that could have major implications for public land policy. These observations do not preclude a collaborative process to address these national issues in a specific locality, but they do suggest that national standards and interests must be part of the process. Not only should these broader interests be represented at the negotiating table, but the debate itself must be confined by national priorities and standards.

This is not to say that community-based conservation and devolved authority on public lands are bad ideas, only that they make sense in limited and well-defined circumstances. The diverse constituencies with interests in public lands are unlikely to support any policy that would have the public land agencies not only relinquishing managerial authority but also eschewing the federal legal protections that ensure public participation, establish resource priorities, and create management standards. Few states or local communities are truly prepared to accept such awesome management responsibilities without considerable federal financial, technical, and other support. The Quincy Library Group experience illustrates the underlying problem: federal ecological goals for public lands can—and often will—conflict with locally defined community goals for these same lands. But when the issue or location does not have national significance, the Quincy experience also suggests the solution: retain federal management authority and federal legal requirements while allowing—even encouraging—representative local groups to pursue their agendas within these parameters. The key to this alternate approach is that any local collaborative group granted a management role over public lands must act within well-defined legal parameters (or sideboards), both substantive and procedural. Without the assurance that enforceable environmental standards are in place, that the group itself is truly representative of the diverse constituencies with interests at stake, that meaningful public involvement procedures are being followed, and that final decisions are reviewable, then this alternative model of public land management lacks both legitimacy and accountability.

INTO THE FUTURE: CONCEIVING A LEGISLATIVE AGENDA

Regardless of which level of government is engaged in managing public lands, the basic questions are the same: Who makes the final decision? Using what standards? And following what procedures? With these questions in mind, it is time to identify and assess alternative legislative models for advancing new ecosystem management and restoration policies. To be sure, much progress has already occurred through administrative reforms, judicial rulings, and local experimental initiatives, but it is Congress that has the final word, and it has yet to be heard from decisively. The models that follow are therefore framed as potential legislative responses to the age of ecology on public lands. They draw on the traditional approaches Congress has employed either to reform basic policy or to resolve specific controversies. On one hand, Congress has used place-based legislation to chart a new course for contentious segments of the public domain—an apt short-term strategy for integrating community conservation ideas with national ecological management and restoration goals. On the other hand, Congress has periodically adopted organic or comprehensive legislation when it wished to revise federal priorities and policies more broadly—an apt response once new ecological management and restoration policies have secured the necessary consensus. Regardless of the approach, Congress is the appropriate institution to reconcile the diverse national and local interests pressing for recognition on public lands.

As we have seen, Congress has regularly employed place-based legislation to protect vulnerable landscapes or to resolve specific controversies. In the new era of ecology and collaboration, similar place-based legislative initiatives represent an attractive short-term method to reconcile national ecological conservation objectives with local economic and social concerns. The Quincy Library Group legislation is one model Congress might follow. It would entail translating collaboratively negotiated ecological and economic management goals into a congressional mandate that also incorporates national environmental legal standards, setting clear limits on how the proposal is to be implemented. A more radical approach is the Valles Caldera trust model, which would have Congress establish a citizen board to oversee the lands at issue, subject to statutorily delineated resource priorities, management standards, and procedural requirements. In essence, Congress would articulate federal standards to define the range of acceptable management options; the collaborative group would then be free to design locally acceptable implementation strategies within the federally defined limits. Such an approach would not only protect national interests in biodiversity and the like but also accommodate local concerns while fostering creativity and experimentation. This is not unlike how federal pollution control laws are designed: federal standards are set, then the states, industry, and others are encouraged to develop implementation strategies to meet the goal. The place-based model is not without problems, however; it could break down the relatively uniform character of our existing public land systems, foreshadowing yet more locally inspired and potentially troublesome legislation. This could transform routine local management controversies into major congressional donnybrooks and actually increase rather than decrease fragmentation on the public domain.

The place-based legislative approach is a particularly appropriate strategy for incorporating ecological principles into the American nature reserve system. Congress has traditionally invoked its legislative power on a site-specific basis to create national parks, wilderness areas, and other protected enclaves. Our protected areas system, however, is neither large enough nor diverse enough to ensure our biological heritage over the long term. We will need larger, better-situated, and interconnected reserves that are adequately buffered to protect against debilitating human disturbances. But to persuade Congress to augment the existing nature reserve system, utilizing such models as the Northern Rockies Ecosystem Protection Act proposal or the Wildlands Project for guidance, the proposals must address affected local communities and economic opportunities on public lands. This argues for selective site-specific modifications to our current hands-off wilderness management policies when creating new reserves to allow compatible economic and restoration activities to occur within carefully delineated protective zones. Such a proposal will not be an easy sell politically with the divisive atmosphere now shrouding the wilderness debate. But without creative provisions melding new ecological reserves with the needs of local communities, and perhaps even providing them with a role in overseeing new reserves, the prospects for passing such legislation are dim. Paradoxically, to address our enlarged biodiversity preservation and restoration needs, we may have to embrace a devolutionary political strategy and management model.[35]

Over time, as ecological management principles become better understood and gain wider acceptance, Congress can be expected to develop comprehensive new ecosystem management and restoration legislation for public lands. The result may well be a new organic statute, perhaps a National Ecosystem Management and Restoration Act. It would contain the usual components of such an organic law: a purpose statement, designated uses, management standards, planning requirements, and public involvement provisions. It would be designed to supplement rather than displace existing legislation, providing each public land agency with new authority and responsibility to engage in ecosystem management. Rather than creating a new super agency, the proposal's proponents may opt not to alter the existing agency structure and to leave established boundaries undisturbed.

The statute should have as its purpose that ecosystems—whether defined in ecological, watershed, or other terms—are to be managed as dynamic entities to conserve and restore biodiversity levels and existing ecological processes in order to ensure a sustainable resource base. It should acknowledge the need for coordinated landscape-scale planning and articulate a new principle of mutual responsibility for addressing transboundary issues. Although designated uses might remain the same, Congress would be well-advised to establish clear priorities among the multiple uses, much as it did in its organic 1997 national wildlife refuge legislation. Without becoming too detailed in deference to the diverse conditions prevailing across public lands, the statute should articulate management standards that require the responsible agencies to conserve and restore ecosystem health and biodiversity levels, while preventing any further ecological impairment. At least two statutory standards are embedded in this proposal: a nonimpairment standard that would establish a threshold basis for evaluating management proposals, and a biodiversity conservation standard that would impose an affirmative obligation on the agencies to protect and restore species diversity. Framed as management standards rather than hard-and-fast rules, the proposal seeks to protect ecological components and processes without placing land managers in a straitjacket, rendering them unable to respond to unique local conditions or exceptional circumstances.[36]

These new ecosystem management standards would be implemented through comprehensive planning, already a critical dimension of modern land management. Each of the public land agencies could be directed to promulgate specific planning regulations to further refine these statutory standards in order to ensure ecosystem integrity and biodiversity levels. Land and resource management plans would be required to adhere to these new standards, first by identifying extant ecosystems and biodiversity resources, then by determining whether contemplated activities might impair them. In the case of transboundary issues, under the principle of mutual responsibility, land managers would be expected to coordinate their ecosystem management efforts with adjacent agencies or landowners. This could be done through an "ecosystem coordination statement" that would document these efforts and respond to any concerns raised during the planning process. The statute also should encourage cooperative planning efforts—perhaps through joint interdisciplinary teams—that would include representatives from other agencies as well as state and local officials with resource management responsibilities within defined ecosystem boundaries. Throughout the planning process, public participation should be encouraged to ensure that all interests are considered and addressed. Provision could also be made for advisory committees to address particular issues, subject to basic FACA fairness requirements. With the agencies already engaged in comprehensive planning and obliged to meet detailed environmental impact analysis requirements, these additional obligations should not prove overly burdensome. Enforcement could be accomplished through existing administrative review processes, and by the courts through citizen suit provisions. Because the proposed standards and procedural mechanisms are already found in various public land laws, both the agencies and courts should be familiar with them and thus not find their institutional capacity taxed.

Given the current degraded condition of many public land ecosystems, the statutory proposal also must address the pressing problem of ecosystem restoration. At a minimum, the new ecosystem planning process should include consideration of whether restoration efforts are necessary and feasible. Where ecological processes or biodiversity levels have been severely affected by past human activities, the agencies should be required to identify the degraded system or depleted population and to undertake affirmative restoration efforts. If the degraded ecosystem crosses administrative boundaries, agencies should be obligated to co-ordinate their restoration efforts, overseen by an interagency restoration commit-tee. Funding for these public land ecological restoration efforts, as in the case of mining reclamation, could be secured by imposing an additional tax or royalty on resource extraction activities or by imposing user fees on currently untaxed activi-ties with a portion of these revenues earmarked for restoration. When restoration involves timber thinning or similar activities, the revenue derived from the sale should be designated for that particular restoration project. In addition, federal Land and Water Conservation Act appropriations should routinely include ear-marked restoration funds. Without an explicit ecological restoration provision and concomitant funding, any new ecosystem management effort may well prove unavailing on those lands that need it most.[37]

No doubt such a comprehensive ecosystem management legislative pro-posal would evoke controversy and opposition. Any change in the status quo that legally redefines management priorities and goals will meet opposition from those who see their traditional access prerogatives threatened. But the deteriorating state of public land ecological resources and the unprecedented levels of conflict now engulfing the land management agencies bespeak the need for change. The statu-tory proposal does not envision a radical restructuring of agencies or boundaries; the proposed legal standards are not new, nor do the procedural or enforcement mechanisms depart from existing law. By linking the nonimpairment standard with an ecosystem restoration obligation, the proposal should help promote truly sustainable resource management policies, thus enhancing community stability and perhaps restoring some peace on the public domain. By establishing coordi-nation processes and requirements, the proposal should better match planning efforts with ecological realities and engage affected communities and others in the process. To the extent that employment opportunities or property-based expecta-tions might be affected, the statute could provide a transitional period as well as economic displacement funds to reduce the hardship.[38] Of course, congressional endorsement of such a statute would signal the final triumph of ecosystem man-agement on public lands, formally opening a new era in public land policy.

NINE

Keeping Faith with Nature

The West has always had to make up its law as it went along.
It has always been both ahead of and behind statutory law
as imported from the seats of power.

BERNARD DEVOTO (1947)

Clearly, the ecosystem . . . has provided a basis for moving beyond
strictly scientific questions into deeper questions of how humans
should live with each other and the environment.

FRANK B. GOLLEY (1993)

This account of contemporary public land policy has been about change, and the processes of change. Slowly but inexorably a new set of resource management priorities is surfacing on public lands, as well as a new way of doing business. The age of ecology is pulling public land policy toward a new era—one where we heed nature's laws more carefully and begin to craft our own in this different image. Just as yesterday's policies served us well in their time, the emerging ecological management regime portends a richer and more sustainable future. To be sure, the processes of change are not coming easily; controversy and conflict have become ever more commonplace across the public domain. But from the clash of competing ideas and priorities, our political institutions have already fostered significant policy reforms, and others are on the horizon. With new devolutionary models helping to reduce the friction, diverse constituencies are coming together in collaborative place-based processes to help chart a course into the age of ecological management. In our efforts to keep faith with nature, we must also keep faith with our democratic heritage.

It may seem curious that a book about public lands has focused so little on the extractive industries and commodity production. After all, mining, logging, and ranching have long been bedrock uses of these lands. Much of the West's history is entwined with the ebb and flow of these regionally important industries. Many long-standing laws and policies governing public lands can be understood only as organic creations rooted in customs derived from these activities. That is changing, however, and in a major way. New biodiversity conservation, ecosystem preservation, and ecological restoration policies are not only challenging but actually displacing the once dominant position of extractive industries in many locations. Some of these new ecological priorities have even been enshrined in law, creating a set of publicly held rights rivaling the private rights that have for so long dictated agency priorities. Other ecological principles have worked their way into agency policies, significantly expanding their planning horizons and prompting new coordination efforts. Expansive new ecologically driven preservation efforts are under way, related proposals are still on the drawing board, and several visionary ideas are gradually garnering scientific credibility and political respectability. This is a whole new way of seeing, understanding, and managing public lands.

As the age of ecology begins to take hold, it would be tempting to think we can now discount the past in deference to a more enlightened future. That would be a serious mistake. Tradition is a powerful force on the western public domain, though any responsible observer would acknowledge that multiple traditions have actually shaped the region's character and policies. Many countervailing tensions inherent in yesterday's public lands conflicts are still very much with us, resurfacing anew as more environmentally sensitive policies take hold. How else to explain the revived conflict over science versus values, preservation versus utilitarianism, or government regulation versus the marketplace? Other perhaps less obvious tensions have surfaced too, creating new schisms among interested constituencies and posing additional resource management challenges. These new schisms can be traced to new public legal rights, an emerging ecological restoration ethic, and the growing irrelevance of conventional boundary lines. How the resulting tensions are ultimately reconciled will tell us much about the role of ecology and collaboration in setting public land policy. It is, in effect, forcing us to harmonize the past with our future needs and aspirations.

We have long struggled to reconcile the relationship between science and human values in public land policy. This tension is once again prominent in the new ecosystem management debate. The science–values conflict has its origins in the early-twentieth-century Progressive conservation movement, which introduced the idea of scientific management and professional resource management agencies. Much of that science, however, was concerned only with single resources and productivity; it did not see the landscape as a whole, nor did it value any resources without economic worth. And while science may have received lip service, basic public land priorities were determined in a political forum, where value judgments are the coin of the realm. In the age of ecology, however, science has taken on new importance and urgency. Reflecting the view that our dwindling biological heritage is both a scarce and supremely important resource, endangered species protection and biodiversity conservation have risen to the fore on our policy agenda. With these urgent new priorities, we have increasingly turned to the ecological sciences to design integrated strategies for protecting and restoring public land ecosystems. But the threshold decision to enshrine biodiversity conservation and ecosystem restoration as policy priorities is still based on a value judgment as to their relative importance in our rapidly changing world. In other words, the science of ecology is now informing our policy choices and management strategies, but it cannot define what is politically feasible, socially acceptable, or economically viable. That is inherently a value judgment, sometimes even a matter of faith.[1]

One tangible dimension of the science–values conflict is the tension that persists between ecology and economics in establishing public land policy. We have long viewed public lands as a national storehouse, stocked with water, timber, minerals, and big game for our consumptive use. Even our national parks have long been seen as attractive playgrounds closely linked to local tourism and recreation industries. Put simply, economic values have historically dominated the public land policy agenda, with environmental concerns at best an afterthought. The age of ecology, however, is moving us to begin reversing these historical priorities. Under the banner of biodiversity conservation, we have protected economically valueless species like the northern spotted owl and put a premium on our remaining old growth forests. Rather than eradicating wolves and suppressing fires, we have begun an active restoration program to re-create functional ecosystems and natural processes. The attendant economic consequences are plainly of consequence, whether measured in diminished timber receipts, livestock production, or other terms. Yet with the aid of the ecological sciences, we are reconceiving public lands as a holistic landscape, not a mere collection of economically exploitable resources. We are also now seeing our resources through a new sustainability lens, not solely in terms of short-range profits but rather as a source of enduring and long-term wealth. Indeed, the new sustainability lens both highlights and strengthens the connection between ecosystems, economics, and our social welfare. The net result is a gradual reordering of our collective values toward more ecologically enlightened policies that more accurately reflect the myriad values and resources that constitute the public lands.[2]

Another tangible dimension of the science–values conflict is the persistent question of the technical expert's role in defining resource priorities. A century ago, when Gifford Pinchot introduced the notion of scientific resource management, he envisioned a cadre of professionally trained managers overseeing our publicly owned resources, making scientific—not political—judgments in allocating our natural capital. But that model has not endured. Not only were the resource management agencies unable to disentangle themselves from politics, but the public eventually realized that the professional managers were themselves making value-based judgments under the guise of science. Science itself was not discredited, but the myth of the omnipotent resource manager was, which has led to the public assuming a more prominent role in agency decision making proesses. With the advent of ecosystem management, science has reclaimed a prominent role in defining resource priorities and related management strategies. Indeed, a quick glance at the detailed planning documents for the Interior Columbia Basin Ecosystem Management Plan (ICBEMP), for example, reveals an extraordinarily complex professional assessment of the region's ecological, economic, and social conditions that set the stage for a major ecosystem restoration proposal. And therein lies the rub: in establishing resource management priorities, how much deference is due the professional judgments of agency scientists and other technical experts, as opposed to the general public's views? Having once rejected the notion of the omnipotent expert, the public may be understandably reluctant to embrace that model now in the fashionable new garb of ecosystem management. Although scientists and technical experts obviously play an important role in ecosystem management, any resource policy priorities that flow from their analyses must be validated in a democratic forum in order to attain legitimacy and garner necessary political consensus. In other words, the new ecosystem management agenda cannot survive simply on its scientific or technical merits; it must be translated into terms that can be understood by the public and that capture its interest and political support.[3]

Few conflicts have been as divisive as the preservation versus utilitarianism schism, long a philosophical fault line in American conservation policy. For more than a century, the preservation–utilitarian debate has played out on the public lands through the struggle over preserving national parks, wilderness areas, and the like. The result has been a steadily expanding network of protected places off-limits to intensive industrial uses that, for the most part, have been relatively small enclaves in deference to the predominant utilitarian philosophy. This has all begun to change, driven largely by our new ecological priorities and understanding. Rather than protecting discrete sites, we have begun trying to preserve entire landscapes and ecosystems, either through expansive new protective designations or with ecosystem management techniques. Our preservation efforts are also no longer limited to land; they now encompass the full range of species that inhabit these landscapes, as reflected in our strict endangered species legislation and related biodiversity conservation policies. All the public land agencies have significant preservation responsibilities too, including the Forest Service with its expansive wilderness network, and the BLM with its own recently inaugurated National Landscape Conservation System. Even on multiple-use lands, new biodiversity conservation priorities have slowed the pace of industrial activity, in deference to the northern spotted owl, red cockaded woodpecker, bull trout, and other previously little-known species. And the innovative ecologically driven preservation proposals that are making their way into the public consciousness are even more breathtaking in scope, stretching across multiple states and even the continent. To be sure, there is staunch resistance to these new initiatives, but the old utilitarian–preservation debate is undeniably being framed in new and more expansive biological terms.[4]

The merits of regulatory versus nonregulatory approaches to public land policy have long been debated. Economists regularly extol the efficiency of the marketplace and decry the inefficiencies associated with planning and related regulatory regimes. Although we have firmly decided to retain the public lands in federal ownership, we still auction off public timber, minerals, and grass, fostering the illusion that market values are at work on the public lands. We have also engaged in a long and acrimonious struggle to curtail corporate abuses through regulatory limitations on the use of public lands, as illustrated by the ongoing controversy over federal hardrock mining regulation. The new ecosystem management regime is not based upon economic principles, however; it is built upon landscape-scale planning and adaptive management strategies designed to promote biodiversity conservation and ecological restoration. Because biodiversity and functioning ecosystems have little immediate market value, some level of government intervention is necessary if these resources are to be acknowledged and protected. Nor are the new regional planning processes, with their extensive environmental analysis and public involvement requirements, particularly efficient; they are technical and political processes that should be measured by accountability and legitimacy standards, not efficiency or profit margins. Given the myriad resources and interests at stake, as well as the absence of ready market values, the trade-off choices required can be determined only through inclusive planning exercises and then enforced through regulatory mechanisms or carefully targeted incentives.[5]

As our values have changed, so too have the laws governing public lands. Our tradition of bestowing individuals with private rights to exploit public resources is now at odds with a new set of publicly held legal rights designed to protect the environment. Indeed, the new ecosystem management policy regime is built squarely on these public legal rights. Perhaps nothing better illustrates this evolution in public land law than comparing the General Mining Law of 1872 with the Endangered Species Act of 1973, which were passed more than a century apart. The Mining Law is based on raw economics, reflecting the frontier era ethic of laissez-faire development. Utilizing the principle of priority, it vests private individuals with property rights in public resources. It manifests no concern for the environment or other resource values, basically opening public lands to mineral exploration and then giving valuable resources away for free. And it relies upon state rather than federal regulation. The Endangered Species Act, on the other hand, seeks to protect vanishing species; it is driven by science, not economics. With its emphasis on preserving ecosystems, it embodies the modern preservationist ethic. It displaces the states' traditional authority over wildlife, replacing it with federal regulatory power both on public and private lands. Its protective provisions can be enforced by citizen suits. Not surprisingly, it is beginning to reshape our conventional understanding of private property rights. A similar public–private dichotomy is now evident in the law across the public domain, whether the resources at issue are water, timber, forage, or recreational uses. As this new countervailing regime of public environmental rights takes hold, it will be up to the courts to resolve these tensions and thus reconcile new priorities with old ones. In short, where a private rights model of public land policy once prevailed, it is now being rivaled—perhaps even displaced—with a new public rights model focused on protecting and restoring our natural heritage.[6]

Ecological restoration is emerging as an important new policy objective on the public lands, rekindling the debate over whether and how to intervene on nature's behalf. The issue is being framed in terms of active versus passive management strategies. On both our multiple-use and preserved lands, we have long pursued interventionist management policies, reflecting our historical commitment to controlling and improving nature for our own benefit. That philosophy, however, has now changed on the preserved lands, owing to revised noninterventionist national park policies and wilderness legislation that protects untrammeled landscapes. As the multiple-use agencies reorder their priorities to accommodate new biodiversity and ecological concerns, they are being urged to curtail logging and other intensive development activities associated with the extractive industries. In the debate over fire policy on the national forests, the Forest Service confronts a Hobbesian choice—either to allow sometimes badly overgrown forests to burn at potentially catastrophic levels or to intercede with selective thinning techniques to re-create the presuppression forest environment and to protect nearby communities. Having battled the timber industry to a standstill, many environmentalists distrust the Forest Service, fearing it will reinstitute an aggressive commercial logging program under an ecological guise. Yet restoration does not always mean nonintervention. The Yellowstone gray wolf reintroduction effort involved extensive intervention to capture, translocate, and then manage the new wolves.

And prescribed burning, riparian corridor recovery, and exotic species control programs all involve interventionist techniques too. Besides, in the case of forest restoration, a carefully regulated timber thinning program could help generate badly needed ecosystem restoration funds. Rather than seeking ideological purity on the intervention question, restoration strategies should be grounded in scientific judgments, recognizing that competing human concerns must also be part of the equation.[7]

A relatively new conflict in public land policy involves the role of boundaries, namely the need to reconcile legal authority with ecological realities on the landscape. Public land policy traditionally has been based on well-defined and sacrosanct boundaries, with quite different management standards prevailing on multiple-use lands and on preserved lands, even when they adjoin each other. On adjacent private lands, the tradition has been near total autonomy, as few regulatory restraints have been placed on private landowners. The new ecosystem management regime, however, conceives of boundaries differently—as permeable and often uncertain lines defined in primarily ecological rather than legal terms. There may even be more than one ecological boundary line, depending on the resource management objective. Of course, any attempt to redraw or transcend boundary lines will invariably provoke hostile reactions from neighboring managers, landowners, and communities, who may perceive that their autonomy, authority, or access prerogatives are at risk. To ameliorate these concerns and thus surmount boundary limitations, the new ecosystem management initiatives have employed interrelated strategies. Several projects have utilized scientific criteria to define the scope of the natural resource problem, but then made pragmatic political compromises over the management strategies used to pursue biodiversity goals and other conservation objectives. Some projects have used a straightforward mix of scientific and political criteria to establish relevant boundaries, as was the case with the Quincy Library Group. Among the public land agencies, coordination is the watchword for defusing potential interagency boundary tensions, though this does not necessarily determine whose management objective prevails in the event of a conflict. Public involvement processes have also been used to educate the public about the need for boundary adjustments and to promote cross-boundary coordination initiatives. When nonfederal lands are at issue, a mix of incentives, cooperative arrangements, and default regulatory strategies have helped overcome boundary obstacles. Reflecting a pragmatic mixture of ecology and politics, the resulting transboundary management policies should prove more functional and durable over the long term.[8]

Finally, the new ecosystem management policy regime has been dogged by an obvious scale problem that also raises age-old political concerns. The scientific evidence pretty clearly demonstrates that biodiversity conservation and ecological restoration goals can be attained only through large-scale, regional management initiatives. In response, the federal public land agencies have undertaken several expansive ecosystem management projects, the most ambitious of which is the multistate ICBEMP. Whether rightly or wrongly, these initiatives have been perceived—particularly by local residents—as large, top-down federal projects divorced from the needs or expectations of the surrounding communities. With the arrival of the Bush administration and with little support in the affected communities, the ICBEMP and other similar initiatives may flounder—tributes to good science but questionable politics. Devolution advocates, who are innately suspicious of big government, eschew such large and top-heavy federal initiatives. They would rather see ecosystem management evolve from local conservation initiatives that then might be nurtured to become broader, regional projects once durable relationships have been established and the participants come to understand the full ecological ramifications of their actions. In essence, they view ecosystem management in political rather than scientific terms, linking its legitimacy to the political theory of devolution rather than the science of ecology. But can we really decouple politics, science, and law, and see ecosystem management only in one-dimensional terms? In the messy real world where the courts stand ready to enforce endangered species mandates and the general populace sees the public domain in preservationist terms, there may be no single truth but rather the recurrent need to strike a compromise—no matter how uneasy—between science and politics.[9]

This excursion through the contemporary fault lines in public land policy should make clear that any effort to keep faith with nature will require us to keep faith with our democratic principles too. Our ecosystem management institutions, laws, and policies should not separate us from the environments we inhabit. People and place are entwined, even on the public lands. That said, however, new ecosystem management ideas are forcing us to acknowledge important biodiversity and ecological restoration concerns that are now being integrated into public land policy. As Aldo Leopold observed over half a century ago: "The land ethic simply enlarges the boundaries of the community to include soils, waters, plants, and animals, or collectively: the land." With our growing understanding of how our lands and resources are interconnected one with another, we are beginning to plan and manage on an ecologically defined scale. These realizations have also changed the terms of the long-standing debates over natural resource policy, expanding them into the realm of ecology and a new human relationship with the natural environment—one without as much hubris or certainty. As these views become instilled ever more deeply in the public consciousness and our political culture, more dramatic shifts in public land policies and practices can be expected.[10]

Change does not occur in a vacuum. As society's values evolve, new priorities are absorbed into our laws and policies—essentially legitimized through our democratic institutions. In the case of the public lands, as we have seen, all our governmental institutions are engaged in the ongoing transition toward an ecosystem management policy regime. They are responding to compelling new scientific insights as well as economic and social changes that are forcing us to reevaluate our collective relationship to the natural environment. Significant reforms are occurring at the federal level, while much of the impetus, both for and against these changes, comes from local pressures exerted by the diverse public land constituencies. No one institution is entirely responsible for the manifold changes afoot; rather, the legislative, executive, and judicial branches are each playing a distinctive role in the ongoing public land policy debate. This is, as has so often been the case, an organic process of reshaping priorities, institutions, and procedures to fit our contemporary needs and values. While we may not know exactly where this process is heading, the participants have eagerly employed our governmental institutions to promote the process of change.

The federal judiciary has played a major role in bringing new ecological management principles to the public lands, and the courts will continue to oversee this new era as it takes shape. The seminal event was the spotted owl litigation, which saw the federal courts enjoin the powerful timber industry from logging the Pacific Northwest's ancient forests to protect the diminutive owl. When it approved the Northwest Forest Plan, the federal judiciary put a stamp of legitimacy on this new ecosystem management approach to public land policy. These decisions can only be regarded as the triumph of publicly held environmental rights over a purely utilitarian policy regime. Since then, in a halting case-by-case fashion, the courts have largely confirmed this new direction in public land policy: affirming federal wolf reintroduction initiatives; enforcing endangered species and biodiversity protections; reading the National Environmental Policy Act (NEPA) cumulative effects requirements broadly; reaffirming the president's expansive national monument designation authority; sustaining the national forest roadless initiative; and mandating various adaptive management strategies. Collectively, the decisions represent a judiciary coming to terms with the full meaning of the 1970s environmental laws and the plethora of new prescriptive standards and ecological obligations they imposed on the public land agencies. Despite the countervailing property rights litigation agenda, the dominant thrust of modern public land litigation has been to impose new environmental obligations on the responsible agencies, while acknowledging that they retain sufficient discretionary authority to experiment with ecosystem management strategies and related collaborative conservation initiatives. With Congress seemingly committed to ever more prescriptive public land legislation, the number and variety of court decisions will only continue to grow, further defining the scope of ecological obligations, collaborative opportunities, and countervailing economic rights. In short, the federal courts are now a major institutional factor in virtually every public land policy controversy, and that role will only deepen as the new ecological policy agenda takes shape and the interested parties test its limits.[11]

The public land agencies are obviously on the front lines in the battle over ecosystem management, devolved management authority, and related controversies. Having long enjoyed broad discretionary authority over the public domain, the agencies have seen that authority eroded by increasingly prescriptive environmental legislation, including extensive public involvement and related procedural requirements. The revered tradition of professional omnipotence that traces to the Progressive Era's scientific management model has given way to a new skepticism toward the agencies, fueled by increasingly diverse constituencies and the values they are called upon to reconcile. The powerful Endangered Species Act (ESA), which gives the U.S. Fish and Wildlife Service regulatory oversight and enforcement responsibilities, has further divested land managers of their discretionary authority, projecting explicit ecological standards onto the public domain and promoting new regional interagency planning initiatives. Although some agency officials have complained about "analysis-paralysis" linked to these complex statutory responsibilities, one wonders what role ecological concerns would play on the public lands without the ESA, NEPA, and other legal requirements. Sensitive to the legacies of the past, the agencies have also each embraced new ecological restoration policies, ranging from making room for fire and reintroduced predators to myriad place-based forest and range restoration projects. Now more constrained and accountable than ever, the agencies cannot ignore the emerging legal sideboards and scientific imperatives that limit their decision space, nor can they afford to overlook the diverse constituencies that are clamoring for change or those that will likely bear the burden of those changes. In this contentious milieu, the collaborative process movement—itself an experiment in devolved governance—offers the agencies an alternative forum where they might better accommodate their constituencies while still meeting their new ecological obligations.[12]

The president has long occupied a central role in protecting the public domain from exploitation and in defining public land policy. The Antiquities Act and other congressionally delegated withdrawal powers bestow the president with the unilateral power to reshape the public domain map, and several of our chief executives have done just that. At the dawn of the twentieth century, Teddy Roosevelt made a major commitment to conservation with his extensive forest reserve designations; a century later, Bill Clinton likewise pursued an expansive new preservation agenda with his national monument designations. Despite much sound and fury, Congress has neither curtailed the president's Antiquities Act powers nor reversed the Clinton national monument designations. The president can also utilize the bully pulpit of his office to influence public land policy. A century ago, President Roosevelt firmly endorsed the Pinchot-inspired utilitarian conservation agenda, which came to dominate public land policy over the ensuing decades. President Clinton sought to do much the same for the new ecological policy agenda, as symbolized by his personal engagement in developing the Northwest Forest Plan. Whether the new ecosystem management and restoration agenda will have the same enduring power as its predecessor remains to be seen, particularly as the Bush administration pursues its own contrary policy agenda. But even during the Clinton administration, an emerging collaborative conservation movement plainly leavened the new ecosystem management agenda, as reflected in the Quincy Library Group's legislative triumph. If history is a guide, perhaps this leavening process could be expected. During the Progressive Era, the western public land constituencies diluted key parts of the Roosevelt-Pinchot conservation agenda, and a similar pattern may be playing out with the new ecological reform initiatives. This does not denigrate the president's role in promoting new public land policies, but it does suggest that presidential protective designations may prove more enduring than related policy shifts.[13]

Congress, of course, has ultimate authority over the public lands, and it will undoubtedly have the final say on new ecological management and collaborative conservation policies. Within the tripartite national governmental structure, Congress may be the most politically accountable branch because its members answer to their constituents at regular elections. This makes Congress a good bellwether of nationwide trends, as well as a barometer for identifying local concerns. In the public lands arena, we have regularly seen these conflicting national–local forces at work, in farsighted organic planning legislation as well as nearsighted appropriations riders. The fact that Congress has not significantly revised any of the groundbreaking 1970s environmental laws that undergird the new ecological policies speaks volumes about how the general public views the public lands. Despite the radical change in its political composition following the 1994 election, Congress has actually written new biodiversity protections into organic statutes, funded several major regional planning initiatives, and rejected most anti-environmental reform proposals. Faced with diverse constituencies and a polarized political landscape, however, Congress has mostly been stalemated over major public land policy reforms, including systemic new biodiversity, ecological management, and landscape preservation proposals. The competing constituencies have each proven able to block objectionable legislative proposals but unable to advance their own agendas, especially on multiple-use public lands. As a result, Congress has tended either to endorse modest procedural reforms, such as the 1995 Federal Advisory Committee Act (FACA) revisions or to adopt place-specific legislation, such as the Quincy Library Group and Valles Caldera trust bills. It has also continued to fund several ecological restoration experiments, including the ICBEMP, Sierra Nevada, and Everglades projects. In this political climate, we can probably expect even more congressional place-based laws, perhaps the outgrowth of local collaborative processes, that will further illuminate what a comprehensive new ecological policy agenda may mean for public lands and the communities dependent on them. As lessons are learned and become clear, Congress will eventually translate them into more uniform legislative mandates that will chart the future course of public land policy.[14]

Although each branch of the federal government has played an obvious role in bringing new ecological management policies to the public lands, it is really the interplay between the various branches that has advanced this agenda to the fore and helped legitimize it. In our system of separated powers, no one branch dominates the nation's internal political life; rather, the branches function symbiotically, both advancing and retarding new ideas when they enter the nation's political consciousness. That symbiotic process has been evident in the advent of new ecosystem management policies. It was Congress that initially passed the environmental laws that were then employed by the federal courts in the spotted owl controversy to block further logging in the Pacific Northwest. The court rulings, in turn, set the stage for executive branch intervention that culminated in the Northwest Forest Plan with its new ecosystem management strategies. Although Congress did not then directly reenter the spotted owl controversy, it did appropriate the funds necessary to implement the plan, and it further directed the same agencies to develop related plans for the Interior Columbia Basin and the Sierra Nevada range. Remarkably similar institutional patterns are evident in the other major ecological management controversies, ranging from wolf reintroduction and endangered species protection to forest management and range restoration. Perhaps the reason is obvious: because the controversies have been intense, the losing side in one forum has regularly escalated the conflict to another forum, seeking either to prevail elsewhere or to gain a tactical advantage. There is little likelihood that this pattern of strategic behavior will end soon, though the collaborative conservation movement may diminish some of it.

While much of the skirmishing over public lands is occurring at the federal level, that should not discount the role states and local communities have long played in the policy process. Our federal structure ensures that local interests are heard within the national government, according sovereign prerogatives to state and local governments. The advent of new ecologically oriented public land policies has not changed our constitutional structure; if anything, basic federalism principles and civic republican ideas have been revitalized in the nation's politics. On the public lands, this revitalization is manifested in the new collaborative conservation movement, which now finds diverse public land constituencies actively engaged in local experimental initiatives that provide a forum for resolving environmental conflicts before they escalate into an overtly political or judicial forum. Operating in the shadow of the law and without express legal sanction, these collaborative processes are bringing natural resource decisions back to a grassroots level, where opportunities for compromise and mitigation are part of the bargaining process. When Congress passed the locally negotiated Quincy Library Group legislation and when the U.S. Fish and Wildlife Service endorsed the locally designed Bitterroot grizzly bear reintroduction plan, the collaborative conservation idea achieved a certain tentative legitimacy. It was, in effect, an acknowledgment that ecological management and restoration goals might coexist with devolved governance institutions and processes. Though both initiatives subsequently experienced significant setbacks, they may still point the way toward more durable models.[15]

Indeed, there is an obvious role for collaborative processes in the new ecosystem management regime. Ecosystems are all quite varied and different, as are the states and our public land communities. Most private landowners more readily identify with local than federal authorities, and their involvement will be critical in many initiatives. A local venue, therefore, offers an attractive forum for negotiating ecological controversies; it is where the ramifications of particular choices will be felt and where important local knowledge resides that can be brought to bear on the problem. That said, however, the local venue notion is subject to important qualifications. First, because public land constituencies are extraordinarily diverse today, collaborative processes must meet basic fairness standards to ensure everyone is represented and heard. Otherwise the process is not local democracy in action; instead, it perpetuates the long-standing agency capture problem. Second, because ecosystems transcend individual locales, any collaborative initiative must incorporate a broader regional perspective, including national standards promoting biodiversity conservation and ecological sustainability. Placed-based proposals, in other words, must be accountable at both a legal and scientific level. Third, some nationally significant issues—endangered species, wilderness, and energy come to mind—may simply not be amenable to locally negotiated resolutions. Having made the decision as a nation to retain public lands for our collective benefit, we should be wary of entrusting them entirely to self-appointed local guardians. The challenge ahead is to identify those issues that lend themselves to collaborative resolution and then to develop workable shared governance models that serve both local and national interests. While clearly not an easy task, it is a necessary one if we are to meld ecological policy with democratic standards on public lands.[16]

This is neither an unqualified endorsement of collaborative processes nor an outright rejection of them. Well-meaning people on all sides of today's public land controversies are heavily invested in these initiatives, many of which have helped resolve, or at least deescalate, long-standing local conflicts. But that has not been the universal experience. The Quincy Library Group initiative has generated as much controversy as it has defused, and the Bitterroot grizzly bear proposal was ultimately scuttled by local, not national, opposition. Our enthusiasm for devolution must therefore be tempered by the realities of parochialism and self-interest that can infect any well-intentioned local entity. There is much lingering distrust over new ecosystem management protocols and toward the public land agencies themselves, who have been captured before by narrow interests. And there are plenty of naysayers from every philosophical perspective who are poised to subvert any collaboratively negotiated proposal that does not advance their own personal agenda. All of this counsels caution, recognizing both the inherent limitations of collaborative processes and the need for further experimentation to better marry ecology and democracy.

The development of federal public land policy is an ongoing and dynamic process. In less than a century, we completely reversed course, moving from a policy of disposal and private ownership to one based on retention and intensive management. Similar evolutionary forces are at work today. Over the past several decades, Congress has imposed ever more rigorous legal constraints on the land management agencies, mostly to ensure that environmental concerns are addressed before resource development proposals are approved. Congress has also elevated the federal commitment to preservation with the passage of powerful laws protecting wilderness areas, river corridors, and endangered species—laws that affect all the public land agencies. These changes, not surprisingly, have fomented intense controversy across the public domain. As the agencies, their constituencies, and the courts have grappled with the full implications of these developments, an ecological perspective has taken center stage in the policy debate. With this new perspective, we may finally have an opportunity to bridge the fault line between our separate utilitarian and preservation traditions, and thus build a more coherent and integrated natural resource policy for the new century. The concurrent emergence of new collaborative conservation initiatives offers a corollary political theory that could help solidify federal ecological initiatives throughout our governmental institutions. It may, in other words, present an opportunity to merge ecology and democracy in public land policy.

The public domain today is rife with change, much of it attributable to progressive new policies and laws designed to protect important environmental values. These changes, as chronicled in the pages of this book, are both substantial and significant. We are no longer madly clear-cutting the nation's forests to meet politically negotiated timber targets, mostly in deference to spotted owls and other forest-dependent species that often have no obvious commercial value. We have reintroduced wolves to restore impoverished ecosystems over the strident objections of the ranching community. We are allowing some fires to burn on public lands, even when they consume valuable timber and other resources in order to restore important ecological processes. We have extended preservation mandates across the public domain, not only creating large new wilderness areas and national monuments but also protecting imperiled species wherever they are located. In fact, with endangered species listings on the upswing and critical habitat designations now blanketing much of the public domain, the land management agencies are exhibiting a new sensitivity to biodiversity concerns. In some locations, they have even expanded their planning efforts to a much broader regional scale. As these new policies have taken hold, the West's many resource-dependent communities have been forced to change their ways. Some have embraced recreation and tourism to forge a new relationship with the public lands; others have bitterly resisted these new federal reform initiatives. But with controversy comes opportunity, namely the chance to engage the full spectrum of public land constituencies in the processes of change, building toward a more enduring and widely accepted set of new policies and priorities.

Given the pervasive nature of change in the modern world, it is clear that public lands will look quite different in a few short decades. If we can merge the new ecosystem management and devolution strategies, we may actually see a more diverse landscape and harmonious populace. Picture the possibilities: Substantial numbers of grizzly bears and wolves roam well beyond the isolated enclaves they now inhabit, once again managed by the states but under new and more enlightened laws. The logging and ranching industries continue to operate on the public domain, but in a more restrained and sustainable manner. Recreation and tourism continue to flourish, but without any new legal entitlements; the worst excesses of industrial-scale recreation have been curbed to fit within the capacity of the landscape. Ecological preservation has taken hold with interconnected nature reserves stretching across the map, designed to both sustain our biological heritage and afford compatible economic opportunities to nearby communities. New jobs are more about restoring the landscape than stripping it bare. Committed to a sustainability policy, the public land agencies have engaged citizen councils to help set local management strategies. Operating under a new set of ecologically sensitive statutory priorities, the responsible agencies are actively coordinating their planning activities, sensitive to the cross-boundary implications of their management decisions. Private landowners have begun to assume biodiversity conservation responsibilities; many are even participating in cooperative landscape conservation projects. Once-critical regulatory strategies have diminished in importance, giving way to new market-based incentives and collaborative planning efforts that reward good ecological stewardship practices. To be sure, there is still controversy aplenty, but more and more disputes are settled through collaborative discourse. By consensus, a biodiversity conservation ethic prevails on the public lands and a new civility has taken hold within the adjacent communities.

Perhaps this picture is far too utopian, given the long history of conflict that has pervaded public land policy. We are certainly in a period of intense controversy; lawsuit filings are on the upswing, congressional lobbyists are busier than ever, and the media bombard us with news about our disagreements. Moreover, the Bush administration is busy trying to undo the most prominent Clinton administration policy reforms, using the specter of energy shortages and catastrophic fire to make its case. But change inevitably follows conflict, and we are seeing significant shifts in policy and attitude. For the reasons set forth throughout this book, the changes afoot point largely in one direction—toward new biodiversity conservation priorities and ecosystem management policies. The collaborative conservation movement is helping to take the sharp edges off the changes ahead, actually shaping the new ecological imperatives to accommodate compelling local concerns and economic needs. Without a contemporary consensus over either ecosystem management or collaborative processes, the skirmishing will continue in every available forum. And because these matters must ultimately be decided in the political arena, we can—and should—expect compromises and detours along the road to a more enlightened policy regime.

The new ecological controversies remind us that it is simply not possible to divorce people and place. Even with all the relevant scientific data, our commitment to ecological conservation on public lands will require a corresponding political commitment, which will inevitably reflect our collective values. If we can manage to conjoin people and place through a communal ecological perspective, then we may begin to comprehend fully the array of species, resources, and communities that depend on our policy decisions and managerial judgments. Ecosystem management and community-based conservation offer an opportunity to begin reconciling our differences and constructing such a new policy order. While some uniformity in public land management goals will undoubtedly be necessary, we should also allow some flexibility in achieving these goals to indulge local needs and interests. Each place is different, capturing the rich social diversity that makes America such a unique and special nation, and these differences should be acknowledged and respected. Each ecosystem is also different, reflecting the incredible natural diversity that is our all-important biological heritage, and these differences also must be acknowledged and valued. On the public lands, the challenge is to respect this diversity on a place-by-place basis while not impairing the overall ecological health of the landscape. Uniform federal legal standards and regional planning efforts can provide a framework (or sideboards) for management, within which local collaborative processes may seek to address particular local interests. Such an approach will not solve all public land controversies, but it does offer a more sensitive strategy for integrating people and place within the ecosystem.

As the new age of ecology takes hold, it is forcing us to surmount the dead weight of tradition that has long pervaded public land policy. Biodiversity conservation and ecosystem restoration are ascendant on the policy agenda, rivaling and in some instances even displacing traditional extractive uses. Lynton Caldwell forecast this development more than 30 years ago when he wrote: "Conservation as a concept has been helpful principally as an intermediary proposition, midway between unrestricted competition among resource users and an ecologically-based view of public responsibility for the self-renewing capabilities of the ecosystem." The immediate challenge is to better integrate these seemingly contradictory policy objectives to achieve both ecological and economic sustainability. Nature and its biological assets are no longer simply commodities to be exploited, but an important part of our national heritage. To better accommodate the natural world, we have begun to think and plan on a regional scale, sensitive to ecological realities and not solely jurisdictional boundaries. Unable to ignore the diverse constituencies deeply concerned with the public lands, we have created new collaborative processes and other opportunities to engage everyone in these planning processes and management decisions. Our commitment to preservation has changed too. We are moving slowly beyond the enclave strategy to pursue ecological preservation objectives on a landscape scale, driven as much by biodiversity protection goals as by aesthetic concerns. With this new commitment to keeping faith with nature, we can begin to reevaluate public lands and their role in our national life. Whatever their original purposes, these lands are now as valuable to us in their natural state as for their resource development potential.[17]

The alluring prospect, of course, is that we may finally overcome the utilitarian–preservation schism that has divided public land policy for more than a century. As ecology has revealed the interdependent nature of the public lands and their resources, we better appreciate the need for integrated policies and management strategies to sustain vital natural systems. Having transformed the bountiful public lands into industrial wastelands, depleted ecosystems, and overcrowded playgrounds, we have come to realize that important resource values have been overlooked, key species lost, and increasingly scarce resources squandered. Even the most diehard adherent to the traditional multiple-use regime will grudgingly acknowledge that adjustments are necessary to preserve our dwindling natural capital. Based upon now widely shared sustainability goals, ecological management does not represent such a sharp break with the past that it would eliminate the human presence from the landscape; rather, it contemplates a more restrained and focused human role—one where we moderate our disparate activities while actively working to restore depleted ecosystems. The new imperative, therefore, is to emulate nature, not to subdue or discount it. And the next step is to build an effective political consensus fully committed to this new ecological ethic. There is no other way in a democracy.

ACRONYMS AND ABBREVIATIONS

ACS	aquatic conservation strategy
ACECs	areas of critical environmental concern
AFSEEE	Association of Forest Service Employees for Environmental Ethics
ANILCA	Alaska National Interest Lands Conservation Act
APA	Administrative Procedures Act
ASQ	allowable sale quantity
ATVs	all-terrain vehicles
AUMs	animal units per month; the standard for measuring public land grazing allotments
bbf	billion board feet
BLM	Bureau of Land Management
CALFED	San Francisco Bay Delta restoration project
CARA	Conservation and Reinvestment Act
CASPO	California spotted owl guidelines
CATS	Californians for Alternatives to Toxics
CCC	Civilian Conservation Corps
C.F.R.	Code of Federal Regulations
CWA	Clean Water Act
DEIS	Draft Environmental Impact Statement
DFPZs	defensible fuel profile zones
Enlibra	Western Governors' Association's principles for environmental management
EPA	Environmental Protection Agency
ESA	Endangered Species Act
ESC	Endangered Species Committee
ESU	evolutionarily significant units
FACA	Federal Advisory Committee Act
FEIS	Final Environmental Impact Statement
FWS	U.S. Fish and Wildlife Service
FLPMA	Federal Land Policy and Management Act
FEMAT	Forest Ecosystem Management Assessment Team
F.2d(3d)	Federal Reporter (circuit court of appeals decisions)
F.Supp.(2d)	Federal Supplement (district court decisions)
FTCA	Federal Tort Claims Act
GAO	U.S. Government Accounting Office
GIS	geographical information systems

GYCC	Greater Yellowstone Coordinating Committee
HCA	habitat conservation area
HCP	habitat conservation plan
IBLA	Interior Board of Land Appeals
ICBEMP	Interior Columbia Basin Ecosystem Management Plan
ICB	Interior Columbia Basin
IGBC	Interagency Grizzly Bear Committee
INFISH	Inland Native Fish Strategy
infra	subsequently referenced
ISC	Interagency Scientific Committee
LWCFA	Land and Water Conservation Fund Act
LS/OG	late-successional–old growth
mbf	million board feet
MOU	Memorandum of Understanding
MWA	Montana Wilderness Association
NAHB	National Association of Home Builders
NBS	National Biological Survey
NEPA	National Environmental Policy Act
NFMA	National Forest Management Act
NMFS	National Marine Fisheries Service
NPS	National Park Service
NREPA	Northern Rockies Ecosystem Protection Act
NSO	northern spotted owl
O and C	Oregon and California; BLM timberlands located in southern Oregon and northern California
OMB	Office of Management and Budget
ONRC	Oregon Natural Resources Council
ORVs	off-road vehicles
PACs	protected activity centers
PACFISH	Interim Strategies for Managing Anadromous Fish-producing Watersheds in Eastern Oregon and Washington, Idaho and Portions of California
PILT	payment in lieu of taxes
PRIA	Public Rangelands Improvement Act
RACs	Resource Advisory Committees
RARE I and II	Roadless Area Review Evaluations
RCAs	riparian conservation areas
R.S. 2477	Revised Statute 2477 (adopted in 1866; repealed by FLPMA in 1976)
ROD	Record of Decision
SCRAP	Students Challenging Regulatory Agency Procedures
SDEIS	Supplemental Draft Environmental Impact Statement
SEIS	Supplemental Environmental Impact Statement
SMCRA	Surface Mining Control and Reclamation Act
SOHAs	spotted owl habitat areas

SNEP	Sierra Nevada Ecosystem Project
supra	previously referenced
TVA	Tennessee Valley Authority
UNESCO	United Nations Educational, Scientific, and Cultural Organization
USDI	U.S. Department of the Interior
UWC	Utah Wilderness Coalition
WSAs	wilderness study areas
Y2Y	Yellowstone to Yukon Conservation Initiative

ONE
Introduction

1. See Donald Worster, Nature's Economy: A History of Ecological Ideas 284, 306 (New York: Cambridge University Press, 1977); see also Hanna J. Cortner and Margaret A. Moote, The Politics of Ecosystem Management (Washington, D.C.: Island Press, 1999); Max Oelschlaeger, The Idea of Wilderness: From Prehistory to the Age of Ecology (New Haven: Yale University Press, 1991).

2. Seattle Audubon Society v. Robertson, 771 F.Supp. 1081 (W.D. Wash. 1991), affirmed sub nom. Seattle Audubon Society v. Evans, 952 F.2d 297 (9th Cir. 1991); see also Kathie Durbin, Tree Huggers: Victory, Defeat and Renewal in the Northwest Ancient Forest Campaign (Seattle: The Mountaineers, 1996); Steven L. Yaffee, The Wisdom of the Spotted Owl: Lessons for a New Century (Washington, D.C.: Island Press, 1994).

3. For the regional forest plan, see Forest Ecosystem Management: An Ecological, Economic, and Social Assessment: Report of the Forest Ecosystem Management Assessment Team (Washington, D.C.: U.S. Government Printing Office, 1993); for Judge Dwyer's decision, see Seattle Audubon Society v. Lyons, 871 F.Supp. 1291, 1311 (W.D. Wash. 1994), affirmed sub nom. Seattle Audubon Society v. Moseley, 80 F.3d 1401 (9th Cir. 1996); see also Chapter 4 for a full examination of the spotted owl controversy.

4. See 50 C.F.R. § 17.84 (2000) (gray wolf listing); Northern Rocky Mountain Wolf Recovery Team, U.S. Fish and Wildlife Service, Northern Rocky Mountain Wolf Recovery Plan (1987); U.S. Fish and Wildlife Service, The Reintroduction of Gray Wolves to Yellowstone National Park and Central Idaho, Final Environmental Impact Statement (1994). Congress added the experimental population provision (16 U.S.C. § 1539(j)) to the Endangered Species Act in 1982 to increase legal flexibility and to facilitate controversial species reintroductions. The wolf reintroduction controversy is described in Hank Fischer, Wolf Wars: The Remarkable Inside Story of the Restoration of Wolves to Yellowstone (Helena, Mt.: Falcon Press 1995); Thomas McNamee, The Return of the Wolf to Yellowstone (New York: Henry Holt, 1997).

5. Wyoming Farm Bureau Federation v. Babbitt, 987 F.Supp. 1349 (D.Wyo. 1997), reversed, 199 F.3d 1224 (10th Cir. 2000); see also Chapter 5 for a more detailed discussion of the Yellowstone wolf reintroduction controversy.

6. For a general description of the evolution of federal fire policy and public attitudes toward fire, see David Carle, Burning Questions: America's Fight with Nature's Fire (Westport, Conn.: Praeger, 2002); Stephen J. Pyne, America's Fires: Management on Wildlands and Forests (Durham, N.C.: Forest History Society, 1997); Stephen J. Pyne, Fire in America: A Cultural History of Wildland and Rural Fire (Princeton, N.J.: Princeton University Press, 1982); see also Chapter 5 for a further discussion of federal fire policy.

7. On the Yellowstone fires, see Micah Morrison, Fire in Paradise: The Yellowstone Fires and the Politics of Environmentalism (New York: Harper Collins, 1993); Norman Christensen et al., Interpreting the Yellowstone Fires of 1988, 39 Bioscience 678 (1989); Paul Schullery, The Fires and Fire Policy, 39 Bioscience 686 (1989).

8. See Dave Forman, Wilderness: From Scenery to Nature, in J. Baird Callicott and Michael P. Nelson, eds., The Great New Wilderness Debate 568 (Athens: University of Georgia Press, 1998); R. Edward Grumbine, Ghost Bears: Exploring the Biodiversity Crisis (Washington, D.C.: Island Press, 1992); Robert B. Keiter and Mark S. Boyce, eds., The Greater Yellowstone Ecosystem: Redefining America's Wilderness Heritage (New Haven: Yale University Press, 1991); Joseph L. Sax, Nature and Habitat Conservation and Protection in the United States, 20 Ecology L. Q. 47 (1993).

9. On new and more expansive approaches to nature conservation, see Reed F. Noss and Allen Y. Cooperrider, Saving Nature's Legacy: Protecting and Restoring Biodiversity 99–177 (Washington, D.C.: Island Press, 1994); Michael E. Soule and John Terborgh, Continental Conservation: Scientific Foundations of Regional Reserve Networks (Washington, D.C.: Island Press, 1999); Edward J. Heisel, Biodiversity and Federal Land Ownership: Mapping A Strategy for the Future, 25 Ecology L. Q. 229 (1998); see also Chapter 6 for a more detailed examination of federal preservation policy. On the debate over active versus passive management, see James A. Pritchard, Preserving Yellowstone's Natural Conditions: Science and the Perception of Nature (Lincoln: University of Nebraska Press, 1999); Frederic H. Wagner et al., Wildlife Policies in the U.S. National Parks (Washington, D.C.: Island Press, 1995); Stephen Budiansky, Nature's Keepers: The New Science of Nature Management (New York: Free Press, 1995); see also Chapter 5 for more on active versus passive management strategies.

10. For data on the federal government's land ownership, see U.S. General Accounting Office, Land Ownership: Information on the Acreage, Management, and Use of Federal and Other Lands 2 (1996); U.S. Department of the Interior, Bureau of Land Management, Public Land Statistics 1998 7–13 (Washington, D.C., 1999); Dale A. Oesterle, Public Land: How Much Is Enough?, 23 Ecology L. Q. 521, 523–24 (1996). Because Alaska's public lands are governed by the unique Alaska National Interest Lands Conservation Act (ANILCA), 16 U.S.C. §§ 3101–3233, the book does not address Alaska public land issues; rather, it focuses on the western states where the public lands are governed by a uniform set of federal laws. For information on ANILCA and the Alaska public lands, see Deborah Williams, ANILCA: A Different Legal Framework for Managing the Extraordinary National Park Units of the Last Frontier, 74 Denver U. L. Rev. 859 (1997); Glenn E. Cravez, ANILCA, Directing the Great Land's Future, 10 UCLA-Alaska L. Rev. 33 (1980).

11. See Douglas S. Kenney et al., Values of the Federal Public Lands 56 (Boulder, Colo.: Natural Resources Law Center, 1998).

12. The relevant laws are found at 16 U.S.C. §§ 428–31 (national forests); 16 U.S.C. § 1 (national parks); 16 U.S.C. §§ 668dd-668ee (national wildlife refuges); 43 U.S.C. §§ 1701–1702 (BLM public lands). A more detailed discussion of the public land agencies can be found in Chapter 2.

13. See Charles F. Wilkinson, Crossing the Next Meridian: Land, Water and the Future of the West (Washington, D.C.: Island Press, 1992); John D. Leshy, The Mining Law: A Study in Perpetual Motion (Baltimore, Md.: Johns Hopkins University Press, 1987); Patricia N. Limerick, Legacy of Conquest: The Unbroken Past of the American West (New York: Norton, 1987). On the implications of the newer planning laws, see Charles F. Wilkinson and H. Michael Anderson, Land and Resource Planning in the National Forests, 64 Oregon L.

Rev. 1 (1985); George C. Coggins, The Law of Public Rangeland Management IV: FLPMA, PRIA, and the Multiple Use Mandate, 14 Envir. L. 1 (1983); Paul J. Culhane and H. Paul Friesema, Land Use Planning for the Public Lands, 19 Natural Resource J. 43 (1979); Robert L. Fischman, The National Wildlife Refuge System and the Hallmarks of Modern Organic Legislation, 29 Ecology L. Q. 457 (2002).

14. See William Perry Pendley, War on the West: Government Tyranny on America's Great Frontier (Washington, D.C.: Regnery Publishing, 1995); Sally Fairfax and Carolyn E. Yale, Federal Lands: A Guide to Planning, Management, and State Revenue (Washington, D.C.: Island Press, 1987).

15. See William E. Riebsame, ed., Atlas of the New West: Portrait of a Changing Region (New York: Norton, 1997); Jan G. Laitos and Thomas A. Carr, The Transformation on Public Lands, 26 Ecology L. Q. 140 (1999). For a further examination of changes in the West's social and economic structure, see Chapters 3 and 7.

16. See R. McGreggor Cawley, Federal Land, Western Anger: The Sagebrush Rebellion and Environmental Politics (Lawrence: University Press of Kansas, 1993); Alan M. Gottlieb, ed., The Wise Use Agenda: The Citizen's Guide to Environmental Resource Issues (Bellevue, Wash.: Free Enterprise Press, 1989); John Echeverria and Raymond Booth Eby, Let the People Judge: Wise Use and the Private Property Rights Movement (Washington, D.C.: Island Press, 1995); see also Chapters 2 and 7 for a more detailed discussion on the Wise Use movement and local reactions to federal public land policies.

17. See U.S. Constitution, Article 4, Section 3, Clause 2 (the property clause); Kleppe v. New Mexico, 426 U.S. 529 (1976). The role and authority of Congress and the states in developing public land policy is examined further in Chapter 2.

18. On the Endangered Species Act, see 16 U.S.C. §§ 1531–1543; Kathryn A. Kohm, ed., Balancing on the Brink of Extinction: Endangered Species Act and Lessons for the Future (Washington, D.C.: Island Press, 1991); Mark Bonnett and Kurt Zimmerman, Politics and Preservation: The Endangered Species Act and the Northern Spotted Owl, 18 Ecology L. Q. 105 (1991). The ESA and its implications for public land management are further explored in Chapters 4 and 7. On the wetlands provisions in the Clean Water Act, see 33 U.S.C. § 1344; United States v. Riverside Bayview Homes, Inc., 474 U.S. 121 (1985); Solid Waste Agency v. Army Corps of Engineers, 531 U.S. 159 (2001); Sam Kalen, Commerce to Conservation: The Call for a National Water Policy and the Evolution of Federal Jurisdiction Over Wetlands, 69 N.D. L. Rev. 873 (1993); Symposium, Wetlands Law and Policy, 7 Va. J. Nat. Resources L. 217 (1988). On the escalating tensions over environmental protection on the public lands, see David Helvarg, The War Against the Greens: The "Wise-Use" Movement, the New Right, and Anti-environmental Violence (San Francisco: Sierra Club Books, 1994); Alston Chase, In a Dark Wood: The Fight over Forests and the Rising Tyranny of Ecology (New York: Houghton Mifflin, 1995); see also Chapter 2.

19. See James Huffman, The Inevitability of Private Rights in Public Lands, 65 U. Colo. L. Rev. 241, 277 (1994); Michael C. Blumm, Public Choice Theory and the Public Lands: Why Multiple Use Failed, 18 Harvard Envir. L. Rev. 405 (1994); Cortner and Moote, supra note 1; Christopher M. Klyza, Who Controls Public Lands?: Mining, Forestry, and Grazing Policies, 1870–1990 (Chapel Hill: University of North Carolina Press, 1996); Bob Pepperman Taylor, Our Limits Transgressed: Environmental Political Thought in America (Lawrence: University Press of Kansas, 1992).

20. For examples of budget riders, see Pub. L. 105–277, § 120(a), 112 Stat. 2681; Pub. L. 106-31, § 3002, 113 Stat. 57 (blocking completion of BLM's proposed § 3809 regulations establish-

ing environmental standards for hardrock mining on public lands, 64 Fed. Reg. 6422 (Feb. 9, 1999)); Pub. L. 104–6, 109 Stat. 73, April 10, 1995 (imposing a moratorium on spending by the U.S. Fish and Wildlife Service for new listings under the Endangered Species Act); see also Chapter 4 for a discussion of budget riders in the spotted owl controversy, including the 1995 salvage logging rider. Ecosystem management critics include Allan K. Fitzsimmons, Defending Illusions: Federal Protection of Ecosystems (Lanham, Md.: Rowman and Littlefield, 1999), and Warren A. Flick and William E. King, Ecosystem Management as American Law, 13(3) Renewable Res. J. 6–11 (1995); see also Chapter 3 for more on ecosystem management. For free market perspectives on public land policy, see Robert H. Nelson, Public Lands and Private Rights: The Failure of Scientific Management (Lanham, Md.: Rowman and Littlefield, 1995); Terry L. Anderson and Donald R. Leal, Free Market Environmentalism (Boulder, Colo.: Westview Press, 1991); see also Chapter 2 for more on market-based approaches to public land policy.

21. See Ronald D. Brunner et al., Finding Common Ground: Governance and Natural Resources in the American West (New Haven: Yale University Press, 2002); Philip Brick et al., eds., Across the Great Divide: Explorations in Collaborative Conservation and the American West (Washington, D.C.: Island Press, 2001); Douglas S. Kenney et al., The New Watershed Source Book: A Directory and Review of Watershed Initiatives in the Western United States (Boulder, Colo.: Natural Resources Law Center, 2000); Steven L. Yaffee et al., Ecosystem Management in the United States: An Assessment of Current Experience (Washington, D.C.: Island Press, 1996); see also Chapters 7 and 8 for an extensive examination of collaborative conservation initiatives.

T W O

Policy and Power on the Public Domain

1. See Christopher M. Klyza, Who Controls Public Lands?: Mining, Forestry, and Grazing Policies, 1870–1990 (Chapel Hill: University of North Carolina Press, 1996); Robert L. Glicksman and George C. Coggins, Hardrock Minerals, Energy Minerals and Other Resources on the Public Lands: The Evolution of Federal Natural Resources Law, 33 Tulsa L. Rev. 765 (1998); see also Marion Clawson, The Federal Lands Revisited 15–62 (Washington, D.C.: Resources for the Future, 1983).

2. See Paul W. Gates, History of Public Land Law Development 341–494, 542–61 (Washington, D.C.: Government Printing Office, 1968); E. Louise Peffer, The Closing of the Public Domain: Disposal and Reservation Policies, 1900–1950 8–14 (New York: Arno Press, 1972); John Wesley Powell, Report on the Lands of the Arid Region of the United States (Cambridge, Mass.: Harvard Common Press, 1983); see also Donald Worster, A River Running West: The Life of John Wesley Powell (New York: Oxford University Press, 2001); Wallace Stegner, Beyond the Hundredth Meridian: John Wesley Powell and the Second Opening of the West (Lincoln: University of Nebraska Press, 1953).

3. Charles F. Wilkinson, Crossing the Next Meridian: Land, Water, and the Future of the West 3–27 (Washington, D.C.: Island Press, 1992).

4. Richard White, "It's Your Misfortune and None of My Own": A New History of the American West 145–47 (Norman: University of Oklahoma Press, 1991); Jon A. Souder and Sally K. Fairfax, State Trust Lands: History, Management, and Sustainable Use 17–36 (Lawrence: University Press of Kansas, 1996); Frederick Jackson Turner, The Significance of the Frontier in American History, Harold P. Simonson, ed. (New York: Frederick Ungar, 1963).

5. On corporate exploitation of the West, see Richard White, supra note 4, at 365–70, 395; Richard D. Lamm and Michael McCarthy, The Angry West: A Vulnerable Land and Its Future 14–17 (Boston: Houghton Mifflin, 1982); William G. Robbins, Colony and Empire: The Capitalist Transformation of the American West (Lawrence: University Press of Kansas, 1994). On Montana's experience with Anaconda Copper, see K. Ross Toole, A History of the Anaconda Copper Mining Company: A Study in the Relationships between a State and Its People and a Corporation, 1880–1950 (unpublished Ph.D. thesis, UCLA, 1954); K. Ross Toole, Montana: An Uncommon Land (Norman: University of Oklahoma Press, 1959). The Great Barbeque's impacts are chronicled in: Gates, supra note 2, at 545–61; William K. Wyant, Westward in Eden: The Public Lands and the Conservation Movement 58–59 (Berkeley: University of California Press, 1982); Peter Matthiessen, Wildlife in America 135–56 (New York: Penguin, 2d edition, 1987).

6. The argument for privatization of the public lands comes in two forms: (1) the strong form contemplates the outright sale and disposal of all (or most) public lands; (2) the weaker form contemplates the application of market principles to public resources but not sale of the public lands themselves. Scott Lehmann, Privatizing Public Lands 15–16, 201–27 (New York: Oxford University Press, 1995); Dale A. Oesterle, Public Lands: How Much Is Enough?, 23 Ecology L. Q. 521 (1996). On the privatization concept generally, see Richard L. Stroup and John A. Baden, Natural Resources: Bureaucratic Myths and Environmental Management (San Francisco: Pacific Institute for Public Policy Research, 1983); Terry L. Anderson and Donald R. Leal, Free Market Environmentalism (Boulder, Colo.: Westview Press, 1991); Robert H. Nelson, Public Lands and Private Rights: The Failure of Scientific Management (Lanham, Md.: Rowman and Littlefield, 1995); Marion Clawson, The Federal Lands Revisited 149–69 (Baltimore, Md.: Resources for the Future, Johns Hopkins University Press, 1983). For a critique of federal public land subsidy programs, see Randal O'Toole, Reforming the Forest Service (Washington, D.C.: Island Press, 1988); Debra L. Donahue, The Western Range Revisited: Removing Livestock from Public Lands to Conserve Native Biodiversity 271–79 (Norman: University of Oklahoma Press, 1999).

7. For critiques of the privatization argument, see Lehmann, supra note 6; Mark Sagoff, The Economy of the Earth: Philosophy, Law, and the Environment (New York: Cambridge University Press, 1988); Joseph L. Sax, The Legitimacy of Collective Values: The Case of the Public Lands, 56 U. Colo. L. Rev. 537 (1985); Clawson, supra note 6, at 123–48.

8. See Peffer, supra note 2, at 14–31; Wyant, supra note 5, at 133–34.

9. Regarding the Progressive Era and conservation policy, see Samuel P. Hays, Conservation and the Gospel of Efficiency: The Progressive Conservation Movement, 1890–1920 (Cambridge, Mass.: Harvard University Press, 1959). On creation of the forest reserves, see Samuel T. Dana and Sally K. Fairfax, Forest and Range Policy: Its Development in the United States 56–63 (New York: McGraw-Hill, 2d edition, 1980). On President Taft's withdrawal decision, see United States v. Midwest Oil Co., 276 U.S. 459 (1915); John D. Leshy, The Mining Law: A Study in Perpetual Motion 31–32 (Baltimore, Md.: Johns Hopkins University Press, 1987). On the Taylor Grazing Act and federal grazing policies, see Dana and Fairfax, id. at 158–65; Phillip O. Foss, Politics and Grass: The Administration of Grazing on the Public Domain 39–72 (Seattle: University of Washington Press, 1960).

10. See Hays, Gospel of Efficiency, supra note 9, at 35–37, 260–76; Gifford Pinchot, Breaking New Ground 261–62 (New York: Harcourt Brace, 1947). Pinchot biographies include: M. Nelson McGeary, Gifford Pinchot (Princeton: Princeton University Press, 1960); Harold T.

Pinkett, Gifford Pinchot: Private and Public Forester (Urbana: University of Illinois Press, 1970); Char Miller, Gifford Pinchot and the Making of Modern Environmentalism (Washington, D.C.: Island Press, 2001). For a general description and critical analysis of Pinchot's conservation philosophy, see Hays, id. at 66–73; Nelson, supra note 6, at 44–51; Douglas H. Strong, Dreamers and Defenders: American Conservationists 61–84 (Lincoln: University of Nebraska Press, 1988); Stewart Udall, The Quiet Crisis 97–108 (New York: Holt, Rinehart and Winston, 1963).

11. On the growth of specialized natural resource agencies, see Curt D. Meine, The Oldest Task in Human History, in Richard L. Knight and Sarah F. Bates, eds., A New Century for Natural Resources 7 (Washington, D.C.: Island Press, 1995); Dana and Fairfax, supra note 9, at 83–86; Michael J. Bean, The Evolution of National Wildlife Law 126–34 (Urbana, Ill.: Praeger, rev. edition, 1983); Thomas Lund, American Wildlife Law 56–79 (Berkeley: University of California Press, 1980); James B. Trefethen, An American Crusade for Wildlife 157 (New York: Winchester Press, 1975); Joseph J. Hickey, Some Historical Phases in Wildlife Conservation, 2(4) Wildlife Society Bulletin 164 (1974).

12. See Paul W. Hirt, A Conspiracy of Optimism: Management of the National Forest since World War Two (Lincoln: University of Nebraska Press, 1994); David A. Clary, Timber and the Forest Service 94–125 (Lawrence: University Press of Kansas, 1986); Dana and Fairfax, supra note 9, at 207–38; Richard W. Behan, The Myth of the Omnipotent Forester, 64 Journal of Forestry 398–407 (1966). For a powerful critique of the doctrine of scientific management in the natural resources field, see Nelson, supra note 6.

13. For the Multiple Use–Sustained Yield Act, see 16 U.S.C. §§ 528–531; George C. Coggins, Of Succotash Syndromes and Vacuous Platitudes: The Meaning of "Multiple Use, Sustained Yield" for Public Land Management (Part I), 53 U. Colo. L. Rev. 229 (1982). For the NFMA legislation, see 16 U.S.C. §§ 1600–14; Charles Wilkinson and H. Michael Anderson, Land and Resource Planning in the National Forests, 64 Ore. L. Rev. 1 (1985). For the FLPMA legislation, see 43 U.S.C. §§ 1701–84; George C. Coggins, The Law of Public Rangeland Management IV: FLPMA, PRIA, and the Multiple Use Mandate, 14 Envir. L. 1 (1983). For NEPA, see 42 U.S.C. §§ 4321–61; Special Focus, Articles and Essays: NEPA at Twenty, 25 Land and Water L. Rev. 1 (1990); Symposium on NEPA at Twenty: The Past, Present and Future of the National Environmental Policy Act, 20 Envir. L. 447 (1990). For the pollution laws, see 33 U.S.C. §§ 1251–1387 (water); 42 U.S.C. §§ 7401–7671q (air); 42 U.S.C. § 7409 (ambient air quality); 42 U.S.C. §§ 7470–79 (prevention of significant deterioration of air quality); 42 U.S.C. § 7491–92 (visibility protection); 33 U.S.C. §1321 (oil and hazardous substance liability); 33 U.S.C. §§1344, 1362(7) (dredge and fill permit requirements); 33 U.S.C. §1342 (National Pollutant Discharge Elimination System); see also William Rodgers, Environmental Law 123–392 (St. Paul, Minn.: West Publishing, 2d edition, 1994).

14. 17 Stat. 32, 16 U.S.C. §§ 21–22; see also Roderick Nash, Wilderness and the American Mind 108 (New Haven: Yale University Press, 3d edition, 1982); Alfred Runte, National Parks: The American Experience 33–47 (Lincoln: University of Nebraska Press, 2d edition, 1987).

15. For a general discussion of the American public's evolving views toward nature, see Hans Huth, Nature and the American: Three Centuries of Changing Attitudes (Lincoln: University of Nebraska Press, 1957); Leo Marx, The Machine in the Garden: Technology and the Pastoral Idea in America (New York: Oxford University Press, 1964); Keith Thomas, Man and the Natural World: A History of the Modern Sensibility (New York: Pantheon,

1983). On Muir and his approach toward nature, see John Muir, Our National Parks (Boston: Houghton Mifflin, 1901); Stephen Fox, John Muir and His Legacy: The American Conservation Movement (Boston: Little, Brown, 1981); Michael Cohen, The Pathless Way: John Muir and American Wilderness (Madison: University of Wisconsin Press, 1984).

16. See Alfred Runte, Yosemite: The Embattled Wilderness 68, 80–82 (Lincoln: University of Nebraska Press, 1990); Nash, supra note 14, at 161–81; Udall, supra note 10, at 116–22.

17. On the Antiquities Act, see 16 U.S.C. §§ 431–33; Dana and Fairfax, supra note 9, at 78–79; John D. Leshy, Putting the Antiquities Act in Perspective, in Robert B. Keiter et al., eds., Visions of The Grand Staircase–Escalante: Examining Utah's Newest National Monument 83–88 (Salt Lake City: University of Utah Press, 1998). On the National Parks Organic Act, see 16 U.S.C. §§ 1–21; Runte, National Parks, supra note 14, at 95–105.

18. See Runte, supra note 14, at 11–47; Richard West Sellars, Preserving Nature in the National Parks: A History 16–27, 70–90 (New Haven: Yale University Press, 1997); see also Joseph L. Sax, Mountains without Handrails: Reflections on the National Parks (Ann Arbor: University of Michigan Press, 1980).

19. On the Wilderness Act, see 16 U.S.C. §§ 1131–36; John Hendee et al., Wilderness Management (Washington, D.C.: U.S. Department of Agriculture, 1978); on the Land and Water Conservation Fund Act, see 16 U.S.C. §§ 460*l*-4 to 460*l*-11; Robert Glicksman and George C. Coggins, Federal Recreational Land Policy: The Rise and Decline of the Land and Water Conservation Fund, 9 Columbia J. Envir. L. 125 (1984); on the National Wildlife Refuge Administration Act, see 16 U.S.C. §§ 668dd–668ee; Nathaniel P. Reed and Dennis Drabelle, The United States Fish and Wildlife Service (Boulder, Colo.: Westview Press, 1984); Richard J. Fink, The National Wildlife Refuges: Theory, Practice, and Prospect, 18d Harvard Envir. L. Rev. 1 (1994); on the Wild and Scenic Rivers Act, see 16 U.S.C. § 1271–87; Sally K. Fairfax, Barbara T. Andrews, and Andrew P. Buchsbaum, Federalism and the Wild and Scenic Rivers Act: Now You See It, Now You Don't, 59 Wash. L. Rev. 417 (1984); on the FLPMA wilderness provision, see 43 U.S.C. § 1782; John D. Leshy, Wilderness and Its Discontents: Wilderness Review Comes to the Public Lands, 1981 Ariz. State L. J. 361; on the Endangered Species Act, see 16 U.S.C. §§ 1531–43; George C. Coggins and Irma S. Russell, Beyond Shooting Snail Darters in Pork Barrels: Endangered Species and Land Use in America, 70 Geo. L. J. 1433 (1982).

20. See Michael E. Soule and John Terborgh, eds., Continental Conservation: Scientific Foundations of Regional Reserve Networks 1–9 (Washington, D.C.: Island Press, 1999); Reed F. Noss and Allen Y. Cooperrider, Saving Nature's Legacy: Protecting and Restoring Biodiversity 142–56 (Washington, D.C.: Island Press, 1994); R. Edward Grumbine, Ghost Bears: Exploring the Biodiversity Crisis 28–63 (Washington, D.C.: Island Press, 1992).

21. See Hope M. Babcock, Dual Regulation, Collaborative Management, or Layered Federalism: Can Cooperative Federalism Models from Other Laws Save Our Public Lands?, 3 Hastings West-Northwest J. Envir. Law and Policy 193 (1996).

22. On the early role of state law on the public lands, see Morton v. Solambo Copper Mining Co., 26 Cal. 527 (1864); Omaechevarria v. Idaho, 246 U.S. 343 (1918). On local resistance to Progressive Era reforms, see Hays, Gospel of Efficiency, supra note 9, at 271–76.

23. See Foss, supra note 9, at 60–61; Dana and Fairfax, supra note 9, at 162–63; Klyza, supra note 1, at 109–140; Jeanne Nienaber Clarke and Daniel C. McCool, Staking Out the Terrain: Power and Performance Among Natural Resource Agencies 161–62 (Albany: State University of New York Press, 2d edition, 1996).

24. On the Administrative Procedures Act, see 5 U.S.C. § 551 et seq.; Bernard Schwartz, Ad-

ministrative Law 28–33 (Boston: Little, Brown, 3d edition, 1991). On NEPA and public involvement processes, see Jonathan Poisner, A Civic Republican Perspective on the National Environmental Policy Act's Process for Citizen Participation, 26 Envir. L. 53 (1996); Reclaiming NEPA's Potential: Can Collaborative Processes Improve Environmental Decision Making? (Missoula, Mt.: O'Connor Center for the Rocky Mountain West, 2000); see also Lynton K. Caldwell, The National Environmental Policy Act: An Agenda for the Future 23–47 (Bloomington: Indiana University Press, 1998). The FLPMA consistency provision is 43 U.S.C. § 1712(c)(9); 43 C.F.R. § 1610.3 (2000).

25. See R. McGreggor Cawley, Federal Land, Western Anger: The Sagebrush Rebellion and Environmental Politics (Lawrence: University Press of Kansas, 1993); John D. Leshy, Unraveling the Sagebrush Rebellion: Law, Politics, and Federal Lands, 14 U.C. Davis L. Rev. 317 (1980); Scott Reed, The County Supremacy Movement: Mendacious Myth Marketing, 30 Idaho L. Rev. 525 (1994); John D. Echevarria and Raymond Booth Eby, eds., Let the People Judge: Wise Use and the Private Property Rights Movement (Washington, D.C.: Island Press, 1995).

26. Collaborative conservation groups and ideas are examined in Steven L. Yaffee et al., Ecosystem Management in the United States: An Assessment of Current Experiences (Washington, D.C.: Island Press, 1996); The Keystone Center, The Keystone National Policy Dialogue on Ecosystem Management (1996) (hereafter, Keystone Ecosystem Management Dialogue); Barb Cestero, Beyond the Hundredth Meeting: A Field Guide to Collaborative Conservation on the West's Public Lands (Tucson, Ariz.: Sonoran Institute, 1999); see also Chapter 7 and the sources cited there. On the state resource advisory councils, see 43 C.F.R. § 1784.6 (2001); Department of the Interior, Bureau of Land Management, Rangeland Reform '94: Draft Environmental Impact Statement 14 (Washington, D.C., 1994); Todd M. Olinger, Public Rangeland Reform: New Prospects for Collaboration and Local Control Using the Resource Advisory Councils, 69 U. Colo. L. Rev. 633, 665–69, 673–92 (1998).

27. See Babcock, supra note 21 at 199–208; Carolyn M. Landever, Whose Home on the Range? Equal Footing, the New Federalism and State Jurisdiction on Public Lands, 47 Florida L. Rev. 557, 604–35 (1995).

28. See U.S. Constitution Art. 1, Sec. 1 (Congress's lawmaking power); id. at Article 1, Section 8 (Congress's enumerated powers). Examples of cases sustaining a broad interpretation of Congress's powers include: Wickard v. Filburn, 317 U.S. 111 (1942); Heart of Atlanta Motel, Inc. v. United States, 379 U.S. 241 (1964); Garcia v. San Antonio Metropolitan Transit Authority, 469 U.S. 528 (1985); see also Jesse Choper, Judicial Review and the National Political Process 175–184 (Chicago: University of Chicago Press, 1980); Herbert Wechsler, The Political Safeguards of Federalism: The Role of the States in the Composition and Selection of the National Government, 54 Columbia L. Rev. 543 (1954). Examples of recent cases limiting Congress's powers include: United States v. Lopez, 514 U.S. 549 (1995); New York v. United States, 505 U.S. 144 (1992); Printz v. United States, 521 U.S. 898 (1997); United States v. Morrison, 529 U.S. 598 (2000); see also Laurence Tribe, American Constitutional Law (Vol. 1) 817, 878 (New York: Foundation Press, 3d edition, 2000).

29. See U.S. Constitution, Article IV, Section 3, Clause 2 (property clause); Camfield v. United States, 167 U.S. 518, 525 (1897); United States v. San Francisco, 310 U.S. 16, 29 (1940).

30. See Kleppe v. New Mexico, 426 U.S. 529, 546 (1976); Minnesota v. Block, 660 F.2d 1240, 1249 (8th Cir. 1981), cert. denied, 455 U.S. 1007 (1982); United States v. Alford, 274 U.S. 264 (1927). On the enclave clause, see U.S. Constitution, Article 1, Section 8, clause 17; George Coggins and Robert Glicksman, Public Natural Resources Law § 3.03[2] (1999) (St. Paul, Minn.: Westgroup, 2000).

31. On the commerce clause and the regulation of private commercial activities or lands, see United States v. Bramble, 103 F.3d 1475 (9th Cir. 1996); United States v. Romano, 929 F. Supp. 502 (D. Mass. 1996); National Association of Home Builders v. Babbitt, 130 F.3d 1041 (D.C. Cir. 1997), cert. denied, 118 S.Ct. 2340 (1998); Gibbs v. Babbitt, 31 F.Supp. 531 (E.D.N.C. 1998), affirmed, 214 F.3d 483 (4th Cir. 2000). On the treaty power and natural resource regulation, see Missouri v. Holland, 252 U.S. 416 (1920). On the spending power, see U.S. Constitution, Article 1, Section 8, clause 1; Tribe, supra note 28, at 833.

32. See Phillip A. Davis, Cry for Preservation, Recreation Changing Public Land Policy, Congressional Quarterly Weekly Edition, Aug. 3, 1991, pp. 2145–51.

33. On the Grand Canyon Protection Act, see Pub. L. 102–575, 106 Stat. 4600 (1992); U.S. Department of the Interior, Bureau of Reclamation, Operation of Glen Canyon Dam Final EIS (Washington, D.C., March 1995); see also Michael Conner, Extracting the Monkey Wrench from Glen Canyon Dam: The Grand Canyon Protection Act—An Attempt at Balance, 15 Public Land L. Rev. 15 (1994). On the Quincy Library Group legislation, see Chapter 8. On the congressional east–west voting split, see Craig W. Allin, The Politics of Wilderness Preservation 119–23, 272–75 (Westport, Conn.: Greenwood Press, 1982); Nash, supra note 14, at 222–24.

34. U.S. Constitution, Article II, Sections 1 and 3.

35. For the sake of convenience and brevity, the text will refer to both cabinet-level offices and administrative agencies as "agencies," which is how both entities are defined in the Administrative Procedures Act. 5 U.S.C. § 551(1). On the presidential appointment power, see U.S. Constitution, Article II, Section 2, Clause 2; Morrison v. Olson, 487 U.S. 654 (1988); see also Tribe, supra note 28, at 677.

36. 16 U.S.C. § 531; Perkins v. Bergland, 608 F.2d 803, 806 (9th Cir. 1979).

37. For an analysis of how presidential administrations have employed their executive powers to reshape public land policy, see Robert F. Durrant, The Administrative Presidency Revisited: Public Lands, the BLM, and the Reagan Revolution (Albany: State University of New York Press, 1992); John D. Leshy, The Babbitt Legacy at the Department of the Interior: A Preliminary View, 31 Envir. L. 199 (2002). Beyond the substantive and procedural standards contained in organic legislation, an assortment of general laws, such as the Administrative Procedures Act, the Federal Advisory Committee Act, and the National Environmental Policy Act, establish uniform procedural standards that all federal agencies must follow.

38. On the APA and public involvement in agency proceedings, see 5 U.S.C. § 552 (rule making); 5 U.S.C. § 552 (Freedom of Information Act); 5 U.S.C. § 552b (Government in the Sunshine Act); 5 U.S.C. Appendix (Federal Advisory Committee Act). On the APA and judicial review, see 5 U.S.C. § 706; Richard J. Pierce, Jr. et al., Administrative Law and Process 361–369 (New York: Foundation Press, 3d edition, 1999); Richard B. Stewart, The Reformation of American Administrative Law, 88 Harv. L. Rev. 1669 (1975). On the Chevron deference doctrine, see Chevron, USA, Inc. v. NRDC, 467 U.S. 837, 842–45 (1984); Pierce et al., id. at 374–80.

39. On the United States' sovereign and proprietorial status on the public lands, see Camfield v. United States, 167 U.S. 518, 524–26 (1897); Kleppe v. New Mexico, 426 U.S. 529, 540 (1976). On the presidential withdrawal power, see Pickett Act of 1910, 43 U.S.C. § 141 (repealed); Federal Land Policy Management Act of 1976, 43 U.S.C. § 1714; Grisar v. McDowell, 6 Wall 364, 381 (1867); United States v. Midwest Oil Co., 236 U.S. 459 (1915); Portland General Electric Co. v. Kleppe, 441 F. Supp. 859 (D. Wyo. 1977); see also David H. Getches, Managing the Public Lands: The Authority of the Executive to Withdraw Lands,

22 Natural Resources J. 279 (1982); Charles F. Wheatley, Withdrawals Under the Federal Land Policy Management Act of 1976, 21 Ariz. L. Rev. 311 (1979).

40. On the General Revision Act, see Dana and Fairfax, supra note 9, at 55–67. In fact, displeased by the aggressive manner in which presidents used the General Revision Act's reservation powers, Congress repealed this legislation in 1907 and thus reserved for itself the power to create national forests. Id. at 372. On the Antiquities Act and its use, see John D. Leshy, Shaping the Modern West: The Role of the Executive Branch, 72 U. Colo. L. Rev. 287 (2001); James R. Rasband, Utah's Grand Staircase: The Right Path to Wilderness Preservation, 70 U. Colo. L. Rev. 483 (1999). On President Clinton's Yellowstone mining claim withdrawal decision, see 60 Fed. Reg. 45732 (Sept. 1, 1995); Kit Miniclier, Mine Firm Grabs Land Near Park: Yellowstone Claims Filed After Clinton Issues Ban, Denver Post, Sept. 6, 1995, p. B-1; see also William J. Lockhart, External Threats to Our National Parks: An Argument for Substantive Protection, 16 Stanford Env. L. J. 3 (1997).

41. On the law of standing, see Sierra Club v. Morton, 405 U.S. 727 (1972); United States v. Students Challenging Regulatory Agency Procedures (SCRAP), 412 U.S. 669 (1973); see Abram Chayes, The Role of the Judge in Public Law Litigation, 89 Harv. L. Rev. 4 (1982); Stewart, supra note 38, at 1723–47. On the "hard look" doctrine, see Citizens to Preserve Overton Park, Inc. v. Volpe, 401 U.S. 402 (1971); Motor Vehicle Manufacturers Association v. State Farm Automobile Insurance Co., 463 U.S. 29 (1983); Pierce et al., supra note 38, at 132–38, 380–93. Citations for the noted cases are: Marbury v. Madison, 5 U.S. (1Cranch) 137, 2 L. Ed. 60 (1803); Brown v. Board of Education, 347 U.S. 483 (1954).

42. The noted early-twentieth-century decisions were: Grimaud v. United States, 220 U.S. 506 (1911); United States v. Light, 220 U.S. 523 (1911); United States v. Midwest Oil Co., 236 U.S. 459 (1915). The noted 1970s decisions were: West Virginia Division of Izaac Walton League of America, Inc. v. Butz, 522 F.2d 945 (4th Cir. 1975); Wilderness Society v. Morton, 479 F.2d 482 (D.C. Cir. 1973) (en banc), cert. denied, 411 U.S. 917 (1973); Natural Resources Defense Council, Inc. v. Morton, 388 F.Supp. 829 (D.D.C. 1974), aff'd per curiam, 527 F.2d 1386 (D.C. Cir. 1976), cert. denied, 427 U.S. 913 (1976). The more recent ESA and NEPA decisions include: Sierra Club v. Peterson, 717 F.2d 1409 (D.C. Cir. 1983); Connor v. Burford, 848 F.2d 1441 (9th Cir. 1988); Thomas v. Peterson, 753 F.2d 754 (9th Cir. 1985); Pacific Rivers Council v. Thomas, 30 F.3d 1050 (9th Cir. 1994). The noted industry and local litigation includes: Bennett v. Spear, 520 U.S. 154 (1997); Catron County Board of Commissioners v. Babbitt, 75 F.3d 1429 (10th Cir. 1996).

43. For a blistering critique of an "activist" federal judiciary, see Robert H. Bork, The Tempting of America: The Political Seduction of the Law (New York: Free Press, 1990). The Chevron doctrine was first articulated in Chevron, USA, Inc. v. NRDC, 467 U.S. 837, 842–45 (1984). On the reality and appearance of judicial activism, compare Joseph L. Sax, The Constitutional Dimensions of Property: A Debate, 26 Loyola L.A. L. Rev. 23 (1992) with Richard A. Epstein, The Seven Deadly Sins of Takings Law: The Dissents in Lucas v. South Carolina Coastal Council, 26 Loyola L.A. L. Rev. 955 (1993).

44. See California Coastal Commission v. Granite Rock Co., 480 U.S. 572 (1987); see also John D. Leshy, Granite Rock and the States' Influence over Federal Land Use, 18 Envir. L. 99 (1987); Eric Freyfogle, *Granite Rock*, Institutional Competence and the State Role in Federal Land Planning, 59 U. Colo. L. Rev. 475 (1988); Richard Cowart and Sally Fairfax, Public Lands Federalism: Judicial Theory and Administrative Reality, 15 Ecology L. Q. 375 (1988).

45. On unfunded federal mandates, see Unfunded Mandates Reform Act of 1995, Pub. L. 104–

4, 109 Stat. 48 (1995), codified at 2 U.S.C. §§ 602, 658 et seq., 1501 et seq.; Robert W. Adler, Unfunded Mandates and Fiscal Federalism: A Critique, 50 Vanderbilt L. Rev. 1137 (1997); Rena L. Steinzor, Unfunded Environmental Mandates and the "New (New) Federalism": Devolution, Revolution, or Reform?, 81 Minn. L. Rev. 97 (1996). On the Clinton adminis-tration's ESA regulatory reforms, see Karin P. Sheldon, Habitat Conservation Planning: Addressing the Achilles Heel of the Endangered Species Act, 6 N.Y.U. Env. L. J. 279 (1998); DeAnne Parker, Natural Community Conservation Planning: California's Emerging Eco-system Management Alternative, 6 U. Balt. J. Envir. L. 107 (1997). On the Supreme Court's recent federalism decisions, see supra note 28 and cases cited therein. On the Enlibra doc-trine, see Western Governors' Association, Enlibra: Principles for Environmental Manage-ment in the West, Policy Resolution 99–013 (June 15, 1999); see also www.westgov.org/wga/policy/99/99013.htm.

46. Daniel Kemmis, Community and the Politics of Place (Norman: University of Oklahoma Press, 1990); see also Daniel Kemmis, This Sovereign Land: A New Vision for Governing the West (Washington, D.C.: Island Press, 2001).

47. On civic republicanism, see H. Jefferson Powell, Reviving Republicanism, 97 Yale L. J. 1703 (1988); Frank Michelmann, Traces of Self-Government, 100 Harv. L. Rev. 4 (1986); Cass Sunstein, Beyond the Republican Revival, 97 Yale L. J. 1539 (1988); Kathleen Sullivan, Rainbow Republicanism, 97 Yale L. J. 1713 (1988). On collaborative processes, see Philip Brick et al., eds., Across the Great Divide: Explorations in Collaborative Conservation and the American West (Washington, D.C.: Island Press, 2001); Keystone Ecosystem Manage-ment Dialogue, supra note 26; Steven L. Yaffee et al., Ecosystem Management in the United States, supra note 26.

48. On the county supremacy ordinances in court, see Boundary County Backpackers v. Boundary County, 913 P.2d 1141 (Idaho 1996); United States v. Nye County, Nevada, 920 F.Supp. 1108 (D. Nev. 1996). See Chapter 7 for further discussion of the advantages and dis-advantages of collaborative processes.

49. The principal laws governing the Forest Service are the Multiple Use–Sustained Yield Act of 1960, 16 U.S.C. §§ 428–31, and the National Forest Management Act of 1976, 16 U.S.C. §§ 1601–141; see also Wilkinson and Anderson, supra note 13. The principal volumes ex-amining the Forest Service and its history are: Harold K. Steen, The United States Forest Service: A History (Seattle: University of Washington Press, 1986); Glen Robinson, The Forest Service: A Study in Public Land Management (Baltimore, Md.: Johns Hopkins Uni-versity Press, 1975); Dana and Fairfax, supra note 9; see also Gifford Pinchot, Breaking New Ground, supra note 10; Herbert Kaufman, The Forest Ranger: A Study in Administrative Behavior (Baltimore, Md.: Johns Hopkins University Press, 1960). The Forest Service's di-minishing bureaucratic reputation is examined in Clarke and McCool, supra note 23, at 49–68.

50. See West Virginia Division of Izaak Walton League of America, Inc. v. Butz, 522 F.2d 945 (4th Cir. 1975); see also Clary, supra note 12, at 183–88, 190–93; Hirt, supra note 12, at 245–51.

51. On the NFMA's environmental constraints, see 16 U.S.C. §§ 1604(g); Wilkinson and An-derson, supra note 13, at 41–44. On the Forest Service's transition to ecosystem manage-ment, see John Fedkiw, The Forest Service's Pathway toward Ecosystem Management, 95(4) Journal of Forestry 30 (April 1997); Margaret Kriz, Fighting Over Forests, 30(22) Na-tional Journal 1232 (May 30, 1998).

52. See Clarke and McCool, supra note 23, at 162–64; Dyan Zaslowsky and T. H. Watkins,

These American Lands: Parks, Wilderness, and the Public Lands 127–28 (Washington, D.C.: Island Press, 1994); Foss, supra note 9, at 198–200; see also Donahue, supra note 6, at 67–113.

53. Before the FLPMA, the Classification and Multiple-Use Act of 1960 provided the BLM with temporary multiple-use management authority over its lands and empowered the agency to classify its lands according to their best uses—a precursor to the FLPMA's more sophisticated land use planning scheme. 43 U.S.C. §§ 1411–18 (expired 1970); see Dana and Fairfax, supra note 9, at 231. On the FLPMA generally, see 43 U.S.C. §§ 1701–84; Coggins, Public Rangeland Management, supra note 13. The FLPMA's resource protection provisions include: 43 U.S.C. § 1732(b) (undue degradation standard); § 1712(c)(3) (ACEC provision); §1782 (wilderness provisions); see also Marla Mansfield, On the Cusp of Property Rights: Lessons from Public Land Law, 18 Ecology L. Q. 43 (1991); Leshy, Wilderness Review, supra note 19.

54. On the Sagebrush Rebellion and the Reagan administration, see Cawley, supra note 25; Leshy, Sagebrush Rebellion, supra note 25; Durant, supra note 37.

55. On the Babbitt range reforms, see 43 C.F.R. §§ 4.77, 1784.0–1 to 1784.62, 4100.0–1 to 4180.2 (2000); Department of the Interior, Bureau of Land Management, Rangeland Reform '94: Final Environmental Impact Statement (Washington, D.C., 1994); Bruce M. Pendery, Reforming Livestock Grazing on the Public Domain: Ecosystem Management–Based Standards and Guidelines Blaze a New Path for Range Management, 27 Envir. L. 513 (1997); Karl N. Arruda and Christopher Watson, The Rise and Fall of Grazing Reform, 32 Land and Water L. Rev. 413 (1997). On the Babbitt mining law reform efforts, see 64 Fed. Reg. 6422 (Feb. 9, 1999); 64 Fed. Reg. 57613 (Oct. 26, 1999); Sam Kalen, An 1872 Mining Law for the New Millennium, 71 U. Colo. L. Rev. 343 (2000); Patrick Garver and Mark Squillace, Mining Law Reform—Administrative Style, 45 Rocky Mountain Mineral Law Institute 14-1 (1999). On the Babbitt wilderness review initiatives, see State of Utah v. Babbitt, 137 F.3d 1193 (10th Cir. 1998); see also Doug Goodman and Daniel McCool, Contested Landscape: The Politics of Wilderness in Utah and the West (Salt Lake City: University of Utah Press, 1999). On the BLM's new national monument responsibilities, see Proclamation No. 6920, 3 C.F.R. 64, 1996 USCCAN A73 (Sept. 18, 1996); see Sanjay Ranchod, The Clinton National Monuments: Protecting Ecosystems with the Antiquities Act, 25 Harvard Envir. L. Rev. 535 (2001). See generally John D. Leshy, The Babbitt Legacy at the Department of the Interior: A Preliminary View, 31 Envir. L. 199 (2001).

56. 16 U.S.C. § 1. The National Parks Organic Act can be found at: 16 U.S.C. §§ 1–18f ; see Robin Winks, The National Park Service Act of 1916: "A Contradictory Mandate"?, 74 Denver U. L. Rev. 575 (1997).

57. On National Park Service resource management policies, see Runte, supra note 14, at 82–105, 111, 168–69; Sellars, supra note 18, at 16–27, 70–90; R. Gerald Wright, Wildlife Research and Management in the National Parks 35–42, 55, 59–69 (Urbana: University of Illinois Press, 1992). On the early history of the National Park Service, see Horace M. Albright and Marian Albright Schenck, Creating the National Park Service: The Missing Years (Norman: University of Oklahoma Press, 1999); Robert Shankland, Steve Mather of the National Parks (New York: Knopf, 1951); Donald C. Swain, Wilderness Defender: Horace M. Albright and Conservation (Chicago: University of Chicago Press, 1970); Horace M. Albright and Robert Cahn, The Birth of the National Park Service: The Founding Years, 1913–33 (Salt Lake City, Utah: Howe Brothers, 1985). The facade management point is made in Sellars, supra note 18, at 4–5, 70, 88–90.

58. Leopold et al., Wildlife Management in the National Parks, in Transactions of the Twenty-Eighth North American Wildlife and Natural Resources Conference 29, 31 (reprinted in Lary M. Dilsaver, America's National Park System: The Critical Documents 237 (Lanham, Md.: Rowman and Littlefield, 1994) (hereafter cited as Dilsaver).

59. See Memorandum from Secretary of the Interior Stuart Udall, on Management of the National Park System to National Park Service Director (July 10, 1964), reprinted in Dilsaver, note 58, at 272; National Park Service, Administrative Policies for Natural Areas (1968), reprinted in Dilsaver, supra note 58, at 354; National Park Service, U.S. Department of the Interior, Management Policies (1988); National Park Service, U.S. Department of the Interior, Management Policies 2001. The Park Service's resource management critics include: Frederic H. Wagner et al., Wildlife Policies in the U.S. National Parks (Washington, D.C.: Island Press, 1995); Alston Chase, Playing God in Yellowstone: The Destruction of America's First National Park (New York: Atlantic Monthly Press, 1986); R. Lee Lyman, White Goats, White Lies: The Abuse of Science in Olympic National Park (Salt Lake City: University of Utah Press, 1998); but for a more sympathetic view, see James A. Pritchard, Preserving Yellowstone's Natural Conditions: Science and the Perception of Nature (Lincoln: University of Nebraska Press, 1999); Robert B. Keiter, Preserving Nature in the National Parks: Law, Policy, and Science in a Dynamic Environment, 74 Denver U. L. Rev. 649 (1997); Mark S. Boyce, Natural Regulation or the Control of Nature, in Robert B. Keiter and Mark S. Boyce, eds., The Greater Yellowstone Ecosystem; Redefining America's Wilderness Heritage 183 (New Haven: Yale University Press, 1991). See Chapter 5 for a further examination of the Park Service's resource management policies and strategies.

60. On the external threats problem, see Office of Science and Technology and National Park Service, U.S. Department of the Interior, State of the Parks 1980: A Report to Congress (1980); U.S. General Accounting Office, Parks and Recreation: Limited Progress Made in Documenting and Mitigating Threats to the Parks (1987) (report to the Chairman, Subcommittee on National Parks and Recreation, House Committee on Interior and Insular Affairs); see also John Freemuth, Islands Under Siege: National Parks and the Politics of External Threats (Lawrence: University of Kansas Press, 1991); Robert B. Keiter, On Protecting the National Parks from the External Threats Dilemma, 20 Land and Water L. Rev. 355 (1985).

61. For background on the national wildlife refuge system, see Zaslowsky and Watkins, supra note 21, at 151–94; Fink, supra note 19. For background on the evolution of wildlife law, see George C. Coggins and Michael E. Ward, The Law of Wildlife Management on the Federal Public Lands, 60 Oregon L. Rev. 59 (1981). On the U.S. Fish and Wildlife Service as an agency, see Reed and Drabelle, supra note 19.

62. The principal early refuge laws were the Refuge Recreation Act of 1962, 16 U.S.C. §§ 460k-460k-4, and the National Wildlife Refuge System Administration Act of 1966, 16 U.S.C. §§ 668dd-668ee; see Fink, supra note 19, at 24–30. Reports documenting deteriorating refuge conditions included: U.S. Fish and Wildlife Service, Department of the Interior, Report on Resource Problems on National Wildlife Refuges, National Fish Hatcheries, Research Centers (1983); U.S. General Accounting Office, Wildlife Management: National Refuge Contamination Is Difficult to Confirm and Clean Up (1987); U.S. General Accounting Office, National Wildlife Refuges: Continuing Problems with Incompatible Uses Call for Bold Action (1989); Defenders of Wildlife, Putting Wildlife First: Recommendations for Reforming Our Troubled Refuge System (Washington, D.C., 1992); see also Fink, id. at 63–82; Zaslowsky and Watkins, supra note 52, at 177–182.

63. The quotation from the 1997 legislation is found at: 16 U.S.C. § 668dd(a)(2). The new compatibility standard is codified at 16 U.S.C. §§ 668dd(a)(3)(C), (d)(3)(A)(I); see also Wilderness Society v. Babbitt, 5 F.3d 383 (9th Cir. 1993); Humane Society v. Lujan, 768 F.Supp. 360 (D.D.C. 1991); Defenders of Wildlife v. Andrus, 455 F.Supp. 446 (D.D.C. 1978). For the new biodiversity provisions, see 16 U.S.C. §§ 668dd(a)(4)(B), (N); for the new conservation planning provisions, see 16 U.S.C. §§ 668dd(e)(1), (4). See also Robert L. Fischman, The National Wildlife Refuge System and the Hallmarks of Modern Organic Legislation, 29 Ecology L. Q. 457 (2002).

64. The crucial 1978 Supreme Court decision was Tennessee Valley Authority v. Hill, 437 U.S. 153 (1978); see also Coggins and Russell, supra note 19; Steven L. Yaffee, Prohibitive Policy: Implementing the Federal Endangered Species Act 32–57 (Cambridge, Mass.: MIT Press, 1982). The various ESA criticisms are recounted and addressed in: Bonnie B. Burgess, Fate of the Wild: The Endangered Species Act and the Future of Biodiversity (Athens: University of Georgia Press, 2001); Brian Czech and Paul R. Krausman, The Endangered Species Act: History, Conservation, Biology, and Public Policy (Baltimore, Md.: Johns Hopkins University Press, 2001). For critical analyses of the FWS's performance as an agency, see Clarke and McCool, supra note 23, at 122–25; Zaslowsky and Watkins, supra note 52, at 186–88; Reed and Drabbelle, supra note 19, at 149–52; Oliver Houck, The Endangered Species Act and Its Implementation by the U.S. Departments of Interior and Commerce, 64 U. Colo. L. Rev. 277 (1993).

65. On the National Biological Survey, see H. Ronald Pulliam, The Birth of a Federal Research Agency, Bio-Science (The Science and Biodiversity Policy Supplement) S-91 (1995); O. J. Reichman and H. Ronald Pulliam, The Scientific Basis for Ecosystem Management, 6(3) Ecological Applications 694 (1996); Karen A. Scanna, The National Biological Survey: A Step Along the Path to Ecosystem Conservation, 4 N.Y.U. Envir. L. J. 134 (1995); Clarke and McCool, supra note 23, at 124–25; Edward J. Heisel, Biodiversity and Federal Land Ownership: Mapping a Strategy for the Future, 25 Ecology L. Q. 229, 254–55 (1998); Frederic H. Wagner, Whatever Happened to the National Biological Survey?, 49(3) BioScience 219 (1999). It should be noted that the Republican-controlled Congress, fearing a further expansion of federal regulatory power in the biological arena, refused to fund the new National Biological Survey agency until it was renamed and relocated within the U.S. Geological Survey and also denied any regulatory powers. Omnibus Consolidated Rescissions and Appropriations Act of 1996, Pub. L. 104–34, 110 Stat. 1321 § 165 (April 26, 1996).

66. On industry's desire for certainty and stability, see Clawson, supra note 6, at 4; Warren A. Flick and William E. King, Ecosystem Management as American Law, 13(3) Renewable Res. J. 6, 10 (1995). For additional background on the philosophy and origins of conservative nonprofit advocacy organizations focused on public land policy, see William Perry Pendley, War on the West: Government Tyranny on America's Great Frontier (Washington, D.C.: Regnery Publishing, 1995); Bruce Yandle, ed., Land Rights: The 1990s' Property Rights Rebellion (Lanham, Md.: Rowman and Littlefield, 1995); Philip D. Brick and R. McGreggor Cawley, A Wolf in the Garden: The Land Rights Movement and Renewing American Environmentalism (Lanham, Md.: Rowman and Littlefield, 1996). For an unsympathetic treatment of the groups and organizations generally opposing the environmental movement and its policy agenda, see David Helvarg, The War Against the Greens (San Francisco: Sierra Club Press, 1994); Carl Deal, The Greenpeace Guide to Anti-environmental Organizations (Berkeley, Calif.: Odonian Press, 1993).

67. On the Wise Use movement and its agenda, see Alan M. Gottlieb, ed., The Wise Use

Agenda: A Task Force Report Sponsored by the Wise Use Movement (Bellevue, Wash.: Free Enterprise Press, 1989); Ron Arnold, Overcoming Ideology, in Brick and Cawley, supra note 66, at 15; Karen Budd-Falen, Protecting Community Stability and Local Economies: Opportunities for County Government Influence in Federal Decision and Policy-Making Process, id. at 73. On the Greater Yellowstone Vision document strategy, see Robert Ekey, Wise Use and the Greater Yellowstone Vision Document: Lessons Learned, in Echeverria and Eby, supra note 25, at 339; see also Chapter 3 for a further description of the Greater Yellowstone vision process.

68. See Dan Daggett, Beyond the Rangeland Conflict: Toward a West that Works (Layton, Utah: Gibbs Smith, 1995); Steve Thompson, After the Fall: As Big Timber Companies Leave the Northern Rockies, A Family Mill Turns to Restoring Forests, 32(9) High Country News 1 (May 8, 2000).

69. Several works provide a sympathetic yet critical analysis of the modern environmental movement, including Philip Shabecoff, A Fierce Green Fire: The American Environmental Movement (New York: Hill and Wang, 1993); Mark Dowie, Losing Ground: American Environmentalism at the Close of the Twentieth Century (Cambridge, Mass.: MIT Press, 1995); Kirkpatrick Sale, The Green Revolution: The American Environmental Movement, 1962–1992 (New York: Hill and Wang, 1993); Robert Gottlieb, Forcing the Spring: The Transformation of the American Environmental Movement (Washington, D.C.: Island Press, 1993).

70. Kathie Durbin, Tree Huggers: Victory, Defeat and Renewal in the Northwest Ancient Forest Campaign 209–15 (Seattle: The Mountaineers, 1996); Dowie, supra note 69, at 212–214; see Chapter 4 for a full description of the spotted owl controversy and accompanying litigation; see Chapter 5 for a description of how environmental organizations also split over litigation strategy on wolf reintroduction. Beginning in the early 1980s, a schism also developed within the environmental community over appropriate political strategies: the long-standing, traditional organizations generally pursued conventional political and social change tactics (i.e., lobbying, litigation, and the like), while a new breed of environmentalists—personified by Earth First!—adopted more confrontational civil disobedience tactics (i.e., sit-ins, demonstrations, and the like) as well as other more aggressive direct action tactics (i.e., monkeywrenching, ecotage, tree spiking, and industrial sabotage) to challenge authority and force policy changes. This book addresses the evolution of ecological policy primarily from the perspective of conventional political change channels; for additional information on the direct action environmental movement, see Christopher Manes, Green Rage: Radical Environmentalism and the Unmaking of Civilization (Boston: Little Brown, 1990); Sale, supra note 69, at 47–69; Gottlieb, supra note 69.

71. On the Muir radical-amateur tradition, see Stephen Fox, John Muir and His Legacy: The American Conservation Movement 333–57 (Boston: Little, Brown, 1981). Critics of the environmental movement's Washington-based strategy include Dowie, supra note 69; Gottlieb, supra note 69. The point about reconnecting with rural working-class constituencies is developed effectively by Donald Snow, who observes that "the central problem with environmentalism is that it lacks a cogent, convincing focus on livelihood, and that has made it vulnerable to wise use attacks." Donald Snow, The Pristine Silence of Leaving It All Alone, in Brick and Cawley, supra note 66, at 27, 37; see also Philip D. Brick, Taking Back the Rural West, in Echeverria and Eby, supra note 25, at 61.

72. On the Defenders of Wildlife wolf compensation fund, see Hank Fischer, Wolf Wars: The Remarkable Inside Story of the Restoration of Wolves to Yellowstone 101–3, 114 (Helena,

Mt.: Falcon Press, 1995). For a brief overview of alternative political strategies available to the modern environmental movement, see Gus diZerega, Environmentalists and the New Political Climate: Strategies for the Future, in Brick and Cawley, supra note 66, at 107. See also Chapter 7 for a critical analysis of collaborative, community-based conservation strategies.

73. On the Bitterroot grizzly bear reintroduction effort, see U.S. Department of the Interior, Fish and Wildlife Service, Bear Recovery in the Bitterroot Ecosystem Final Environmental Impact Statement (Missoula, 2000); Sarah B. van de Wetering, Bears, People, Power, 2(2) Chronicle of Community 15 (1998). On new rangeland collaborative efforts, see Teresa Jordan, The Truth of the Land, in Robert B. Keiter, Reclaiming the Native Home of Hope: Community, Ecology and the American West 43 (Salt Lake City: University of Utah Press, 1998); William E. Riebsame and Robert G. Woodmansee, Mapping Common Ground on Public Rangelands, in Echeverria and Eby, supra note 25, at 9; Daggett, supra note 68.

THREE
Ecology and the Public Domain

1. Frederic E. Clements, Bio-ecology (New York: Wiley, 1939); Frederic E. Clements, Plant Succession: An Analysis of the Development of Vegetation (Washington, D.C.: Carnegie Institute of Washington, 1916). For a general discussion of Clements's contributions to ecological science, see Frank B. Golley, A History of the Ecosystem Concept in Ecology: More than the Sum of the Parts 11–27, 201 (New Haven: Yale University Press, 1993).

2. A. G. Tansley, The Use and Abuse of Vegetational Concepts and Terms, 16 Ecology 284, 306 (1935).

3. For a general historical overview of the ecosystem concept from a scientific and social-cultural perspective, see Golley, supra note 1. For additional historical perspective on the evolution of ecology as a science, see Joel B. Hagen, An Entangled Bank: The Origins of Ecosystem Ecology (New Brunswick, N.J.: Rutgers University Press, 1992); Donald Worster, Nature's Economy: A History of Ecological Ideas (New York: Cambridge University Press, 1985).

4. Golley, supra note 1, at 60 (explaining Raymond L. Lindemann, The Trophic-Dynamic Aspect of Ecology, 23 Ecology 399–418 (1942)); see also Hagen, supra note 3, at 87–99.

5. Hagen, supra note 3, at 99.

6. Golley, supra note 1, at 69; see also Hagen, supra note 3, at 126. For a general description of Odums's ecological principles, see Golley, id. at 62–69.

7. The Facts on File Dictionary of Biology 97 (1988); see also Daniel B. Botkin, Discordant Harmonies: A New Ecology for the Twenty-first Century 230 (New York: Oxford University Press, 1990) (defining an ecosystem as "a set of interacting species and their local, nonbiological environment, functioning together to sustain life"). The nonliving or inorganic components of an ecosystem consist of such materials as air, water, and soil.

8. Edward O. Wilson, The Biological Diversity Crisis, 35 BioScience 700, 703 (1985); see also Edward O. Wilson, The Diversity of Life (Cambridge, Mass.: Harvard University Press, 1992); Paul Ehrlich and Ann Ehrlich, Extinction: Causes and Consequences of the Disappearances of Species (New York: Random House, 1981); O. H. Frankel and Michael Soule, Conservation and Evolution 22–30 (New York: Cambridge University Press, 1981).

9. On our evolving conception of ecological change, see Botkin, supra note 7, at 188–92;

S. T. A. Pickett and Richard S. Ostfeld, The Shifting Paradigm in Ecology, in Richard L. Knight and Sarah F. Bates, A New Century for Natural Resources Management 261 (Washington, D.C.: Island Press, 1995); Worster, The Wealth of Nature, supra note 3, at 156–70. On the connection between ecosystems and species conservation, see O. H. Frankel, The Place of Management in Conservation, in Christine M. Schonewald-Cox et al., eds., Genetics and Conservation: A Reference for Managing Wild Animal and Plant Populations 2 (Reading, Mass.: Benjamin/Cummings, 1983).

10. On island biogeography, see Robert H. McArthur and Edward O. Wilson, The Theory of Island Biogeography (Princeton, N.J.: Princeton University Press, 1967); David Quammen, The Song of the Dodo: Island Biogeography in an Age of Extinctions (New York: Scribners, 1996). The science of species extinctions is explored more fully in: Raymond F. Dasmann, Wildlife Biology 169–74 (New York: Wiley, 2d edition, 1981); Larry D. Harris, The Fragmented Forest: Island Biogeography Theory and the Preservation of Biotic Diversity (Chicago: University of Chicago Press, 1984); John Terborgh et al., The Role of Top Carnivores in Regulating Terrestrial Ecosystems, in Michael E. Soule and John Terborgh, eds., Continental Conservation: Scientific Foundations of Regional Reserve Networks 39–64 (Washington, D.C.: Island Press, 1999). On the loss of species in U.S. national parks, see William D. Newmark, Legal and Biotic Boundaries of Western North American National Parks: A Problem of Congruence, 33 Biological Conservation 197–208 (1985); William D. Newmark, Extinction of Mammal Populations in Western North American National Parks, 9(3) Conservation Biology 512–26 (1995); Harris, id. at 72 (noting species losses at Mt. Rainier National Park).

11. On the nonequilibrium concept of ecology and its consequences for nature conservation, see Pickett and Ostfeld, supra note 9, at 261; Worster, The Wealth of Nature, supra note 3, at 156–70; N. Christensen et al., The Report of the Ecological Society of America Committee on the Scientific Basis for Ecosystem Management, 6(3) Ecological Applications 665, 672–73 (1996); Nels C. Johnson et al., eds., Ecological Stewardship: A Common Reference for Ecosystem Management, vol. 1, 33–35 (Oxford: Elsevier Science, 1999) (hereafter, Ecological Stewardship); Reed F. Noss and Alan Y. Cooperrider, Saving Nature's Legacy: Protecting and Restoring Biodiversity 43–49, 165–67 (Washington, D.C.: Island Press, 1994); see also Botkin, supra note 7. On the principles and practice of adaptive management, see Kai N. Lee, Compass and Gyroscope: Integrating Science and Politics for the Environment (Washington, D.C.: Island Press, 1993); Carl J. Walters, Adaptive Management of Renewable Resources (New York: Macmillan, 1986); see also Paul L. Ringold, Adaptive Monitoring Design for Ecosystem Management, 6(3) Ecological Applications 745 (1996); Gene Lessard, An Adaptive Approach to Planning and Decision-making, 40 Landscape and Urban Planning 81 (1998).

12. Office of Technology Assessment, Technologies to Maintain Biological Diversity 3 (Washington, D.C., 1987); see also The Keystone Center, Final Consensus Report on Biological Diversity on Federal Lands (Keystone, Colo., 1991).

13. For a more detailed description of the hierarchical dimensions of biodiversity, see National Research Council, Perspectives on Biodiversity: Valuing Its Role in an Ever Changing World 20–42 (Washington, D.C.: National Academy Press, 1999); Noss and Cooperrider, supra note 11, at 3–13; R. Edward Grumbine, Ghost Bears: Exploring the Biodiversity Crisis 22–28 (Washington, D.C.: Island Press, 1992). Ecosystem complexity is discussed in Christensen et al., supra note 11, at 671–72; Ariel E. Lugo et al., Ecosystem Processes and Functioning, in Ecological Stewardship, vol. 2, supra note 11, at 219. On the filter approach

to conservation, see Noss and Cooperrider, id. at 105–7; David W. Crumpacker, Prospects for Sustainability of Biodiversity Based on Conservation Biology and US Forest Service Approaches to Ecosystem Management, 40 Landscape and Urban Planning 47, 54–57 (1998); Malcolm L. Hunter, Jr., Coping with Ignorance: The Coarse-Filter Strategy for Maintaining Biodiversity, in Kathryn A. Kohm, ed., Balancing on the Brink of Extinction: The Endangered Species Act and Lessons for the Future 266–81 (Washington, D.C.: Island Press, 1991); J. Michael Scott et al., Gap Analysis: A Geographic Approach to Protection of Biological Diversity, 123 Wildlife Monographs 1, 7–9 (1993). On the "hot spots" concept, see National Research Council, id. at 27–28; Noss and Cooperrider, id. at 100. An "endemic species" is one native to a particular place and found only in that location; the term "endemism" is defined as "the relative number of endemic species found within a particular geographic area or region." Id. at 392.

14. Michael Soule and Daniel Simberloff, What Do Genetics and Ecology Tell Us about the Design of Nature Reserves, 35(1) Biological Conservation 19 (1986); William L. Baker, The Landscape Ecology of Large Disturbances in the Design and Management of Nature Reserves, 7(3) Landscape Ecology 181 (1992).

15. For further discussion of the "fine filter" strategy concept, see Noss and Cooperrider, supra note 11, at 101–2; Hann et al., Landscape Management Assessment Using Hierarchical Principles of Landscape Ecology, in M. E. Jensen and P. S. F. Bourgeron, eds., Ecosystem Management: Principles and Applications, vol. II, p. 285 (Portland, Ore.: U.S. Forest Service Pacific Northwest Research Station, 1994).

16. See Mark S. Boyce, Population Viability Analysis: Adaptive Management for Threatened and Endangered Species, in Mark S. Boyce and Alan Haney, eds., Ecosystem Management: Applications for Sustainable Forest and Wildlife Resources 226 (New Haven: Yale University Press, 1997); Frank D'Erchia, Geographical Information Systems and Remote Sensing Applications for Ecosystem Management, in Ecosystem Management, id. at 201, 217–21; William T. Sexton et al., Implementing Ecosystem Management: A Framework for Remotely Sensed Information at Multiple Scales, 40 Landscape and Urban Planning 173 (1998); see also Remote Sensing Articles in 98(6) Journal of Forestry 4–63 (June 2000).

17. See J. Lubchenco et al., The Sustainable Biosphere Initiative: An Ecological Research Agenda, 72(2) Ecology 371, 373 (1991); N. Christensen et al., supra note 11; Society of American Foresters, Task Force Report on Sustaining Long-term Forest Health and Productivity xiii (Bethesda, Md., 1993).

18. See Michael E. Soule, What Is Conservation Biology?, 35 (11) BioScience 727 (December 1985); Reed F. Noss, Some Principles of Conservation Biology, as They Apply to Environmental Law, 69 Chi-Kent L. Rev. 893, 895 (1994).

19. See Roderick F. Nash, The Rights of Nature: A History of Environmental Ethics (Madison: University of Wisconsin Press, 1989). Nash's book provides a detailed historical account of the gradual extension of philosophical thinking to the realm of the environment, including descriptions of the primary individuals and ideas that have influenced this development.

20. Lynn White, Jr., The Historical Roots of Our Ecological Crisis, 155 Science 1203 (1967); Keith Thomas, Man and the Natural World: A History of the Modern Sensibility 17–50 (New York: Pantheon, 1983). See also Nash, Rights of Nature, supra note 19, at 87–92.

21. See Nash, Rights of Nature, supra note 19, at 23–43; Charles Darwin, The Origin of Species: By Means of Natural Selection, or, The Preservation of Favored Races in the Struggle for Life (New York: Modern Library, 1993). Biographies examining Darwin's life and his

contribution to the science of evolution include Michael White and John Gribbin, Darwin: A Life in Science (New York: Dutton, 1995); L. Robert Stevens, Charles Darwin (Boston: Twayne Publishers, 1978); Gertrude Himmelfarb, Darwin and the Darwinian Revolution (Garden City, N.Y.: Anchor Books, 1962).

22. The Marsh quotations are found at George Perkins Marsh, Man and Nature; Or, Physical Geography as Modified by Human Action 40, 36 (Cambridge, Mass.: Harvard University Press, 1965).

23. For a brief description of Muir's views on the human relationship to nature, see Nash, Rights of Nature, supra note 19, at 39–41; see also Stephen R. Fox, The American Conservation Movement: John Muir and His Legacy (Madison: University of Wisconsin Press, 1985); Frederick W. Turner, Rediscovering America: John Muir in His Time and Ours (New York: Viking, 1985); Michael W. Cohen, The Pathless Way: John Muir and the American Wilderness (Madison: University of Wisconsin Press 1984).

24. Worster, Nature's Economy, supra note 3, at 316–338.

25. The principal Leopold biographies are Curt Meine, Aldo Leopold: His Life and Work (Madison: University of Wisconsin Press, 1988); Susan Flader, Thinking Like a Mountain: Aldo Leopold and the Evolution of an Ecological Attitude toward Deer, Wolves, and Forests (Columbia: University of Missouri Press, 1974); see also Eric Freyfogle, The Land Ethic and Pilgrim Leopold, 61 U. Colo. L. Rev. 217 (1990). For compilations and analyses of Leopold's essays, see Susan L. Flader and J. Baird Callicott, eds., River of the Mother of God and Other Essays by Aldo Leopold (Madison: University of Wisconsin Press, 1991); J. Baird Callicott and Eric T. Freyfogle, eds., Aldo Leopold for the Health of the Land: Previously Unpublished Essays and Other Writings (Washington, D.C.: Island Press, 1999); Curt Meine and Richard L. Knight, eds., The Essential Aldo Leopold: Quotations and Commentaries (Madison: University of Wisconsin Press, 1999).

26. The quoted passages can be found in Aldo Leopold, A Sand County Almanac with Essays on Conservation from Round River (New York: Ballantine, 1966); they are from Leopold's "Round River," "The Land Pyramid," and "Thinking Like a Mountain" essays. See also Aldo Leopold, A Sand County Almanac and Sketches Here and There (New York: Oxford University Press, 1949); Luna Leopold, Round River: From the Journals of Aldo Leopold (New York: Oxford University Press, 1953).

27. The quotations are from Leopold's "Land Ethic" essay, which can be found in the Ballantine version of A Sand County Almanac, id., at 237–64. On the idea of a biocentric ethic, see Holmes Rolston III, Philosophy Gone Wild (Buffalo, N.Y.: Prometheus Books, 1986); Max Oelschlaeger, The Idea of Wilderness: From Prehistory to the Age of Ecology (New Haven: Yale University Press, 1991); Lewis P. Hinchman, Aldo Leopold's Hermeneutic of Nature, 57 The Review of Politics 225 (1995).

28. J. Baird Callicott, The Scientific Substance of the Land Ethic, in Thomas Tanner, ed., Aldo Leopold: The Man and His Legacy 87 (Ankeny, Iowa: Soil Conservation Society of America, 1987) (emphasis in original); see also Flader, Thinking Like a Mountain, supra note 25, at 5.

29. The quotations are from Rachel Carson, Silent Spring 5–6, 296–97 (Boston: Houghton Mifflin, 1962). Biographical information about Rachel Carson and her work is available in Linda Lear, Rachel Carson: Witness for Nature (New York: Henry Holt, 1997); Mary A. McCay, Rachel Carson (New York: Twayne Publishers, 1993); see also Nash, The Rights of Nature, supra note 19, at 82; Yaakov Garb, Rachel Carson's Silent Spring, Dissent 539 (Fall 1995).

30. See Christopher D. Stone, Should Trees Have Standing? Toward Legal Rights for Natural Objects, 45 S. Calif. L. Rev. 450 (1972); Christopher D. Stone, Should Trees Have Standing? Toward Legal Rights for Natural Objects (Los Altos, Calif.: W. Kaufman, 1974) (subsequently published book version); Sierra Club v. Morton, 405 U.S. 727, 741–42 (1972) (J. Douglas, concurring opinion); Bill Devall and George Sessions, Deep Ecology 67 (Salt Lake City, Utah: Gibbs Smith, 1985); William D. Newmark, Selected Contributions of Aldo Leopold to Conservation Biology, 19 J. Land, Resources, and Envir. L. 211 (1999); Eric Katz, The Traditional Ethics of Natural Resources Management, in Knight and Bates, supra note 9, at 101.

31. For philosophical arguments justifying biodiversity conservation, see National Research Council, supra note 13, at 43–86; Stephen R. Kellert, The Value of Life: Biological Diversity and Human Society 9–26 (Washington, D.C.: Island Press, 1996); Bryan G. Norton, Why Preserve Natural Variety? 3–22 (Princeton, N.J.: Princeton University Press, 1987); Holmes Rolston III, supra note 27; Oelschlaeger, supra note 27, at 292–301. The 25 percent figure comes from Office of Technology Assessment, Technologies to Maintain Biological Diversity 6 (1987). The economic valuation arguments for ecosystems are made in Robert Costanza et al., The Value of the World's Ecosystem Services and Natural Capital, 387 Nature 253 (1997); Gretchen P. Daily, ed., Nature's Services: Societal Dependence on Natural Ecosystems (Washington, D.C.: Island Press, 1997); Symposium, 20 Stanford Envir. L. J. 309–536 (2001); but see National Research Council, supra note 13, at 87–117.

32. See Hans Huth, Nature and the American: Three Centuries of Changing Attitudes (Lincoln: University of Nebraska Press, Bison Edition, 1990); Alfred Runte, National Parks: The American Experience (Lincoln: University of Nebraska Press, 2d edition, 1987); Thomas Lund, American Wildlife Law (Berkeley: University of California Press, 1980); Michael Bean, The Evolution of National Wildlife Law 89–98, 262–67 (New York: Praeger, 1983); Norton, supra note 31, at 98–118; see also Jack Turner, The Abstract Wild (Tucson: University of Arizona Press, 1996); Joseph Sax, Mountains without Handrails: Reflections on the National Parks (Ann Arbor: University of Michigan Press, 1980).

33. Aldo Leopold, A Sand County Almanac, supra note 26, at 139, 240.

34. On animal rights, see Kellert, supra note 31, at 22–24; Norton, supra note 31, at 151–82; Nash, Rights of Nature, supra note 19, at 146–52; Lisa Mighetto, Wild Animals and American Environmental Ethics (Tucson: University of Arizona Press, 1991). On the Native American relationship to nature, see Klara Bonsack Kelley and Harris Francis, Navajo Sacred Places 97–100 (Bloomington: Indiana University Press, 1994); Deward E. Walker, Jr., Blood of the Monster: The Nez Perce Coyote Cycle (Worland, Wyo.: High Plains Publishing Co., 1994); B. L. Driver et al., eds., Nature and the Human Spirit: Toward an Expanded Land Management Ethic (State College, Pa.: Venture Publishing, 1996). For examples of statutory damages provisions, see, e.g., Comprehensive Environmental Response, Compensation, and Liability Act of 1980, 42 U.S.C. § 9607(a)(4)(C); Clean Water Act, 33 U.S.C. § 1321(f)(4); Oil Pollution Act of 1990, 33 U.S.C. § 2702(b)(2)(A); 43 C.F.R. § 11.83 (2000); 15 C.F.R. §§ 990.10–990.66 (2000); General Electric Co. v. U.S. Department of Commerce, 128 F.3d 767 (D.C. Cir. 1997); Ohio v. U.S. Department of the Interior, 880 F.2d 432 (D.C. Cir. 1989).

35. Samuel P. Hays, Beauty, Health, and Permanence: Environmental Politics in the United States, 1955–1985 3–4, 26–34, 88, 111 (New York: Cambridge University Press, 1987); see also Brent S. Steel and Nicholas P. Lovrich, An Introduction to Natural Resource Policy and the Environment: Changing Paradigms and Values, in Brent S. Steel, ed., Public Lands

Management in the West: Citizens, Interest Groups, and Values 3–15 (Westport, Conn.: Praeger, 1997).

36. Charles F. Wilkinson, Fire on the Plateau: Conflict and Endurance in the American Southwest xii, 178–229 (Washington, D.C.: Island Press, 1999). See also Peter Wiley and Robert Gottlieb, Empires in the Sun: The Rise of the New American West (Tucson: University of Arizona Press, 1985); William Dietrich, Northwest Passage: The Great Columbia River (New York: Simon and Schuster, 1995); Richard D. Lamm and Michael McCarthy, The Angry West: A Vulnerable Land and Its Future 122–23 (Boston: Houghton Mifflin, 1982).

37. On the economic history of the American West, see Richard White, "It's Your Misfortune and None of My Own": A History of the American West 236–69 (Norman: University of Oklahoma Press, 1991); William G. Robbins, Colony and Empire: The Capitalist Transformation of the American West (Lawrence: University Press of Kansas, 1994). On the fading utility of the multiple-use concept, see Michael Blumm, Public Choice Theory and the Public Lands: Why "Multiple Use" Failed, 18 Harv. Env. L. Rev. 405 (1994); Note, Managing Federal Lands: Replacing the Multiple-Use System, 82 Yale L. J. 787 (1973); see also Charles F. Wilkinson, Crossing the Next Meridian: Land, Water, and the Future of the West (Washington, D.C.: Island Press, 1992).

38. See Jan G. Laitos and Thomas A. Carr, The Transformation on Public Lands, 26 Ecology L. Q. 140, 167–72 (1999); Kerri S. Smith, Global Mining Helps Local Firms, Denver Post, Feb. 2, 1997, at D9:1; James Brooke, U.S. Gold Mining Companies Increasingly Look Abroad, New York Times, Aug. 13, 1996; James Brooke, For U.S. Miners, The Rush Is On to Latin America, New York Times, April 17, 1994, at C9. To be sure, the world is less stable in the aftermath of the September 11, 2001, tragedy, but that has not significantly altered U.S. corporate interest in international business opportunities. It has, however, prompted more interest in U.S. domestic energy sources, including those located on public lands.

39. U.S. Census Bureau, Statistical Abstract of the United States: 1999 28 (Population) (1999) (noting population trends); id. at 7 (noting state growth rates). On western urbanization, see Christopher McGrory Klyza, Reform at a Geological Pace: Mining Policy on the Federal Lands, 1964–1994, in Charles E. Davis, ed., Western Public Lands and Environmental Politics 103 (Boulder, Colo.: Westview, 1997); see also Lamm and McCarthy, supra note 36, at 117–33. On growth in the Yellowstone region, see Dennis Glick et al., eds., An Environmental Profile of the Greater Yellowstone Ecosystem 92 (Bozeman, Mt.: Greater Yellowstone Coalition, 1991); on Colorado Plateau growth, see Walter E. Hecox, Charting the Colorado Plateau: An Economic and Demographic Exploration 23–27 (Flagstaff, Ariz.: Grand Canyon Trust, 1996); Richard L. Knight, Field Report from the New American West, in Curt Meine, ed., Wallace Stegner and the Continental Vision: Essays on Literature, History, and Landscape 183 (Washington, D.C.: Island Press, 1997).

40. For a concise summary of the socioeconomic changes affecting the post–World War Two West and the resulting shift in values concerning the West's public lands, see Steel and Lovrich, supra note 35, at 3–15. The "new West" concept is explored in John A. Baden and Donald Snow, eds., The Next West: Public Lands, Community, and Economy in the American West (Washington, D.C.: Island Press, 1997); The Changing Needs of the West, Oversight Hearing before the Committee on Natural Resources, House of Representatives, 103d Cong., 2d Sess., Serial No. 103–80 (Washington, D.C.: U.S. Government Printing Office, 1994); Richard White, supra note 37, at 541–73; William E. Riebsame, ed., Atlas of the New West: Portrait of a Changing Region (New York: Norton, 1997). The West's evolving economy is examined in Raymond Rasker, A New Look at Old Vistas: The Economic

Role of Environmental Quality in Western Public Lands, 65 U. Colo. L. Rev. 369, 37786 (1994); see also Thomas M. Power and Richard Barrett, Cowboy Economics: Pay and Prosperity in the New American West (Washington, D.C.: Island Press, 2001); Thomas M. Power, Lost Landscapes and Failed Economies: The Search for a Value of Place (Washington, D.C.: Island Press, 1996).

41. On general trends in traditional extractive uses on public lands, see Laitos and Carr, supra note 38, at 15062; John D. Leshy, Public Lands at the Millennium, 46 Rocky Mt. Min. L. Inst. 1-1 (2000); Annual Energy Review, Fossil Fuel Production on Federally Administered Lands, at Table 1.14 (1998). On the issue of federal subsidies, see John A. Baden, The True-Mann's West: Endangered and Forsaken, in Baden and Snow, supra note 40, at 107; Dyan Zaslowsky and T. H. Watkins, These American Lands: Parks, Wilderness, and the Public Lands 137–41 (Washington, D.C.: Island Press, 1994).

42. See Hays, supra note 35, at 110–17; Samuel T. Dana and Sally K. Fairfax, Forest and Range Policy: Its Development in the United States, 179–206 (New York: McGraw-Hill, 2d edition, 1980); Steel and Lovrich, supra note 35, at 7–14; Laitos and Carr, supra note 38, at 143–66; see also BLM Shifting Role Towards Recreation, Preservation, 19(5) Land Letter 1 (March 28, 1999); H. Ken Cordell et al., Outdoor Recreation in American Life: A National Assessment of Demand and Supply Trends (1999).

43. John Hunt, Tourism on Our Terms: Tourism in the Western United States 14 (Denver, Colo.: Western Governors' Association, 1991). But for more disquieting accounts of tourism's impact on western communities, see Hal K. Rothman, Devil's Bargains: Tourism in the Twentieth Century West (Lawrence: University Press of Kansas, 1998); Scott Norris, ed., Discovered Country: Tourism and Survival in the American West (Albuquerque, N.M.: Stone Ladder Press, 1994); Power, Lost Landscapes, supra note 40, at 214–30.

44. The private employer statistic is from Klyza, in Western Public Lands and Environmental Politics, supra note 39, at 106; the "few opportunities" quotation is from Patrick T. Long, Tourism on Our Terms: Rural Community Tourism Development Impacts and Policies 2 (Denver, Colo.: Western Governors' Association, 1991); see also David I. Kass and Sumiye Okubo, U.S. Travel and Tourism Satellite Accounts for 1996 and 1997, Survey of Current Business 8 (July 2000) (noting substantial growth in the tourism and travel sectors, while also noting the difficulty in computing actual expenditures and employment in these industries).

45. For accounts of local community transformations related to tourism and recreation, see Rasker, supra note 40, at 380–84; Tara Gunter, Beyond Green and Brown, 4(1) Chronicle of Community 5 (2000); Jim Howe et al., Balancing Nature and Commerce in Gateway Communities (Washington, D.C.: Island Press, 1997). On public support for the environment, see Steele and Lovrich, supra note 35, at 7–14; Philip Shabecoff, A Fierce Green Fire: The American Environmental Movement, 247–48 (New York: Hill and Wang, 1993).

46. On western rural communities and local attitudes, see U.S. Department of the Interior, Bureau of Land Management, Economic and Social Conditions of Communities, ICBEMP (Portland, Ore., 1998); Dorothy Nelson, Rural Life and Social Change in the Modern West, in R. Douglas Hurt, ed., The Rural West Since World War II 38–57 (Lawrence: University Press of Kansas, 1998). On urban versus rural values, see Gundars Rudzitis, Amenities Increasingly Draw People to the Rural West, 14(2) Rural Development Perspectives 9 (1999); Grand Canyon Trust, Beyond the Boundaries: The Human and Natural Communities of the Greater Grand Canyon 1819 (1997); but see Leslie R. Alm and Stephanie Witt, The Rural-Urban Environmental Conflict in the American West: A Four-State Study, 69(4) Spectrum: The Journal of State Government 26 (1996).

47. On tribal fishing rights on the Columbia River, see The Boldt Decision, 384 F. Supp. 312 (W.D. Wash. 1974), aff'd, 520 F.2d 676 (9th Cir. 1985), cert. denied, 423 U.S. 1086 (1976); United States v. Washington, 459 F. Supp. 1020 (W.D. Wash. 1978), aff'd sub nom. Puget Sound Gillnetters Association v. United States Dist. Court, 573 F.2d 1123 (9th Cir. 1987), aff'd in part, vacated in part, and remanded sub nom. Washington v. Washington State Commercial Passenger Fishing Vessel Association, 443 U.S. 658 (1979). On Blackfeet treaty rights on the ceded strip, see Agreement with the Indians of the Blackfeet Indian Reservation in Montana, Sept. 26, 1895, ch. 398, § 9, 29 Stat. 321, 353–54 (1896); Jay Hansford C. Vest, Traditional Blackfeet Religion and the Sacred Badger–Two Medicine Wildlands, 6 J. of Law and Religion 455 (1988); Walt Borges, Law Is the Weapon in Tribal War Against Drilling; Legal Grounds v. Sacred Grounds, 17 (19) Legal Times, Sept. 26, 1994, at p. 7. On tribal rights to sacred sites, see Bear Lodge Multiple Use Association v. Babbitt, 175 F.3d 814 (10th Cir. 1999); Sandra B. Zellmer, Sustaining Geographies of Hope: Cultural Resources on Public Lands, 73 U. Colo. L. Rev. 413 (2002). On tribal authority to establish environmental standards, see City of Albuquerque v. Browner, 97 F.3d 415 (10th Cir. 1996); Arizona Public Service Co. v. Environmental Protection Agency, 211 F.3d 1280 (D.C. Cir. 2000); Bugenig v. Hoopa Valley Tribe, 266 F.3d 1201 (9th Cir. 2001); Joe W. Stuckey, Tribal Nations: Environmentally More Sovereign than States, 31 Envir. L. Rpt. News & Analysis 11198 (2001); Allison M. Dussias, Asserting a Traditional Environmental Ethic: Recent Developments in Environmental Regulation Involving Native American Tribes, 33 New Eng. L. Rev. 653 (1999).

48. For a discussion of the Redwood National Park controversy and the accompanying litigation, see Dale A. Hudson, Sierra Club v. Department of the Interior. The Fight to Preserve Redwood National Park, 7 Ecology L. Q. 781 (1979); Tom Turner, Wild by Law: The Sierra Club Legal Defense Fund and the Places It Has Saved 65 (San Francisco: Sierra Club Books, 1990). On the national parks' external threats problem, see Office of Science and Technology, National Park Service, U.S. Department of the Interior, State of the Parks 1980: A Report to Congress (1980); U.S. General Accounting Office, Parks and Recreation, Limited Progress Made in Documenting and Mitigating Threats to the Parks (Washington, D.C., 1987); John Freemuth, Islands Under Siege: National Parks and the Politics of External Threats (Lawrence: University of Kansas Press, 1991); David Simon, ed., Our Common Lands: Defending the National Parks (Washington, D.C.: Island Press, 1988). The national wildlife refuge external threats problem is documented and discussed in U.S. General Accounting Office, Wildlife Management: National Refuge Contamination Is Difficult to Confirm and Clean Up (Washington, D.C., 1987); Richard J. Fink, National Wildlife Refuges: Theory, Practice, and Prospect, 18 Harv. Env. L. Rev. 1, 77–82 (1994).

49. On Yellowstone region grizzly bear management policy, see Frank C. Craighead, Jr., Track of the Grizzly Bear (San Francisco: Sierra Club Books, 1979); see also Frank C. Craighead, Jr., et al., The Grizzly Bears of Yellowstone: Their Ecology in the Yellowstone Ecosystem (Washington, D.C.: Island Press, 1995); Thomas McNamee, The Grizzly Bear (New York: Knopf, 1982). For a historical overview of the Greater Yellowstone ecosystem concept, see Paul Schullery, Searching for Yellowstone: Ecology and Wonder in the Last Wilderness 197–207 (New York: Houghton Mifflin, 1997); Douglas B. Houston, Ecosystems of National Parks, 172 Science 648, 651 (1971); Douglas B. Houston, The Northern Yellowstone Elk 1–2, 196–97 (New York: Macmillan, 1982). Management of the Greater Yellowstone region as an ecological entity is discussed in: Greater Yellowstone Ecosystem: Oversight Hearing Before the Subcommittee on Public Lands and the Subcommittee on National

Parks and Recreation of the House Committee on Interior and Insular Affairs, 99th Cong., 1st Sess. 94–103 (1985); see also Rick Reese, Greater Yellowstone: The National Park and Adjacent Wildlands 55–99 (Helena: Montana Magazine, 1991); Robert B. Keiter and Mark S. Boyce, The Greater Yellowstone Ecosystem: Redefining America's Wilderness Heritage 3–18, 19–26 (New Haven: Yale University Press, 1991); Tim W. Clark and Steven C. Minta, Greater Yellowstone's Future: Prospects for Ecosystem Science, Management, and Policy 13–36 (Moose, Wyo.: Homestead Publishing, 1994).

50. Greater Yellowstone Coordinating Committee, Visions for the Future: A Framework for Coordination in the Greater Yellowstone Area. Draft (Billings, Mt.: National Park Service and U.S. Forest Service, 1990). For the critical report that triggered the so-called vision document, see Congressional Research Service, Library of Congress, Greater Yellowstone Ecosystem: An Analysis of Data Submitted by Federal and State Agencies, 99th Cong., 2d Sess. (Comm. Print No. 6, Dec. 1986).

51. Greater Yellowstone Coordinating Committee, A Framework for Coordination in the Greater Yellowstone Area (Billings, Mt.: National Park Service and U.S. Forest Service, 1991). For a critical analysis of the Greater Yellowstone vision document process, see Bruce Goldstein, Can Ecosystem Management Turn an Administrative Patchwork into a Greater Yellowstone Ecosystem?, 8 Northwest Envir. J. 285–324 (1992); Robert B. Keiter, Greater Yellowstone: Managing a Charismatic Ecosystem, 3 Utah State Univ. Nat. Res. and Envir. Issues 75–86 (1995); Pamela Lichtman and Tim W. Clark, Rethinking the "Vision" Exercise in the Greater Yellowstone Ecosystem, 7 Society and Natural Resources 459 (1994); John Freemuth and R. McGreggor Cawley, Science, Expertise and the Public: The Politics of Ecosystem Management in the Greater Yellowstone Ecosystem, 40 Landscape and Urban Planning 211 (1998).

52. On the Region One forest supervisors' letter, see Association of Forest Service Employees for Environmental Ethics, Inner Voice 11 (Winter 1990); Wilkinson, Crossing the Next Meridian, supra note 37, at 151–52. On AFSEEE, see Jeff DeBonis, Natural Resource Agencies: Questioning the Paradigm, in Knight and Bates, supra note 9, at 159. On the New Perspectives program, see Jerry F. Franklin, Creating a Forestry for the 21st Century: The Science of the Ecosystem Management (Washington, D.C.: Island Press, 1997); Winifred Kessler, New Perspectives for Sustainable Natural Resources Management, 2(3) Ecological Applications 221 (1992); James J. Kennedy and Thomas M. Quigley, Evolution of USDA Forest Service Organizational Culture and Adaptation Issues in Embracing an Ecosystem Management Paradigm, 40 Landscape and Urban Planning 113 (1998). On the Forest Service's shift to an ecosystem management policy, see Memorandum from Dale Robertson, U.S. Forest Service Chief to Regional Foresters (June 4, 1992); see also Crumpacker, supra note 13, at 50–51; Kennedy and Quigley, id. at 113.

53. On the Sagebrush Rebellion, see R. McGreggor Cawley, Federal Land, Western Anger: The Sagebrush Rebellion and Environmental Politics (Lawrence: University Press of Kansas, 1993); John D. Leshy, Unraveling the Sagebrush Rebellion: Law, Politics, and Federal Lands, 14 U.C. Davis L. Rev. 317 (1980). On the BLM grazing litigation, see Natural Resources Defense Council v. Hodel, 388 F. Supp. 829 (D.D.C. 1974), aff'd per curiam, 527 F.2d 1386 (D.C. Cir. 1976), cert. denied, 427 U.S. 913 (1976); but see Natural Resource Defense Council v. Hodel, 624 F. Supp. 1045 (D. Nev. 1985), aff'd, 819 F.2d 927 (9th Cir. 1987). On federal energy development policies, see Kleppe v. Sierra Club, 427 U.S. 390 (1976); Natural Resources Defense Council v. Berklund, 609 F.2d 553 (D.C. Cir. 1979); Conner v. Burford, 848 F.2d 1441 (9th Cir. 1988); National Wildlife Federation v. Burford, 871 F.2d 849

(9th Cir. 1989); see also Robert H. Nelson, The Making of Federal Coal Policy (Durham, N.C.: Duke University Press, 1983).

54. See Tennessee Valley Authority v. Hill, 437 U.S. 153 (1978); George C. Coggins and Irma Russell, Beyond Shooting Snail Darters in Pork Barrels: Endangered Species and Land Use in America 70 Geo. L. J. 1433 (1982); see also Catherine J. Tinker, Introduction to Biological Diversity: Law, Institutions, and Science, 1 Buffalo J. of International Law 1 (1994) for a description of the International Biodiversity Convention. Cases illustrating the U.S. Fish and Wildlife Service's regulatory power include: Thomas v. Peterson, 753 F.2d 754 (9th Cir. 1985); Sierra Club v. Yeutter, 926 F.2d 429 (5th Cir. 1991); Pacific Rivers Council v. Thomas, 873 F. Supp. 365 (D. Idaho, 1995).

55. See James K. Agee and Daryll R. Johnson, eds., Ecosystem Management for Parks and Wilderness (Seattle: University of Washington Press, 1988). On the biosphere reserve concept and its application in the U.S. public land context, see John D. Peine, ed., Ecosystem Management for Sustainability: Principles and Practices Illustrated by a Regional Biosphere Reserve Cooperative (Boca Raton, Fla.: Lewis Publishers, 1999); Vernon C. Gilbert, Cooperation in Ecosystem Management, in Agee and Johnson, id. at 180; Joseph L. Sax and Robert B. Keiter, Glacier National Park and Its Neighbors: A Study in Federal Interagency Relations, 14 Ecology L. Q. 207, 253 (1987).

56. On national park protection legislation, see Robert B. Keiter, On Protecting the National Parks from the External Threats Dilemma, 20 Land and Water L. Rev. 355, 396–408 (1985). The Pacific Northwest logging controversy is addressed in Chapter 4.

57. See Interagency Ecosystem Management Task Force, The Ecosystem Approach: Healthy Ecosystems and Sustainable Economies, vol. I, Overview 3 (Washington, D.C., 1995) (hereafter, The Ecosystem Approach); see also U.S. General Accounting Office, Ecosystem Management: Additional Actions Needed to Adequately Test a Promising Approach (Washington, D.C., 1994) (hereafter, GAO Ecosystem Management); Congressional Research Service, Ecosystem Management: Federal Agency Activities (Washington, D.C.: Library of Congress, 1994); Susan M. Stein and Diane Gelburd, Healthy Ecosystems and Sustainable Economies: The Federal Interagency Ecosystem Management Initiative, 40 Landscape and Urban Planning 73 (1998).

58. Hanna J. Cortner and Margaret A. Moote, The Politics of Ecosystem Management 40 (Washington, D.C.: Island Press, 1999); Christensen et al., supra note 11, at 669–70; Robert C. Szaro et al., The Emergence of Ecosystem Management as a Tool for Meeting People's Needs and Sustaining Ecosystems, 40 Landscape and Urban Planning 1, 3–4 (1998).

59. In order of appearance, the quoted definitions can be found in: Christensen et al., supra note 11, at 668–69; R. Edward Grumbine, What Is Ecosystem Management, 8 Conservation Biology 27, 31 (1994); Society of American Foresters, supra note 17, at iv–v; Robert Lackey, Seven Pillars of Ecosystem Management, 40 Landscape and Urban Planning 21, 29 (1998). For other ecosystem management definitions, see Agee and Johnson, supra note 55, at 7; Szaro et al., supra note 58, at 2–3; Crumpacker, supra note 13, at 48–49; Richard L. Knight, Ecosystem Management and Conservation Biology, 40 Landscape and Urban Planning 41, 44 (1998).

60. See The Ecosystem Approach, vol. 1, supra note 57, at 17. A compendium of these and other federal agency ecosystem management definitions can be found in Richard Haeuber, Ecosystem Management and Environmental Policy in the United States: Open Window or Closed Door?, 40 Landscape and Urban Planning 221, 224 (1998); see also Congressional Research Service, Ecosystem Management, supra note 57.

61. Several if not most of these six principles can be found in the following sources, where the authors define and examine the ecosystem management concept: Cortner and Moote, supra note 58, at 40–45; Christensen et al., supra note 11, at 670–82; The Ecosystem Approach, vol. 1, supra note 57, at 3, 6, 9–10, 46–47; Haeuber, supra note 60, at 222–25; Grumbine, Ecosystem Management, supra note 59, at 29–31; Lackey, supra note 59, at 28; Crumpacker, supra note 13, at 47, 48–49; Robert B. Keiter, Beyond the Boundary Line: Constructing a Law of Ecosystem Management, 65 U. Colo. L. Rev. 293, 301–2 (1994); Moote et al., Principles of Ecosystem Management 5 (Tucson: University of Arizona Water Resources Center, 1994); D. S. Slocomb, Environmental Planning, Ecosystem Science, and Ecosystem Approaches for Integrating Environment and Development, 17 Environmental Management 289, 297 (1993); Peter F. Brussard et al., Ecosystem Management: What Is It Really?, 40 Landscape and Urban Planning 9, 11–14 (1998); Gene Lessard, An Adaptive Approach to Planning and Decision-making, 40 Landscape and Urban Planning 81 (1998).

62. For a more detailed comparison between ecosystem management and traditional resource management approaches, see Cortner and Moote, supra note 58, at 37–38; Douglas W. MacCleery and Dennis C. LeMaster, The Historical Foundations and Evolving Context for Natural Resources Management on Federal Lands, in Ecological Stewardship, vol. 2, supra note 11, at 519–21; Kessler, supra note 52; see also Knight and Bates, supra note 9.

63. 16 U.S.C. § 1604(g)(3)(B); 36 C.F.R. §§ 219.19, 219.27(g) (1999); Seattle Audubon Society v. Evans, 952 F.2d 297, 301 (9th Cir. 1991) (relying on the NFMA biological diversity requirement to conclude that forest plans must consider the entire biological community); see also Charles F. Wilkinson and H. Michael Anderson, Land and Resource Planning in the National Forests, 64 Ore. L. Rev. 1, 296–304 (1985). Although FLPMA does not contain an equivalent biodiversity provision, the BLM has adhered to the Forest Service's biodiversity obligations in the Northwest Forest Plan, both to meet its ESA and other legal obligations and to avoid future legal challenges to its forest management plans. See U.S. Department of Agriculture, Forest Service, and U.S. Department of the Interior, Bureau of Land Management, Record of Decision for Amendments to Forest Service and Bureau of Land Management Planning Documents within the Range of the Northern Spotted Owl 26–28, 44 (April 13, 1994).

64. See 16 U.S.C. §§ 1604(b), (f)(3), (h)(1); 43 U.S.C. § 1712(c)(2) (interdisciplinary planning provisions);16 U.S.C. § 1604(d); 43 U.S.C. § 1712(f) (public participation provisions); 16 U.S.C. § 1604(a); 43 U.S.C. § 1712(c)(9) (transboundary coordination provisions).

65. See 16 U.S.C. § 1536(a)(2), § 1538(a)(1)(B); Coggins and Russell, supra note 54; Keiter, Beyond the Boundary Line, supra note 61, at 307–9; see also Oliver Houck, The Endangered Species Act and Its Implementation by the U.S. Departments of Interior and Commerce, 64 U. Colo. L. Rev. 277 (1993).

66. To ensure analysis of environmental impacts and public disclosure of the analysis, the NEPA requires preparation of an environmental impact statement (EIS) whenever a major federal action significantly affecting the quality of the human environment is contemplated. 42 U.S.C. § 4332(2)(c); see Robert B. Keiter, NEPA and the Emerging Concept of Ecosystem Management on the Public Lands, 25 Land and Water L. Rev. 43, 50–59 (1990). The NEPA procedural requirements have been rigorously enforced by the courts. See Neighbors of Cuddy Mountain v. U.S. Forest Service, 137 F.3d 1372 (9th Cir. 1998); Seattle Audubon Soc'y v. Espy, 998 F.2d 699 (9th Cir. 1993), affirming Seattle Audubon Soc. v. Moseley, 798 F. Supp. 1473 (W.D. Wash. 1992); Thomas v. Peterson, 753 F.2d 754 (9th Cir. 1985). See generally Symposium, Special Focus, Articles and Essays: NEPA at Twenty, 25

Land and Water L. Rev. 1 (1990); Symposium on NEPA at Twenty: The Past, Present and Future of the National Environmental Policy Act, 20 Envir. L. 447 (1990).

67. On NEPA and cumulative effects, see 40 C.F.R. §§ 1508.7, 1508.8(b), 1508.25(a) (2000); Neighbors of Cuddy Mountain v. U.S. Forest Service, 137 F.3d 1372 (9th Cir. 1998); Thomas v. Peterson, 753 F.2d 754 (9th Cir. 1985); Conner v. Burford, 848 F.2d 1441 (9th Cir. 1988); Thomas v. Peterson, 753 F.2d 754 (9th Cir. 1985); see also Keiter, NEPA, supra note 66, at 50–54. On NEPA and conservation biology, see Marble Mountain Audubon Society v. Rice, 914 F.2d 179 (9th Cir. 1990); Seattle Audubon Society v. Moseley, 798 F. Supp. 1473, 1483 (W.D. Wash. 1992); Dinah Bear, Using the National Environmental Policy Act to Protect Biological Diversity, 8 Tulane Envir. L. J. 77 (1984). On NEPA and interagency coordination see 42 U.S.C. § 4332(2)(C); 40 C.F.R. § 1501.7; 40 C.F.R. § 1502.16(c) (2000); Keiter, NEPA, supra note 66, at 47–50.

68. On the related topics of biodiversity conservation, ecosystem management, and the law, see William L. Snape III, ed., Biodiversity and the Law (Washington, D.C.: Island Press, 1996); Robert B. Keiter et al., Legal Perspectives on Ecosystem Management: Legitimizing a New Federal Land Management Policy, in Ecological Stewardship, vol. 3, supra note 11, at 9; Oliver Houck, On the Law of Biodiversity and Ecosystem Management, 81 Minn. L. Rev. 869 (1997); Keiter, Beyond the Boundary Line, supra note 61.

69. The spotted owl litigation reference is Seattle Audubon Society v. Lyons, 871 F.Supp. 1291, 1311 (D. Wash. 1994), aff'd sub nom., Seattle Audubon Society v. Moseley, 80 F.3d 1401 (9th Cir. 1996); see Chapter 4 for a detailed analysis of the spotted owl court decisions. The ESA reconsultation decision is: Pacific Rivers Council v. Thomas, 30 F.3d 1050 (9th Cir. 1994); see also Pacific Rivers Council v. Thomas, 873 F.Supp. 365 (D. Idaho 1995); Pacific Rivers Council v. Thomas, 897 F.Supp. 454 (D. Idaho 1995) (dissolving earlier injunction).

70. On evolving utilitarian notions concerning public land policy, see Bruce Babbitt, Address at the Annual Meeting of the Sierra Club (1985), in George C. Coggins et al., Federal Public Land and Resources Law 1080–81 (Mineola, N.Y.: Foundation Press, 3d edition 1993); Hays, supra note 35; Mark Sagoff, Where Ickes Went Right, Or Reason and Rationality in Environmental Law, 14 Ecology L. Q. 265 (1987). On the evolution in preservationist thought, see Runte, supra note 32, at 197–208; Richard W. Sellars, Preserving Nature in National Parks: A History (New Haven: Yale University Press, 1997). On the arguments for biodiversity conservation, see Mighetto, supra note 34; Nash, The Rights of Nature, supra note 19; Norton, supra note 31.

71. Useful discussions addressing how ecological management principles might be reconciled with existing law include: A. Dan Tarlock, The Non-equilibrium Paradigm in Ecology and the Partial Unraveling of Environmental Law, 27 Loyola L.A. L. Rev. 1121 (1994); Keiter, Beyond the Boundary Line, supra note 61; Joseph P. Sax, Property Rights and the Economy of Nature: Understanding Lucas v. South Carolina Coastal Council, 45 Stanford L. Rev. 1412 (1993); see also Fred P. Bosselmann and A. Dan Tarlock, eds., Symposium on Ecology and the Law, 69 Chicago-Kent L. Rev. 843–985 (1994).

72. See Allan K. Fitzsimmons, Sound Policy or Smoke and Mirrors: Does Ecosystem Management Make Sense?, 32(2) J. American Water Res. Ass'n 217, 220 (1996) (hereafter, Fitzsimmons, Sound Policy); Warren A. Flick and William E. King, Ecosystem Management as American Law, 13(3) Renewable Res. J. 6, 16 (1995); Lackey, supra note 59, at 26–27; GAO Ecosystem Management, supra note 57, at 37–39.

73. Allan K. Fitzsimmons, Why a Policy of Federal Management and Protection of Ecosystems Is a Bad Idea, 40 Landscape and Urban Planning 195, 196–98 (1998) (hereafter, Fitzsim-

mons, Bad Idea); Fitzsimmons, Sound Policy, supra note 72, at 218–19; Rebecca W. Thomson, Ecosystem Management: Great Idea, But What Is It, Will It Work, and Who Will Pay?, Natural Resources and the Environment 70–71 (Winter, 1995); Allen K. Fitzsimmons, Federal Ecosystem Management: A "Train Wreck" in the Making, 217 Cato Institute Policy Analysis 1–33 (Washington, D.C., 1994) (hereafter, Fitzsimmons, Train Wreck); GAO Ecosystem Management, supra note 57, at 57–59; Lackey, supra note 59, at 24.

74. See Ronald N. Johnson, Ecosystem Management and Reinventing Government, in Terry L. Anderson, ed., Breaking the Environmental Policy Gridlock 22, 24, 38–41 (Stanford, Calif.: Hoover Institution Press, 1996); Fitzsimmons, Train Wreck, supra note 73, at 198–99; Fitzsimmons, Sound Policy, supra note 72, at 222–24; Thomson, supra note 73, at 72; Lackey, supra note 59, at 27–28; Roger A. Sedjo, Toward an Operational Approach to Public Forest Management, 94(8) Journal of Forestry 24–27 (1996).

75. Fitzsimmons, Sound Policy, supra note 72, at 220–21 (citing Ed Grumbine, Protecting Biological Diversity through the Greater Ecosystem Concept, 10(3) Natural Areas J. 117 (1990).

76. On the scientific expertise point, see John Freemuth, The Emergence of Ecosystem Management: Reinterpreting the Gospel?, 9(4) Society and Natural Resources 411 (1996); Christopher M. Klyza, Who Controls Public Lands? Mining, Forestry, and Grazing Policies, 1870–1990, 148–49 (Chapel Hill: University of North Carolina Press, 1996). On the coordination point, see Daryll R. Johnson and James K. Agee, Introduction to Ecosystem Management, in Agee and Johnson, supra note 55, at 3, 7.

77. On the increasing tendency among anti-environmental forces to discount and even attack scientific theories, information, and data, see Paul Ehrlich and Ann Ehrlich, Betrayal of Science and Reason: How Anti-Environmental Rhetoric Threatens Our Future (Washington, D.C.: Island Press, 1996). On the limits of collaborative processes, see Michael McCloskey, The Skeptic: Collaboration Has Its Limits, 28(9) High Country News 7 (1995); George C. Coggins, Of Californicators, Quislings and Crazies, 2(2) Chronicle of Community 27 (1998). On the need for legal standards in ecosystem management, see Houck, Law of Biodiversity, supra note 68; Oliver Houck, Are Humans Part of Ecosystems, 28 Envir. L. 1 (1998).

78. See Flick and King, supra note 72, at 10–11; Sedjo, supra note 74, at 26–27; Wayne A. Morrissey, An Ecosystem-Based Approach to Managing America's Resources: A View from the U.S. Capitol Hill, 40 Landscape and Urban Planning 203, 205–7 (1998).

FOUR
Ecology Triumphant?

1. In 1990, approximately 14 million acres of national forest and BLM lands were identified as suitable for timber production. The Scientific Panel on Late-Successional Forest Ecosystems, Alternative for Management of Late-Successional Forests of the Pacific Northwest: A Report to the Agriculture Committee and The Merchant Marine and Fisheries Committee of the U.S. House of Representatives 33 (1991) (hereafter, 1991 Late-Successional Forests Report). The affected forests and BLM resource districts are noted in: U.S. Forest Service and Bureau of Land Management, Draft Supplemental Environmental Impact Statement on Management of Habitat for Late-Successional and Old-Growth Forest Related Species within the Range of the Northern Spotted Owl 2–4 to 2–6 (Portland, Ore., 1993) (hereafter, 1993 Draft SEIS on Old Growth Species).

2. The Forest Service's management obligations are set forth in the Multiple Use–Sustained Yield Act of 1960, 16 U.S.C. §§ 528–31; National Forest Management Act of 1976, 16 U.S.C. § 1604(e); see also Charles F. Wilkinson and H. Michael Anderson, Land and Resource Planning in the National Forests, 64 Ore. L. Rev. 1 (1984). The BLM is responsible for managing Oregon and California ("O and C") forestlands in southwestern Oregon; these lands, which total approximately 3 million acres, reverted to the federal government in the mid-1930s when the Oregon and California railroad went bankrupt. The BLM's management obligations are set forth in the Oregon and California Act of 1937, 43 U.S.C. § 1181a–j; Headwaters, Inc. v. BLM, Medford District, 914 F.2d 1174 (9th Cir. 1990).

3. See Wilkinson and Anderson, supra note 2, at 137–38; see also Paul W. Hirt, A Conspiracy of Optimism: Management of the National Forests since World War Two (Lincoln: University of Nebraska Press, 1994).

4. On the timber sale figures, see E. Thomas Tuchman et al., The Northwest Forest Plan: A Report to the President and Congress 101–2 (Portland, Ore.: U.S. Department of Agriculture, Office of Forestry and Economic Assistance, 1996) (hereafter, 1996 Forest Plan Report); Harriet H. Christensen et al., Atlas of Human Adaptation to Environmental Change, Challenge, and Opportunity: Northern California, Western Oregon, and Western Washington 1 (Portland, Ore.: U.S. Forest Service Pacific Northwest Research Station, Gen. Tech. Rpt. PNW-GTR-478, 2000). Not all the region's timber was destined for domestic consumption; approximately 3 million board feet of timber were shipped overseas annually from 1983 to 1990. With high timber prices prevailing on the international market during the late 1980s, major timber companies like Weyerhauser and Plum Creek exported much of their high-quality timber to Japan and other emerging Asian markets. But when this unprocessed timber was exported abroad, it deprived local mills of "value-added" processing opportunities, and thus had an indirect impact on regional timber mill employment levels. Congress finally intervened in 1990 to prohibit the export of unprocessed timber from public lands and to limit the export of timber sold from state and local government lands. But even then, as raw logs cut on state and private lands were exported to meet international demand, additional pressure was placed on the federal agencies to allow more cutting on the public lands to meet domestic consumption needs. See Forest Resources Conservation and Shortage Relief Act of 1990 as amended, 16 U.S.C. §§ 620–20(j); Board of Natural Resources v. Brown, 992 F.2d 937 (9th Cir. 1993); see also Steven L. Yaffee, The Wisdom of the Spotted Owl: Lessons for a New Century 161–62 (Washington, D.C.: Island Press, 1994) (hereafter, Yaffee, Wisdom); 1996 Forest Plan Report, id., at 209.

5. On the Forest Service's community stability policy, see Sustained Yield Forest Management Act of 1944, Act of March 29, 1944, ch. 146, 58 Stat. 132 (1944); Samuel T. Dana and Sally K. Fairfax, Forest and Range Policy 167 (New York: McGraw-Hill, 2d edition, 1980); Sarah F. Bates, Public Lands Communities: In Search of a Community of Values, 14 Public Land L. Rev. 81, 92–94 (1993); Con H. Schallu and Richard M. Alston, Principles of Decisionmaking: The Commitment to Community Stability: A Policy or Shibboleth?, 17 Envir. L. 429 (1987). On the sociology of rural logging communities, see Matthew S. Carroll, Community and the Northwestern Logger: Continuities and Change in the Era of the Spotted Owl (Boulder, Colo.: Westview, 1995); Robert G. Lee et al., eds., Community and Forestry: Continuities in the Sociology of Natural Resources (Boulder, Colo.: Westview, 1990).

6. See Elliott A. Norse, Ancient Forests of the Pacific Northwest 4 (Washington, D.C.: Island Press, 1990); Kathie Durbin, Tree Huggers: Victory, Defeat and Renewal in the Northwest Ancient Forest Campaign 23–24, 49 (Seattle, Wash.: The Mountaineers, 1996).

7. See Jerry Franklin et al., Ecological Characteristics of Old-Growth Douglas-Fir Forests (Portland, Ore.: U.S. Forest Service, Pacific Northwest Experiment Station, Gen. Tech. Rpt. PNW-118, 1981); Victor M. Sher and Andy Stahl, Spotted Owls, Ancient Forests, Courts, and Congress: An Overview of Citizens' Efforts to Protect Old-Growth Forests and the Species That Live in Them, 6 Northwest Envir. J. 361, 363 (1990).

8. For a description of the Northwest's old growth forests, see Norse, supra note 6, at 17–32; Durbin, Tree Huggers, supra note 6, at 49–53.

9. See Endangered and Threatened Wildlife and Plants: Determination of Threatened Status for the Northern Spotted Owl, 55 Fed. Reg. 26,114 (June 26, 1990); Interagency Scientific Committee to Address the Conservation of the Northern Spotted Owl, A Conservation Strategy for the Northern Spotted Owl 20, 197 (1990) (hereafter, ISC Report).

10. These recommendations were the product of the Oregon Threatened and Endangered Species Task Force, created in 1973 to help the state address endangered species issues and listings. The interagency task force consisted of biologists from the Oregon State Game Commission, Forest Service, BLM, U.S. Fish and Wildlife Service, and Oregon State University. The 300-acre set-aside recommendation represented the task force's best guess, based on limited scientific knowledge at the time, of how much habitat an owl pair requires. ISC Report, supra note 9, at 51–53; Yaffee, Wisdom, supra note 4, at 33–35, 58–82.

11. Steven L. Yaffee, Lessons about Leadership from the History of the Spotted Owl Controversy, 35 Nat. Res. J. 381, 390–92 (1995) (hereafter, Yaffee, Lessons); E. Charles Meslow, Spotted Owl Protection: Unintentional Evolution Toward Ecosystem Management, 10 (3 & 4) Endangered Species UPDATE 34, 35 (1993).

12. See U.S. Forest Service, Draft Supplement to the Environmental Impact Statement for an Amendment to the Pacific Northwest Regional Guide, Spotted Owl Guidelines S-7 to S-15 (Portland, Ore., 1986); U.S. Forest Service, Final Supplement to the Environmental Impact Statement for an Amendment to the Pacific Northwest Regional Guide, Spotted Owl Guidelines II-1 to II-6 (Portland, Ore., 1988); see also Yaffee, Lessons, supra note 11, at 395–97.

13. ISC Report, supra note 9, at 56.

14. The quotations and points noted can be found at ISC Report, supra note 9, at 47, 1, 3, 36–39, 18, 40–43, 45. The committee members were wildlife biologists or ecologists, employed by either the Forest Service, BLM, or U.S. Fish and Wildlife Service, though other scientists with state or university affiliations also worked closely with the committee. True to its scientific mission, the report set forth very specific quantitative guidelines to implement its conservation strategy, including HCAs designated to support 20 or more owl pairs, 7- to 12-mile spacing between HCAs, a 50–11–40 rule to govern connecting habitat, and no further logging in designated HCAs. Id. at 30. The so-called 50–11–40 rule was designed to address fragmentation concerns and to promote landscape connectivity between HCAs; it required that at least 50 percent of the forest landscape with a mean tree diameter at breast height of 11 inches and 40 percent of the canopy cover should be retained. Id. at 26–28. The recommended HCA strategy represented a major shift from the existing conservation strategy that focused on protecting much smaller and more fragmented areas for a few owls rather than larger, connected areas for 20 or more owl pairs. The existing Spotted Owl Habitat Area strategy protected only 1 to 3 pairs of owls on habitat ranging from 1,000 to 3,000 acres in size, while the proposed strategy sought to protect 15 to 20 pairs of owls on contiguous habitat that could extend over more than 50,000 acres. Id. at 17, 24–30.

15. See the Scientific Panel on Late-Successional Forest Ecosystems, Alternative for Manage-

ment of Late-Successional Forests of the Pacific Northwest: A Report to the Agriculture Committee and The Merchant Marine and Fisheries Committee of the U.S. House of Representatives 12 (1991). The Gang of Four consisted of: Norman Johnson, a forest management professor from Oregon State University; Jerry Franklin, Bloedel Professor of Ecosystem Analysis from the University of Washington; Jack Ward Thomas, chief research wildlife biologist with the Forest Service's Pacific Northwest Research Station and chair of the just completed Interagency Scientific Committee Report; and John Gordon, dean of the Yale University School of Forestry and Environmental Studies.

16. In 1984, Congress passed the Washington Wilderness Act of 1984, Pub. L. 98–339, 98 Stat. 302 (1984), and the Oregon Wilderness Act of 1984, Pub. L. 98–328, 98 Stat. 275 (1984), both of which designated large blocks of national forestland as wilderness in each state. See Durbin, Tree Huggers, supra note 6, at 54–63. On the potential impact of the NEPA, ESA, and NFMA, see Tennessee Valley Authority v. Hill, 437 U.S. 153 (1978); Thomas v. Peterson, 753 F.2d 754 (9th Cir. 1985); Wilkinson and Anderson, supra note 2, at 296.

17. See Tennessee Valley Authority v. Hill, 437 U.S. 153 (1978); George C. Coggins and Irma S. Russell, Beyond Shooting Snail Darters in Pork Barrels: Endangered Species and Land Use in America, 70 Geo. L. J. 1433 (1982).

18. See Endangered and Threatened Wildlife and Plants: Finding on Northern Spotted Owl Petition, 52 Fed. Reg. 48552, 48554 (Dec. 23, 1987); Northern Spotted Owl v. Hodel, 716 F.Supp. 479, 483 (W.D. Wash. 1988); Endangered and Threatened Wildlife and Plants: Determination of Threatened Status for the Northern Spotted Owl, 55 Fed. Reg. 26, 114 (June 26, 1990).

19. On the ESA's critical habitat designation provision, see 16 U.S.C. § 1533(a)(3); Thomas F. Darin, Designating Critical Habitat Under the Endangered Species Act: Habitat Protection versus Agency Discretion, 24 Harvard Envir. L. Rev. 209 (2000); Cynthia Yagerman, Protecting the Critical Habitat Under the Federal Endangered Species Act, 20 Envir. L. 811 (1990). On FWS's critical habitat designation process for the northern spotted owl, see Endangered and Threatened Wildlife and Plants: Determination of Threatened Status for the Northern Spotted Owl, 55 Fed. Reg. 26,114 (June 26, 1990); Northern Spotted Owl v. Lujan, 758 F.Supp. 621, 629 (W.D. Wash. 1991); U.S. Fish and Wildlife Service, Critical Habitat for the Northern Spotted Owl (1992). For the timber industry litigation challenging the critical habitat designation, see Douglas County v. Lujan, 810 F.Supp. 1470 (D. Ore. 1992), reversed sub nom., Douglas County v. Babbitt, 48 F.3d 1495 (9th Cir. 1995); see also Chapter 7. Additional information on this phase of the spotted owl litigation can be found in Victor M. Sher, Travels with Strix: The Spotted Owl's Journey through the Federal Courts, 14 Public Land L. Rev. 41, 48 (1993); Alyson C. Flournoy, Beyond the "Spotted Owl Problem": Learning from the Old-Growth Controversy, 17 Harvard Envir. L. Rev. 261, 298–300 (1993).

20. On the marbled murrelet, see 1993 Draft SEIS on Old Growth Species, supra note 1, at 3 & 4–61 to -62; Marbled Murrelet v. Babbitt, 83 F.3d 1060 (9th Cir. 1996), cert. denied, 519 U.S. 1108 (1997). The aquatic conservation strategy and the NFMA's biodiversity conservation mandate are further discussed later in this chapter.

21. See U.S. Department of Agriculture, Forest Service, Final Supplement to the Environmental Impact Statement for an Amendment to the Pacific Northwest Regional Guide (Pacific Northwest Region, 1988); Seattle Audubon Society v. Robertson, No. C89–160-WD (W.D. Wash. March 24, 1989) (order on motions for preliminary injunction and change of venue).

22. See Department of the Interior and Related Agencies Appropriations Act for Fiscal Year 1990, Pub. L. 101–21, § 318, 103 Stat. 701, 745–50 (1989); Seattle Audubon Society v. Lujan, No. C89-160-WD (W.D. Wash. Nov. 6, 1989) (order dissolving injunction), reversed sub nom. Seattle Audubon Society v. Robertson, 914 F.2d 1311 (9th Cir. 1990), reversed, 503 U.S. 429 (1992). On the Section 318 rider litigation, see Michael C. Blumm, Ancient Forests and the Supreme Court: Issuing a Blank Check for Appropriations Riders, 43 J. Urban and Contemp. Law 35 (1993); Victor M. Ster and Carol Sue Hunting, Eroding the Landscape, Eroding the Laws: Congressional Exemptions from Judicial Review of Environmental Laws, 15 Harvard Envir. L. Rev. 435 (1991).

23. See ISC Report, supra note 9, at 18; Northern Spotted Owl Habitat Management: Notice of Vacation of Northern Spotted Owl Guidelines, 55 Fed. Reg. 40,412–13 (Oct. 3, 1990); see also Flournoy, supra note 19, at 294–300.

24. Seattle Audubon Society v. Robertson, No. C89–160WD, 1991 WL 180099, pp. 6, 13 (W.D. Wash. March 7, 1991) (order on motions for summary judgment and dismissal); Seattle Audubon Society v. Robertson, 771 F.Supp. 1081, 1090, 1096 (W.D. Wash. 1991) (injunction), aff'd sub nom. Seattle Audubon Society v. Evans, 952 F.2d 297, 302 (9th Cir. 1991). In noting "a remarkable series of violations of environmental laws," the district court cited testimony from Dr. Eric Forsman, the Forest Service research wildlife biologist responsible for originally documenting the spotted owl's plight, who observed: "On all of those plans . . . there was a considerable—I would emphasize considerable—amount of political pressure to create a plan which was an absolute minimum. That is, which had a very low probability of success and which had a minimum impact on timber harvest." 771 F. Supp. at 1089.

25. Seattle Audubon Society v. Moseley, 798 F.Supp. 1473, 1478–83, 1490 (W.D. Wash. 1992), aff'd sub nom. Seattle Audubon Society v. Espy, 998 F.2d 699, 704–5 (9th Cir. 1993).

26. See Seattle Audubon Society v. Moseley, 798 F.Supp. 1484, 1490 (W.D. Wash. 1992); Dale F. Robertson, Ecosystem Management of the National Forests and Grasslands, Memo to Regional Foresters and Station Directors (Washington, D.C.: U.S. Forest Service, June 4, 1992). On the Forest Service's evolution toward ecosystem management, see Winifred B. Kessler and Hal Salwasser, Natural Resource Agencies: Transforming from Within, in Richard L. Knight and Sarah F. Bates, eds., A New Century for Natural Resources Management 171–87 (Washington, D.C.: Island Press, 1995).

27. On the Oregon and California Act, see 43 U.S.C. § 1181a; Headwaters, Inc. v. BLM, Medford District, 914 F.2d 1174 (9th Cir. 1990).

28. The recently available scientific information included the FWS's 1982 status review of the spotted owl, the Forest Service's 1986 draft SEIS analyzing spotted owl habitat requirements, a 1986 National Audubon Society study of the spotted owl by a "blue ribbon" panel of scientists, Dr. Russell Lande's 1985 and 1987 analyses of spotted owl population demographics and viability, and two BLM analyses of the spotted owl completed in 1986 and 1987. See Portland Audubon Society v. Lujan, 795 F.Supp. 1489, 1492 (D. Ore. 1992).

29. See Continuing Resolution, H.J. Res. 395, Pub. L. 100–202, § 314, 101 Stat. 1329–214, 1329–254 (1987); Pub. L. 100–446, § 314, 102 Stat. 1774, 1825–26 (1988); Portland Audubon Society v. Lujan, 712 F.Supp. 1456, 1485, 1488–89 (D. Ore. 1989). The rider was ultimately renewed annually for three consecutive years, and the courts consistently interpreted it to preclude all challenges to the BLM's timber program. Portland Audubon Society v. Lujan, 712 F. Supp. 1456, 1488 (D. Ore. 1989), aff'd, 884 F.2d 1233 (9th Cir. 1989), cert. denied, 110 S. Ct. 1470 (1990); Oregon Natural Resources Council v. Mohla, 19 Envir. L. Rep. 21,177

(D. Ore. May 25, 1989), aff'd, 895 F.2d 627 (9th Cir. 1990), cert. denied, 110 S.Ct. 2621 (1990). For perceptive critiques of these decisions, see Sher and Hunting, supra note 22, at 452–70; Flournoy, supra note 19, at 284–94.

30. Seattle Audubon Society v. Evans, 952 F.2d 297 (9th Cir. 1991); see also Sher and Hunting, supra note 22, at 452–60.

31. Portland Audubon Society v. Lujan, 795 F.Supp. 1489, 1501, 1505 (D. Ore. 1992), aff'd sub nom. Portland Audubon Society v. Babbitt, 998 F.2d 705 (9th Cir. 1993).

32. On the Jamison strategy and its implementation, see Bureau of Land Management, Management Guidelines for the Conservation of the Northern Spotted Owl FY 1991 through FY 1992 (1990); U.S. Fish and Wildlife Service, Formal Consultation on 174 Bureau of Land Management Fiscal Year 1991 Timber Sales (June 17, 1991, as amended July 3, 1991). The ensuing litigation is found at: Lane County Audubon Society v. Jamison, No. 91–6123-JO, 1991 WL 354885, at p. 2 (D. Ore. Sept. 11, 1991), reversed, 958 F.2d 290 (9th Cir. 1992), on remand, Lane County Audubon Society v. Jamison, No. 91–6123-JO (D. Ore. Jan. 21, 1993) (opinion and order).

33. On the Endangered Species Committee, see 16 U.S.C. §§ 1536(e)–(o); Jared des Rosiers, The Exemption Process Under the Endangered Species Act: How the "God Squad" Works and Why, 66 Notre Dame L. Rev. 825 (1991). The ESC is composed of seven members: the secretary of agriculture, secretary of the army, chairman of the Council of Economic Advisors, administrator of the Environmental Protection Agency, secretary of the interior, administrator of the National Oceanic and Atmospheric Administration, and one member from the affected state appointed by the president. 16 U.S.C. § 1536(e)(3). For a description of the two previous ESC exemption request proceedings (involving the Tellico dam in Tennessee and the Grayrocks dam on the Platte River), see Comment, The 1978 Amendments to the Endangered Species Act: Evaluating the New Exemption Process Under § 7, 9 Envir. L. 10031 (1979).

34. See Endangered Species Committee, Application for Exemption by the Bureau of Land Management to Conduct 44 Timber Sales in Western Oregon (May 15, 1992); Scott Fonner, Bush Prods "God Squad" to OK Timber Sales, Sources Say, The Oregonian, May 6, 1992; Portland Audubon Society v. Endangered Species Committee, 984 F.2d 1534 (9th Cir. 1993); see also Sher, Travels with Strix, supra note 19, at 51–57.

35. See Forest Ecosystem Management: An Ecological, Economic, and Social Assessment Report of the Forest Ecosystem Management Assessment Team (hereafter, FEMAT), Appendix VII-A, at p. 101, for a list of the individual conference participants. To maintain the conference's problem solving focus, the speakers were admonished against voicing extremist positions and instructed to avoid related east-side timber harvesting and log export issues. See also Durbin, Tree Huggers, supra note 6, at 199–200; Yaffee, Wisdom, supra note 4, at 142.

36. See FEMAT, supra note 35, at Appendix VII-A, pp. 4–100 for a topical summary of the speaker presentations; id. at pp. 9, 87–88 for Thomas's quoted remarks; see also Durbin, Tree Huggers, supra note 6, at 201.

37. FEMAT, supra note 35, at ii. More specifically, the FEMAT members were instructed to focus on the region's federal lands, to develop an array of alternative management strategies that addressed environmental, economic, and social concerns, to analyze biodiversity in multispecies terms that included the northern spotted owl, marbled murrelet, anadromous fish species, and other old growth species, and to suggest adaptive management strategies for monitoring impacts on the region's old growth forests. Id. at ii–iv. Significantly, al-

though neither the FLPMA nor the O and C legislation that governs BLM timber management on its Oregon and California lands mandates biodiversity conservation, the FEMAT was expressly instructed to apply the NFMA population viability regulations to BLM lands to formulate consistent management options for biodiversity purposes across the region's federal forestlands. FEMAT (Excerpts), supra note 35, at 5, 11.

38. See FEMAT, supra note 35, at v–xi for a list of the FEMAT team members. Besides FEMAT itself, the White House created two other committees to address related dimensions of the timber controversy: one was charged with preparing an economic development plan to provide transitional assistance to the affected communities, and the other was directed to address interagency coordination issues. 1996 Forest Plan Report, supra note 4, at 33. (1996).

39. The secretive environment did not, however, prevent leaks to the press, nor did it prevent environmentalists, the timber industry, and other antagonists in the controversy from seeking to influence the process by lobbying their executive, congressional, and other allies. See Durbin, Tree Huggers, supra note 6, at 203, 205; Yaffee, Wisdom, supra, note 4 at 147–50.

40. Although the FEMAT team initially evaluated 54 previously identified forest management strategies, its original 10 forest plan alternatives did not include Option 9. However, when it became evident that the original proposed alternatives would not assure even a 1-bbf annual timber harvest, the committee reassessed its strategic assumptions and created Option 9, which enabled it to squeeze 1.2 bbf from the regional forests. See Durbin, Tree Huggers, supra note 6, at 202–4.

41. Forest Service and Bureau of Land Management, Record of Decision for Amendments to Forest Service and Bureau of Land Management Planning Documents within the Range of the Northern Spotted Owl, at Table ROD-1, B-4–B-5, 14–17, 116–23, C-4–C-6 (1994) (hereafter, NSO Record of Decision). In a controversial move, however, the FEMAT permitted some logging of younger trees in the late-successional reserves to accomplish ecological objectives. See FEMAT (Excerpts), supra note 35, at 11–14, 115–16; NSO Record of Decision, id. at p. B-5. The survey and manage system was specifically intended to reach all nondispersing species that would not receive protection through the dispersal corridors and other allocations associated with the plan's landscape-level designations. See id. at C-49–C-61 for an enumeration of the protected species.

42. See NSO Record of Decision, supra note 41, at B-9–34. According to Option 9, Tier 1 Key Watersheds (8.1 million acres) were selected to protect at-risk fish species and habitat, while Tier 2 Key Watersheds (1 million acres) were selected to protect sources of high-quality waters that might not contain any fish. Id. at B-18–20. For a comprehensive and critical analysis of the Option 9 aquatic conservation strategy, see Henry B. Lacey, New Approach of Business as Usual: Protection of Aquatic Ecosystems Under the Clinton Administration's Westside Forests Plans, 10 J. Envir. Law and Litigation 309 (1995).

43. FEMAT (Excerpts), supra note 35, at 48–68 (see p. 53 for the opening quotation). After employing approximately 144,900 persons in 1990, the region's timber industry employment had dropped to 125,400 employees in 1992 as the old growth controversy deepened. The Option 9 harvesting forecasts anticipated a timber-based employment level at 119,800 workers. Southwest Oregon would be most heavily affected, because that area's local mills depended primarily on federal timber and could not draw upon state or private timberlands to make up the federal timber deficit. The other FEMAT options anticipated employment levels ranging from 123,700 to 112,900. Id. at 60–61. Within the region, recreation and tourism employment totaled 50,000–80,000 full-time jobs—only half the number of jobs that were in timber-based employment. Id. at 60–63.

44. Id. at 70–80 (the opening quotation is found at p. 70). The impacts, however, would vary widely among communities depending upon such factors as the individual community's leadership, economic diversification, and linkage to markets. Because of limited information, FEMAT social scientists were unable to evaluate the plan's impact on Native American tribes, but they noted that timber harvesting limitations could affect those tribes that depended on logging for employment while the same limitations could protect cultural and spiritual values linked to particular forestlands. Id. at 80–81.

45. The Northwest Economic Adjustment Initiative was overseen by a federal Economic Development Assistance Team, which included representatives from the National Economic Council; Council of Economic Advisors; Office of Management and Budget; Domestic Policy Council; Departments of Agriculture, Commerce, Interior, and Labor; Environmental Protection Agency; and Small Business Administration. 1996 Forest Plan Report, supra note 4, at 33. It established a three-tier federal, regional, and state administrative structure to deliver assistance to the affected communities. Id. at 48–49, 157–58.

46. See 1996 Forest Plan Report, supra note 4, at 209; Yaffee, Wisdom, supra note 4, at 161–62.

47. 1996 Forest Plan Report, supra note 4, at 43.

48. For a description of the interagency coordination structure, see id. at 44–49. The Regional Interagency Executive Committee included regional executives from the Forest Service, Bureau of Land Management, U.S. Fish and Wildlife Service, National Park Service, Bureau of Indian Affairs, National Marine Fisheries Service, and Environmental Protection Agency; the separate Intergovernmental Advisory Committee was composed of 20 members, with one official from local, state, and tribal governments in Oregon, Washington, and California, and federal officials from each of the region's participating agencies. The Provincial Interagency Executive Committees were composed of the federal managers responsible for lands or resources in each province; they were responsible for implementing the Forest Plan in each province. The Provincial Advisory Committees included representatives from federal, state, county, and local governments, as well as the timber industry, environmental groups, recreation and tourism organizations, and at-large members, who advised the provincial executive committees on local conditions, concerns, and related matters.

49. Id. at 56–63; see also U.S. General Accounting Office, Ecosystem Planning: Northwest Forest and Interior Columbia River Basin Plans Demonstrate Improvements to Land Use-Planning Appendix I:2.3, 2.4 (Washington, D.C., 1999) (hereinafter, 1999 GAO Ecosystem Planning Report).

50. Even before the final plan was announced, the timber industry mounted a peremptory legal challenge, arguing that the FEMAT process violated the Federal Advisory Committee Act (FACA), 5 U.S.C. App. 2. Northwest Forest Resource Council v. Espy, 846 F. Supp. 1009 (D.D.C. 1994); see also Chapter 7 for a further description of this FACA litigation. The environmental community's reaction to the Forest Plan is described in Durbin, Tree Huggers, supra note 6, at 209–15.

51. Seattle Audubon Society v. Lyons, 871 F.Supp. 1291 (W.D. Wash. 1994) (the quotations respectively can be found at pp. 1311 (emphasis in original) and 1300). In addition, the court found that the agencies had objectively assessed alternative management approaches and properly utilized the FEMAT report in their decision processes. Id. at 1318–19.

52. Id. at 1320–24 (the monitoring quotation is found at p. 1324). While acknowledging that plaintiffs had persuasively argued that the agencies had skewed their economic analysis by exaggerating the negative effects of lower timber harvests and by discounting the benefits

associated with unlogged forests, the court nonetheless felt compelled to defer to the government's economic assumptions and approach. Id. at 1324–25.

53. Seattle Audubon Society v. Moseley, 80 F.3d 1401, 1404–6 (9th Cir. 1996).

54. The Forest Service's Western Forest Health Initiative is found at 60 Fed. Reg. 95 (Jan. 3, 1995); 60 Fed. Reg. 32485 (June 22, 1995); see Durbin, Tree Huggers, supra note 6, at 235–40; see also Chapter 5 for a further discussion of federal fire policy and forest health problems.

55. Emergency Supplemental Appropriations for Additional Disaster Assistance, for Anti-terrorism Initiatives, for Assistance in the Recovery from the Tragedy that Occurred at Oklahoma City, and Recisions Act, Pub. L. 104–19, §§ 2001–2, 109 Stat. 194, 240–47 (1995) (codified at 16 U.S.C. § 1611 (hereafter, Recisions Act); Conference Report to H.R. 1158, H.R. 104–24, 104th Cong., 1st Sess. (1994); see also Michael Axeline, Forest Health and the Politics of Expediency, 26 Envir. L. 613 (1996); Slade Gorton and Julie Kays, Legislative History of the Timber and Salvage Amendments Enacted in the 104th Congress: A Small Victory for Timber Communities in the Pacific Northwest, 26 Envir. L. 641 (1996).

56. See Memorandum from President William J. Clinton, to Secretaries of Interior, Agriculture, Commerce and EPA Administrator, Re: Implementing Timber-Related Provisions to Public Law 104–19 (Aug. 1, 1995); Letter from Jack Ward Thomas, Chief of U.S. Forest Service et al., to the Regional Foresters of the U.S. Forest Service et al., Re: Salvage Sale Provisions of Pub. L. 104–19 and Memorandum of Agreement (Aug. 9, 1995). For an overview of the environmental problems and the ensuing litigation accompanying the 1995 Recisions Act's salvage sale provisions, see Patti A. Goldman and Kristen L. Boyles, Forsaking the Rule of Law: The 1995 Logging without Laws Rider and Its Legacy, 27 Envir. L. 1035 (1997).

57. On the Kootenai litigation, see Inland Empire Public Lands Council v. Glickman, 911 F.Supp. 431 (D. Mont. 1995), aff'd, 88 F.3d 697 (9th Cir. 1996). On the Thunderbolt litigation, see Idaho Conservation League v. Thomas, 917 F.Supp. 1458 (D. Idaho 1995), aff'd, 91 F.3d 1345 (9th Cir. 1996); see also Kristen L. Boyles, Making the Case for Enforceable Standards, 13 J. Envir. Law and Litigation 1, 5–6 (1998).

58. Northwest Forest Resource Council v. Glickman, No. 95–6244-HO (D. Ore. Jan 19, 1996) (order granting injunctive and declaratory relief). The court, however, excluded those sale units where threatened or endangered bird species were known to be nesting. And in another minor environmental victory, the timber industry failed to convince the courts that timber sales predating the Section 318 rider were resurrected by the new rider. See Northwest Forest Resource Council v. Pilchuck Audubon Society, 97 F.3d 1161 (9th Cir. 1996).

59. On the North Umpqua litigation, see Oregon Natural Resources Council v. Thomas, 92 F.3d 792 (9th Cir. 1996); Goldman and Boyles, supra note 56, at 1066–68.

60. On the post-plan timber sale figures, see 1999 GAO Ecosystem Planning Report (Letter Report), supra note 49, at 5.1, Appendix I:4, I:5. On the small mill closures, see Chris Carrel, Renewed Controversy Threatens the Truce of Clinton's Northwest Forest Plan, 30(22) High Country News 1, 8, 12 (Nov. 23, 1998). On the drop in county revenues, see "Budget Would Increase, Fix Timber Receipts," The Bulletin (Bend, Ore.), March 23, 1998, at p. A2; Jim Barnett and Dave Hogan, Senators Offer Plan to Rescue Forest Counties, Portland Oregonian, Sept. 8, 2000, at p. A20. On the Bush administration's potential Forest Plan revisions, see Michael Milstein, Bush Will Overhaul Northwest Forest Plan, Portland Oregonian, April 8, 2002.

61. On the post-plan ESA listings, see 1999 GAO Ecosystem Planning Report, supra note 49, at Appendix I:5.2. For a description of the ESA salmon listing decisions, see Michael C.

Blumm and Greg D. Corbin, Salmon and the Endangered Species Act: Lessons from the Columbia Basin, 74 Wash. L. Rev. 519, 525–48 (1999); see also Lacey, supra note 42, at 316–33. On the federal agencies' continuing uncertainty over spotted owl trends, see Gerry Jackson, assistant director for Ecological Services, U.S. Fish and Wildlife Service, Capitol Hill Hearing Testimony, Federal Document Clearing House Congressional Testimony, March 19, 1998; James Lyons, undersecretary, U.S. Department of Agriculture, Capitol Hill Hearing Testimony, Federal Document Clearing House Congressional Testimony, March 19, 1998. For the timber industry's listing challenge, see Timber Industry Asks for Owl Recount; Will Bush Deal? Public Land News, Aug. 2, 2002, at p. 6; Owl, Murrelet Listings Need Review, Group Says, Endangered Species & Wetlands Report 10 (February 2002). On salmon listing challenges, see Alsea Valley Alliance v. Evans, 161 F. Supp.2d 1154 (D. Ore. 2001), which is discussed further in Chapter 7.

62. The initial litigation is found at Oregon Natural Resources Council Action v. U.S. Forest Service, 59 F. Supp.2d 1085 (W.D. Wash. 1999). For the plan's survey and management requirements, see Forest Plan Record of Decision, pp. 36, C-4, C-49 (1994). On Congress's reaction, see Congress Targets Survey Requirements for Forests, 4(11) Endangered Species & Wetlands Report 1 (August 1999). Discretion to Conduct Surveys in Forests May Be Left to FS, BLM, 4(12) Endangered Species & Wetlands Report 1 (September 1999). On the Forest Plan amendment, see 63 Fed. Reg. 65167 (Nov. 25, 1998); U.S. Forest Service and Bureau of Land Management, Final Supplemental Environmental Impact Statement for Amendment to the Survey and Manage, Protection Buffer, and Other Mitigating Measures Standards and Guidelines (Portland, Ore., 2000). On the next round of litigation, see Brian Stempeck, "Survey and Manage" Lawsuits Go Head to Head in Northwest, Land Letter, Feb. 28, 2002, at p. 9; "Survey and Manage" Suits Filed by Industry, Enviros, Endangered Species & Wetlands Report, February 2002, at p. 16; Michael Milstein, Bush's Revised Rule May Lead to More Logging, The Oregonian, Oct. 1, 2002, at p. 1.

63. See Pacific Coast Federation of Fishermen's Associations v. National Marine Fisheries Service, No. C97–775R, Order Granting . . . Plaintiff's Motion for Summary Judgment on Count V (W.D. Wash., April 28, 1998); Pacific Coast Federation of Fishermen's Associations v. National Marine Fisheries Service, 71 F.Supp. 2d 1063, 1073 (W.D. Wash. 1999), aff'd, 253 F.3d 1137 (9th Cir. 2001). See also Dan Berman, Feds May Gut Northwest Forest Plan Rule Protecting Endangered Salmon, 21(21) Land Letter 5 (Dec. 12, 2002).

64. See Scientists Seek Changes in Northwest Forest Plan: Recommend Protection for All Remaining Old Growth Forests, at www.biodiversitynw.org/oldgrowth/ogpetition.htm; Chris Carrel, Renewed Controversy Threatens the Truce of Clinton's Northwest Forest Plan, 30(22) High Country News 1 (Nov. 23, 1998); Douglas Gantenbein, Old Growth for Sale, Audubon, May/June, 1998, at pp. 76–80; Brent Foster, The Failure of Watershed Analysis Under the Northwest Forest Plan: A Case Study of the Gifford Pinchot National Forest, 5 Hastings West-Northwest J. Envir. L. and Policy 337 (1999); Lacey, supra note 42.

65. See 1999 GAO Ecosystem Planning Report, supra note 49; James Pipkin, Northwest Forest Plan Revisited (Washington, D.C.: Department of the Interior Office of Policy Analysis, 1998); Kathie Durbin, An Uneasy Compromise, 10(4) Inner Voice 10–11, 15–16 (July/August 1998); Joan Smith, Supervisor—Elect Siskiyou County, Federal Documents Clearing House Congressional Testimony, July 23, 1996. The GAO report, however, was completed before Judge Dwyer enjoined the agencies for failing to meet the Forest Plan's survey and management requirements.

66. See U.S. Forest Service, 1998 Northwest Forest Plan Accomplishment Report 2 (Portland,

Ore., 1998); U.S. Forest Service, Northwest Economic Adjustment Initiative 3 (Portland, Ore., 1998); Michael Milstein, Study Says Federal Aid Was Little Boon to Loggers, The Oregonian, January 23, 2003, at p. 1. See also Kathie Durbin, Life after Lumber, 10(4) Inner Voice 20 (July/August 1998).

67. See Chapter 2 for additional information on the National Biological Survey proposal.

68. On the HCP "no surprises" policy, see John D. Leshy, The Babbitt Legacy at the Department of the Interior: A Preliminary View, 31 Envir. L. 199, 212–14 (2001); Graham M. Lyons, Habitat Conservation Plans: Restoring the Promise of Conservation, 23 Environs Envir. Law and Policy J. 83 (1999); Karin P. Sheldon, Habitat Conservation Planning: Addressing the Achilles Heel of the Endangered Species Act, 6 N.Y.U. Envir. L. J. 279 (1998).

69. Interagency Ecosystem Management Task Force, The Ecosystem Approach: Healthy Ecosystems and Sustainable Economies, vol. 1, at p. 3 (Washington, D.C., 1995) (hereafter, The Ecosystem Approach).

70. The papers from the 1995 conference are available in Nels Johnson et al., eds., Ecological Stewardship: A Common Reference for Ecosystem Management (3 vols.) (Oxford: Elsevier Science, 1999). On federal agency cooperative efforts, see www.cesu.org/cesu (announcing and describing the cooperative Ecosystem Studies Units); 65 Fed. Reg. 8834–40 (Feb. 22, 2000) (notice of a proposed unified federal agency watershed management approach).

71. On the 2000 NFMA planning regulations, see 36 C.F.R. part 219 (2001); 65 Fed. Reg. 67514 (Nov. 9, 2000); see also 97(5) Journal of Forestry 5–46 (May 1999), which analyzes the Committee of Scientists' report). The Bush administration's proposed revisions to the NFMA planning regulations are found at 67 Fed. Reg. 72770-816 (Dec. 6, 2002); see also Brian Stempeck, Bush Administration Proposes Easing Planning Regs., Land Letter, Dec. 5, 2002, at p. 1.

72. On the road construction moratorium, see 36 C.F.R. § 212 (2001); 64 Fed. Reg. 7290 (Feb. 12, 1999); 66 Fed. Reg. 3206 (Jan. 12, 2001). On the roadless area review initiative, see William J. Clinton, Memorandum on Protection of Forest "Roadless" Areas, Oct. 13, 1999; U.S. Forest Service, Roadless Area Conservation Final Environmental Impact Statement (2000); Special Areas: Roadless Area Conservation Final Rule, 36 C.F.R. part 294 (2002); 66 Fed. Reg. 3244, 3277 (Jan. 12, 2001). On the ensuing litigation, see infra note 80 and accompanying text; Counterattacks Launched Against FS Roadless Area Policy, 26(2) Public Land News 5 (Jan. 19, 2001); Forest Service Weakening Roadless Policy via Road Policy, Enviros Say, 27(5) Public Lands News 5 (March 1, 2002); see also Pamela Baldwin, The National Forest System Roadless Areas Initiative (Congressional Research Service, 2001); U.S. General Accounting Office, Forest Service Roadless Areas: Potential Impact of Proposed Regulations on Ecological Sustainability (Washington, D.C., 2000). These Forest Service road-related initiatives are explored further in Chapter 6.

73. On the BLM and ecosystem management, see Bureau of Land Management, Ecosystem Management in the BLM: From Concept to Commitment (1994). On rangeland reform regulations, see 43 C.F.R. § 4180 (2001) (standards); § 1784.6–1 (2001) (resource advisory councils); see also Bruce Pendery, Reforming Livestock Grazing on the Public Domain: Ecosystem Management–Based Standards and Guidelines Blaze a New Path for Range Management, 27 Envir. L. 513 (1997). On the mining law regulations, see Mining Claims Under the General Mining Laws; Surface Management, 65 Fed. Reg. 69,998 (Nov. 21, 2000); 43 C.F.R. § 3809.420, § 3809.50 et seq. (2001). On the BLM planning revisions, see Bureau of Land Management, Land Use Planning Handbook (BLM Handbook H-1601–

1, 2000). On the Grand Staircase monument and the National Landscape Conservation System, see Proclamation No. 6920, 3 C.F.R. 64 (1997); Bureau of Land Management, National Landscape Conservation System, at www.blm.gov/nhp/what/nlcs/index.html.

74. On the Park Service's ecosystem management initiatives, see National Park Service, Management Policies 2001 12, 18–20, 28 (2001); see also id. at 30 (committing to "reestablish natural functions and processes in human-disturbed components of natural systems in parks"). On the Yellowstone snowmobile decision, see Chapter 7. On the FWS's ecosystem management initiatives, see Office of the Director, U.S. Fish and Wildlife Service, Ecosystem Approach to Fish and Wildlife Conservation, Fish and Wildlife Service Manual Part 052, 1.12E, 1.4, 1.10 (1996); see also id. at 1.8B(4) (monitor and evaluate provision). See Chapter 5 for an analysis of the FWS's wolf restoration efforts.

75. Cases illustrating the impact of an ESA listing on public land management practices include: Sierra Club v. Yeutter, 926 F.2d 429 (5th Cir. 1991) (red cockaded woodpecker); Silver v. Babbitt, 924 F.Supp. 976 (D. Ariz. 1995) (Mexican spotted owl); Pacific Rivers Council v. Thomas, 30 F.3d 1050 (9th Cir. 1994) (salmon); see also Mark Bonnett and Kurt Zimmerman, Politics and Preservation: The Endangered Species Act and the Northern Spotted Owl, 18 Ecology L. Q. 105 (1991). On ESA listing litigation, see Northern Spotted Owl v. Hodel, 716 F.Supp. 479 (W.D. Wash. 1988); Friends of the Wild Swan v. U.S. Fish and Wildlife Service, 945 F.Supp. 1388 (D. Ore. 1996) (bull trout); Forest Guardians v. Babbitt, 164 F.3d 1261 (10th Cir. 1998), amended, 174 F.3d 1178 (10th Cir. 1999) (Rio Grande silvery minnow); Defenders of Wildlife v. Babbitt, 958 F.Supp. 670 (D.D.C. 1997) (Canadian lynx); see also George C. Coggins and Robert L. Glicksman, Public Natural Resources Law § 15C.02[1][a] (St. Paul, Minn.: West Group, 2000). On critical habitat designation litigation, see Northern Spotted Owl v. Lujan, 758 F.Supp. 621 (W.D. Wash. 1991); Silver v. Babbitt, 924 F.Supp. 972 (D. Ariz. 1995) (Mexican spotted owl); Natural Resources Defense Council v. U.S. Department of the Interior, 113 F.3d 1121 (9th Cir. 1997) (California gnatcatcher). On recovery plan litigation, see Defenders of Wildlife v. Babbitt, 130 F.Supp.2d 121 (D.D.C. 2001) (Sonoran pronghorn); Fund for Animals v. Babbitt, 903 F.Supp. 96 (D.D.C. 1995) (grizzly bear).

76. Judge Dwyer's NFMA biodiversity decisions are: Seattle Audubon Society v. Robertson, 771 F.Supp. 1081 (W.D. Wash. 1991), aff'd sub nom. Seattle Audubon Society v. Evans, 952 F.2d 297 (9th Cir. 1991). The relevant Ninth Circuit biodiversity rulings are: Seattle Audubon Society v. Lyons, 80 F.3d 1401 (9th Cir. 1996); Oregon Natural Resources Council v. Lowe, 109 F.3d 521 (9th Cir. 1997). On the principle of deference to agency expertise, see Sierra Club v. Marita, 46 F.3d 606 (7th Cir. 1995); Colorado Environmental Coalition v. Dombeck, 185 F.3d 1162 (10th Cir. 1999); Inland Empire Public Lands Council v. U.S. Forest Service, 88 F.3d 754 (9th Cir. 1996); Sierra Club v. Espy, 38 F.3d 792 (5th Cir. 1994); Sierra Club v. Robertson, 28 F.3d 753 (8th Cir. 1994), affirming, 810 F. Supp. 1021 (W.D. Ark. 1992). On the revised NFMA regulations, see 36 C.F.R. § 219.20 (2002) (addressing species diversity as a dimension of ecological sustainability); see also Oliver A. Houck, On the Law of Biodiversity and Ecosystem Management, 81 Minn. L. Rev. 869 (1997) (criticizing the Forest Service's initial proposed NFMA regulation revisions).

77. On the Forest Service and NEPA cumulative effects analysis, see Neighbors of Cuddy Mountain v. U.S. Forest Service, 137 F.3d 1372 (9th Cir. 1998); Blue Mountains Biodiversity Project v. Blackwood, 161 F.3d 1208 (9th Cir. 1998); Resources Limited, Inc. v. Robertson, 8 F.3d 1394 (9th Cir. 1993); Thomas v. Peterson, 753 F.2d 754 (9th Cir. 1985); see also Robert B. Keiter, NEPA and the Emerging Concept of Ecosystem Management on the Public Lands,

25 Land and Water L. Rev. 43 (1990). On the role of NEPA cumulative effects analysis requirements in regional assessments, see U.S. Forest Service and Bureau of Land Management, Report to the Congress on the Interior Columbia Basin Ecosystem Management Project 13 (Boise, Idaho, 2000); U.S. Forest Service and Bureau of Land Management, Interior Columbia Basin Draft Environmental Impact Statement Summary 2 (Boise, Idaho, 2000); see also The Ecosystem Approach, vol. 2, supra note 69, at 85–88. On NEPA and wildlife corridors, see Marble Mountain Audubon Society v. Rice, 914 F.2d 179 (9th Cir. 1990); see also Western Land Exchange Project v. Dombeck, 47 F.Supp.2d 1216 (D. Ore., 1999) (acknowledging the value of minimizing landscape fragmentation for biodiversity purposes, and thus sustaining proposed land exchanges designed to consolidate public ownership into large habitat blocks). On ESA reconsultation and forest planning obligations, see Pacific Rivers Council v. Thomas, 30 F.3d 1050 (9th Cir. 1994); see also Bob Marshall Alliance v. Hodel, 852 F.2d 1223 (9th Cir. 1988), cert. denied, 489 U.S. 1066 (1989) (requiring preparation of a comprehensive biological opinion assessing the effects of oil and gas leasing and post-leasing activities on "listed" species in the affected area).

78. The relevant Judge Dwyer inventory and monitoring decisions are: Seattle Audubon Society v. Lyons, 871 F.Supp. 1291, 1324 (W.D. Wash. 1994); Oregon Natural Resources Council Action v. U.S. Forest Service, 59 F.Supp.2d 1085 (W.D. Wash. 1999). The other citations are: Sierra Club v. Martin, 168 F.3d 1 (11th Cir. 1999), affirming, 71 F.Supp.2d 1268 (N.D. Ga. 1996); Friends of South East's Future v. Morrison, 153 F.3d 1059 (9th Cir. 1998); see also Sierra Club v. Peterson, 185 F.3d 349 (5th Cir. 1999), reversed on other grounds, 228 F.3d 559 (5th Cir. 2000) (en banc); Utah Environmental Congress v. Zieroth, 190 F.Supp.2d 1265 (D. Utah 2002).

79. On the role of public opinion in agency decisions under NEPA, see Christianson v. Hauptman, 991 F.2d 59, 63–64 (2d Cir. 1993); Conservation Law Foundation v. Secretary of the Interior, 864 F.2d 954, 959 (1st Cir. 1989); Paul J. Culhane, NEPA's Impact on Federal Agencies, Anticipated and Unanticipated, 20 Envir. L. 681 (1990). On Rocky Mountain front leasing, see Rocky Mountain Oil and Gas Association v. U.S. Forest Service, No. CV 98–22-H-CCL, Opinion and Order, March 7, 2000 (D. Mont. 2000), aff'd, 2001 U.S. App. LEXIS 8758 (9th Cir. 2001), cert. denied sub nom. Independent Petroleum Association v. U.S. Forest Service, 122 S.Ct. 541 (2002); Judge: FS May Bar Oil Leasing to Protect 'Value of Place', Public Land News, March 17, 2000, at p. 6. On the role of citizen management committees, see National Parks and Conservation Association v. Stanton, 54 F.Supp.2d 7 (D.D.C. 1999); see also Chapter 7. On FACA litigation, see Northwest Forest Resource Council v. Espy, 846 F.Supp. 1009 (D.D.C. 1994); Aluminum Company of America v. National Marine Fisheries Service, 92 F.3d 902 (9th Cir. 1996); Alabama-Tombigbee Rivers Coalition v. Department of the Interior, 26 F.3d 1103 (11th Cir. 1994); see also Chapter 7.

80. On the ESA reintroduction litigation, see Gibbs v. Babbitt, 214 F.3d 483 (4th Cir. 2000); Wyoming Farm Bureau v. Babbitt, 199 F.3d 1224 (10th Cir. 2000); United States v. McKittrick, 143 F.3d 1170 (9th Cir. 1998); New Mexico Cattle Growers Association v. U.S. Fish and Wildlife Service, 81 F.Supp.2d 1141 (D.N.M. 1999); on the national monument designation decisions, see Tulare County v. Bush, 306 F.3d 1138 (D.C. Cir. 2002); Mountain States Legal Foundation v. Bush, 306 F.3d 1132 (D.C. Cir. 2002); on rangeland reform litigation, see Public Lands Council v. Babbitt, 529 U.S. 728 (2000); on critical habitat designation, see Biodiversity Legal Foundation v. Badgley, 284 F.3d 1046 (9th Cir. 2002); Forest Guardians v. Babbitt, 174 F.3d 1178 (10th Cir. 1999); Natural Resources Defense Council v. U.S. Department of the Interior, 113 F.3d 1121 (9th Cir. 1997). On national forest roadless

area litigation, see Kootenai Tribe of Idaho v. Veneman, 313 F.3d 1094 (9th Cir. 2002), reversing, Kootenai Tribe v. Veneman, 142 F. Supp.2d 1231 (D. Idaho 2001) and Idaho ex rel. Kempthorne v. U.S. Forest Service, 142 F.Supp.2d 1248 (D. Idaho 2001); see also Wyoming Timber Industry Association v. U.S. Forest Service, 80 F.Supp.2d 1245 (D. Wyo. 2000).

81. On BLM-related litigation, see ONRC Action v. Bureau of Land Management, 150 F.3d 1132 (9th Cir. 1998); Natural Resources Defense Council v. Hodel, 624 F.Supp. 1045 (D. Nev. 1985); Coggins and Glicksman, supra note 75, at § 19.03[4][a]; but see Idaho Watersheds Project v. Hahn, 187 F.3d 1035 (9th Cir. 1999); Oregon Natural Desert Association v. Singleton, 47 F.Supp.2d 1182 (D. Ore. 1998). On ESA critical habitat designation and economic considerations, see Catron County Board of Commissioners v. U.S. Fish and Wildlife Service, 75 F.3d 1429 (10th Cir. 1996); see also Chapter 7, which discusses the Wise Use movement's litigation strategy.

82. See U.S. House of Representatives, Committee on Natural Resources, Oversight Hearing on the Changing Needs of the West (1994); U.S. General Accounting Office, Ecosystem Management: Additional Actions Needed to Adequately Test a Promising Approach (1994); Congressional Research Service, Ecosystem Management: Status and Potential (S. Prt. 103–98, 1994); "Ecosystem Management Act of 1994," S. 2190, 103d Cong., 2d Sess. (1994); "Ecosystem Management Act of 1995," S. 93, 104th Cong., 1st Sess. (1995); spotted owl legislative proposals included: H.R. 1164 103d Cong., 1st Sess. (1993); S. 1156, 102d Cong., 1st Sess. (1992); H.R. 4899, 102d Cong., 2d Sess. (1992). For a more detailed summary of Congress's pre-1994 actions supporting ecosystem management concepts, see Robert B. Keiter, Beyond the Boundary Line: Constructing a Law of Ecosystem Management, 65 U. Colo. L. Rev. 293, 314–16 (1994).

83. See Unfunded Mandates Reform Act of 1995, Pub. L. 104-4, 109 Stat. 48 (1995), codified at 2 U.S.C. §§ 602, 658 et seq., 1501 et seq.; see also Robert W. Adler, Unfunded Mandates and Fiscal Federalism: A Critique, 50 Vanderbilt L. Rev. 1137 (1997); Rena L. Steinzor, Unfunded Environmental Mandates and the "New (New) Federalism": Devolution, Revolution, or Reform?, 81 Minn. L. Rev. 97 (1996). The FACA amendments are found at Unfunded Mandates Reform Act of 1995, Pub. L. 104-4, 109 Stat. 48, § 204 (1995), codified at 2 U.S.C. §§ 658 et seq., 1501 et seq.

84. On the mining regulation controversy, see Pub. L. 106–31, § 3002 (1998) (blocking BLM's proposed mining regulations, 64 Fed. Reg. 6422 (Feb. 9, 1999)); Sam Kalen, An 1872 Mining Law for the New Millennium, 71 U. Colorado L. Rev. 33, 390–92 (2002). On the R.S. 247 controversy, see Coggins and Glicksman, supra note 75, at § 10E.19; Mitchell R. Olson, The R.S. 2477 Right of Way Dispute: Constructing a Solution, 27 Envir. L. 289 (1997); Symposium, 14 J. Energy, Nat. Res., and Envir. Law 295–354 (1994). The National Biological Survey agency is discussed in Chapter 2.

85. See National Wildlife Refuge System Improvement Act of 1997, Pub. L. 105–57, 111 Stat. 1252 (codified at 16 U.S.C. § 668dd); National Parks Omnibus Management Act of 1998, Pub. L. 105–391, 112 Stat. 3497 (codified as amended in scattered sections of 16 U.S.C.); CARA-Lite Beefs Up LWCF and PILT, Subject to Money Bills, 25(20) Public Lands News 6 (Oct. 13, 2000); Secure Rural Schools and Community Self Determination Act of 2000, Pub. L. 106–393, 114 Stat. 1607 (2000); see also Robert L. Fischman, The National Wildlife Refuge System and the Hallmarks of Modern Organic Legislation, 29 Ecology L. Q. 457 (2002).

86. See Herger-Feinstein Quincy Library Group Forest Recovery Act, Pub. L. 105–277, Div. A, § 101(e) Title IV, § 401, 112 Stat. 2681–305, 105th Cong., 2d Sess. (1998), codified at 16

U.S.C. § 2104 note; Valles Caldera Preservation Act, Pub. L. 106–248, 114 Stat. 598, 106th Cong., 2d Sess. (2000). The Quincy Library Group legislation and the underlying collaborative process are examined in detail in Chapter 8. The Valles Caldera legislation was modeled on earlier legislation establishing similar trust and sustainability requirements to govern the Park Service's management of the scenic Presidio lands adjacent to San Francisco Bay. See Donald J. Hellmann, The Path of the Presidio Trust Legislation, 28 Golden Gate U. L. Rev. 319 (1998); Johanna H. Wald, The Presidio Trust and Our National Parks: Not a Model to Be Trusted, 28 Golden Gate U. L. Rev. 369 (1998).

F I V E

Making Amends with the Past

1. The quotation is from Bruce Hampton, The Great American Wolf 255 (New York: Henry Holt, 1997), which also describes the wolf extermination campaign. Id. at 102–46; see also Barry Holstun Lopez, Of Wolves and Men (New York: Scribners, 1978); Rick McIntyre, ed., War Against the Wolf: America's Campaign to Exterminate the Wolf (Stillwater, Minn.: Voyageurs Press, 1995).

2. At roughly the same time as the bison slaughter, market hunters also decimated the region's big game populations, which left wolves and other predators with little prey base and caused them to begin depredating on domestic livestock. Hampton, supra note 1, at 112–19. The Roosevelt statement is quoted in McIntyre, supra note 1, at 108; Lopez, supra note 1, at 142. For accounts of the early state bounty programs, see Hampton, id. at 102–26; McIntyre, id. at 75–132. For examples of state bounty laws, see McIntyre, id. at 118–21; for an account of Montana's bounty program, see Hank Fischer, Wolf Wars: The Remarkable Inside Story of the Restoration of Wolves to Yellowstone 17–18 (Helena, Mont.: Falcon Press, 1995).

3. See Peter Matthiessen, Wildlife in America 195 (New York: Viking, 1987); Stanley Young and Edward Goldman, The Wolves of North America 383 (Washington, D.C.: American Wildlife Institute, 1944); Hampton, supra note 1, at 127–45; McIntyre, supra note 1, at 149–215; Fischer, supra note 2, at 19–23.

4. On the Park Service's evolving view toward wolves, see Paul Schullery, Searching for Yellowstone: Ecology and Wonder in the Last Wilderness 160–61 (New York: Houghton Mifflin, 1997); Richard West Sellars, Preserving Nature in the National Parks: A History 119–23 (New Haven: Yale University Press, 1997); A. Starker Leopold et al., Wildlife Management in the National Parks, in Transactions of the Twenty-Eighth North American Wildlife and Natural Resources Conference 29, 29–44 (1963), reprinted in Larry M. Dilsaver, ed., America's National Park System: The Critical Documents 237, 237–52 (Lanham, Md.: Rowman and Littlefield, 1994) (hereafter, NPS Leopold Report); Memorandum from Secretary of the Interior Stewart Udall, on Management of the National Park System to National Park Service Director (July 10, 1964), reprinted in Dilsaver, id. at 272, 273 (hereafter, 1964 Udall Memo).

5. The Aldo Leopold quote is from Aldo Leopold, Review of The Wolves of North America, 42(12) Journal of Forestry 928–29 (1944); for more on Leopold's evolving view of wolves, see McIntyre, supra note 1, at 187–91; Aldo Leopold, Thinking Like a Mountain, in A Sand County Almanac 137–41 (New York: Oxford University Press, 1949). For more on the Olson, Errington, and Murie studies, see Fischer, supra note 2, at 24–34; Hampton, supra note 1, at 152, 155–57; McIntyre, supra note 1, at 312–21. For the other noted reports and

studies, see A. Starker Leopold et al., Predator and Rodent Control in the United States, report submitted to Stewart Udall, Secretary of the Interior (March 9, 1964), reprinted in Transactions of the Twenty-Ninth North American Wildlife Conference 27, 35 (1964); L. David Mech, The Wolf: The Ecology and Behavior of an Endangered Species (New York: American Museum of Natural History, 1970); for more recent information on wolf ecology, see Ludwig N. Carbyn et al., eds., Ecology and Conservation of Wolves in a Changing World (Edmonton: Canadian Circumpolar Institute, 1995).

6. The wolf is "listed" at 50 C.F.R. § 17.11 (2001). The relevant ESA provisions are: 16 U.S.C. § 1536(a)(1) (federal conservation duty); 16 U.S.C. § 1533(f) (recovery plan requirement); 16 U.S.C. § 1539(j) (experimental population provision); see also 16 U.S.C. § 1532(3), which defines "conserve" as using "all methods and procedures which are necessary to bring any endangered species or threatened species to the point at which the measures provided [in this act] are no longer necessary." For the recovery plan, see U.S. Fish and Wildlife Service, Northern Rocky Mountain Wolf Recovery Plan (1987).

7. On Canadian wolf dispersal, see McIntyre, supra note 1, at 357–66; Hampton, supra note 1, at 202–203. The extensive political maneuvering behind the wolf reintroduction proposal is recounted in Fischer, supra note 2; Thomas McNamee, The Return of the Wolf to Yellowstone (New York: Henry Holt, 1997); see also Robert B. Keiter and Patrick T. Holscher, Wolf Recovery Under the Endangered Species Act: A Study in Contemporary Federalism, 11 Public Land Law Rev. 19 (1990). Examples of wolf reintroduction reports include Wolf Management Committee (U.S. Fish and Wildlife Service), Reintroduction and Management of Wolves in Yellowstone National Park and the Central Idaho Wilderness Area: Report to the United States Congress (Washington, D.C., 1991); National Park Service et al., Wolves for Yellowstone? A Report to the United States Congress (Washington, D.C., 1990). The wolf reintroduction EIS funding was provided for in H.R. Rep. No. 102–256, at 16 (1991).

8. U.S. Fish and Wildlife Service, The Reintroduction of Gray Wolves to Yellowstone National Park and Central Idaho Draft Environmental Impact Statement (1993) (hereafter, Wolf Reintroduction DEIS); U.S. Fish and Wildlife Service, The Reintroduction of Gray Wolves to Yellowstone National Park and Central Idaho Final Environmental Impact Statement (1994); see also Steven H. Fritts et al., Planning and Implementing a Reintroduction of Wolves to Yellowstone National Park and Central Idaho, 5 Restoration Ecology 7–27 (vol. 5, March 1997).

9. Wyoming Farm Bureau Federation v. Babbitt, No. 94-CV-286-D (D. Wyo., filed Nov. 25, 1994) Order Denying Plaintiff's Motion for Preliminary Injunction (entered Jan. 3, 1995); see also McNamee, supra note 7, at 13–55; Edward E. Bangs and Steven H. Fritts, Reintroducing the Gray Wolf to Central Idaho and Yellowstone National Park, 24(3) Wildlife Society Bulletin 402 (1996).

10. See Wyoming Farm Bureau Federation v. Babbitt, 987 F. Supp. 1349 (D. Wyo. 1997); 16 U.S.C. § 1539(j)(2)(A) ("outside the current range" provision). For more detailed accounts of this litigation, see Fischer, supra note 2, at 143–57; McNamee, supra note 7, at 45–47, 211–27. During the pending appeal, in a separate criminal prosecution of a Montana resident accused of illegally killing one of the Yellowstone wolves, the Court of Appeals for the Ninth Circuit, interpreted the same Section 10(j) "outside the current range" language quite differently from the way Judge Downes had interpreted it, finding no legal problem with the Yellowstone wolf reintroduction program. United States v. McKittrick, 143 F.3d 1170 (9th Cir. 1998).

11. The figures cited in this paragraph are from the U.S. Fish and Wildlife Service, Rocky Mountain Wolf Recovery 1999 Annual Report (1999), at www.r6.fws.gov/wolf/annualrpt99; see also Edward E. Bangs et al., Status of Gray Wolf Restoration in Montana, Idaho, and Wyoming, 26(4) Wildlife Society Bulletin 785 (1998). By the year 2000, over 50 wolves were roaming free in northwest Montana, whereas 134 cattle or sheep have been lost to wolves resulting in compensation of $35,000. In total, Defenders of Wildlife had paid nearly $100,000 in compensation to area livestock producers. 1999 Annual Report, id. at 4, table 5. On the economic benefits from the wolves, see Wolf Reintroduction DEIS, supra note 8, at 4–25; John Duffield, An Economic Analysis of Wolf Recovery in Yellowstone: Park Visitor Attitudes and Values, in Wolves for Yellowstone?, supra note 7, at 2–35 to 2–87. On the Nez Perce tribe's role in Idaho wolf reintroduction, see Patrick I. Wilson, Wolves, Politics, and the Nez Perce: Wolf Recovery in Central Idaho and the Role of Native Tribes, 39 Nat. Res. J. 543 (1999); Michelle Nijhuis, Return of the Natives, 33(4) High Country News 1 (Feb. 26, 2001).

12. Wyoming Farm Bureau Federation v. Babbitt, 199 F.3d 1224 (10th Cir. 2000). The court also rejected the argument that the Section 10(j) provision could not be used to reintroduce Canadian wolves into the recovery areas because it was a different wolf subspecies from the original wolves that inhabited the area, which were still present and which would suffer genetic mutations by intermingling with the different reintroduced wolves. And the court deferred to the FWS's conclusion that no original wolves were still present in the recovery areas and that biologists now lumped the two subspecies for classification purposes, which avoided any ESA problems.

13. See U.S. Fish and Wildlife Service, Rocky Mountain Wolf Recovery 2001 Annual Report (Helena, Mont. 2002); Ray Ring, Wolf at the Door, 34(10) High Country News 1 (May 27, 2002); April Reese, Sawtooth Wolves Take Precedence over Livestock, Judge Rules, Land Letter, June 20, 2002, at 1.

14. See Idaho Legislative Wolf Oversight Comm., Idaho Wolf Conservation and Management Plan (March 2002); Montana Dept. of Fish, Wildlife and Parks, Draft Montana Wolf Conservation and Management Planning Document (January 2002); Wyoming Game and Fish Dept., Draft Wyoming Gray Wolf Management Plan (November 2002); see also Jeff Gearino, Wolf Delisting May Hinge on Reclassification, Casper Star Tribune, June 28, 2002; Associated Press, Wolf-Management Bill Heads to Governor, The Spokesman Review, March 15, 2002; Jim Mann, State [Montana] Releases Plan for Wolves, The Daily Inter Lake (Kalispell, Mont.), January 17, 2002; Gray Wolf Recovery "Countdown" Started Last Year, FWS Says, Endangered Species & Wetlands Report 3 (September 2001); Fish and Wildlife Service, Proposal to Reclassify and Remove the Gray Wolf from the List of Endangered and Threatened Wildlife in Portions of the Conterminous United States, 65 Fed. Reg. 43450 (July 13, 2000).

15. John Wesley Powell, Report on the Lands of the Arid Region of the United States, 45th Cong., 2d Sess., H.R. Exec. Doc. 73 (1878), reprinted by Harvard Common Press, 1983; quoted in Stephen J. Pyne, Fire in America: A Cultural History of Wildland and Rural Fire 80 (Princeton, N.J.: Princeton University Press, 1982) (hereafter, Pyne, Fire in America); see also Thomas M. Bonnicksen et al., Native American Influences on the Development of Forest Ecosystems, in Robert C. Szaro et al., eds., Ecological Stewardship: A Common Reference for Ecosystem Management, vol. 2, 442–45 (Oxford: Elsevier Science, 1999) (hereafter, Ecological Stewardship). For historical analyses of fire and fire policy, see David Carle, Burning Questions: America's Fight with Nature's Fire (Westport, Conn.: Praeger, 2002);

Pyne, Fire in America, id.; for a concise review of fire policy focusing on public lands, see Stephen J. Pyne, America's Fires: Management on Wildlands and Forests (Durham, N.C.: Forest History Society, 1997) (hereafter, Pyne, America's Fires).

16. Because the Forest Service has historically played such a major role in federal fire policy on public lands, the textual discussion focuses largely on the Forest Service and its policies. For the most part, the other federal land management agencies followed the Forest Service's lead in managing fire on the public lands. For the BLM's experience with fire and fire policy, see Pyne, Fire in America, supra note 15, at 307–14.

17. The quotation is from Pyne, Fire in America, supra note 15, at 103; see also Stephen J. Pyne, Year of the Fires: The Story of the Great Fires of 1910 (New York: Viking, 2001).

18. On the Forest Service's "10 a.m. policy," see Stephen J. Pyne et al., Introduction to Wildland Fire 256 (New York: Wiley, 2d edition, 1996).

19. The Park Service's new fire policy was set forth in National Park Service, Administrative Policies for Natural Areas (1968), reprinted in Dilsaver, supra note 4, at 354–55; the Forest Service's fire policy changes are described in Pyne, Fire in America, supra note 15, at 260–94. During the 1970s, as federal agencies consolidated their fire programs, the Forest Service shifted from being the dominant player in the national fire program to being merely a key participant. Other agencies and jurisdictions began assuming more visible and prominent roles in fire management circles. Pyne, id. at 321.

20. Prior to 1988, the burned acreage had exceeded 200,000 acres annually only once in over a dozen years. Pyne et al., supra note 18, at 264–65. For a recounting of the Yellowstone fires, see Micah Morrison, Fire in Paradise: The Yellowstone Fires and the Politics of Environmentalism (New York: Harper Collins, 1993).

21. The ecological impacts of the Yellowstone fires are analyzed in Mary Ann Franke, Yellowstone in the Afterglow: Lessons from the Fires (Mammoth, Wyo.: Yellowstone Center for Resources, 2000); Dennis H. Knight, The Yellowstone Fire Controversy, in Robert B. Keiter and Mark S. Boyce, eds., The Greater Yellowstone Ecosystem: Redefining America's Wilderness Heritage 87 (New Haven: Yale University Press, 1991); Norman Christensen et al., Interpreting the Yellowstone Fires of 1988, 39 Bioscience 678 (1989); Paul Schullery, The Fires and Fire Policy, 39 Bioscience 686 (1989). The political fallout from the Yellowstone fires is discussed and analyzed in Pamela Lichtenstein, The Politics of Wildfire: Lessons from Yellowstone, 96(5) Journal of Forestry 4 (May 1998); Schullery, id.; Chris Elfring, Yellowstone: Fire Storm over Fire Management, 39(10) BioScience 667 (November 1989); see also Morrison, supra note 20.

22. U.S. Dept. of Agriculture and U.S. Dept. of the Interior, Fire Policy Management Review Team, Final Report on Fire Management Policy (Washington, D.C., 1989), reprinted in 54 Fed. Reg. 25666 (June 16, 1989). More specifically, the report noted that many agency fire management plans did not meet current policies, planned burning could reduce hazardous fuel buildups, social and economic concerns must be balanced against the ecological benefits of fire, and the public should be more involved in developing fire management plans.

23. See Pyne, America's Fires, supra note 15, at 28–30, 36; U.S. General Accounting Office, Federal Fire Management: Limited Progress in Restarting the Prescribed Fire Program (Washington, D.C., 1990); see also John N. Maclean, Fire on the Mountain: The True Story of the South Canyon Fire (New York: William Morrow, 1999).

24. U.S. Dept. of the Interior and U.S. Dept. of Agriculture, Federal Wildland Fire Management Policy and Program Review: Final Report (Washington, D.C., 1995).

25. See U.S. Dept. of the Interior et al., Review and Update of the 1995 Federal Wildland Fire

Management Policy (2001); U.S. Dept. of the Interior, U.S. Dept. of Agriculture, and Western Governors' Association, A Collaborative Approach for Reducing Wildland Fire Risks to Communities and the Environment: Ten Year Comprehensive Strategy (2001); Mark Blaine, Fixing Fire, Forest Magazine 18 (July/August 2001); Larisa Epatko, List of Communities at Risk of Wildfires Balloons, Land Letter, May 8, 2001, at 3; Timothy Egan, Idea of Fighting Fire with Fire Wins Converts, New York Times, June 30, 2002, at p. 1; Rocky Barker, Fire Prevention Efforts Stall: Risk in West Grows as New Policy Catches on Slowly, Idaho Statesman, June 30, 2002, at p. 1; Dan Gallagher, With Fire Season Over, 2003 Looks Just as Bad, Idaho Post Register, Oct. 15, 2002.

26. Good discussions of fire ecology can be found in Pyne et al., supra note 18, at 171; Stephen F. Arno and Steven Allison-Bunnell, Flames in Our Forests: Disaster or Renewal? 65-89 (Washington, D.C.: Island Press, 2002).

27. See R. Neil Sampson et al., Overview, in R. Neil Sampson and David L. Adams, eds., Assessing Forest Ecosystem Health in the Inland West 3–10 (Binghamton, N.Y.: Food Products Press, 1994) (hereafter, Assessing Forest Ecosystem Health). Another telling statistic highlights the urban–wildland interface problem: between 1985 and 1992, over 1,300 homes were destroyed by pine forest wildfires. Bonnicksen et al., supra note 15, at 455. The relationship between fire and forest ecosystems is discussed further in the last section of this chapter.

28. Under the Federal Tort Claims Act, 28 U.S.C. § 2671 et seq., the federal government has waived its sovereign immunity from liability for damages except when the claim is "based upon the exercise or performance or the failure to exercise or perform a discretionary function or duty . . . whether or not the discretion involved is abused." 28 U.S.C. § 2680(a). Because the formulation of public land fire policy inevitably involves discretionary judgments, liability is unlikely to attach based on the policy itself. See United States v. Varig Airlines, 467 U.S. 797, 814 (1984); Berkovitz by Berkovitz v. United States, 486 U.S. 531, 537 (1988); McDougal v. U.S. Forest Service, 195 F.Supp.2d 1229 (D. Ore. 2002). However, if government officials negligently implement the policy, then tort liability might attach. See Rayonier, Inc. v. United States, 352 U.S. 315 (1957) (U.S. liable for negligently allowing a contained fire on public lands to escape and damage nearby property); see also Peter H. Froelicher, Issues of Liability Surrounding Fire Management in the Greater Yellowstone Area, 27 Land and Water L. Rev. 123 (1992); Pyne et al., supra note 18, at 329–39.

29. On federal criminal sanctions and public land fire, see 18 U.S.C. § 1855; 18 U.S.C. §§ 3571(b)(3), (c)(3); see also United States v. Alford, 274 U.S. 264 (1927); United States v. Lindsey, 595 F.2d 5 (9th Cir. 1979). On fire and NEPA, see Rhodes v. Johnson, 153 F.3d 785 (7th Cir. 1998); Laura Sweedo, Where There Is Fire, There Is Smoke: Prescribed Burning in Idaho's Forests, 8 Dick. J. Envir. L. Pol. 121 (1999). On fire and air pollution, see 42 U.S.C. § 7418(a) (requiring federal agencies to comply with state Clean Air Act standards); United States v. Tennessee Air Pollution Control Board, 185 F.3d 529 (6th Cir. 1999). Clean Air Act provisions potentially applicable to prescribed burning on public lands are those governing visibility, carbon monoxide, and particulates. 42 U.S.C. §§ 7491–92, 7512, 7513; see also Sweedo, id.; Pyne et al., supra note 18, at 554–66.

30. On mining's legacy, see U.S. General Accounting Office, Federal Land Management: An Assessment of Hardrock Mining Damage (Washington, D.C., 1988); U.S. General Accounting Office, Public Lands: Interior Should Ensure Against Abuses from Hardrock Mining (Washington, D.C., 1986); John D. Leshy, The Mining Law: A Study in Perpetual Motion 183–228 (Baltimore, Md.: Johns Hopkins University Press, 1987); on the legacy of

livestock grazing, see Debra L. Donahue, The Western Range Revisited: Removing Live-stock from Public Lands to Conserve Native Biodiversity 42–66 (Norman: University of Oklahoma Press, 1999); U.S. General Accounting Office, Public Rangelands: Some Ripar-ian Areas Restored but Widespread Improvement Will Be Slow (Washington, D.C., 1988) (hereafter, GAO Riparian Areas); on the impact of logging and road construction on the national forests, see U.S. Forest Service, Administration of the Forest Development Sys-tem: Temporary Suspension of Road Construction and Reconstruction in Unroaded Areas (proposed rule making), 63 Fed. Reg. 4351 (Jan. 28, 1998); 64 Fed. Reg. 7290 (Feb. 12, 1999); see also Special Section: Ecological Effects of Roads, 14(1) Conservation Biology 16–94 (2000). See also Dan Flores, The Natural West: Environmental History in the Great Plains and Rocky Mountains (Norman: University of Oklahoma Press, 2001), for a historical overview of the human impact on the western environment.

31. On early-twentieth-century wildlife restoration efforts, see Daniel A Poole and James B. Trefethen, The Maintenance of Wildlife Populations, in Howard P. Brokaw, ed., Wildlife and America 339 (Washington, D.C.: Council on Environmental Quality, 1978); see also James B. Trefethen, An American Crusade for Wildlife (New York: Winchester Press, 1975). The Yellowstone bison restoration effort is recounted in H. Duane Hampton, How the U.S. Cavalry Saved Our National Parks, 165–67 (Bloomington: Indiana University Press, 1971); Aubrey L. Haines, The Yellowstone Story, vol. 2, 54–77 (Boulder: Colorado Associ-ated University Press, 1977). The Park Service's early predator policies are examined in Sel-lars, supra note 4, at 71–75, 119–24.

32. The opening quotation is from William E. Shands and Robert G. Healy, The Lands No-body Wanted xiv (Washington, D.C.: Conservation Foundation, 1977), which generally examines eastern national forest restoration policy; see also Bill McKibben, An Explosion of Green, Atlantic Monthly 61 (April 1995). On the Weeks Act, see Samuel T. Dana and Sally K. Fairfax, Forest and Range Policy: Its Development in the United States 111–14 (New York: McGraw-Hill, 2d edition, 1980); Martha Carlson, Private Lands–Public For-est: The Story of the Weeks Act, Forest Notes 3 (Summer 1986). The navigable streams re-quirement was written into the Weeks Act to address lingering constitutional concerns over whether the federal government had the authority to purchase private land. By requiring that the purchased land encompass navigable streams, Congress's interstate commerce powers could be used to justify the federal purchases. The Clarke-McNary Act of 1924 sub-sequently extended the Forest Service's acquisition authority to include lands "for the pro-duction of timber." Shands and Healy, id. at 120.

33. The history of ecological restoration is briefly recounted in William R. Jordan III et al., Restoration Ecology: Ecological Restoration as a Technique for Basic Research, in William R. Jordan III et al., eds., Restoration Ecology: A Synthetic Approach to Ecological Research 3–4 (New York: Cambridge University Press, 1987). On the University of Wisconsin Arboretum prairie restoration project and the Leopold family restoration efforts, see Stephanie Mills, In Service of the Wild: Restoring and Reinhabiting Damaged Land 93–129 (Boston: Beacon Press, 1995).

34. Aldo Leopold, The Arboretum and the University, 18(2) Park and Recreation 59–60 (1934), quoted in Curt Meine and Richard L. Knight, eds., The Essential Aldo Leopold: Quota-tions and Commentaries 123 (Madison: University of Wisconsin Press, 1999). On the work of early ecologists, see A. D. Bradshaw, Restoration: An Acid Test for Ecology, in Jordan et al., supra note 33, at 25–27.

35. Society for Ecological Restoration Mission Statement (1993), reprinted in Wallace Coving-

ton et al., Ecosystem Restoration and Management: Scientific Principles and Concepts, in Ecological Stewardship, vol. 2, supra note 15, at 601. On the expanding scale of ecological restoration efforts, see Peter S. White, Spatial and Biological Scales in Reintroduction, in Donald A. Falk et al., eds., Restoring Diversity: Strategies for Reintroduction of Endangered Plants 49–86 (Washington, D.C.: Island Press, 1996). For a comprehensive overview of environmental laws with restoration components, see Celia Campbell-Mohn et al., Environmental Law: From Resources to Recovery (St. Paul, Minn.: West Publishing Co., 1993).

36. The principal books on ecological restoration include: William R. Jordan III et al., eds., Restoration Ecology: A Synthetic Approach to Ecological Research (New York: Cambridge University Press, 1987); Research Council Committee on Restoration of Aquatic Ecosystems, Restoration of Aquatic Ecosystems: Science, Technology, and Public Policy (Washington, D.C.: National Academy Press, 1992) (hereafter, Restoration of Aquatic Ecosystems); A. Dwight Baldwin et al., eds., Beyond Preservation: Restoring and Inventing Landscapes 32 (Minneapolis: University of Minnesota Press, 1994); Falk et al., eds., supra note 35; Donald Harker et al., Landscape Restoration Handbook (Boca Raton, Fla.: Lewis Publishers, 2d edition, 1999); Steven G. Whisenant, Restoring Damaged Wildlands: A Process-Orientated, Landscape-Scale Approach (New York: Cambridge University Press, 1999). The two primary journals addressing ecological restoration are: Restoration and Management Notes (published by the University of Wisconsin Arboretum); Restoration Ecology (published by Blackwell Science).

37. The first quotation is from John J. Berger, Ecological Restoration Comes of Age, Forum for Applied Research and Public Policy 90 (Summer 1995); see also Restoration of Aquatic Ecosystems, supra note 36, at 2; William R. Jordan III, "Sunflower Forest": Ecological Restoration as the Basis for a New Environmental Paradigm, in Baldwin et al., supra note 36, at 32 (defining ecological restoration as "the active attempt to compensate for human influence on an ecological system in order to return the system to its historic condition"). The second quotation is from Society for Ecological Restoration Mission Statement (1993), reprinted in Wallace Covington et al., supra note 35, at 601; see also Restoration of Aquatic Ecosystems, id. at 2. The third quotation is from Wallace Covington, id. at 601; see also Restoration of Aquatic Ecosystems, id. at 17–21; Andy P. Dobson et al., Hopes for the Future: Restoration Ecology and Conservation Biology, 277 Science 515 (July 1997).

38. For a more detailed discussion of these various restoration techniques, see James G. Kenna et al., Ecosystem Restoration: A Manager's Perspective, in Ecological Stewardship, vol. 2, supra note 15, at 620–21; see also Restoration of Aquatic Ecosystems, supra note 36, at 17–18; Harker et al., supra note 36, at 19–40, 63–90.

39. See Ecological Restoration, in Ecological Stewardship, vol. 1, supra note 15, at 96; see also James A. Harris et al., Land Restoration and Reclamation: Principles and Practice 16–18 (Essex: Addison Wesley, 1996); John Cairns, Jr., Eco-Societal Restoration: Rehabilitating Human Society's Life-support System, in B. C. Rana, ed., Damaged Ecosystems and Restoration 1–23 (Singapore: World Scientific Publishing, 1998); Restoration of Aquatic Ecosystems, supra note 36, at 19–21. On mitigation, see 40 C.F.R. 1508.20 (2001) (defining mitigation for NEPA purposes); see also Falk et al., supra note 35, at 273–394 (discussing reintroduction and mitigation).

40. For general descriptions of ecological restoration strategies, processes, and practices, see Kenna et al., in Ecological Stewardship, vol. 2, supra note 15, at 619–76; Harker et al., supra note 36, at 63–90; Leslie Jones Sauer, The Once and Future Forest: A Guide to Forest Restoration Strategies 89–129 (Washington, D.C.: Island Press, 1998).

41. To get the full flavor of this debate, compare Jordan, Sunflower Forest, in Baldwin et al., eds., supra note 36, at 17–34, with G. Stanley Kane, Restoration or Preservation? Reflections on a Clash of Environmental Philosophies, id. at 69–84; see also Alastair S. Gunn, The Restoration of Species and Natural Environments, 13 Envir. Ethics 213 (1991); C. Mark Cowell, Ecological Restoration and Environmental Ethics, 15 Envir. Ethics 19 (1993); Robert Elliot, Extinction, Restoration, Naturalness, 16 Envir. Ethics 135 (1994).

42. See NPS Leopold Report, in Dilsaver, supra note 4, at 237, 239; 1964 Udall Memo, in Dilsaver, supra note 4, at 272; National Park Service, Management Policies 2001 (Washington, D.C., 2001). On the evolution of the National Park Service's natural resource management policies, see Sellars, supra note 4; R. Gerald Wright, Wildlife Research and Management in the National Parks (Urbana: University of Illinois Press, 1991); Alfred Runte, National Parks: The American Experience (Lincoln: University of Nebraska Press, 1987); Robert B. Keiter, On Preserving Nature in the National Parks: Law, Policy, and Science in a Dynamic Environment, 74 Denver U. Law Review 649 (1997).

43. The principal works criticizing the Park Service's resource management policies, particularly in the Yellowstone setting, are: Frederic H. Wagner et al., Wildlife Policies in the U.S. National Parks (Washington, D.C.: Island Press, 1995); Stephen Budiansky, Nature's Keepers: The New Science of Nature Management 131–58 (New York: Free Press, 1995); Karl Hess, Jr., Rocky Times in Rocky Mountain National Park: An Unnatural History (Niwot: University of Colorado Press, 1993); Alston Chase, Playing God in Yellowstone: The Destruction of America's First National Park (Boston: Atlantic Monthly Press, 1986). For a thoughtful rebuttal, see James A. Pritchard, Preserving Yellowstone's Natural Conditions: Science and the Perception of Nature (Lincoln: University of Nebraska Press, 1999); see also National Research Council Committee on Ungulate Management in Yellowstone National Park, Ecological Dynamics on Yellowstone's Northern Range (Washington, D.C.: National Academy Press, 2002).

44. Compare Covington et al., supra note 35, at 602–3, with Bonnicksen et al., supra note 15, at 461–64.

45. On the trajectory of evolutionary change, see Covington et al., supra note 35, at 602–3; Nancy Langston, Forest Dreams, Forest Nightmares: The Paradox of Old Growth in the Inland West 274–78 (Seattle: University of Washington Press, 1995). As an alternative to the "trajectory of evolutionary change" standard, the Forest Service and other commentators have suggested a "natural range of variability" standard. See Penelope Morgan et al., Historical Range of Variability: A Useful Tool for Evaluating Ecosystem Change, in Assessing Forest Ecosystem Health, supra note 27, at 87. Potential sources of information about the precontact environment include historical records, local knowledge, photographic evidence, and comparative adjacent pristine areas. See Covington et al., id. at 604; Morgan et al., id. at 96–100.

46. On intervening in the Yellowstone ecosystem, see Budiansky, supra note 43, at 131–55; Wagner et al., supra note 43, at 49–91, 371–75; Chase, supra note 43, at 49–91, 371–75; see also Hess, supra note 43, making the same point about elk management at Rocky Mountain National Park.

47. For the argument that nature should be left alone, see Bill Willers, ed., Unmanaged Landscapes: Voices for Untamed Nature (Washington, D.C.: Island Press, 1999).

48. For a critical analysis of species reintroduction strategies, see Holly Doremus, Restoring Endangered Species: The Importance of Being Wild, 23 Harvard Envir. L. Rev. 1 (1999); see also Federico Cheever, From Population Segregation to Species Zoning: The Evolution of

Reintroduction Law Under Section 10(j) of the Endangered Species Act, 1 Wyoming L. J. 287 (2001). On the Columbia River salmon restoration controversy, see William Dietrich, Northwest Passage: The Great Columbia River 323–53 (New York: Simon and Schuster, 1995); Blaine Harden, A River Lost: The Life and Death of the Columbia 213–38 (New York: Norton, 1996); Michael C. Blumm et al., Saving Snake River Water and Salmon Simultaneously: The Biological, Economic, and Legal Case for Breaching the Lower Snake River Dams, Lowering John Day Reservoir, and Restoring Natural River Flows, 28 Envir. L. 997 (1998).

49. On the question of scale for ecological restoration purposes, see Jonathan B. Haufler et al., Scale Considerations for Ecosystem Management, in Ecological Stewardship, vol. 2, supra note 15, at 331; Peter S. White, supra note 35, at 49; Michael E. Soule and John Terborgh, eds., Continental Conservation: Scientific Foundations of Regional Reserve Networks (Washington, D.C.: Island Press, 1999) (hereafter, Continental Conservation). On linkage strategies, see Daniel J. Simberloff et al., Regional and Continental Restoration, in Continental Conservation, id. at 70–71.

50. Examples of ecosystem management projects, including those focused primarily on ecological restoration, can be found in Steven L. Yaffee et al., Ecosystem Management in the United States: An Assessment of Current Experience 16–17, 73–75 (Washington, D.C.: Island Press, 1996); see also Douglas S. Kenney, The New Watershed Sourcebook: A Directory and Review of Watershed Initiation in the Western United States (Boulder, Colo.: Natural Resources Law Center, 2000); The Keystone Center, The Keystone National Policy Dialogue on Ecosystem Management Final Report B-1 to B-39 (Keystone, Colo., 1996).

51. See 16 U.S.C. § 1604 (NFMA planning provision); 43 U.S.C. § 1712 (FLPMA planning provision); 30 U.S.C. § 1258 (SMCRA reclamation plan requirements); 16 U.S.C. § 1604 (g)(3)(B) (NFMA biodiversity provision); 16 U.S.C. § 1604 (g)(3)(E)(ii) (NFMA five-year tree restocking requirement); 33 U.S.C. § 1251(a) (CWA water quality restoration); 42 U.S.C. § 4332(2)(C) (NEPA EIS provision); see also Campbell-Mohn et al., supra note 35; Robert B. Keiter, NEPA and the Emerging Concept of Ecosystem Management, 25 Land and Water L. Rev. 43 (1990); Dinah Bear, Using the National Environmental Policy Act to Protect Biological Diversity, 8 Tulane Envir. L. J. 77 (1994). On the role of litigation in ecological restoration, see Daniel F. Luecke, An Environmental Perspective on Large Ecosystem Restoration Processes and the Role of the Market, Litigation, and Regulation, 42 Ariz. L. Rev. 395 (2000).

52. See Yaffee et al., supra note 50, at 55–56.

53. See GAO Riparian Areas, supra note 30, at 2, 4, 35; 43 C.F.R. §§ 4.77, 1784.0–1 to 1784.62, 4100.0–1 to 4180.2 (2001) (range reform regulations); Bureau of Land Management, Rangeland Reform '94: Final Environmental Impact Statement (Washington, D.C., 1994); Public Lands Council v. Babbitt, 167 F.3d 1287 (10th Cir. 1999), aff'd, 529 U.S. 728 (2000). For an overview of the conflict over public land grazing, see Charles F. Wilkinson, Crossing the Next Meridian: Land, Water, and the Future of the West 75–113 (Washington, D.C.: Island Press, 1992); see also George Wuerthener and Mollie Matteson, eds., Welfare Ranching: The Subsidized Destruction of the American West (Washington, D.C.: Island Press, 2002); Donahue, supra note 30; Dan Daggett, Beyond the Rangeland Conflict: Toward a West that Works (Salt Lake City, Utah: Gibbs Smith, 1995); Karl Hess, Jr., Visions upon the Land: Man and Nature on the Western Range (Washington, D.C.: Island Press, 1992).

54. On range restoration strategies, see Kenna et al., supra note 38, at 644, 646; GAO Riparian Areas, supra note 30, at 18–35. On federal authority to control livestock numbers, see Nat-

ural Resources Defense Council v. Morton, 388 F.Supp. 829 (D.D.C. 1974); Oregon Natural Desert Association v. Singleton, 47 F.Supp.2d 1182 (D. Ore. 1998); Idaho Watersheds Project v. Hahn, 187 F.3d 1035 (9th Cir. 1999); see also McKinley v. United States, 828 F.Supp. 888 (D. N.M. 1993) (sustaining the Forest Service's authority to reduce livestock numbers); Donahue, supra note 30, at 193–228. The Comb Wash litigation can be found at National Wildlife Federation v. Bureau of Land Management, 140 IBLA 85 (1997); see also Joseph Feller, What Is Wrong with the BLM's Management of Livestock Grazing on the Public Land?, 30 Idaho L. Rev. 555 (1993–94). On ESA litigation involving range conditions, see Southwest Center for Biological Diversity v. U.S. Forest Service, 82 F.Supp.2d 1070 (D. Ariz. 2000); Pacific Rivers Council v. Thomas, 936 F.Supp. 738 (D. Idaho 1996); see also Oregon Natural Desert Association v. Green, 953 F. Supp. 1133 (D. Ore. 1997). On application of the Clean Water Act to range conditions, see Pronsolino v. Nastri, 291 F.3d 1123 (9th Cir. 2002); Oregon Natural Desert Association v. Dombeck, 151 F.3d 945 (9th Cir. 1998); see also Debra L. Donahue, The Untapped Power of the Clean Water Act Section 401, 23 Ecology L. Q. 201 (1996); Richard L. Braun, Emerging Limits on Federal Land Management Discretion: Livestock, Riparian Ecosystems, and Clean Water Law, 17 Envir. L. 43 (1986).

55. See Doc and Connie Hatfield, Trout Creek Mountain Working Group, www.mtnvisions.com/Aurora/tcmwghat (visited Nov. 15, 2000); Ron Rhew, Oregon Lahontan Cutthroat Trout, and Results of Six Years of Management Developed by the Trout Creek Working Group, www.mtnvisions.com/Aurora/ronrhew (visited Nov. 15, 2000); Tom Knudson, The Ranch Restored: An Overworked Land Comes Back to Life, High Country News, March 1, 1999, at p. 13; David E. Brown, The Trout Creek Mountain Experience, 58 Wilderness 28 (1995); see also Ecological Stewardship, vol. 2, supra note 15, at 644.

56. See William McDonald and Ronald J. Bemis, Community Involvement and Sustainability: The Malpai Borderlands Effort, in Peter H. Raven, ed., Nature and Human Society: The Quest for a Sustainable World 596 (Washington, D.C.: National Academy Press, 1997); Kelly Cash, Malpai Borderlands: The Searchers for Common Ground, in Philip Brick et al., eds., Across the Great Divide: Explorations in Collaborative Conservation and the American West 112 (Washington, D.C.: Island Press, 2001); see also Yaffee et al., supra note 50, at 183–84; www.malpaiborderlandsgroup.org (visited July 11, 2002).

57. The interior West's forest health problems have been documented in several publications. See U.S. General Accounting Office, Western National Forests: A Cohesive Strategy Is Needed to Address Catastrophic Wildfire Threats (Washington, D.C., 1999); Assessing Forest Ecosystem Health, supra note 27; Raymond G. Jaindl and Thomas M. Quigley, eds., Search for a Solution: Sustaining the Land, People, and Economy of the Blue Mountains (Washington, D.C.: American Forests, 1996); Langston, supra note 45.

58. On the environmental community's general views on forest restoration, see Michael Axline, Forest Health and the Politics of Expediency, 26 Envir. L. 613 (1996); Gregory H. Aplet, Beyond Even- vs. Uneven-Aged Management: Toward a Cohort-Based Silviculture, in Assessing Forest Ecosystem Health, supra note 27, at 423; Langston, supra note 45, at 278–80; Reed F. Noss and Alan Y. Cooperrider, Saving Nature's Legacy: Protecting and Restoring Biodiversity 209–19 (Washington, D.C.: Island Press, 1994); Reed F. Noss, The Wildlands Project: Land Conservation Strategy, Wild Earth 10–25 (Special Issue 1992). Alternative forest restoration goals and strategies are outlined in Sampson and Adams, supra note 27, at 7–8; Russell T. Graham, Silviculture, Fire, and Ecosystem Management, in Assessing Forest Ecosystem Health, supra note 27, at 339; Chadwick D. Oliver et al., Manag-

ing Ecosystems for Forest Health: An Approach and the Effects on Uses and Values, id., at 113; Covington et al., in Ecological Stewardship, vol. 2, supra note 15, at 603–6 (describing Flagstaff area Ponderosa pine restoration project); see also Langston, id. at 278–80.

59. On the Knudsen-Vandenburg Act, see Randal O'Toole, Reforming the Forest Service 4, 132–33, 135–36 (Washington, D.C.: Island Press, 1988); Wilkinson, Crossing the Next Meridian, supra note 53, at 169–74 (also noting that 1976 amendments to the act allowed the Forest Service to use its K-V funds for virtually any timber-related purposes, not just reforestation). The NFMA salvage sale exceptions are found at 16 U.S.C. §§ 1604(g)(3)(F)(iv), (m)(1). Examples of environmental challenges to Forest Service salvage timber sales include: The Wilderness Society v. Rey, 180 F.Supp.2d 1141 (D. Mont. 2002); Utah Environmental Congress v. Zieroth, 190 F.Supp.2d 1265 (D. Utah 2002); Sierra Club v. Bosworth, 199 F.Supp.2d 971 (N.D. Cal. 2002); Heartwood, Inc. v. U.S. Forest Service, 73 F.Supp.2d 962 (S.D. Ill. 1999); Blue Mountains Biodiversity Project v. Blackwood, 161 F.3d 1208 (9th Cir. 1998); Krichbaum v. U.S. Forest Service, 17 F.Supp.2d 549 (W.D. Va. 1998); Kettle Range Conservation Group v. U.S. Forest Service, 971 F.Supp. 480 (D. Ore. 1997); see also April Reese, Debate Over Post-Fire Salvage Logging Heats Up, Land Letter, Jan. 31, 2002, at 5.

60. On the 1995 timber salvage sale rider and related litigation, see Patti A. Goldman and Kristen L. Boyles, Forsaking the Rule of Law: The 1995 Logging without Laws Rider and Its Legacy, 27 Envir. L. 1035 (1997); see also Chapter 4 for a more detailed discussion of this controversy. On the Bush administration's forest health and fire initiative, see President of the United States, Healthy Forests: An Initiative for Wildfire Prevention and Stronger Communities (2002); Brian Stempeck, Bush Moves to Speed Up Forest-Thinning Projects, 21(21) Land Letter 1 (Dec. 12, 2002).

61. According to scientists, the number of Ponderosa pine trees per acre had grown from 24.3 trees in 1876 to 1,254 trees in 1992, while grassy forest openings had declined from 81 percent of the forest in 1876 to only 7 percent in 1992. Covington et al., in Ecological Stewardship, vol. 2, supra note 15, at 606; see also Wallace W. Covington and M. M. Moore, Postsettlement Changes in Natural Fire Regimes and Forest Structure: Ecological Restoration of Old-Growth Ponderosa Pine Forests, in Assessing Forest Ecosystem Health, supra note 27, at 153.

62. The Greater Flagstaff Forests Partnership's members include the city of Flagstaff, the Flagstaff Fire Department, Northern Arizona University, Grand Canyon Trust, The Nature Conservancy, U.S. Fish and Wildlife Service, Arizona Game and Fish Department, and others. For a description of the partnership and its restoration agenda, see Michelle Nijhuis, Flagstaff Searches for Its Forests' Future, High Country News, March 1, 1999, at 8–12; www.gffp.org (visited July 11, 2002); for a scientific analysis of Ponderosa pine forest restoration, see Covington et al., Ecological Stewardship, vol. 2, supra note 15, at 603–6; Covington and Moore, supra note 61, at 153; see also John Herron, Wildfire Policy and Suppression in the American Southwest, in Christopher J. Huggard and Arthur R. Gomez, eds., Forests Under Fire: A Century of Ecosystem Mismanagement in the Southwest 181–210 (Tucson: University of Arizona Press, 2001).

63. See the first part of this chapter for a more detailed description of the Section 10(j) provision.

64. See 50 C.F.R. § 17.84(j) (2001) (California condor special regulations); 50 C.F.R. § 17.84(g) (2001) (black-footed ferret special regulations); U.S. Fish and Wildlife Service, Pacific Region, Recovery Plan for the California Condor (Portland, Ore., 1996); Bureau of Land

Management, Black-footed Ferret Reintroduction Conata Basin/Badlands, South Dakota Final Environmental Impact Statement (Washington, D.C., 1994); see also Les Line, Phantom of the Plains: The Continuing Saga of the Black-Footed Ferret, 100(4) Wildlife Conservation 20 (July/August 1997); Tim W. Clark, Averting Extinction: Reconstructing Endangered Species Recovery (New Haven: Yale University Press, 1997); Brian Miller et al., Prairie Night: Black-Footed Ferrets and the Recovery of Endangered Species (Washington, D.C.: Smithsonian Institution Press, 1996); Jane Hendron, Condor Soars toward Recovery, Endangered Species Bulletin 26 (March, 1998); Frank Graham, Jr., Day of the Condor, 102(6) Audubon 46 (January/February 2000). On the condor litigation, see National Audubon Society v. Hester, 627 F.Supp. 1419 (D.D.C. 1986), reversed, 791 F.2d 210 (D.C. Cir. 1986) (challenging the initial captive breeding decision); West's Legal News, Federal Judge Denies Request to Halt Condor Release in Arizona, 10–17–96 WLN 11047 (Oct. 17, 1996); Greenwire, Condors: Court Ruling Opens the Way for Bird Release in AZ, Oct. 15, 1996; Jim Woolf, San Juan in Court Fighting Condor, Salt Lake Tribune, Oct. 8, 1996, at B1. On captive breeding generally, see Michael E. Soule and Bruce A. Wilcox, eds., Conservation Biology: An Evolutionary-Ecological Perspective 197–270 (Sunderland, Mass.: Sinauer Associates, 1980).

65. On the North Carolina wolf restoration initiative, see Gibbs v. Babbitt, 31 F.Supp.2d 531 (E.D.N.C. 1998), affirmed, 214 F.3d 483 (4th Cir. 2000); 50 C.F.R. § 17.84 (c) (2001); 51 Fed. Reg. 41,790 (1986); see also Jennifer Gilbreath, A Bright Decade for the Red Wolf, 15(4) Endangered Species Update E18 (July/August 1998); T. Edward Nickens, North Carolina Wolf Country, 102 Wildlife Conservation 64 (January/February 1999). On the Southwestern wolf restoration program, see 50 C.F.R. § 17.84(k) (2001); New Mexico Cattle Growers Association v. U.S. Fish and Wildlife Service, 81 F.Supp.2d 1141 (D. N.M. 1999); U.S. Fish and Wildlife Service, Reintroduction of the Mexican Wolf within Its Historic Range in the Southwestern United States Final Environmental Impact Statement (Albuquerque, N.M., 1996); see also Heidi Ridgley, Opening the Door to Wolf Recovery, 74(4) Defenders 6 (Fall 1999); U.S. Fish and Wildlife Service, Mexican Wolf Returns to the Wild, Endangered Species Bulletin 12 (March 1998).

66. Department of the Interior, U.S. Fish and Wildlife Service, Proposal to Reclassify and Remove the Gray Wolf from the List of Endangered and Threatened Wildlife in Portions of the Conterminous United States; Proposal to Establish Three Special Regulations for Threatened Gray Wolves, 65 Fed. Reg. 43450 (July 13, 2000). On the illegal wolf killings in the Southwest, see U.S. Fish and Wildlife Service News Releases, Wolves to Be Moved into Gila Wilderness, March 21, 2000; Endangered Mexican Wolves May Be Moved to Gila National Forest, U.S. Newswire, March 2, 2000. On the Montana, Idaho, and Wyoming reaction to wolf reintroduction, see Bangs et al., supra note 11, at 794. Specifically, Idaho, Montana, and Wyoming signed a memorandum of understanding that effectively precluded their involvement in wolf recovery management until wolves were removed from ESA coverage. However, the states agreed to coordinate their wolf management and planning efforts once they are delisted, including the control of problem wolves. They have also asserted that federal funds must be provided to manage wolves both before and after recovery. See Memorandum of Understanding among the States of Montana, Idaho and Wyoming Concerning Monitoring and Management of a Recovered Wolf Population in the Central Rocky Mountain Region, Dec. 2, 1999. The Wyoming statute classifying the wolf as a predator is found at Wyo. Stat. Ann. §§ 23–1–101(viii), 23–3–103(a) (2001).

67. On grizzly bear management and policy, see generally Thomas McNamee, The Grizzly

Bear (New York: Knopf, 1984); John Craighead et al., The Grizzly Bears of Yellowstone: Their Ecology in the Yellowstone Ecosystem, 159–92 (Washington, D.C.: Island Press, 1995); Brian L. Kuehl, Conservation Observations Under the Endangered Species Act: A Case Study of the Yellowstone Grizzly Bear, 64 U. Colo. L. Rev. 607 (1993); see also Symposium on Large Carnivore Conservation, 10(9) Conservation Biology 936–1055 (1996). The case for grizzly bear linkage corridors is made in Noss and Cooperrider, supra note 58, at 150–56; Tim W. Clark and Steven C. Minta, Greater Yellowstone Ecosystem: Prospects for Ecosystem Science, Management and Policy 84–85 (Moose: Homestead Press, 1994); Mark Schaffer, Keeping the Grizzly Bear in the American West: A Strategy for Real Recovery (Washington, D.C.: The Wilderness Society, 1992). On the Cabinet-Yaak and Selkirk litigation, see Carlton v. Babbitt, 900 F. Supp. 526 (D.D.C. 1995); Carlton v. Babbitt, 26 F. Supp.2d 102 (D.D.C. 1998).

68. U.S. Fish and Wildlife Service, Grizzly Bear Recovery Plan 140–42 (Missoula, Mont., 1993). Because the grizzly bear was listed under the ESA before 1978 (when Congress amended the ESA to require that critical habitat be designated as part of the listing decision), the FWS is not legally obligated to designate critical habitat for the bear. 16 U.S.C. § 1532(5)(B); Fund for Animals v. Babbitt, 903 F.Supp. 96, 115–17 (D.D.C. 1995).

69. On the accomplishments and shortcomings of the federal grizzly bear recovery effort, see Louisa Willcox, The Last Grizzlies of the American West: The Long Hard Road to Recovery, 14 Endangered Species UPDATE 11–16 (Nos. 9 & 10, 1997); Kuehl, supra note 67, at 612–22; Shaffer, supra note 67, at 10–13; see also Todd Wilkinson, Science Under Siege: The Politicians' War on Nature and Truth 109 (Boulder, Colo.: Johnson Books, 1998) for a critical description of the grizzly bear recovery effort. Examples of grizzly bear habitat–focused litigation include Resources Ltd., Inc. v. Robertson, 35 F.3d 1300 (9th Cir. 1993); Swan View Coalition, Inc. v. Turner, 824 F.Supp. 923 (D. Mont. 1992); Bob Marshall Alliance v. Hodel, 852 F.2d 1223 (9th Cir. 1988), cert. denied, 489 U.S. 1066 (1989). The grizzly bear recovery plan decision is in Fund for Animals v. Babbitt, 903 F. Supp. 96 (D.D.C. 1995).

70. See U.S. Fish and Wildlife Service, Grizzly Bear Recovery in the Bitterroot Ecosystem Final Environmental Impact Statement (Missoula, Mont., 2000). On the Bitterroot ecosystem grizzly bear reintroduction citizen management committee proposal, see Sarah B. Van de Wetering, Bears, People, Power, 2(2) Chronicle of Community 15 (Winter 1998); Brenda L. Hall, Subdelegation of Authority Under the Endangered Species Act: Secretarial Authority to Subdelegate His Duties to a Citizen Management Committee as Proposed for the Selway-Bitterroot Wilderness Grizzly Bear Reintroduction, 20 Public Land and Resources L. Rev. 81 (1999); see also Hank Fischer, Moving Past the Polarization: Wolves, Grizzly Bears, and Endangered Species Recovery (supporting the committee proposal) and Doug Honnold, Wolves, Bears, and the Spirit of the Wild: Asking the Right Questions (opposing the committee proposal), in Robert B. Keiter, ed., Reclaiming the Native Home of Hope: Community, Ecology and the American West 121, 129 (Salt Lake City: University of Utah Press, 1998); Willcox, supra note 69, at 15–16 (opposing the committee proposal).

71. The ESA listing/delisting criteria are found at 16 U.S.C. § 1533(a); 50 C.F.R. § 424.11 (2001); see also Philip Kline, Grizzly Bear Blues: A Case Study of the Endangered Species Act's Delisting Process and Recovery Plan Requirements, 31 Envir. L. 371 (2001); Robert B. Keiter, Observations on the Future Debate over "Delisting" the Grizzly Bear in the Greater Yellowstone Ecosystem, 13 Envir. Professional 248 (1991); Willcox, supra note 69, at 15; Todd Wilkinson, Grizzly War, 30(21) High Country News 1 (Nov. 9, 1998).

72. See U.S. Fish and Wildlife Service, Interagency Conservation Strategy Team, Draft Con-

servation Strategy for the Grizzly Bear in the Yellowstone Area (Missoula, Mont., 2000); see also Holly Doremus, Delisting Endangered Species: An Aspirational Goal, Not a Realistic Expectation, 30 Envir. Law Reporter 10434 (2000), and Federico Cheever, The Rhetoric of Delisting Species Under the Endangered Species Act, 31 Envir. Law Reporter 11302 (2001), which argue against an aggressive ESA delisting program because of practical differences between species recovery and delisting.

73. On the role of salmon in the Pacific Northwest regional management initiatives, see U.S. Forest Service and Bureau of Land Management, Integrated Scientific Assessment for Ecosystem Management in the Interior Columbia Basin 68 (Portland, Ore., 1996). At the same time, it is clear that salmon recovery depends on more than modifications in public land management policies; it will also require modifications in river management, hatchery policy, and harvest levels as well as improvements in salmon habitat on the public lands. Id.; see also Michael C. Blumm, Sacrificing the Salmon: A Legal and Policy History of the Decline of Columbia Basin Salmon (Bookworld Publications, 2002).

74. U.S. Forest Service and Bureau of Land Management, Status of the Interior Columbia Basin, Summary of Scientific Findings 9 (Portland, Ore., 1996) (hereafter, 1996 ICB Summary of Scientific Findings).

75. On the Snake River chinook salmon listing, see 50 C.F.R. § 17.11(h) (2001); 57 Fed. Reg. 14,635 (April 22, 1992). Ensuing court decisions were: Pacific Rivers Council v. Thomas, 30 F.3d 1050 (9th Cir. 1994); Pacific Rivers Council v. Thomas, 873 F.Supp. 365 (D. Idaho 1995); Pacific Rivers Council v. Thomas, 897 F.Supp. 454 (D. Idaho 1995) (dissolving earlier injunction). The interim management standards can be found at: U.S. Forest Service and Bureau of Land Management, Interim Strategies for Managing Anadromous Fish-producing Watersheds in Eastern Oregon and Washington, Idaho, and Portions of California (PACFISH) (1995);U.S. Forest Service, Inland Native Fish Strategy (INFISH), Decision Notice and Finding of No Significant Impact (1995); U.S. Forest Service, Interim Management Direction Establishing Riparian, Ecosystem, and Wildlife Standards for Timber Sales (East-side Screens) (1994, amended 1995). Other Columbia River Basin salmon listings include the Snake River sockeye salmon, 15 evolutionarily significant units (ESUs) of West Coast chinook salmon, 5 ESUs of West Coast steelhead, 2 ESUs of chum, the bull trout, and the Oregon Coho salmon. See Michael C. Blumm and Greg D. Corbin, Salmon and the Endangered Species Act: Lessons from the Columbia Basin, 74 Wash. L. Rev. 519, 525–47 (1999). After much litigation, the bull trout was finally listed in 1998 as a threatened species. 50 C.F.R. §§ 17.11(h), 17.44(w) (2001); Friends of Wild Swan v. U.S. Fish and Wildlife Service, 945 F.Supp. 1388 (D. Ore. 1996). Subsequent litigation after these listings included: Friends of Wild Swan v. United States Forest Service, 966 F.Supp. 1002 (D. Ore. 1997); Prairie Wood Products v. Glickman, 971 F.Supp. 457 (D. Ore. 1997). On the Columbia River Basin salmon recovery effort generally, see Symposium on Salmon Recovery, 74 Wash. L. Rev. 511 (1999).

76. The noted statistics can be found at 1996 ICB Summary of Scientific Findings, supra note 74, at 119, 131. On Congress's additional socioeconomic study mandate, see Department of the Interior and Related Agencies Appropriations Act of 1998, Pub. L. 105–83 § 323(b), 111 Stat. 1543, 1597 (1998) (the ICBEMP "to the extent practicable, shall analyze the economic and social conditions, and culture and customs, of the communities at the subbasin level within the Project area and the impact the alternatives in the draft EISs will have on those communities"). On the study itself, see Bureau of Land Management, Economic and Social Conditions of Communities, ICBEMP (Portland, Ore., 1998).

77. The ecosystem stress quotation can be found at U.S. Forest Service and Bureau of Land Management, Considering All Things: Summary of the Draft Environmental Impact Statements 22 (Boise, Idaho, 1997). The proposed new ecosystem restoration management policy is found at U.S. Forest Service and Bureau of Land Management, East-side Draft Environmental Impact Statement, vol. 1, ch. 3, p. 28 (Portland, Ore., 1996). Initially, the ICBEMP agencies decided to prepare two EISs, basically dividing the region at the western border of Idaho, in an effort to address perceived ecological, economic, and social differences between the two areas. But later concluding that the differences were more imaginary than real, the agencies prepared only one EIS to develop and analyze the supplemental draft alternative management proposals. U.S. Forest Service and Bureau of Land Management, Interior Columbia Basin Supplemental Draft Environmental Impact Statement, ch. 1, pp. 4–5 (Boise, Idaho, 2000) (hereafter, Interior Columbia Basin Supplemental DEIS).

78. Public comments on the Draft EISs are summarized in the Interior Columbia Basin Supplemental DEIS, vol. 2, supra note 77, at Appendix 4; U.S. Forest Service and Bureau of Land Management, Final Analysis of Public Comment for the East-side and Upper Columbia Basin Draft Environmental Impact Statements Executive Summary (1998). On the regulatory agencies' response to the ICBEMP DEISs, see U.S. General Accounting Office, Ecosystem Planning: Northwest Forest and Interior Columbia River Basin Plans Demonstrate Improvements in Land-Use Planning, Appendix II:4.3.1 (Washington, D.C., 1999). For an analysis of the ICBEMP DEIS process and proposals, see the articles in 96(10) Journal of Forestry 4–46 (1998); see also Lorna Jorgensen, The Move toward Participatory Democracy in Public Land Management Under NEPA: Is It Being Thwarted by the ESA?, 20 J. Land, Resources, and Envir. L. 311 (2000).

79. See U.S. Forest Service and Bureau of Land Management, Interior Columbia Basin Supplemental DEIS Summary 17 (Boise, Idaho, 2000); id. at 22 (asserting that the "first priority is restoration of ecosystems and watersheds") (hereafter, Interior Columbia Basin Supplemental DEIS Summary); see also Interior Columbia Basin Supplemental DEIS, supra note 77, at ch. 3, pp. 92–124. See also U.S. Forest Service and Bureau of Land Management, Interior Columbia Basin Final EIS Proposed Decision 1–8 (Boise, Idaho, 2000).

80. Information on the T habitat, A1/A2 subwatersheds, and RCAs can be found in the ICBEMP Proposed Decision, supra note 79, at 3–4, 41–69, 104–12, 117, 126. The noted timber harvest and job projections can be found respectively in U.S. Forest Service and Bureau of Land Management, Report to the Congress on the Interior Columbia Basin Ecosystem Management Project 21–33 (Walla Walla, Wash., 2000) (hereafter, ICBEMP Report to Congress); Interior Columbia Basin Supplemental DEIS Summary, supra note 79, at 23 (explaining that an active restoration emphasis would create local employment opportunities, projecting that 40 percent of the new jobs would be in stewardship timber harvesting while 60 percent would be associated with prescribed fire and fuels management activities).

81. The $67 million funding figure is found in 2000 ICBEMP Report to Congress, supra note 80, at 18; see also Becky Kramer, Aid Sought to Offset U.S. Laws, The Spokesman-Review, April 25, 2001, noting that ICBEMP-area rural communities seek $100 million in federal assistance to offset job losses attributed to changing public land policies. On ICBEMP-related litigation, see ONRC Action v. Bureau of Land Management, 150 F.3d 1132 (9th Cir. 1998) (action seeking to enjoin BLM grazing and other resource decisions pending completion of the ICBEMP plan properly dismissed on ripeness grounds).

82. See U.S. Forest Service et al., The Interior Columbia Basin Strategy: A Strategy for Applying the Knowledge Gained by the Interior Columbia Basin Ecosystem Management Proj-

ect to the Revision of Forest and Resource Management Plans and Project Implementation (2003); Natalie M. Henry, Feds Abandon Regional [ICBEMP] Ecosystem Management Project, Land Letter, February 27, 2003, at p. 6. See also Allan K. Fitzsimmons, Defending Illusions: Federal Protection of Ecosystems 185–89 (Lanham, Md.: Rowman and Littlefield, 1999).

83. On the Sierra Nevada initiative, see Center for Water and Wildland Resources, Status of the Sierra Nevada: Assessment Summaries and Management Strategies, 2 vols. (Davis: University of California, Davis, 1996); U.S. Forest Service, Sierra Nevada Forest Plan Amendment Environmental Impact Statement (Vallejo, Calif., 2001); see also Chapter 8. On other similar western regional ecological restoration initiatives, see Greater Yellowstone Coalition, Sustaining Greater Yellowstone, a Blueprint for the Future (Bozeman, Mont., 1994); Grand Canyon Trust, Beyond the Boundaries: The Human and Natural Communities of the Greater Grand Canyon (Flagstaff, Ariz., 1997); Richard L. Knight et al., eds., Forest Fragmentation in the Southern Rocky Mountains (Boulder: University Press of Colorado, 2000); Bureau of Land Management, The Great Basin: Healing the Land (2000). On other regional ecological restoration initiatives, see John C. Tucker, Biodiversity Conservation and Ecosystem Management in Florida: Obstacles and Opportunities, 13 Fordham Envir. L. J. 1 (2001); Mary Doyle, Implementing Everglades Restoration, 17 J. Land Use and Envir. Law 59 (2001); Cory W. Borish et al., Conducting Regional Environmental Assessments: The Southern Appalachian Experience, in John D. Peine, ed., Ecosystem Management for Sustainability: Principles and Practices Illustrated by a Regional Biosphere Reserve Cooperative 117 (Boca Raton, Fla.: Lewis Publishers, 1999); Elizabeth Ann Rieke, The Bay-Delta Accord: A Stride Toward Sustainability, 67 U. Colo. L. Rev. 341 (1996); Robert J. Glennon and John E. Thorson, Federal Environmental Restoration Initiatives: An Analysis of Agency Performance and the Capacity for Change, 7 Ariz. L. Rev. 483, 516–21 (2000).

84. On river restoration generally, see Ludwik A. Teclaff and Eileen Teclaff, Restoring River and Lake Ecosystems, 34 Natural Resources J. 905 (1994); Christine A. Klein, Dam Policy: The Emerging Paradigm of Restoration, 31 Envir. Law Reporter 10486 (2001); Denise D. Fort, Restoring the Rio Grande: A Case Study in Environmental Federalism, 28 Envir. L. 1 (1998). On the Glen Canyon dam experiment, see Grand Canyon Protection Act of 1992, Pub. L. 102–575, 106 Stat. 4669; U.S. Bureau of Reclamation, Operation of Glen Canyon Dam Final Environmental Impact Statement (Salt Lake City, Utah, 1995); Michael Conner, Extracting the Monkey Wrench from Glen Canyon Dam: The Grand Canyon Protection Act: An Attempt at Balance, 15 Public Land L. Rev. 135 (1994). On the Elwha River restoration proposal, see Public Law 102–495, 106 Stat. 3173 (1992); National Park Service, Elwha River Ecosystem Restoration Implementation Final Environmental Impact Statement (Denver, Colo., 1996); Phillip M. Bender, Restoring the Elwha, White Salmon, and Rogue Rivers: A Comparison of Dam Removal Proposals in the Pacific Northwest, 17 J. Land, Resources and Envir. Law 189 (1997).

S I X

Shaping a New Heritage

1. See Craig W. Allin, The Politics of Wilderness Preservation (Westport, Conn.: Greenwood Press, 1982), for a detailed analysis of the Wilderness Act and its passage.

2. An excellent synopsis of the history of Montana wilderness legislation through 1992 can be

found in Lorna Naegele, A History of Wilderness Designation in Montana (unpublished 1992 draft paper prepared under the auspices of the Missoula Rotary Club). In preparing this discussion of the Montana national forest wilderness process, the author spoke with John Gatchell of the Montana Wilderness Association and Professor Bill Chaloupka at the University of Montana, both of whom provided helpful information and insights.

3. For an area-by-area description of Montana's wilderness lands, see Steve Woodruff and Don Schwennesen, Montana Wilderness: Discussing the Heritage (Kansas City, Kan.: Lowell Press, 1984). WSAs, however, are open to limited motorized recreational access—a political concession that has generated much controversy over the intervening years. See Montana Wilderness Association v. U.S. Forest Service, 146 F.Supp.2d 1118 (D. Mont. 2001).

4. See Montana Wilderness Act of 1994, H.R. 2473 (103rd Cong., 2d Sess., 1994).

5. The relevant cases are Montana Wilderness Association v. U.S. Forest Service, 655 F.2d 951 (9th Cir. 1981), cert. denied, 455 U.S. 989 (1982) (Gallatin right-of-way dispute); Bob Marshall Alliance v. Hodel, 852 F.2d 1223 (9th Cir. 1988), cert. denied sub nom. Kohlman v. Bob Marshall Alliance, 489 U.S. 1066 (Rocky Mountain Front oil and gas leasing); Montana Wilderness Association v. U.S. Forest Service, 146 F.Supp.2d 1118 (D. Mont. 2001), re-manded, 314 F.3d 1146 (9th Cir. 2003) (ATV use); see also Conner v. Burford, 848 F.2d 1441 (9th Cir. 1988) (Flathead and Gallatin National Forest oil and gas leasing). Curiously, during the Gallatin litigation, the environmental challengers were thwarted by an obscure right-of-way provision clandestinely inserted into the 1980 Alaska National Interest Lands Conservation Act (ANILCA) by a friendly Montana senator.

6. See 36 C.F.R. part 294 (2002); U.S. Forest Service, Roadless Area Conservation Final Environmental Impact Statement (Washington, D.C.: Dept. of Agriculture, 2000); see also infra note 56 for citations to the roadless rule litigation. See Chapter 4 for additional information on the Bush administration and the roadless rule.

7. On the Forest Service's new road construction rules, see 36 C.F.R. Part 212 (2001); 66 Fed. Reg. 3206 (Jan. 12, 2001); U.S. Forest Service, National Forest System Road Management Strategy Environmental Assessment (Washington, D.C., 2000).

8. Utah Wilderness Act of 1984, Pub. L. 98–428, 98 Stat. 1657 (1984). On the events leading to passage of the Utah Wilderness Act of 1984, see Richard Warnick, The Year of Utah Wilderness: Lessons from 1984, in Doug Goodman and Daniel McCool, eds., Contested Landscape: The Politics of Wilderness in Utah and the West 217–20 (Salt Lake City: University of Utah Press, 1999) (hereafter, Contested Landscape).

9. On the Escalante National Monument proposal, see T. H. Watkins, Righteous Pilgrim: The Life and Times of Harold L. Ickes, 1874–1952, 582–83 (New York: Henry Holt, 1990); Elmo R. Richardson, Federal Park Policy in Utah: The Escalante National Monument Controversy of 1935–1940, 33 Utah Hist. Q. 109 (1965).

10. Examples of the literature that helped popularize the desert Southwest include Wallace Stegner, Beyond the Hundredth Meridian: John Wesley Powell and the Second Opening of the West (Lincoln: University of Nebraska Press, 1982 edition); Edward Abbey, Desert Solitaire: A Season in the Wilderness (New York: Ballantine, 1968).

11. 43 U.S.C. § 1782; John D. Leshy, Wilderness and Its Discontents—Wilderness Review Comes to the Public Lands, 1981 Ariz. State L. J. 361 (hereafter, Leshy, Wilderness). The key wilderness study area court ruling is State of Utah v. Andrus, 486 F. Supp. 995 (D. Utah 1979). On the BLM's wilderness evaluation criteria, see Bureau of Land Management, Wilderness Inventory Handbook (Washington, D.C., 1978); see also H. Michael Anderson and Aliki Moncrief, America's Unprotected Wilderness, 76 Denver U. L. Rev. 413, 426–31 (1999).

12. See Sierra Club v. Watt, 608 F. Supp. 305 (E.D. Cal. 1985).

13. The various BLM wilderness study area figures can be found at 45 Fed. Reg. 75,602 (Nov. 14, 1980) (2.5 million acres); 1 Utah State Office, U.S. Bureau of Land Management, Utah BLM Statewide Wilderness Draft Environmental Impact Statement 35 (1986) (1.89 million acres); 1 Utah State Office, U.S. Bureau of Land Management, Utah BLM Statewide Wilderness Final Environmental Impact Statement 62–63 (1990) (1.97 million acres). The IBLA administrative appeal process and related developments are discussed in James R. Rasband, Utah's Grand Staircase: The Right Path to Wilderness Preservation?, 70 U. Colo. L. Rev. 483, 492–98 (1999); Sara McCormick and Bet Osiek, Weighted in the Balance: The Bureau of Land Management Wilderness Inventory, in Contested Landscape, supra note 8, at 35; Ricky S. Torrey, The Wilderness Inventory of the Public Lands: Purity, Pressure, and Procedure, 12 J. Energy, Natural Resources and Envir. Law 453 (1992). The original legislative proposals were H.R. 1501, 101st Cong. (1989) (1.4 million acres, by Rep. Hansen); H.R. 1500, 101st Cong. (1989) (5.4 million acres, by Rep. Owens). For an overview of the Utah wilderness controversy, see The Utah Wilderness Coalition, ed., Wilderness at the Edge: A Citizen Proposal to Protect Utah's Canyons and Deserts (Salt Lake City, Utah, 1990); Kevin Hayes, History and Future of the Conflict Over Wilderness Designations of BLM Land in Utah, 16 J. Envir. Law and Litigation 203 (2001).

14. See H.R. 1745, 104th Cong. (1995); S. 884, 104th Cong. (1995); H.R. Rep. No. 104–396. For descriptions of the 1995 events, see Ray Wheeler, Utah Hearings Misfire, 27(24) High Country News 1 (Dec. 25, 1995); Ann Nilson, Wilderness Backers Pack U. Meeting: Spirited Remarks Touch on Religion, Shame, Family Values, Salt Lake Tribune, April 16, 1995, at p. B1; T. H. Watkins, The Redrock Chronicles: Saving Wild Utah 104 (Baltimore, Md.: Johns Hopkins University Press, 2000); Lucas Smart and Amber Ayers, Don't Cry Out Acreage Figures in a Crowded Theater, in Contested Landscape, supra note 8, at 81–94.

15. See Interior Department Review and Budget: Oversight Hearing before the House Committee on Resources 27 (104th Cong., 1996) (statement of Rep. Hansen).

16. See Establishment of the Grand Staircase–Escalante National Monument, Proclamation No. 6920, 3 C.F.R. 64 (1997), 1996 USCCAN A73 (Sept. 18, 1996). The Antiquities Act is found at 16 U.S.C. §§ 431–33. For a succinct account of the various legal challenges to the president's creation of the monument, see Rasband, supra note 13, at 514–32. See generally Robert B. Keiter et al., eds., Visions of the Grand Staircase–Escalante: Examining Utah's Newest National Monument (Salt Lake City: University of Utah Press, 1998) (hereafter, Visions).

17. Under the FLPMA, the BLM has the authority periodically to inventory its lands "to reflect changes in conditions and to identify new and emerging resource and other values." 43 U.S.C. § 1711(a). It also has the authority periodically to revise its land use management plans. Id. at § 1712(a). On the reinventory controversy, see Jim Woolf, Babbitt vs. Hansen: Wilderness Debate in Utah Goes Back to Drawing Board, Salt Lake Tribune, Aug. 1, 1996, at A1; Jim Woolf, BLM Experts Set to Reassess Wilderness, Salt Lake Tribune, Aug. 29, 1996, at B1; State of Utah v. Babbitt, 137 F.3d 1193, 1197–1201 (10th Cir. 1998), reversing, State of Utah v. Babbitt, No. 2:96-CV-870B (Order, Nov. 18, 1996). The BLM's reinventory recommendations are found at U.S. Dept. of the Interior, Bureau of Land Management, Utah Wilderness Inventory (1999). On the UWC's revised wilderness recommendations, see Wild Utah Unveiled, 15(2) Southern Utah Wilderness Alliance 1 (Summer 1998); America's Redrock Wilderness Act, H.R. 1732, S. 861 (106th Cong., 2000). Biodiversity-based wilderness arguments can be found in Dinah Davidson et al., Selecting Wilderness Areas to

Conserve Utah's Biological Diversity, 56(2) Great Basin Naturalist 95 (1996); Science on the Wild Side: Where Wilderness Advocacy Meets Conservation Biology, 16(2) Southern Utah Wilderness Alliance 5 (Summer 1999); see also Wilderness at the Edge, supra note 13, at 25–27; Grand Canyon Trust, Beyond the Boundaries: The Human and Natural Communities of the Greater Grand Canyon 43–63 (Flagstaff, Ariz., 1997).

18. Utah Schools and Lands Exchange Act of 1998, Pub. L. No. 105–335, 112 Stat. 3139 (1998). In addition, the exchange included school trust land inholdings in the state's national parks, recreation areas, forests, and Indian reservations, which resolved a checkerboard problem that had plagued management of these lands for years. And the state received an extra $50 million in compensation from the federal government for its schools. The decision to include state trust land inholdings in Utah's national parks, national recreation areas, national forests, and Indian reservations stemmed from earlier legislation that provided for the exchange of these inholdings, but valuation problems had stalled negotiations. See Utah Schools and Lands Improvement Act of 1993, 31 U.S.C. § 6902 (codifying Pub. L. No. 103–93, 107 Stat. 995 (1993)). On the West Desert wilderness bill, see Utah National Parks and Public Lands Wilderness Act, H.R. 3035 106th Cong. (1999); Brent Israelsen, Leavitt, Babbitt Talk Wilderness; Two Meet in Washington to Discuss Land Issues. Areas of West Desert to Be Tackled First; Wilderness Changes to Be Piecemeal, Salt Lake Tribune, Feb. 3, 1999, at A1; Editorial, West Desert Wilderness, Salt Lake Tribune, Feb. 8, 1999, at A6.

19. On the San Rafael Swell proposal, see San Rafael Western Legacy District and National Conservation Area Act, H.R. 3605, S. 2048 (106th Cong., 2d Sess., 2000); Jim Woolf, Bill Would "Conserve" San Rafael Swell, But Not as Wilderness, Salt Lake Tribune, Feb. 9, 2000, at A1; Judy Fahys, San Rafael Land-Use Bill Derailed, Salt Lake Tribune, June 8, 2000, at A1; Jim Woolf, Cannon Gives Up on San Rafael Swell Bill, Salt Lake Tribune, July 7, 2000; Federico Cheever, Talking about Wilderness, 76 Denver U. L. Rev. 335, 337–40 (1999); see also Brent Israelsen, Leavitt Halts San Rafael Plan, Salt Lake Tribune, Dec. 5, 2002, at A1 (describing the demise of a subsequent San Rafael Swell National Monument designation plan). On the West Desert exchange, see West Desert Land Exchange Act, H.R. 4579, S. 2754 (106th Cong., 1999).

20. On the loss of species in national parks, see William D. Newmark, Legal and Biotic Boundaries of Western North American National Parks: A Problem of Congruence, 33 Biological Conservation 197–208 (1985) (hereafter, Newmark I); William D. Newmark, Extinction of Mammal Populations in Western North American National Parks, 9(3) Conservation Biology 512–26 (1995) (hereafter, Newmark II); Larry Harris, The Fragmented Forest: Island Biogeography Theory and the Preservation of Biotic Diversity 72 (Chicago: University of Chicago Press, 1984). On the ecological shortcomings of our nature reserves generally, see Dyan Zaslowsky and T. H. Watkins, These American Lands: Parks, Wilderness, and the Public Lands 226–28 (Washington, D.C.: Island Press, 1994); Reed Noss and Allen Cooperrider, Saving Nature's Legacy: Protecting and Restoring Biodiversity 172–73 (Washington, D.C.: Island Press, 1994). References on national park and national wildlife refuge management policy can be found in Chapters 2 and 5; references on the Greater Yellowstone Ecosystem can be found in Chapter 3. On the Everglades National Park problems, see U.S. House of Representatives, Committee on Interior and Insular Affairs, Modifying the Boundaries of the Everglades National Park and to Provide for the Protection of Lands, Waters, and Natural Resources within the Park, H.R. Rpt. No. 101-182 (1989); John C. Tucker, Biodiversity Conservation and Ecosystem Management in Florida: Obstacles and Opportunities, 13 Fordham Envir. L. J. 1 (2001).

21. The cited goals are set forth in Noss and Cooperrider, supra note 20, at 89. The quotation is from Michael Soule and Daniel Simberloff, What Do Genetics and Ecology Tell Us about the Design of Nature Reserves?, 35(1) Biological Conservation 19, 32 (1986).

22. See Jack Ward Thomas et al. (Interagency Scientific Committee to Address the Conservation of the Northern Spotted Owl), A Conservation Strategy for the Northern Spotted Owl 23 (Portland, Ore.: U.S. Dept. of Agriculture et al., 1990); David Wilcove and Dennis Murphy, The Spotted Owl Controversy and Conservation Biology, 5 Conservation Biology 261 (1991). On the need for roadless habitat, see Noss and Cooperrider, supra note 20, at 141; Reed Noss, Some Principles of Conservation Biology as They Apply to Environmental Law, 69 Chicago-Kent L. Rev. 893, 900–904 (1994); see also Special Section: Ecological Effects of Roads, 14(1) Conservation Biology 16 (2000). The hot spots concept is discussed in Noss and Cooperrider, id. at 104–7.

23. See Noss and Cooperrider, supra note 20, at 144–74; Reed Noss, The Wildlands Project: Land Conservation Strategy, Wild Earth (Special Issue) 15 (1992) (hereafter, Noss, The Wildlands Project); see also Michael E. Soule and John Terborgh, Continental Conservation: Scientific Foundations of Regional Reserve Networks (Washington, D.C.: Island Press, 1999).

24. See Soule and Terborgh, supra note 23, at 39–64; Dave Foreman, The Wildlands Project and the Rewilding of North America, 76 Denver U. L. Rev. 535, 546–48 (1999). The acreage estimates and implications are discussed in Noss and Cooperrider, supra note 20, at 168–72. The Noss quotation can be found at Reed Noss, The Wildlands Project, supra note 23, at 15.

25. See Noss and Cooperrider, supra note 20, at 147; Dave Foreman, Wilderness: From Scenery to Nature, in J. Baird Callicott and Michael P. Nelson, eds., The Great New Wilderness Debate 582 (Athens: University of Georgia Press, 1998) (hereafter, The Great New Wilderness Debate).

26. Studies analyzing the need to protect diverse ecosystems include: John Loomis and J. Chris Echohawk, Using GIS to Identify Under-Represented Ecosystems in the National Wilderness Preservation System in the U.S.A., 26(1) Envir. Conservation 53 (1999); Edward T. Laroe et al., eds., Our Living Resources: A Report to the Nation on the Distribution, Abundance, and Health of U.S. Plants, Animals, and Ecosystems 295 (Washington, D.C.: U.S. Dept. of the Interior, National Biological Survey, 1995); Reed F. Noss et al., Endangered Ecosystems of the United States: A Preliminary Assessment of Loss and Degradation (Washington, D.C.: U.S. Dept. of the Interior, National Biological Survey, 1995); Michael J. Mac et al., Status and Trends of the Nation's Biological Resources (Washington, D.C.: Dept. of the Interior, U.S. Geological Survey, 1999). Examples of Congress's recent national monument or conservation area designations include: Steens Mountain Cooperative Management and Protection Act of 2000, Pub. L. 106–399, 114 Stat. 1655 (2000); Santa Rosa and San Jacinto Mountains National Monument Act of 2000, Pub. L. 106–351, 114 Stat. 1362 (2000); see also John D. Leshy, The Babbitt Legacy at the Department of the Interior: A Preliminary View, 31 Envir. L. 199, 210–11 (2001) (hereafter Leshy, Babbitt's Legacy). On the Gallatin and Utah land exchanges, see supra notes 5, 18 and accompanying text.

27. On the "greater ecosystem" concept, see R. Edward Grumbine, Ghost Bears: Exploring the Biodiversity Crisis 57–60 (Washington, D.C.: Island Press, 1992). Specific greater ecosystem proposals are discussed in: Robert B. Keiter and Mark S. Boyce, eds., The Greater Yellowstone Ecosystem: Redefining America's Wilderness Heritage (New Haven: Yale University Press, 1991); Glacier National Park, The Crown of the Continent Ecosystem, at

www.nps.gov/glac/resources/bio1/html; Grand Canyon Trust, Beyond the Boundaries: The Human and Natural Communities of the Greater Grand Canyon (Flagstaff, Ariz., 1997); Grumbine, id. at 72–75, 166–67 (North Cascades). Critics of this concept include Allan K. Fitzsimmons, Defending Illusions: Federal Protection of Ecosystems 6–8 (Lanham, Md.: Rowman and Littlefield, 1999); William Perry Pendley, War on the West: Government Tyranny on America's Great Frontier 136 (Washington, D.C.: Regnery Publishing, 1995).

28. See H.R. 488, 106th Cong., 2d Sess. (1999); H.R. 852, 104th Cong., 1st Sess. (1995); Tara Weinman, The Northern Rockies Ecosystem Protection Act: In Support of Enactment, 5 Public Interest J. 287 (1995); Mike Bader, The Need for an Ecosystem Approach for Endangered Species Protection, 13 Public Land L. Rev. 137 (1992); Mike Bader, The Northern Rockies Ecosystem Protection Act: A Citizen Plan for Wilderness Management, 17(2) Western Wildlands 22 (Summer 1991).

29. See Harvey Locke, Preserving the Wild Heart of North America, 5(1) Borealis 20 (Spring 1994); Gary M. Tabor, Yellowstone-to-Yukon: Canadian Conservation Efforts and Continental Landscape/Biodiversity Strategy (Boston: Henry P. Kendall Foundation, 1996); Michael Finkel, From Yellowstone to Yukon, 101 Audubon 44–53 (July/August 1999); Douglas H. Chadwick, The Yellowstone to Yukon Initiative (Washington, D.C.: National Geographic, 2000); The Wilderness Society, The New Challenge: People, Commerce and the Environment in the Yellowstone to Yukon Region (1997); Symposium, Large Carnivore Conservation in the Northern Rockies, 10(4) Conservation Biology 939–1058 (August 1996).

30. On the Wildlands Project generally, see The Wildlands Project Mission Statement, Wild Earth (Special Issue) 3 (1992); Reed F. Noss, The Wildlands Project: Land Conservation Strategy, Wild Earth (Special Issue) 15 (1992); Greg Hanscum, Visionaries or Dreamers? A Team of Activists and Scientists Tries to Stitch the West's Wilderness Back Together, 31(4) High Country News 9 (April 26, 1999); Dave Foreman, Developing a Regional Wilderness Recovery Plan, Wild Earth (Special Issue) 26 (1992); on the "Sky Islands" proposal, see Dave Foreman et al., The Elements of a Wildlands Network Conservation Plan: An Example from the Sky Islands, 10(1) Wild Earth 17 (Spring 2000); Landmark Rewilding Plan Unveiled for Southwestern "Sky Islands," 10(2) Wild Earth: The Wildlands Project 1 (Fall 2000).

31. See U.S. General Accounting Office, Federal Lands: Information on Land Owned and on Acreage with Conservation Restrictions (Washington, D.C., 1995).

32. See David N. Cole, Management Dilemmas That Will Shape Wilderness in the 21st Century, 99(1) J. of Forestry 4 (January 2001); William S. Alverson et al., Wild Forests: Conservation Biology and Public Policy 242–43 (Washington, D.C.: Island Press, 1994) (proposing diversity maintenance areas); Holly Doremus, Nature, Knowledge and Profit: The Yellowstone Bioprospecting Controversy and the Core Purposes of America's National Parks, 26 Ecology L. Q. 401 (1999).

33. See The Great New Wilderness Debate, supra note 25; Michael McCloskey, Changing Views of What the Wilderness System Is All About, 76 Denver U. L. Rev. 369 (1999); Gregory H. Aplet, On the Nature of Wildness: Exploring What Wilderness Really Protects, 76 Denver U. L. Rev. 347 (1999).

34. On the evolution of nature preservation rationales, see Allin supra note 1, at 157–60; Zaslowsky and Watkins, supra note 20, at 213; John C. Hendee et al., Wilderness Management 119–20 (Golden, Colo.: North American Press, 2d edition, 1990); see also Max Oelschlae-

ger, The Idea of Wilderness: From Prehistory to the Age of Ecology (New Haven: Yale University Press, 1991); The Great New Wilderness Debate, supra note 25; William Cronon, ed., Uncommon Ground: Toward Reinventing Nature (New York: Norton, 1995).

35. On the evolution of the national park system, see Alfred Runte, National Parks: The American Experience (Lincoln: University of Nebraska Press, 2d edition, 1987).

36. Antiquities Act of 1916, 16 U.S.C. §§ 431–33; Ronald F. Lee, The Antiquities Act of 1906 (Washington, D.C.: National Park Service, 1970). Cases interpreting the Antiquities Act include: Cameron v. United States, 252 U.S. 450 (1920); Wyoming v. Franke, 58 F. Supp. 890 (D. Wyo. 1945); Mountain States Legal Foundation v. Bush, 306 F.3d 1132 (D.C. Cir. 2002); Tulare County v. Bush, 306 F.3d 1138 (D.C. Cir. 2002). On President Clinton's national monument designations, see Sanjay Ranchod, The Clinton National Monuments: Protecting Ecosystems with the Antiquities Act, 25 Harvard Envir. L. Rev. 535 (2001); Special Issue: Learning from the Monument, 21 J. Land, Resources, and Envir. Law 509–634 (2001).

37. On the evolution of the national wildlife refuge system, see Robert L. Fischman, The National Wildlife Refuge System and the Hallmarks of Modern Organic Legislation, 29 Ecology L. Q. 457 (2002); Zaslowsky and Watkins, supra note 20, at 151–77.

38. On loss of species in national parks, see Newmark I, supra note 20; Newmark II, supra note 20; Harris, supra note 20. On Everglades, Alaska, and California Desert park boundaries, see Runte, supra note 35, at 128–37; Roderick Nash, Wilderness and the American Mind 307 (New Haven: Yale University Press, 3d edition, 1982). References to the external threats problem can be found in Chapter 2.

39. Valles Caldera Preservation Act, Pub. L. No. 106–248, 114 Stat. 598 (2000); Sen. Rpt. 106–267, 106th Cong., 2d Sess. (2000); H. Rpt. 106–724, 106th Cong., 2d Sess. (2000); see also Chapter 4 for additional discussion of the Valles Caldera trust legislation.

40. 16 U.S.C. § 1131(c).

41. See Allin, supra note 1, at 83–84; Samuel T. Dana and Sally K. Fairfax, Forest and Range Policy: Its Development in the United States 157–58 (New York: McGraw-Hill, 1980); Hendee et al., supra note 34, at 101–2.

42. The "untrammeled" statutory language is found at 16 U.S.C. § 1131 (c). On the history of the Wilderness Act, see Allin, supra note 1; Nash, supra note 38, at 220–27; Michael Frome, Battle for the Wilderness (Salt Lake City: University of Utah Press, revised edition, 1997); Michael McCloskey, The Wilderness Act of 1964: Its Background and Meaning, 45 Oregon L. Rev. 288 (1966).

43. On western opposition to the Wilderness Act, see Zaslowsky and Watkins, supra note 20, at 209. The mining provisions are found at 16 U.S.C. § 1133(d). The grazing provision was contained in the Colorado Wilderness Act of 1980, 94 Stat. 3271, § 108 (1980); see also H.Rpt. 96–617. On the National Park Service's original opposition to the Wilderness Act, see Allin, supra note 1, at 110–18; Dana and Fairfax, supra note 41, at 198.

44. Under its wilderness purity policy, the Forest Service seized upon the Wilderness Act's definition of wilderness as a large pristine or untrammeled area to conclude that few other national forestlands merited protection under these strict criteria, which thus enabled the agency to limit its recommendations for new wilderness areas. See Allin supra note 1, at 157–60; Zaslowsky and Watkins, supra note 20, at 213. The Gore Range case established the important precedent that the Wilderness Act protected roadless lands contiguous to existing wilderness study areas, which enabled environmental activists to focus their attention elsewhere. Parker v. United States, 448 F.2d 793 (10th Cir. 1971), cert. denied, 405 U.S. 989

(1972); Wyoming Outdoor Coordinating Council v. Butz, 484 F.2d 1244 (10th Cir. 1973). On the RARE process, see U.S. Forest Service, Final Environmental Statement: Roadless Area Review and Evaluation (Washington, D.C., 1973); U.S. Forest Service, Final RARE II Environmental Impact Statement (Washington, D.C., 1979); Allin, id. at 160–65; Hendee et al., supra note 34, at 123–56. The Forest Service based its ecosystem representation criteria on the Wilderness Act, which included in its definition of "wilderness" an undeveloped area that "may also contain ecological, geological, or other features of scientific, educational, scenic, or historical value." 16 U.S.C. § 1131(c)(4); Hendee et al., id. at 130. On the RARE litigation, see Sierra Club v. Butz, 3 Envir. Law Reporter 20071 (N.D. Cal. 1972); Allin, id. at 160–65; Hendee et al., id. at 131, 138–39.

45. On the Eastern Wilderness Areas Act, see Pub. L. 93–622, 88 Stat. 2096 (1975); Allin, supra note 1, at 186–92; Hendee et al., supra note 34, at 132–33. The Endangered American Wilderness Act of 1978, which both created new wilderness areas and added to existing ones, designated 17 wilderness areas in Arizona, California, Colorado, Idaho, New Mexico, Oregon, Utah, Washington, and Wyoming. Pub. L. 95–237, 92 Stat. 40 (1978); Allin, id. at 192–93; Hendee et al., id. at 133–34. The 1980 statewide wilderness bills included the Colorado Wilderness Act of 1980, Pub. L. 96–560, 94 Stat. 3266 (1980); New Mexico Wilderness Act of 1980, Pub. L. 96–550, 94 Stat. 3222 (1980); Central Idaho Wilderness Act of 1980, Pub. L. 96–312, 94 Stat. 948 (1980); see also Hendee et al., id. at 140–41.

46. The quoted statutory language is found at 16 U.S.C. § 3101(b). On ANILCA and its passage in Congress, see Allin, supra note 1, at 207–65; Runte, supra note 35, at 236–58.

47. On the wilderness oil and gas leasing controversy, see Pacific Legal Foundation v. Watt, 529 F. Supp. 982 (D. Mont. 1981); Zaslowsky and Watkins, supra note 20, at 216–18; Hendee et al., supra note 34, at 141–42. On the RARE II litigation, see California v. Block, 690 F.2d 753 (9th Cir. 1982); Allin, supra note 1, at 164; Hendee et al., id. at 134–39.

48. Examples of the 1984 statewide wilderness bills include: Arizona Wilderness Act of 1984, Pub. L. 98–406, 98 Stat. 1485 (1984); California Wilderness Act of 1984, Pub. L. 98–425, 98 Stat. 1620 (1984); Washington Wilderness Act of 1984, Pub. L. 98–339, 98 Stat. 300 (1984); Wyoming Wilderness Act of 1984, Pub. L. 98–550, 98 Stat. 2808 (1984). On the subject of wilderness "release" language, see Meredith M. Graff, Babel in the Wilderness: The Rhetoric of Sufficiency/Release and Excepted Activities, in Contested Landscape, supra note 8, at 95; Hendee et al., supra note 34, at 141–42.

49. Examples of national park wilderness legislation include: Pub. L. 100–668, 102 Stat. 3965 (1988) (Mount Rainier, Olympic); Pub. L. 98–425, 98 Stat. 1626 (1984) (Sequoia–Kings Canyon, Yosemite).

50. In fact, the political compromise that paved the way for passage of the Wilderness Act included an agreement to create a Public Land Law Review Commission to reevaluate the laws and policies governing management of the BLM's public lands. The commission's final report was the genesis of the FLPMA legislation and its wilderness review provisions. See Dana and Fairfax, supra note 41, at 229–35.

51. See Federal Land Policy and Management Act of 1976, 43 U.S.C. §§ 1701–84; George C. Coggins, The Law of Public Rangeland Management IV: FLPMA, PRIA, and the Multiple Use Mandate, 14 Envir. L. 1 (1983). On the FLPMA wilderness provision, see 43 U.S.C. § 1782; Leshy, Wilderness, supra note 11; on the California Desert Conservation Area, see 43 U.S.C. § 1781; Zaslowsky and Watkins, supra note 20, at 220–22.

52. See Arizona Desert Wilderness Act of 1990, Pub. L. 101–628, 104 Stat. 4469 (1990); California Desert Protection Act of 1994, Pub. L. 103–433, 108 Stat. 4478 (1994); Zaslowsky and

Watkins, supra note 20, at 219–22. Notably, despite strong national support for the California Desert bill, the legislation's very expansiveness triggered strenuous local opposition, prodding the district's congressman into using the appropriations process to deny the Park Service funding to manage its newly acquired park lands. Michelle Nijhuis, So Much Land, So Little Money, 31(10) High Country News, May 24, 1999, at 12.

53. For examples of the Clinton national monument proclamations, see: Establishment of the Agua Fria National Monument, Presidential Proclamation No. 7263, 65 Fed. Reg. 2817 (Jan. 11, 2000); Establishment of the Grand Canyon–Parashant National Monument, Presidential Proclamation No. 7265, 65 Fed. Reg. 2825 (Jan. 11, 2000); Upper Missouri River Breaks National Monument, Presidential Proclamation No. 7398, 66 Fed. Reg. 7359 (Jan. 17, 2001). On the BLM's National Landscape Conservation System, see U.S. Dept. of the Interior, Bureau of Land Management, National Landscape Conservation System, at www.blm.gov/nhp/what/nlcs/index.html (site visited June 20, 2002). See also Christine A. Klein, Preserving Monumental Landscapes Under the Antiquities Act, 87 Cornell L. Rev. 1333 (2002).

54. On federal agency ESA responsibilities, see 16 U.S.C. §§ 1536(b), 1539(a)(2); Sierra Club v. Glickman, 156 F.3d 606 (5th Cir. 1998); J. B. Ruhl, Section 7(a)(1) of the "New" Endangered Species Act: Rediscovering and Redefining the Untapped Power of Federal Agencies' Duty to Conserve Species, 25 Envir. L. 1107 (1995); Thomas France and Jack Tuholske, Stay the Hand: New Directions for the Endangered Species Act, 7 Public Land L. Rev. 1 (1986). For federal statutory biodiversity provisions, see 16 U.S.C. § 1604(g)(3)(B) (national forests); 16 U.S.C. § 668dd (a)(4)(B) (national wildlife refuges). The Northwest Forest Plan is discussed in Chapter 4. The FLPMA ACEC provisions are found at 43 U.S.C. §§ 1702(a), 1712(c)(3).

55. On the Forest Service's Roadless Area Conservation initiative, see Special Areas; Roadless Area Conservation; Final Rule, 36 C.F.R. pt. 294 (2001); see also 66 Fed. Reg. 3244, 3272 (Jan. 12, 2001); U.S. Forest Service, Forest Service Roadless Area Conservation Final Environmental Impact Statement S-7, S-26–27 (Washington, D.C., 2000). On the Forest Service's road construction rule, see 36 C.F.R. § 212 (2001); 66 Fed. Reg. 3206 (Jan. 12, 2001); 64 Fed. Reg. 7290 (Feb. 12, 1999). On the Montana national forest withdrawals, see Kathie Durbin, Drawing the Line in Montana, 11(3) Inner Voice 10 (May/June 1999); Bureau of Land Management, Public Land Order No. 7480; Withdrawal of National Forest System Lands in the Rocky Mountain Front, Montana, 66 Fed. Reg. 6657 (Jan. 22, 2001).

56. On the Clinton national monument designation litigation, see Mountain States Legal Foundation v. Bush, 306 F.3d 1132 (D.C. Cir. 2002); Tulare County v. Bush, 306 F.3d 1138 (D.C. Cir. 2002); see also supra note 36 for related precedent. On legal challenges to the Forest Service's roadless area rule, see Counterattacks Launched Against FS Roadless Area Policy, 26(2) Public Land News 5 (Jan. 19, 2001); Kootenai Tribe of Idaho v. Veneman, 313 F.3d 1094 (9th Cir. 2002), reversing, Kootenai Tribe v. Veneman, 142 F.Supp. 2d 1231 (D. Idaho 2001) and Idaho ex rel. Kempthorne v. U.S. Forest Service, 142 F. Supp. 2d 1248 (D. Idaho 2001). On the Rocky Mountain Front withdrawal legal challenge, see Rocky Mountain Oil and Gas Association v. U.S. Forest Service, No. CV 98–22-H-CCL, Opinion and Order, March 7, 2000 (D. Mont. 2000), affirmed, 2001 U.S. App. LEXIS 8758 (9th Cir. 2001), cert. denied sub nom. Independent Petroleum Association v. U.S. Forest Service, 122 S.Ct. 541 (2002). The George W. Bush administration's contrary policies are discussed at the end of Chapter 4.

57. See Steens Mountain Cooperative Management and Protection Area Act, Pub. L. 106–

399, 114 Stat. 1655 (2000); see also Pub. L. 106–351, 114 Stat. 1362 (2000) (Santa Rosa–San Jacinto); Pub. L. 106–353, 114 Stat. 1374 (2000) (Colorado Canyons); Pub. L. 106–538, 114 Stat. 2563 (2000) (Las Cienegas); Pub. L. 106–554, app. D-1, 114 Stat. 2763 (2000) (Black Rock); Pub. L. 106–530, 114 Stat. 2527 (2000) (Great Sand Dunes); Leshy, Babbitt's Legacy, supra note 26, at 210–11. On the idea of devolution in recent congressional preservation debates, see "Hansen to Address Wilderness by Area; 'Hard' Release Likely," Public Lands News, Aug. 3, 2001, at 3 (reporting that "House Resources Committee Chairman James Hansen (R-Utah) has adopted an informal policy of addressing potential wilderness legislation through bills addressing individual areas, rather than statewide or regional bills").

58. Prior to 1970, however, Congress decommissioned 10 presidentially created national monuments that were relatively small and unimportant. These decommissioning decisions have never been regarded as a broader precedent for such congressional action. See Lee, supra note 36, at 93–97; John D. Leshy, Putting the Antiquities Act in Perspective, in Visions, supra note 16, at 85.

59. See U.S. General Accounting Office, Land Ownership: Information on the Acreage, Management, and Use of Federal and Other Lands (Washington, D.C., 1996); see also Bradley C. Karkkainen, Biodiversity and Land, 83 Cornell L. Rev. 1 (1997).

60. See David W. Crumpacker et al., A Preliminary Assessment of the Status of Major Terrestrial and Wetland Ecosystems on Federal and Indian Lands in the United States, 2 Conservation Biology 103, 11–14 (1988); U.S. General Accounting Office, Endangered Species Act: Information on Species Protection on Nonfederal Lands 4–5 (Washington, D.C., 1994); David Farrier, Conserving Biodiversity on Private Land: Incentives for Management or Compensation for Lost Expectations?, 19 Harvard Envir. L. Rev. 303, 310–11 (1995); Helen E. Hunt, ed., Conservation on Private Lands: An Owner's Manual 4 (Washington, D.C.: World Wildlife Fund, 1997).

61. See Dale D. Goble et al., Local and National Protection of Endangered Species: An Assessment, 2 Envir. Science & Policy 43 (1999); Center for Wildlife Law, State Endangered Species Acts: Past, Present and Future (1998); Barton H. Thompson, Jr., The Endangered Species Act: A Case Study in Takings and Incentives, 49 Stanford L. Rev. 303 (1997); see also Symposium: Biodiversity and Its Effect on Private Property, 38 Idaho L. Rev. 291–520 (2002).

62. For more on this view of property and ownership rights, see Lynda L. Butler, The Pathology of Property Norms: Living within Nature's Boundaries, 73 S. Calif. L. Rev. 927, 934–53 (2000); Joseph L. Sax, Property Rights and the Economy of Nature: Understanding Lucas v. South Carolina Coastal Council, 45 Stanford L. Rev. 1412 (1993). The principal recent Supreme Court decisions addressing Fifth Amendment regulatory takings claims are Nollan v. California Coastal Commission, 483 U.S. 825 (1987); Lucas v. South Carolina Coastal Council, 505 U.S. 1003 (1992); Dolan v. City of Tigard, 512 U.S. 374 (1994); Palazzolo v. Rhode Island, 533 U.S. 606 (2001); Tahoe-Sierra Preservation Council v. Tahoe Regional Planning Agency, 535 U.S. 302 (2002).

63. The Leopold quotations can be found in Aldo Leopold, A Sand County Almanac 258, 262 (New York: Ballantine Books, 1966); see Chapter 3 for more on Leopold's "land ethic." Among the extensive recent literature on evolving conceptions of property and ownership rights, see Eric T. Freyfogle, Bounded People, Boundless Lands (Washington, D.C.: Island Press, 1998); Lynton K. Caldwell and Kristin Shrader-Frechette, Policy for Land: Land and Ethics (Lanham, Md.: Rowman and Littlefield, 1993); Henry Diamond and Patrick Noo-

nan, Land Use in America (Washington, D.C.: Island Press, 1996); Carol M. Rose, Property Rights and Responsibilities, in Marian R. Chertow and Daniel C. Esty, eds., Thinking Ecologically: The Next Generation of Environmental Policy 49–59 (New Haven: Yale University Press, 1997).

64. For the National Academy of Science report, see Committee on Scientific and Technical Criteria for Federal Acquisition of Lands for Conservation, Setting Priorities for Land Conservation (Washington, D.C.: National Research Council, 1993).

65. For more information on federal acquisition authority or the Land and Water Conservation Fund, see George C. Coggins and Robert L. Glicksman, Public Natural Resources Law § 10C.06 (St. Paul, Minn.: West Group, 2001 updated edition). Federal condemnation authority was established by the Supreme Court in United States v. Gettysburg Electric Railroad Co., 160 U.S. 668 (1896). On the Valles Caldera legislation, see Valles Caldera Preservation Act, Pub. L. 106–248, 114 Stat. 598 (106th Cong., 2000); see also Chapter 4. The Nevada exchange legislation is the Southern Nevada Public Land Management Act of 1998, Pub. L. 105–263, 112 Stat. 2343 (2000). Under the CARA proposal, federal off-shore oil and gas revenues would be directly transferred to the Land and Water Conservation Fund, and the public land agencies could use these funds to acquire private lands, applying such criteria as "important or special resource attributes [or] threats to resource integrity." In the 107th Congress, the CARA legislation was introduced as H.R. 701 and S. 1328 (2001).

66. The FLPMA exchange provision can be found at 43 U.S.C. § 1716. The Federal Land Exchange Facilitation Act can be found at Pub. L. 100–409, 102 Stat. 1086 (1988). For a detailed description of the law governing federal land exchanges, see Coggins and Glicksman, supra note 65, at § 10C.05; see also Ryan M. Beaudoin, Federal Ownership and Management of America's Public Lands through Land Exchanges, 4 Great Plains Nat. Res. J. 228 (2000); Amy Stengel, "Insider's Game" or Valuable Land Management Tool? Current Issues in the Federal Land Exchange Program, 14 Tulane Envir. L. J. 567 (2001).

67. See U.S. General Accounting Office, BLM and the Forest Service: Land Exchanges Need to Reflect Appropriate Value and Serve the Public Interest (Washington, D.C., GAO/RCED-00–73, 2000); George Draffan and Janine Blaeloch, Commons or Commodity? The Dilemma of Federal Land Exchanges 17–34 (Seattle, Wash.: Western Land Exchange Project, 2000); Deborah Nelson et al., Trading Away the West: How the Public Is Losing Trees, Land and Money. Weyerhaeuser Gets Forest Land, but What Do the Taxpayers Get?, Seattle Times, Sept. 27, 1998, at A1.

68. The principal federal land exchange cases include: Desert Citizens Against Pollution v. Bisson, 231 F.3d 1172 (9th Cir. 2000); Muckleshoot Indian Tribe v. U.S. Forest Service, 177 F.3d 800 (9th Cir. 1999); RESTORE: The North Woods v. United States Dept. of Agriculture, 968 F.Supp. 168 (D. Vt. 1997); see also Kettle Range Conservation Group v. BLM, 150 F.3d 1083 (9th Cir. 1998); National Coal Association v. Hodel, 825 F.2d 523 (D.C. Cir. 1987); National Coal Association v. Hodel, 675 F.Supp. 1231 (D. Mont. 1989), aff'd, 874 F.2d 661 (9th Cir. 1989); Lodge Tower Condominium Association v. Lodge Properties, Inc., 880 F. Supp. 1370 (D. Colo. 1995), aff'd, 85 F.3d 476 (10th Cir. 1996) (addressing the public interest requirement).

69. For further discussion of the Utah and Montana land exchanges, including their relationship to the wilderness designation process and relevant legislation, see supra notes 5, 18 and accompanying text.

70. See Snowbasin Land Exchange Act of 1995, Pub. L. 104–333, 110 Stat. 4093 (1995); Christopher Smith, Earl Holding: The Inside Story: Holding Gave Forest Service Employees Free-

bies, Salt Lake Tribune, Feb. 13, 2000, at A15; Mike Gorrell, Ex-Federal Official's Job Raises Questions; Official's New Job Raising Ethical Questions, Salt Lake Tribune, March 16, 1997, at C1; Rick Egan, Land Grab or Fair Trade? Snowbasin: Small Resort Has Big Plans, Salt Lake Tribune, Nov. 12, 1995, at A1; Lee Davidson, Hansen Bill Slammed as "Land Grab," Deseret News, September 30, 1995, at B1; Warren Cornwall, Olympic-sized Rip-off, 27(20) High Country News 5 (Oct. 30, 1995). For a critique of large-scale public land exchange transactions that require congressional involvement, see Murray D. Feldman, The New Public Land Exchanges: Trading Development Rights in One Area for Public Resources in Another, 43 Rocky Mt. Min. L. Inst. 2-1 (1997).

71. The relevant cases are Minnesota v. Block, 660 F.2d 1240 (8th Cir. 1981), cert. denied, 455 U.S. 1007 (1982) (motorboats in wilderness); Columbia Gorge United v. Yeutter, 960 F.2d 110 (9th Cir. 1992) (Columbia River Gorge legislation); Gibbs v. Babbitt, 214 F.3d 483 (4th Cir. 2000) (red wolf reintroduction); Christy v. Hodel, 857 F.2d 1324 (9th Cir. 1988), cert. denied, 490 U.S. 1114 (1989) (wildlife damages claim); Mountain States Legal Foundation v. Hodel, 799 F.2d 1423 (10th Cir. 1986), cert. denied, 480 U.S. 951 (1987) (wildlife damages claim). Articles addressing the scope of congressional power to regulate beyond public land boundaries include Peter Appel, The Power of Congress "Without Limitation": The Property Clause and Federal Regulation of Private Property, 86 Minn. L. Rev. 1 (2001). Joseph L. Sax, Helpless Giants: The National Parks and the Regulation of Private Lands, 75 Mich. L. Rev. 239 (1976).

72. See Graham M. Lyons, Habitat Conservation Plans: Restoring the Promise of Conservation, 23 Environs Envir. Law and Policy J. 83 (1999); Karin P. Sheldon, Habitat Conservation Planning: Addressing the Achilles Heel of the Endangered Species Act, 6 N.Y.U. Envir. L. J. 279 (1998); Albert C. Lin, Participants' Experiences with Habitat Conservation Plans and Suggestions for Streamlining the Process, 23 Ecology L. Q. 369 (1996); see also U.S. Fish and Wildlife Service and Plum Creek Timber Company, Final Environmental Impact Statement and Native Fish Habitat Conservation Plan (Boise, Idaho, 2000).

73. The Nature Conservancy's acquisition and natural heritage programs are described in William Stolzenburg, Detectives of Diversity, 42(1) Nature Conservancy 22 (January/February 1992); Landmarks: A Report on Conservancy Actions and Accomplishments, 40(1) Nature Conservancy 21 (January/February 1990); Dennis H. Grossman and Kathleen Lemon Goodin, Rare Terrestrial Ecological Communities of the United States, in Our Living Resources, supra note 26, at 218. Conservation easements are discussed in Peter M. Morrisette, Conservation Easements and the Public Good: Preserving the Environment on Private Lands, 41 Natural Resources J. 373 (2001); Melissa W. Baldwin, Conservation Easements: A Viable Tool for Land Preservation, 32 Land and Water L. Rev. 89 (1997); Liz Rosan, ed., Preserving Working Ranches in the West (Tucson, Ariz.: Sonoran Institute, 1997). On the tax consequences associated with conservation easements, see Nancy A. McLaughlin, Tax Benefits of Conservation Easements, 23(6) Tax Management Estates, Gifts and Trusts J. 253 (1998).

74. Defenders of Wildlife wolf compensation and bounty programs are described in Robert C. Moore, The Pack Is Back: The Political, Social, and Ecological Effects of the Reintroduction of the Gray Wolf to Yellowstone National Park and Central Idaho, 12 T. M. Cooley L. Rev. 647, 659–60 (1995); Defenders of Wildlife, Wolf Compensation Trust, www. defenders.org/wolfcomp.html. See Farrier, supra note 60, at 399–405, for a discussion of the federally funded private land stewardship proposal. See also Barton H. Thompson, Jr., Providing Biodiversity Through Policy Diversity, 38 Idaho L. Rev. 355 (2002) and James L.

Huffman, Marketing Biodiversity, 38 Idaho L. Rev. 421 (2002) for additional proposals for integrating market strategies into private biodiversity conservation efforts.

75. The emergence of public–private collaborative groups is explored more fully in Chapter 7.

S E V E N

Collaborative Conservation

1. This historical overview of Moab and the surrounding region is derived primarily from the following sources: Fawn McConkie Tanner, A History of Moab, Utah (Moab, Press of the Times Independent, 1937); Richard A. Firmage, A History of Grand County (Salt Lake City: Utah State Historical Society, 1996); Raye C. Ringholz, Little Town Blues: Voices from the Changing West (Salt Lake City, Utah: Peregrine Smith Books, 1992); Raye C. Ringholz, Paradise Paved: The Challenge of Growth in the New West (Salt Lake City: University of Utah Press, 1996); Elmo R. Richardson, Federal Park Policy in Utah: The Escalante National Monument Controversy of 1935–1940, 33(2) Utah Historical Quarterly 109 (1965).

2. See Raye C. Ringholz, Uranium Frenzy: Boom and Bust on the Colorado Plateau (New York: Norton, 1989).

3. For an example of local national park support, see Statement of Norman Boyd, Mayor of Moab, Utah, in Hearings on H.R. 6925 and S. 27 (Canyonlands National Park), Subcommittee on National Parks, Committee on Interior and Insular Affairs, House of Representatives, 88th Cong., 2d Sess., June 20 and 29, 1964, at p. 30. The Canyonlands National Park legislation is codified at 16 U.S.C. § 271. In 1971, Congress added another 80,000 acres to the park, recognizing that the original boundaries were too small to protect at-risk resources. One year later, Congress protected another 150,000 acres contiguous to the park by adding it to the newly created Glen Canyon National Recreation Area. Robert B. Keiter, Completing Canyonlands, 74 National Parks 26 (March/April 2000). For biographical information on Edward Abbey, see James M. Cahalen, Edward Abbey: A Life (Tucson: University of Arizona Press, 2001).

4. Firmage, supra note 1, at 397. Furthermore, the state's economic statistics revealed that by 1984 government salaries had displaced mining salaries as the major source of personal income in the Grand County. Id. at 383.

5. Ringholz, Paradise Paved, supra note 1, at 145–46; see also Charles Wilkinson, Paradise Revised, in William E. Riebsame, ed., Atlas of the New West 18–22 (New York: Norton, 1997).

6. Canyon Country Partnership Charter (Nov. 26, 1993) (emphasis in original). The partnership's original membership included local federal agency representatives (from the BLM, Forest Service, and Park Service), state agency representatives (from the wildlife, mining, parks, and forestry agencies), and local government officials (from Grand, San Juan, Carbon, and Emery counties). For a more detailed account of the Sand Flats controversy and the ensuing agreement, see Sarah Van de Wetering, Doing It the Moab Way: A Public Land Partnership at Sand Flats, 1(1) Chronicle of Community 5 (1996).

7. For insightful analyses of the Canyon Country Partnership, see R. Eric Smith, The Canyon Country Partnership and Ecosystem-Based Management on the East-Central Colorado Plateau, 19 J. Land, Resources, and Envir. Law 19 (1999); Barb Cestero, Beyond the Hundredth Meeting: A Field Guide to Collaborative Conservation on the West's Public Lands 56–61 (Tucson, Ariz.: Sonoran Institute, 1999).

8. Firmage, supra note 1, at 407–8.

9. General information about the Catron County public land controversies is available in: Sam Burns, Catron County, New Mexico: Mirroring the West, Healing the Land, Rebuilding Community, in Jonathan Kusel and Elisa Adler, eds., Forest Communities, Community Forests: A Collection of Case Studies of Community Forestry 63 (Taylorsville, Calif.: Forest Community Research, 2001); Timothy Egan, Lasso the Wind: Away to the New West 11–32 (New York: Knopf, 1998); Tony Davis, Catron County's Politics Heat Up as Its Land Goes Bankrupt, 28(12) High Country News 1 (June 24, 1996); D'Lyn Ford, The Catron Way, www.cahe.nmsu.edu/pubs/resourcesmag/fall95/catron.html (site visited May 30, 2001). The author also conducted interviews in southwestern New Mexico during July 2001 with Delbert Griego, deputy supervisor for the Gila National Forest; Andrea Martinez, Gila National Forest information officer; Stephen Libby, Gila National Forest staff officer; Todd Schulke, environmental activist; and Danny Fryar, local businessman, rancher, and former county manager.

10. See U.S. Forest Service, Gila National Forest Plan (Silver City, N.M., 1986); U.S. Forest Service, Gila National Forest Plan Environmental Impact Statement 70–71 (Silver City, N.M., 1986) (listing the spotted owl as an indicator species for ponderosa pine and mixed conifer forest types); 58 Fed. Reg. 14248–01 (March 16, 1993) (ESA Mexican spotted owl listing decision); Silver v. Babbitt, 924 F. Supp. 976 (D. Ariz. 1995); Silver v. Babbitt, 924 F. Supp. 972 (D. Ariz. 1995). On the legal and biological status of the northern goshawk, see James Peck, Seeing the Forest through the Eyes of a Hawk: An Evaluation of Recent Efforts to Protect Northern Goshawk Populations in Southwestern Forests, 40 Nat. Res. J. 125 (2000); see also 63 Fed. Reg. 35183 (June 29, 1998) (FWS's decision refusing to list the northern goshawk under the ESA); Center for Biological Diversity v. Badgley, 2001 WL 844399 (D. Ore.) (sustaining the no-listing decision); "No-list Decision on Northern Goshawk Upheld," Endangered Species & Wetlands Report 6 (July 2001). For a general overview and history of southwestern forest policy, see Christopher J. Huggard and Arthur R. Gomez, Forests Under Fire: A Century of Ecosystem Mismanagement in the Southwest (Tucson: University of Arizona Press, 2001).

11. On the environmentalists' unsuccessful challenge to the Catron County "custom and culture" ordinance, see Gila Watch et al. v. Catron County, No. 94–1312-M-Civil (D. N.M., filed Jan. 16, 1996); Tony Davis, Catron County Wins in Court, Loses on the Ground, 28 High Country News 2 (Feb. 5, 1996); Anita Miller, The War for the West: At Issue, 28 Urban Lawyer 861, 868–71 (1996). On Catron County's successful NEPA challenge to ESA critical habitat designations, see Catron County Board of County Commissioners v. U.S. Fish and Wildlife Service, 75 F.3d 1429 (10th Cir. 1996).

12. The Catron County Citizens Group is described in Burns, supra note 9; Melinda Smith, The Catron County Citizens Group: A Case Study in Community Collaboration (Albuquerque, N.M.: New Mexico Center for Dispute Resolution, 1998).

13. See Forest Guardians v. U.S. Forest Service, No. CIV. 97–2562 PHX-SMN (D. Ariz., April 17, 1998); Forest Guardians v. Dombeck, 131 F.3d 1309 (9th Cir. 1997) (noting an earlier temporary injunction on grazing activities); Southwest Center for Biological Diversity v. Clark, 90 F.Supp.2d 1300 (D. N.M. 1999) (ordering the FWS to designate critical habitat for the spikedace and loach minnow); Keith Bagwell, Judges Void Hundreds of Grazing Leases, Extend Logging Halt, Arizona Daily Star, July 26, 1997, at A1; Tony Davis, Healing the Gila, 33(20) High Country News 1 (Oct. 22, 2001); Forest Guardians, Dislodging the Sacred Cow: A Summary of Ongoing and Past Litigation, at www.fguardians.org/cowcourt

(site visited Aug. 8, 2001); see also Forest Guardians v. U.S. Forest Service, No. 00-612-TUC-RCC (D. Ariz., Nov. 22, 2002) (ordering the Forest Service to consult with the FWS over the effect of grazing on the Mexican spotted owl); Judge Orders End to Grazing on National Forests in Ariz., N.M., Land Letter, Dec. 5, 2002, at p. 13; Center for Biological Diversity v. U.S. Forest Service, No. 97–666 TUC JMR, (D. Ariz., March 30, 2001) (ordering the Forest Service to reconsult with the FWS over the effect of grazing on the spikedace and loach minnow); Reconsultation Ordered for Handful of Grazing Allotments, Endangered Species & Wetlands Report 11 (April 2001); April Reese, Enviros Turning to ESA to Ban Cattle from Public Lands, Land Letter, Jan 2, 2003, at p. 4.

14. See New Mexico Cattle Growers Association v. U.S. Fish and Wildlife Service, 248 F.3d 1277 (10th Cir. 2001) (ordering the FWS to integrate local economic concerns into its critical habitat determinations); Final Designation of Critical Habitat for the Spikedace and the Loach Minnow, 65 Fed. Reg. 24328 (April 25, 2000) (increasing the stream mileage considered critical habitat from 1,600 miles under the FWS's 1986 designation to 1,800 miles in 2000); Arizona Cattle Growers' Association v. Cartwright, 29 F.Supp.2d 1100 (D. Ariz. 1998) (rejecting a ranching industry challenge to the Forest Service's grazing allotment reductions). For more information on the Laney controversy, see Timothy Egan, supra note 9, at 11–32; Diamond Bar Cattle Co. v. United States, 168 F.3d 1209 (10th Cir. 1999).

15. On the southwestern wolf reintroduction litigation, see New Mexico Cattle Growers' Association v. U.S. Fish and Wildlife Service, 81 F.Supp.2d 1141 (D. N.M. 1999). The Gila National Forest's elk problems are addressed in Karl Hess, Jr., Incentive-Based Conservation for the Sky Island Complex: A Draft Report to the Wildlands Project on Livestock, Elk, and Wolves (Feb. 20, 1998); Brian Thomas, The Multiplication of Fractional Elk: New Mexico's New Math, 36 Nat. Res. J. 793 (1996).

16. Board of County Commissioners, Catron County Resolution No. 006–2001; Rene Romo, Thin Forests or We Will, Counties Say, Albuquerque Journal, Sept. 13, 2000; Jennifer McKee, U.S. Forests Called Public Safety Threat, Albuquerque Journal, Jan. 30, 2001; New Mexico Senate Approves County Fire Work in Fed Forests, Public Land News, March 2, 2001, p. 6.

17. See William Pendley Perry, War on the West: Government Tyranny on America's Great Frontier (Washington, D.C.: Regnery Publishing, 1995).

18. The Sagebrush Rebellion and Wise Use movement are also addressed in Chapter 2, where the primary sources for these two political movements are also referenced.

19. See Catron County Interim Land Use Policy Plan, adopted by Catron County, New Mexico, May 21, 1991, reprinted in National Federal Lands Conference Update (August 1992); Scott W. Reed, The County Supremacy Movement: Mendacious Myth Making, 30 Idaho L. Rev. 525 (1993/94); Florence Williams, Sagebrush Rebellion II: Some Rural Counties Seek to Influence Federal Land Use, 24(3) High Country News 1, 9 (Feb. 24, 1992). The argument for state ownership can be found in Wayne Hage, Storm Over Rangelands: Private Rights in Federal Land (Bellevue, Wash.: Free Enterprise Press, 1989).

20. Citations to the court decisions are: United States v. Nye County, Nevada, 920 F. Supp. 1108 (D. Nev. 1996); United States v. Gardner, 107 F.3d 1314 (9th Cir. 1997), aff'ing, 903 F. Supp. 1394 (D. Nev. 1995); Boundary Backpackers v. Boundary County, 128 Idaho 371, 913 P.2d 1141 (1996). The extensive legal literature on these cases and Wise Use litigation includes: Peter D. Coppelman, The Federal Government's Response to the County Supremacy Movement, Natural Resources and the Environment 30 (Summer 1997); Elizabeth M. Osenbaugh and Nancy K. Stoner, The County Supremacy Movement, 28 Urban

Lawyer 497 (1996); Andrea Hungerford, "Custom and Culture" Ordinances: Not a Wise Move for the Wise Use Movement, 8 Tulane Envir. L. J. 457 (1995); Patrick A. Austin, Law West of the Pecos: The Growth of the Wise Use Movement and the Challenge to Federal Public Land Use Policy, 30 Loyola L.A. L. Rev. 275 (1996); Anita P. Miller, America's Public Lands: Legal Issues in the New War for the West, 24 Urban Lawyer 895 (1992); Anita P. Miller, All Is Not Quiet on the Western Front, 25 Urban Lawyer 827 (1993); Anita P. Miller, The Western Front Revisited, 26 Urban Lawyer 845 (1994); Anita P. Miller, The War for the West: At Issue, 28 Urban Lawyer 861 (1996).

21. Jon Christensen, Nevadans Drive out Forest Supervisor, 31 High Country News 2 (Nov. 22, 1999); Rebels Roust a Forester, Forest Magazine, January/February 2000, at 10.

22. See Jim Carlton, Bitter Battle over Rural West, Wall Street Journal, Feb. 16, 2001, at p. B1; David Foste and Scott Sonner, Nevadans, U.S. Make Much Ado about Old Road; Shovel Wielding Residents Protest Government's Land Regulations, Chicago Tribune, March 1, 2000, at p. 8; Jarbidge Agreement Reached, Alpine Enterprise, April 11, 2001.

23. The relevant cases are: Mountain States Legal Foundation v. Andrus, 499 F. Supp. 383 (D. Wyo. 1980) (oil and gas leasing); Mountain States Legal Foundation v. Hodel, 668 F. Supp. 1466 (D. Wyo. 1987) (oil and gas leasing); Intermountain Forest Industry Association v. Lyng, 683 F. Supp. 1330 (D. Wyo. 1988) (national forest timber sale levels); Mountain States Legal Foundation v. Hodel, 799 F.2d 1423 (10th Cir. 1986), cert. denied, 480 U.S. 951 (1987) (BLM forage); Christy v. Hodel, 857 F.2d 1324 (9th Cir. 1988), cert. denied, 490 U.S. 1114 (1989) (grizzly bear depredation). Examples of state litigation funds can be found at Wyo. Stat. Ann. § 9–4–218 (1999); Nev. Rev. Stat. Ann. § 328.500 (1999).

24. The Supreme Court's recent takings decisions include: Nollan v. California Coastal Commission, 483 U.S. 825 (1987); Lucas v. South Carolina Coastal Council, 505 U.S. 1003 (1992); Dolan v. City of Tigard, 512 U.S. 374 (1994); Palazzolo v. Rhode Island, 533 U.S. 606 (2001); Tahoe-Sierra Preservation Council v. Tahoe Regional Planning Agency, 535 U.S. 302 (2002). On the application of these takings decisions to public land resources, see George C. Coggins and Robert L. Glicksman, Public Natural Resources Law § 4.05[4] (St. Paul, Minn.: West Group, 2001); Barton H. Thompson, Jr., Resource Use and the Emerging Law of "Takings": A Realistic Appraisal, 42 Rocky Mt. Min. L. Inst. 2–1 (1996); Jan Laitos, Regulation of Natural Resources Use and Development in Light of the "New" Takings Clause, 34 Rocky Mt. Min. L. Inst. 1–1 (1989); Jan G. Laitos and Richard A. Westfall, Government Interference with Private Interests in Public Resources, 11 Harvard Envir. L. Rev. 1 (1987). The principal grazing permit cases are: McKinley v. United States, 828 F. Supp. 888 (D. N.M. 1993); Nevada Land Action Association v. U.S. Forest Service, 8 F.3d 713 (9th Cir. 1993); United States v. Gardner, 107 F.3d 1314 (9th Cir. 1997). On the Hage litigation, see Hage v. United States, 51 Fed Cl. 570 (2002); Hage v. United States, 42 Fed. Cl. 249 (1998); see also Diamond Bar Cattle Co. v. United States, 168 F.3d 1209 (10th Cir. 1999) (rejecting the argument that grazing rights are attached to water rights). On the question of whether endangered species limitations constitute a taking, see Croman Corporation v. United States, 49 Fed. Cl. 776 (2001); Buse Timber and Sales, Inc. v. United States, 45 Fed. Cl. 258 (1999); Klamath Water Users Protective Association v. Patterson, 204 F.3d 1206 (9th Cir. 2000) (ESA takes precedence over water rights in operation of Klamath Basin federal reclamation project); Pacific Coast Federation of Fishermen's Associations v. U.S. Bureau of Reclamation, 138 F.Supp.2d 1228 (N.D. Cal. 2001) (enjoining federal water deliveries in the Klamath Basin until ESA consultation is completed); Tulare Lake Basin Water Storage District v. United States, 49 Fed. Cl. 313 (April 30, 2001) (finding ESA compliance require-

ments to be a taking of water rights); Coggins and Glicksman, id. at § 4.16 et seq.; Mark Sagoff, Muddle or Muddle Through? Takings Jurisprudence Meets the Endangered Species Act, 38 William and Mary L. Rev. 825 (1997); Barton H. Thompson, Jr., The Endangered Species Act: A Case Study in Takings and Incentives, 49 Stanford L. Rev. 305 (1997).

25. The relevant cases are: Public Lands Council v. Babbitt, 529 U.S. 728 (2000) (range reforms); Wyoming Farm Bureau Federation v. Babbitt, 199 F.3d 1224 (10th Cir. 2000), reversing, 987 F. Supp. 1349 (D. Wyo. 1997) (wolf reintroduction); State of Utah v. Babbitt, 137 F.3d 1193 (10th Cir. 1998), reversing, State of Utah v. Babbitt, Civ. No. 2:96-CV-870B (Order, Nov. 18, 1996) (BLM wilderness reinventory); Mountain States Legal Foundation v. Glickman, 92 F.3d 1228 (D.C. Cir. 1996) (timber sales); United States v. Garfield County, 122 F. Supp.2d 1201 (D. Utah 2000) (R.S. 2477); Southern Utah Wilderness Alliance v. Bureau of Land Management, 147 F. Supp.2d 1130 (D. Utah, 2001) (R.S. 2477); Tulare County v. Bush, 306 F.3d 1138 (D.C. Cir. 2002) (national monuments); Mountain States Legal Foundation v. Bush, 306 F.3d 1132 (D.C. Cir. 2002) (national monuments); Kootenai Tribe of Idaho v. Veneman, 313 F.3d 1094 (9th Cir. 2002) (Forest Service roadless area rule).

26. The citation for the 1982 critical habitat amendment is: Pub. L. 97–304, § 2(a), 96 Stat. 1411 (1982) (codified at 16 U.S.C. § 1533(b)(2). For an overview of Congress's 1978 and 1982 ESA amendments, see P. Stephanie Easley et al., The Endangered Species Act: A Stanford Environmental Law Society Handbook 22–26 (Palo Alto, Calif.: Stanford University Press, 2001). The congressional ESA reform efforts included: H.R. 3159, 102nd Cong. (1992); H.R. 3092, 102nd Cong. (1992); H.R. 6134, 102nd Cong. (1992); H.R. 2043, 103rd Cong. (1993); H.R. 1490, 103rd Cong. (1993); H.R. 2207, 103rd Cong. (1993); see also Laura Spitzberg, The Reauthorization of the Endangered Species Act, 13 Temple Envir. Law and Tech. J. 193 (1994); Jon A. Souder, Chasing Armadillos Down Yellow Lines: Economics in the Endangered Species Act, 33 Nat. Res. J. 1095, 1107–13 (1993). On Congress's timber harvest appropriations riders, see Pub. L. 101–121, 103 Stat. 701, 745 (1989); Robertson v. Seattle Audubon Society, 503 U.S. 429 (1992); and Chapter 4 for a more detailed discussion of these timber sale riders.

27. Douglas County v. Babbitt, 48 F.3d 1495 (9th Cir. 1995). Under the ESA, the secretary of the interior is required to designate critical habitat for listed species concurrently with the listing decision, factoring both scientific and economic considerations into the determination (unlike the initial listing decision, which is based solely on scientific criteria). 16 U.S.C. § 1533. But see Catron County Board of Commissioners v. U.S. Fish and Wildlife Service, 75 F.3d 1429 (10th Cir. 1996), which reaches the opposite conclusion.

28. Babbitt v. *Sweet Home* Chapter of Communities, 515 U.S. 687 (1995). The relevant ESA statutory provisions and regulations are: 16 U.S.C. § 1538(a)(1)(B) (the taking prohibition); 16 U.S.C. § 1532(19) (defining "take" as "to harass, harm, pursue, hunt, shoot, wound, kill, trap, capture, or collect, or to attempt to engage in any such conduct"); 50 C.F.R. § 17.3 (2000) (defining "harm" as an act of "significant habitat modification or degradation where it actually kills or injures wildlife by significantly impairing essential behavioral patterns, including breeding, feeding, or sheltering"). In Sweet Home, the Court's 6–3 majority invoked the so-called Chevron doctrine, which counsels courts to defer to agency legal interpretations absent clear evidence of a contrary congressional intent. See Chevron, USA, Inc. v. Natural Resources Defense Council, 467 U.S. 837 (1984). The Court then examined the ESA's statutory language, purpose, structure, and legislative history and also expressed reluctance to second-guess the secretary's interpretation because he was faced with a "complex policy choice" that could affect "a vast range of economic and social enterprises and

endeavors." The three dissenting justices rejected the linkage between the statutory "take" prohibition and habitat alteration, arguing that the term *take* was limited to direct impacts such as the killing or capturing of individual animals.

29. For a brief synopsis of these reports and anecdotes, see Karin P. Sheldon, Habitat Conservation Planning: Addressing the Achilles Heel of the Endangered Species Act, 6 N.Y.U. Envir. L. J. 279–81 (1998).

30. As examples of the 104th Congress's anti-ESA legislative proposals, see S. 768, 104th Cong. (1995); S. 1364, 104th Cong. (1995); H.R. 2275, 104th Cong. (1995); H.R. 2364, 104th Cong. (1995) H.R. 2374, 104th Cong. (1995); see also Jeffrey S. Kopf, Slamming Shut the Ark's Doors: Congress's Attack on the Listing Process of the Endangered Species Act, 3 Animal Law 103 (1997); Tanya L. Godfrey, The Reauthorization of the Endangered Species Act: A Hotly Contested Debate, 98 W.V. L. Rev. 979 (1996). In addition, several stringent property rights–based takings bills were also introduced that session; they would have saddled the government with mandatory compensation requirements for any regulatory-based takings claim and opened the federal courts to such claims. See S. 609, 104th Cong. (1995); H.R. 9, 104th Cong. (1995); Godfrey, id., at 1013–15; Bonnie B. Burgess, Fate of the Wild: The Endangered Species Act and the Future of Biodiversity 77–102 (Athens: University of Georgia Press, 2001). On the Young-Pombo bill, see H.R. 2275, 104th Cong. (1995); Eva Tompkins, Reauthorization of the Endangered Species Act—A Comparison of Two Bills that Seek to Reform the Endangered Species Act: Senate Bill 768 and House Bill 2275, 6 Dick. J. Envir. Law and Policy 119 (1997). The ESA listing moratorium was contained in the Emergency Supplemental Appropriations and Rescissions for the Department of Defense to Preserve and Enhance Military Readiness Act of 1995, Pub. L. 104–6, tit. II, ch. 4, 109 Stat. 73, 86 (1995). On the Interior Department's "no surprises" and "safe harbor" policies, see 50 C.F.R. §§ 13.23–.25, 13.28, 13.50 (safe harbor), 50 C.F.R. §§ 17.22–.23 (no surprises) (2000); Final Addendum to the Handbook for Habitat Conservation Planning and Incidental Take Permitting Process, 65 Fed. Reg. 35242–47 (June 1, 2000); 64 Fed. Reg. 32705 (June 17, 1999) (safe harbor policy); 63 Fed. Reg. 8859 (Feb. 23, 1998) (no surprises policy); see also Douglas L. Huth, Endangered Species Act Reauthorization: Congress Proposes a Rewrite with Private Landowners in Mind, 48 Okl. L. Rev. 383 (1995). Since the 104th Congress, the Wise Use movement's congressional supporters have regularly introduced legislation that would give the states a greater role in conserving endangered species, interject new economic factors into the process, and carve loopholes in the consultation process. See, e.g., S. 1180, 105th Cong. (1997); H.R. 3160, 106th Cong. (1999); S. 747, 107th Cong. (2001).

31. The relevant cases are: Bennett v. Spear, 520 U.S. 154 (1997) (sustaining ESA standing for ranchers); Catron County Commissioners v. U.S. Fish and Wildlife Service, 75 F.3d 1429 (10th Cir. 1996) (NEPA applies to ESA critical habitat designation process); Douglas County v. Babbitt, 48 F.3d 1495 (9th Cir. 1995) (NEPA does not attach to the ESA critical habitat designation process); New Mexico Cattle Growers Association v. U.S. Fish and Wildlife Service, 248 F.3d 1277 (10th Cir. 2001) (requiring an economic analysis in the critical habitat designation process); FWS, NMFS Want to Re-do Critical Habitat Economic Analyses, 7(2) Endangered Species & Wetlands Report 1 (Nov. 2001); Arizona Cattle Growers Association v. U.S. Fish and Wildlife Service, 273 F.3d 1229 (9th Cir. 2001) (incidental take permit and livestock grazing). On the salmon listing controversy, see Alsea Valley Alliance v. Evans, 161 F.Supp.2d 1154 (D. Ore. 2001). The court order, however, has been stayed pending an appeal by environmental groups; the United States chose not to appeal

the ruling, but to review its hatchery policy. See National Marine Fisheries Service to Review 23 Ecologically Significant Units, Endangered Species & Wetlands Report 3 (November 2001); Appeals Court Keeps Oregon Coho Listing in Place for Now, Endangered Species & Wetlands Report 3 (December 2001). In related litigation, the courts have ruled that federal species recovery and reintroduction efforts do not violate constitutional commerce clause limitations, even though the federal government has assumed traditional state wildlife management responsibilities. Gibbs v. Babbitt, 214 F.3d 483 (4th Cir. 2000) (red wolf restoration does not violate the commerce clause); National Association of Home Builders v. Babbitt, 130 F.3d 1041 (D.C. Cir. 1997) (Delphi Sand Flower-Loving Fly recovery plan does not violate the commerce clause).

32. Pendley, supra note 17 at xviii.

33. A. Dan Tarlock, Can Cowboys Become Indians? Protecting Western Communities as Endangered Cultural Remnants, 31 Arizona State L. J. 539, 543–44 (1999). On the evolving nature of property rights, see Chapter 6, note 63 and accompanying text.

34. See Tarlock, supra note 33, at 550, for a discussion of the ethnic white population as a cultural minority. On the Native American analogy, see James R. Rasband, The Rise of Urban Archipelagos in the West: A New Reservation Policy?, 31 Environmental Law 1 (2001). See generally Paula M. Nelson, Rural Life and Social Change in the Modern West, in R. Douglas Hirt, ed., The Rural West Since World War II 38–57 (Lawrence: University Press of Kansas, 1998).

35. The community-held property rights idea is discussed in Tarlock, supra note 33, at 566–79; see also Charles Geisler and Gail Daneker, eds., Property and Values: Alternatives to Public and Private Ownership (Washington, D.C.: Island Press, 2000).

36. On NEPA's potential role in facilitating coordinated planning and collaborative public involvement, see Center for the Rocky Mountain West and Institute for Environment and Resources, Reclaiming NEPA's Potential: Can Collaborative Processes Improve Environmental Decisionmaking? (Missoula: University of Montana, 2000); Robert B. Keiter, NEPA and the Emerging Concept of Ecosystem Management on the Public Lands, 25 Land and Water L. Rev. 43 (1990). For information on third world conservation, see David Western and R. Michael Wright, eds., Natural Connections: Perspectives in Community-Based Conservation (Washington, D.C.: Island Press, 1994).

37. See Daniel Kemmis, Community and the Politics of Place 44–63 (Norman: University of Oklahoma Press, 1990) (hereafter Kemmis, Politics of Place); Douglas S. Kenney, Arguing about Consensus: Examining the Case Against Western Watershed Initiatives and Other Collaborative Groups Active in Natural Resources Management 19 (Boulder, Colo.: Natural Resources Law Center, 2000).

38. For information on the origins and evolution of these new collaborative processes, see Julia M. Wondolleck and Steven L. Yaffee, Making Collaboration Work: Lessons from Innovation in Natural Resource Management (Washington, D.C.: Island Press, 2000); Philip Brick et al., eds., Across the Great Divide: Explorations in Collaborative Conservation and the American West (Washington, D.C.: Island Press, 2001) (hereafter, Across the Great Divide); Barb Cestero, Beyond the Hundredth Meeting: A Field Guide to Collaborative Conservation on the West's Public Lands (Tucson, Ariz.: Sonoran Institute, 1999); Douglas S. Kenney, The New Watershed Sourcebook: A Directory and Review of Watershed Initiatives in the Western United States (Boulder, Colo.: Natural Resources Law Center, 2000); see also Edward P. Weber, Pluralism by the Rules: Conflict and Cooperation in Environmental Regulation (Washington, D.C.: Georgetown University Press, 1998); Daniel Press,

Democratic Dilemmas in the Age of Ecology: Trees and Toxins in the American West (Durham, N.C.: Duke University Press, 1994).

39. On the various benefits to be derived from collaborative initiatives, see Wondolleck and Yaffee, supra note 38, at 23–46; Kemmis, Politics of Place, supra note 38 at 64–83. On the relationship between ecosystem management and collaborative processes, see Bradley C. Karkkainen, Collaborative Ecosystem Governance: Scale, Complexity, and Dynamism, 21 Virginia Envir. L. J. 189 (2002); Environmental Protection Agency, Community Based Environmental Protection: A Resource Book for Protecting Ecosystems and Communities (Washington, D.C., 1997); Symposium: The Ecosystem Approach: New Departures for Land and Water, 24 Ecology L. Q. 619–886 (1997); Symposium: Power, Politics, and Place: Who Holds the Reins of Environmental Regulation, 25 Ecology L. Q. 559–763 (1999); The Keystone Center, The Keystone National Policy Dialogue on Ecosystem Management Final Report (Keystone, Colo., 1996).

40. The Forest Service's community stability policy is examined in Sarah F. Bates, Public Land Communities: In Search of a Community of Values, 14 Public Land L. Rev. 81 (1993); Con H. Schallau and Richard M. Alston, The Commitment to Community Stability: A Policy or Shibboleth?, 17 Envir. L. 429 (1987).

41. On the community of interests concept, see Bates, supra note 40, at 90–91, 104–11; Daniel S. Reimer, The Role of "Community" in the Pacific Northwest Logging Debate, 66 U. Colo. L. Rev. 222, 247 (1995); Timothy P. Duane, Community Participation in Ecosystem Management, 24 Ecology L. Q. 771 (1997). The idea of a sense of place is developed in Kemmis, Politics of Place, supra note 38, at 116–19; Charles F. Wilkinson, The Eagle Bird: Mapping a New West 132–86 (New York: Pantheon, 1992); see also Daniel Kemmis, This Sovereign Land: A New Vision for Governing the West 111–16 (Washington, D.C.: Island Press, 2001).

42. See Richard Halvey and Karen Deike, Unleashing Enlibra, 17(5) The Environmental Forum 20–31 (2000) and related articles; see also www.westgov.org/wga/initiatives/enlibra.

43. The basic elements or attributes of successful collaborative processes are described in more detail in Wondolleck and Yaffee, supra note 38; Cestero, supra note 38; The Keystone Center, supra note 39; see also Susan L. Carpenter and W. J. D. Kennedy, Managing Public Disputes: A Practical Guide to Handling Conflict and Reaching Agreements (San Francisco: Jossey-Bass, 1988); Robert H. Mnookin and Lewis Kornhauser, Bargaining in the Shadow of the Law, 88 Yale L. J. 950 (1979).

44. See Wondolleck and Yaffee, supra note 38, at 229–46; Matthew J. McKinney, What Do We Mean by Consensus? Some Defining Principles, in Across the Great Divide, supra note 38, at 33.

45. K. Norman Johnson et al. (Committee of Scientists), Sustaining the People's Lands: Recommendations for Stewardship of the National Forests and Grasslands into the Next Century xiv–xv (Washington, D.C.: U.S. Dept. of Agriculture, 1999); see also 36 C.F.R. §§ 219.19–.21 (2001). On the concept of sustainability, see Donald Snow, Coming Home: An Introduction to Collaborative Conservation, in Across the Great Divide, supra note 38, at 7–8; Charles F. Wilkinson, Crossing the Next Meridian: Land, Water, and the Future of the West 298–99 (Washington, D.C.: Island Press, 1992); Daniel Sitarz, ed., Sustainable America: America's Environment, Economy and Society in the 21st Century (Carbondale, Ill.: Earthpress, 1998); President's Council on Sustainable Development, Natural Resources Management and Protection Task Force Report (Washington, D.C.: 1998); The World Commission on Environment and Development, Our Common Future 43–66 (New York: Oxford University Press, 1987).

46. Jim Burchfield, Finding Science's Voice in the Forest, in Across the Great Divide, supra note 38, at 236; Wondolleck and Yaffee, supra note 38, at 134–37; see also 36 C.F.R. §§ 219.22–.25 (2001) (explaining the role of science in national forest planning); Daniel Sarewitz et al., eds., Prediction: Science, Decision Making, and the Future of Nature (Washington, D.C.: Island Press, 2000).

47. National Parks Conservation Association v. Stanton, 54 F.Supp.2d 7, 20 (D.D.C. 1999). See Sydney Cook, Revival of Jeffersonian Democracy or Resurgence of Western Anger? The Emergence of Collaborative Decision Making, 2000 Utah Law Rev. 575; Brenda L. Hall, Subdelegation of Authority Under the Endangered Species Act: Secretarial Authority to Subdelegate His Duties to a Citizen Management Committee as Proposed for the Selway-Bitterroot Wilderness Grizzly Bear Reintroduction, 20 Public Land and Res. L. Rev. 81 (1999).

48. Natural Resources Defense Council v. Hodel, 618 F.Supp. 848, 871 (E.D. Cal. 1985).

49. 5 U.S.C. App. 2, §§ 9, 10. According to FACA, an "advisory committee" means "any committee, board, commission, council, conference, panel, task force, or other similar group, or any subcommittee or other subgroup thereof . . . established or utilized by the President, or . . . by one or more agencies." 5 U.S.C. App. 2, § 3(2). On the Northwest Forest Plan FACA litigation, see Northwest Forest Resource Council v. Espy, 846 F. Supp. 1009 (D.D.C. 1994), reversed sub nom., Northwest Forest Resource Council v. Dombeck, 107 F.3d 897 (D.C. Cir. 1997); Seattle Audubon Society v. Lyons, 871 F. Supp. 1291, 1309–10 (W.D. Wash. 1994) (rejecting a similar FACA challenge to the FEMAT process, because the public comment and review opportunities accompanying release of the draft and final EISs effectively cured any earlier procedural infirmities in the process). For cases on FACA judicial remedies, see California Forestry Association v. U.S. Forest Service, 102 F.3d 609 (D.C. Cir. 1996); Aluminum Company of America v. National Marine Fisheries Service, 93 F.3d 902 (9th Cir. 1996); Natural Resources Defense Council v. Pena, 147 F.3d 1012 (D.C. Cir. 1998); Cargill, Inc. v. United States, 173 F.3d 323 (5th Cir. 1999); but see Alabama-Tombigbee Rivers Coalition v. Department of the Interior, 26 F.3d 1103 (11th Cir. 1994) (enjoining the National Marine Fisheries Service from using scientific advice from an improperly convened advisory committee in making its endangered species listing decision).

50. The 1995 FACA amendments are found at Unfunded Mandates Reform Act of 1995, Pub. L. 104-4, 109 Stat. 48, § 204 (1995), codified at 2 U.S. C. §§ 658 et seq., 1501 et seq. On FACA generally, see Sheila Lynch, The Federal Advisory Committee Act: An Obstacle to Ecosystem Management by Federal Agencies, 71 Washington L. Rev. 431 (1996); Herb McHarg, The Federal Advisory Committee Act: Keeping Interjurisdictional Ecosystem Management Groups Open and Legal, 15 J. Energy, Natural Resources, and Envir. Law 437 (1995).

51. See Oregon Natural Resources Council v. Daley, 6 F.Supp.2d 1139 (D. Ore. 1998); Carlotta Collette, Oregon's Plan for Salmon and Watersheds: The Basics of Building a Recovery Plan, in Across the Great Divide, supra note 38, at 140–49; Carlotta Collette, The Oregon Way: Governor John Kitzhaber Casts for Consensus in the Northwest's Troubled Waters, 30(2) High Country News 1, Oct. 26, 1998; see also Valerie Ann Lee and Jaelith Hall-Rivera, Breathing New Life into the ESA: The Pacific Northwest's Endangered Species Act Experiment in Devolution, 31 Envir. L. Reporter 10102 (January 2001); 4(d) Rules Challenged by Kittitas County, NAHB, Endangered Species & Wetlands Report 13 (February 2001).

52. See, e.g., Idaho Farm Bureau Federation v. Babbitt, 58 F.3d 1392 (9th Cir. 1995) (Bruneau Hot Springs snail); City of Las Vegas v. Lujan, 891 F.2d 927 (D.C. Cir. 1989) (desert tor-

toise); National Association of Homebuilders v. Babbitt, 990 F.Supp. 1 (D.D.C. 1997) (karst invertebrate species); but see Alsea Valley Alliance v. Evans, 161 F.Supp.2d 1154 (D. Ore. 2001) (salmon runs); Dan Berman, FWS Agrees to Review Northern Spotted Owl, Marbled Murrelet Listings, Land Letter, January 16, 2003, at 3. See also Leslie M. Lewallen and Russell C. Brooks, Alsea Valley v. Evans and the Meaning of "Species" Under the Endangered Species Act: A Return to Congressional Intent, 25 Seattle U. L. Rev. 731 (2002).

53. The ESA listing and delisting criteria are the same; see 16 U.S.C. § 1533(a)(1). There is yet but little literature on delisting. See Todd Wilkinson, Science Under Siege: The Politicians' War on Nature and Truth 65–112 (Boulder, Colo.: Johnson Books, 1998); Philip Kline, Grizzly Bear Blues: A Case Study of the Endangered Species Act's Delisting Process and Recovery Plan Requirements, 31 Envir. L. 317 (2001); Holly Doremus, Delisting Endangered Species: An Aspirational Goal, Not a Realistic Expectation, 30 Envir. Law Reporter 10434 (2000); Federico Cheever, The Rhetoric of Delisting Species Under the Endangered Species Act, 31 Envir. Law Reporter 11302 (2001).

54. George C. Coggins, Of Californicators, Quislings, and Kings: Some Perils of Devolved Collaborations, in Across the Great Divide, supra note 38, at 167; see also Michael McCloskey, The Skeptic: Collaboration Has Its Limits, High Country News, May 13, 1996, at 7.

55. Richard Behan, Plundered Promise: Capitalism, Politics, and the Fate of the Federal Lands 212–15 (Washington, D.C.: Island Press, 2001).

56. See Coggins, in Across the Great Divide, supra note 38, at 167; John D. Echeverria, No Success Like Failure: The Platte River Collaborative Watershed Planning Process, 25 William and Mary Envir. Law and Policy Rev. 559 (2001); but see Joseph L. Sax, Environmental Law at the Turn of the Century: A Reportorial Fragment of Contemporary History, 88 Calif. L. Rev. 2377 (2000).

57. See David H. Getches, Some Irreverent Questions about Watershed-Based Efforts, in Across the Great Divide, supra note 38, at 180; Douglas S. Kenney, Are Community Watershed Groups Effective? Confronting the Thorny Issue of Measuring Success, in Across the Great Divide, supra note 38, at 188; Wondolleck and Yaffee, supra note 38, at 213.

58. See Wondolleck and Yaffee, supra note 38, at 229–45.

59. For an overview of public land visitation figures, see Jan G. Laitos and Thomas A. Carr, The Transformation on Public Lands, 26 Ecology L. Q. 140, 160–62 (1999). The Colorado Plateau visitation data can be found at Colorado Plateau Forum Town Hall Research Committee, The Devil's Bargain? Community and Tourism on the Colorado Plateau 1 (1997). See generally H. Ken Cordell et al., Outdoor Recreation in American Life: A National Assessment of Demand and Supply Trends (Champaign, Ill.: Sagamore Publishing, 1999).

60. For further historical perspectives on recreation and tourism, see Mark D. Barringer, Selling Yellowstone: Capitalism and the Construction of Nature (Lawrence: University Press of Kansas, 2002); Hal K. Rothman, Devil's Bargains: Tourism in the Twentieth Century American West (Lawrence: University Press of Kansas, 1998); Samuel T. Dana and Sally K. Fairfax, Forest and Range Policy: Its Development in the United States 179–238 (New York: McGraw-Hill, 2d edition, 1980).

61. On gateway communities, see Jim Howe et al., Balancing Nature and Commerce in Gateway Communities (Washington, D.C.: Island Press, 1997); on the growth of the ski industry, see Rothman, supra note 60, at 202–27. For a brief history of public land outfitting, see Arthur D. Smith, Jr., Outfitting on Public Lands: A Study of Federal and State Regulation, 26 Idaho L. Rev. 9, 14–20 (1989).

62. See United States v. Curtis-Nevada Mines, Inc., 611 F.2d 1277 (9th Cir. 1980).

63. For a sample of the court cases addressing recreation-based conflicts on public lands, see Northwest Motorcycle Association v. U.S. Department of Agriculture, 18 F.3d 1468 (9th Cir. 1994) (sustaining Forest Service closure of ORV trail); Wilderness Public Rights Fund v. Kleppe, 608 F.2d 1250 (9th Cir. 1979) (sustaining Park Service limits on Grand Canyon river trips); Hells Canyon Alliance v. U.S. Forest Service, 227 F.3d 1170 (9th Cir. 2000) (sustaining Forest Service limitations on motorized water craft); Bicycle Trails Council v. Babbitt, 82 F.3d 1445 (9th Cir. 1996) (sustaining Park Service limits on mountain bikes).

64. Patrick T. Long, For Residents and Visitors Alike: Seeking Tourism's Benefits, Minimizing Tourism's Costs, in David M. Wrobel and Patrick T. Long, eds., Seeing and Being Seen: Tourism in the American West 82 (Lawrence: University Press of Kansas, 2001).

65. Christopher M. Klyza, Reform at a Geological Pace: Mining Policy on the Federal Lands, 1964–1994, in Charles Davis, ed., Western Public Lands and Environmental Politics 106 (Boulder, Colo.: Westview, 1997).

66. For a more detailed examination of the positive and negative implications of recreation and tourism, see Thomas Michael Power, Lost Landscapes and Failed Economies: The Search for a Value of Place 213–34 (Washington, D.C.: Island Press, 1996); see also Rothman, supra note 60; Allen Best, Vail and the Road to a Recreational Empire, 30(23) High Country News 1 (Dec. 7, 1998); Walter E. Hecox, Charting the Colorado Plateau: An Economic and Demographic Exploration (Flagstaff, Ariz.: Grand Canyon Trust, 1996).

67. The figures cited are derived from Keith Easthouse, Out of Control, Forest Magazine 16 (May/June 2000); Southern Utah Wilderness Alliance, Overriding Utah's Wilderness: The Search for Balance and Quiet in Utah's Wilderness 2 (Salt Lake City, Utah, 1999); National Park Service, Winter Use Plans: Final Environmental Impact Statement 184–90 (Moose, Wyo., 2000); National Park Service, Winter Use Plans: Draft Supplemental Environmental Impact Statement 132 (Moose, Wyo., 2002). For a sample of the California Desert Conservation Area ORV litigation, see Sierra Club v. Clark, 756 F.2d 686 (9th Cir. 1985); Sierra Club v. Clark, 774 F.2d 1406 (9th Cir. 1985); American Motorcycle Association v. Watt, 543 F. Supp. 789 (C.D. Cal. 1982); Center for Biological Diversity v. Bureau of Land Management, 2001 WL 777088 (N.D. Cal.); see also Daniel B. Wood, Land Battle Heats Up Dunes, Christian Science Monitor, May 20, 2002. See generally David G. Havlick, No Place Distant: Roads and Motorized Recreation on America's Public Lands (Washington, D.C.: Island Press, 2002).

68. See U.S. Forest Service, White River National Forest Land and Resource Management Plan—2002 Revision (Glenwood Springs, Colo., 2002); U.S. Forest Service, Rocky Mountain Region, Record of Decision for the White River National Forest Land and Resource Management Plan—2002 Revision (Denver, Colo., 2002); April Reese, Ski Area Expansions, Logging in Roadless Areas Part of White River Plan, Land Letter, June 13, 2002, at 1–3; Allen Best, Stop: A National Forest Tries to Rein in Recreation, 32(1) High Country News 1 (Jan. 17, 2000).

69. See Southern Utah Wilderness Alliance, supra note 67; Southern Utah Wilderness Alliance v. Dabney, 7 F. Supp. 2d 1205 (D. Utah 1998), remanded, 222 F.3d 1819 (10th Cir. 2000) (Canyonlands National Park ORV litigation); Southern Utah Wilderness Alliance v. Norton, 301 F.3d 1217 (10th Cir. 2002).

70. See National Park Service, Winter Use Plans: Final Environmental Impact Statement 184–90 (Moose, Wyo., 2000); Ray Ring, Move Over! Will Snowmobile Tourism Relax Its Grip on a Gateway Town, 34(6) High Country News 1 (April 1, 2002); Dan Berman, NPS Increases Number of Snowmobiles Allowed in Yellowstone, Land Letter 7 (Nov. 14, 2002).

71. U.S. General Accounting Office, Federal Lands: Information on the Use and Impact of Off-Highway Vehicles (Washington, D.C., 1995); David Sheridan, Off Road Vehicles on Public Land (Washington, D.C.: Council on Environmental Quality, 1979); Keith Easthouse, Out of Control, Forest Magazine, May/June 2000, at 14; see also Joseph L. Sax, Mountains without Handrails: Reflections on the National Parks (Ann Arbor: University of Michigan Press, 1980), for a powerful argument endorsing contemplative recreation as a primary purpose of the national parks. On efforts to protect natural quiet and solitude on public lands, see the Wilderness Act, 16 U.S.C. at § 1131(c); National Park Service, Management Policies 4.9 (Washington, D.C., 2001); U.S. Air Tour Association v. Federal Aviation Administration, 298 F.3d 997 (D.C. Cir. 2002); Grand Canyon Air Tour Coalition v. Federal Aviation Administration, 154 F.3d 455 (D.C. Cir. 1998).

72. The access arguments from the ORV user perspective are perhaps best captured in Blue Ribbon Magazine, published by the Idaho Falls–based Blue Ribbon Coalition. The magazine is available at www.sharetrails.org/mag.

73. See Executive Order 11,644, 37 Fed. Reg. 2877 (1972); Executive Order 11,989, 42 Fed. Reg. 26,959 (1977); Northwest Motorcycle Association v. U.S. Department of Agriculture, 18 F.3d 1468 (9th Cir. 1994); Washington Trails Association v. U.S. Forest Service, 935 F. Supp. 1117 (W.D. Wash. 1996); Southern Utah Wilderness Alliance v. Dabney, 7 F. Supp.2d 1205 (D. Utah 1998), remanded, 222 F.3d 1819 (10th Cir. 2000); Friends of Boundary Waters Wilderness v. Robertson, 978 F.2d 1484 (8th Cir. 1992), cert denied, 508 U.S. 972 (1993); Montana Wilderness Association v. U.S. Forest Service, 146 F. Supp.2d 1118 (D. Mont. 2001); see also Utah Shared Access Alliance v. U.S. Forest Service 288 F.3d 1205 (10th Cir. 2002); Jeffrey Bleich, Chrome on the Range: Off Road Vehicles on Public Lands, 15 Ecology L. Q. 159 (1988).

74. The Symms Act is found at 23 U.S.C. § 206. The mountain bike controversy is examined in Bicycle Trails Council v. Babbitt, 82 F.3d 1335 (9th Cir. 1996); Scott Havlick, Mountain Bicycles on Federal Lands: Over the River and through Which Woods?, 7 Energy Law and Policy 123 (1986).

75. For more information on the R.S. 2477 controversy, see Southern Utah Wilderness Alliance v. Bureau of Land Management, 147 F. Supp.2d 1130 (D. Utah 2001); United States v. Garfield County, 122 F. Supp. 2d 1201 (D. Utah 2000); Sierra Club v. Hodel, 848 F.2d 1068 (10th Cir. 1988); Mitchell R. Olson, The R.S. 2477 Right of Way Dispute: Constructing a Solution, 27 Envir. L. 289 (1997); Harry R. Bader, Potential Legal Standards for Resolving the R.S. 2477 Right of Way Crisis, 11 Pace Envir. L. Rev. 485 (1994). On the San Rafael Swell National Monument proposal, see Christopher Smart, The Proposal for San Rafael Monument Fails in Emery, Salt Lake Tribune, Nov. 6, 2002, at A15; Brent Israelsen, Leavitt Halts San Rafael Plan, Salt Lake Tribune, Dec. 5, 2002, at A1.

76. See Outfitter Policy Act of 2001, S. 978, H.R. 2386, 107th Cong., 1st Sess. (2001); Smith, supra note 61. On the Forest Service litigation involving timber sale modifications, see Scott Timber Co. v. United States, 40 Fed. Cl. 492 (1998), vacated on reconsideration, 44 Fed. Cl. 170 (1999); Precision Pine and Timber, Inc. v. United States, 50 Fed. Cl. 35 (2001); Westel-Oviatt Lumber Co. v. United States, 38 Fed. Cl. 563 (1997); see also U.S. General Accounting Office, Timber Management: Forest Service Has Considerable Liability for Suspended or Canceled Timber Sales (Washington, D.C., 2000). The legal and policy problems involved in extending legal rights to public land users are examined in Wilkinson, Crossing the Next Meridian, supra note 45.

77. 16 USC § 4601–6a et seq.; Pub. L. 107–63, 115 Stat. 466 (2001); see also H.R. 104–259,

104th Cong., 2d Sess. (1996); U.S. General Accounting Office, Recreation Fees: Demonstration Fee Program Successful in Raising Revenues but Could Be Improved (Washington, D.C., 1999); Kira Dale Pfisterer, Foes of Forest Fees: Criticism of the Recreation Fee Demonstration Project at the Forest Service, 22 J. Land, Resources, and Envir. Law 309 (2002); Matt Rasmussen, Booming Business, Forest Magazine, September/October 1999, at 34; Hal Clifford, Land of the Fee, 32(3) High Country News 1 (Feb. 14, 2000).

78. See Greater Yellowstone Coordinating Committee, Winter Visitor Management: A Multi-Agency Assessment (1999); Bureau of Land Management, Montana State Office, and U.S. Forest Service, Northern Region, Off-Highway Vehicle Environmental Impact Statement and Proposed Plan Amendment for Montana, North Dakota and Portions of South Dakota (Billings, Mont., 2001).

79. Thomas Michael Power, The Changing Economic Role of Natural Landscapes in the West: Moving Beyond an Extractive and Tourist Perspective, 31 Envir. Law Reporter News & Analysis 10438 (2001).

80. See John W. Ragsdale, Jr., National Forest Land Exchanges and the Growth of Vail and Other Gateway Communities, 31 Urban Lawyer 1 (1999); see also Chapter 6 for a detailed examination of the intermixed public–private land ownership problem.

E I G H T
Toward a New Order

1. Just in the Quincy area, the timber harvest levels dropped from 430 mbf in 1990 to 187 mbf in 1998. See U.S. Forest Service, Herger-Feinstein Quincy Library Group Forest Recovery Act Final Environmental Impact Statement 3-117–24 (Quincy, Calif., 1999) (hereafter, Quincy Library Group FEIS). Across the Sierra Nevada forests, total harvest levels dropped from approximately 1.1 bbf in 1988 to 300 mbf in 1994. See Timothy P. Duane, Community Participation in Ecosystem Management, 24 Ecology L. Q. 771, 785 n.70 (1997) (hereafter, Duane, Community Participation).

2. For basic information about Sierra Nevada ecology, evolution of the region's timber industry, and related socioeconomic information, see Centers for Water and Wildland Resources, Sierra Nevada Ecosystem Project: Final Report to Congress (Davis, Calif., 1996) (hereafter, SNEP Report); Timothy P. Duane, Shaping the Sierra: Nature, Culture, and Conflict in the Changing West (Berkeley: University of California Press, 1998) (hereafter, Duane, Shaping the Sierra). For more specific information about the Quincy region, see Quincy Library Group FEIS, supra note 1, at 3-117–24.

3. Much of the information contained in this section came from on-site interviews with officials and individuals involved with the Quincy Library Group and related issues. The author conducted interviews during the 2001 summer with Jamie Rosen, Department of Agriculture General Counsel's Office; Erin Noel, a local environmental activist and attorney; Linda Blum and Harry Reeves, Quincy Library Group members representing local environmental interests; Mike Yost, a local forestry professor and environmental representative on the Quincy Library Group; Todd Bryan, a University of Michigan graduate student studying the Quincy Library Group process; Dave Peters, the U.S. Forest Service district ranger responsible for implementing the Herger-Feinstein Quincy Library Group legislation. In addition, the author attended the Aug. 8, 2001, Quincy Library Group meeting as an observer.

4. See Julia M. Wondolleck and Steven L. Yafee, Making Collaboration Work: Lessons from Innovation in Natural Resource Management 71–3 (Washington, D.C.: Island Press, 2000).

5. For more detailed descriptions of the Quincy Library Group's evolution, see Ronald D. Brunner et al., Finding Common Ground: Governance and Natural Resources in the American West 159–76 (New Haven: Yale University Press, 2002); Ed Marston, The Quincy Library Group: A Divisive Attempt at Peace, in Philip Brick et al., eds., Across the Great Divide: Explorations in Collaborative Conservation and the American West 79 (Washington, D.C.: Island Press, 2001) (hereafter, Across the Great Divide); Duane, Community Participation, supra note 1 at 784–96.

6. See Quincy Library Group Community Stability Proposal (August 1993), available at www.qlg.org.

7. See Tom Knudsen, Majesty and Tragedy: The Sierra in Peril, Sacramento Bee, June 9–13, 1991 (5-part series), at p. A1; Jared Verner et al., The California Spotted Owl: A Technical Assessment of Its Current Status 3–25 (Albany, Calif.: U.S. Forest Service Pacific Southwest Research Station, 1992); U.S. Forest Service, California Spotted Owl Sierran Province Interim Guidelines Environmental Assessment (San Francisco, Calif., 1993); U.S. Forest Service, Decision Notice and Finding of No Significant Impact for California Spotted Owl Sierran Province Interim Guidelines Environmental Assessment (San Francisco, Calif., 1993); U.S. Forest Service, CASPO Interim Guidelines (San Francisco, Calif., 1993); see also California Forestry Association v. Thomas, 936 F. Supp. 13 (D.D.C. 1996).

8. Ed Marston, The Quincy Library Group: A Divisive Attempt at Peace, in Across the Great Divide, supra note 5, at 85.

9. See Herger-Feinstein Quincy Library Group Forest Recovery Act, Pub. L. 105–277, Div. A, § 101(e) Title IV, § 401, 112 Stat. 2681-305 (105th Cong., 1998), codified at 16 U.S.C. § 2104 note (2000); Lawrence Ruth, Conservation on the Cusp: The Reformation of National Forest Policy in the Sierra Nevada, 18 UCLA J. Envir. Law and Policy 1, 72–81 (1999/2000).

10. See Quincy Library Group FEIS, supra note 1; U.S. Forest Service, Herger-Feinstein Quincy Library Group Forest Recovery Act FEIS Record of Decision 6 (Quincy, Calif., 1999). Only the Californians for Alternatives to Toxics (CATS) went to court, arguing that the Quincy EIS had failed to properly analyze the environmental effects of herbicide spraying in the proposed DFPZs. The CATS litigants eventually won when the court ordered the Forest Service to prepare a supplemental EIS on this point.Californians for Alternatives to Toxics v. U.S. Forest Service, Docket No. Civ. S-00-2016 LKK/JFM (E.D. Cal., Aug. 28, 2001).

11. See U.S. Forest Service, Sierra Nevada Forest Plan Amendment Final Environmental Impact Statement (San Francisco, Calif., 2001); SNEP Report, supra note 2, at 100; see also Ed Marston et al., Restoring the Range of Light, 33(16) High Country News 1, 8–14 (Aug. 27, 2001).

12. See Wondolleck and Yaffee, supra note 4, at 117; Brunner et al., supra note 5, at 172.

13. On ESA listing for the California spotted owl, see Dan Berman, FWS Rejects Petition to List California Spotted Owl, Land Letter, February 13, 2003, at p. 7; 68 Fed. Reg. 7580–7608 (February 14, 2003). Memorandum to U.S. Fish and Wildlife Service Director from Region One Acting Director re. Administrative 90-Day Finding on Petition to List the California Spotted Owl (July 12, 2000) (finding that listing "may be warranted"). On California wilderness, see California Wild Heritage Act of 2002, S. 2535 (107th Cong., 2002).

The Bush administration national forest policy proposals are discussed in Chapters 4 and 5.

14. See Dale Bosworth, Decision for Appeals of the Record of Decision for the Sierra Nevada Forest Plan Amendment and Its Final Environmental Impact Statement (Nov. 16, 2001); Sierra Nevada Framework Review Team, Focus Area Summaries (June 16, 2002); Summary of Sierra Nevada Forest Plan Review (Jan. 24, 2002), at www.fs.fed.us/r5/snfpa/library/archives/review/review-summary-01-24-02.htm (site visited Dec. 27, 2002); U.S. Forest Service, Plumas and Lassen National Forests: California Administrative Study 4202-02-01, 67 Fed. Reg. 72136-38 (Dec. 4, 2002); see also Brian Stempeck, Panel to Recommend Changes to Sierra Nevada Logging Plan Next Week, Land Letter, February 27, 2003, at p. 7; Jane Braxton Little, Coming of Age for Quincy Experiment, Forest Magazine 38 (Spring 2002); Bob Dale, Sideways Attack in the Sierra, Forest Magazine 29 (Spring 2002); Enviros Feel They're on Short End of Sierra Nevada Stick, Public Land News 8 (March 15, 2002).

15. See note 13 and accompanying text; see also Restoring the Range of Light, High Country News, Aug. 27, 2001, at p. 9.

16. See Herger-Feinstein Quincy Library Group Forest Recovery Act, Pub. L. No. 105–277, Div. A, § 101(d)(2), 112 Stat. 2681 (106th Cong., 1998), codified at 16 U.S.C. 2104 note (d)(2) (providing for fuel break construction on "not less than 40,000, but not more than 60,000 acres per year").

17. 36 C.F.R. § 219.1(b)(3) (2001); see also Chapter 7 for further discussion of the sustainability concept.

18. See Marston, supra note 8, at 86; Duane, Community Participation, supra note 1, at 792; Erin Noel, Legislating Localism: The Quincy Library Group and the History of Community Involvement in the Management of Public Lands 39–45 (1998) (unpublished paper); Quincy Library Group ROD Draws Criticism from Environmentalists, Public Land News, Sept. 17, 1999, at 7.

19. See Louis Blumberg and Darrell Knuffke, Count Us Out, 2(2) Chronicle of Community 41, 43 (Winter 1998); Marston, supra note 5, at 86; Duane, Community Participation, supra note 1, at 792; Brunner et al., supra note 5, at 183.

20. On the concept of hierarchical nesting for ecosystem management purposes, see Jonathan B. Haufler et al., Scale Considerations for Ecosystem Management, in Robert C. Szaro et al., eds., Ecological Stewardship: A Common Reference for Ecosystem Management (vol. 2) 331 (Oxford: Elsevier Science, 1999) (hereafter, Ecological Stewardship); Duane, Shaping the Sierra, supra note 2, at 450. On the interagency coordination statement idea, see Robert B. Keiter, Beyond the Boundary Line: Constructing a Law of Ecosystem Management, 65 U. Colo. L. Rev. 293, 330 (1994).

21. See Chapter 6 for a detailed examination of private land conservation strategies and issues.

22. On the temporal dimensions of ecosystems, see Peter S. White et al., Disturbance and Temporal Dynamics, in Ecological Stewardship, supra note 20, at 281,

23. Adaptive management concepts are addressed in Bernard T. Bormann et al., Adaptive Management, in Ecological Stewardship (vol. 3), supra note 20, at 505; Gene Lessard, An Adaptive Approach to Planning and Decision-Making, 40 Landscape and Urban Planning 81 (1998); Kai N. Lee, Compass and Gyroscope: Integrating Science and Politics for the Environment (Washington, D.C.: Island Press, 1993). For the Quincy Library Group appeal, see Notice of Appeal Pursuant to 36 CFR Part 217, at www.qlg.org/pub/miscdoc/appeal (site visited June 6, 2001).

24. These collaborative conservation process principles are examined more fully in Chapter 7; see also Wondolleck and Yaffee, supra note 4.

25. See Duane, Community Participation, supra note 1, at 789; Noel, Legislating Localism, supra note 18, at 40–45; Charles Davis and M. Dawn King, The Quincy Library Group and Collaborative Planning within U.S. National Forests (2000), at www.qlg.org/pub/Perspectives/daviskingcasestudy.pdf (site visited June 27, 2002).

26. Jane Braxton Little, Quincy Library Group Bars Outsiders, 31(8) High Country News 4 (April 26, 1999).

27. See Marston, supra note 5, at 85; Blumberg and Knuffke, supra note 19, at 41.

28. On the analysis-paralysis point, see U.S. Forest Service, The Process Predicament: How Statutory, Regulatory, and Administrative Factors Affect National Forest Management (Washington, D.C., 2002); Brian Stempeck, Forest Service Releases Management Gridlock Report, 21(12) Land Letter 3 (June 20, 2002); U.S. General Accounting Office, Forest Service: Appeals and Litigation of Fuel Reduction Projects (Washington, D.C., 2001). On revising the multiple-use concept, see Jack Ward Thomas and Jory Ruggiero, Politics and the Columbia Basin Assessment: Learning from the Past and Moving to the Future, 19 Public Land and Resources L. Rev. 33, 47 (1998); Michael C. Blumm, Public Choice Theory and the Public Lands: Why "Multiple Use" Failed, 18 Harvard Envir. L. Rev. 405 (1994).

29. The quotations are from: Sierra Nevada Ecosystem Project, Final Report to Congress: Status of the Sierra Nevada (vol. 1) 45 (Davis, Calif.: Center for Water and Wildland Resources, 1996); Don Snow, Coming Home, 1(1) Chronicle of Community 43 (1996).

30. See Derek Kauneckis et al., Tahoe Regional Planning Agency: The Evolution of Collaboration (Bloomington: Indiana University School of Public and Environmental Affairs, 2000); Bowen Blair, Jr., The Columbia River Gorge National Scenic Area: The Act, Its Genesis and Legislative History, 17 Envir. L. 863 (1987); Carl Abbott et al., Planning a New West: The Columbia River Gorge National Scenic Area (Corvallis: Oregon State University Press, 1997); see also Michael C. Blumm and Brad L. Johnson, Promising a Process for Parity: The Pacific Northwest Electric Power Planning and Conservation Act and Andromous Fish Protection, 11 Envir. L. 497 (1981), for a description of the Northwest Power Planning Council.

31. Daniel Kemmis, This Sovereign Land: A New Vision for Governing the West 181–93 (Washington, D.C.: Island Press, 2001); Richard Behan, Plundered Promise: Capitalism, Politics, and the Fate of the Federal Lands 202–15 (Washington, D.C.: Island Press, 2001); Donald Snow et al., The Lubrecht Conversations, 3(1) Chronicle of Community 15 (1998); April Reese, Bush Administration Tests Waters for "Charter Forests" Initiative, Land Letter, Feb. 28, 2002, at 8–9.

32. The proponents of such devolution of management authority include: Kemmis, This Sovereign Land, supra note 31; Behan, supra note 31; Sally K. Fairfax, State Trust Lands Management: A Promising New Application for the Forest Service?, in Roger A. Sedjo, ed., A Vision for the U.S. Forest Service: Goals for Its Next Century 105 (Washington, D.C.: Resources for the Future, 2000). Critics and skeptics of such devolution include: Rena I. Steinzor, EPA and Its Sisters at Thirty: Devolution, Revolution, or Reform?, 31 Envir. Law Reporter 11086 (2001); Rena I. Steinzor, The Corruption of Civic Environmentalism, 30 Envir. Law Reporter 10909 (2000); Hope M. Babcock, Dual Regulation, Collaborative Management, or Layered Federalism: Can Cooperative Federalism Models from Other Laws Save Our Public Lands?, 3 Hastings West-Northwest J. Envir. Law and Policy 193 (1996).

33. See Dale D. Goble et al., Local and National Protection of Endangered Species: An Assessment, 2 Envir. Science and Policy 43 (1999); Center for Wildlife Law, State Endangered Species Acts: Past, Present, and Future (1998); Daniel R. Mandelker, NEPA Law and Litigation § 12.01 et seq. (1992).

34. The Bitterroot grizzly bear reintroduction proposal and process is discussed in U.S. Fish and Wildlife Service, Grizzly Bear Recovery in the Bitterroot Ecosystem Final Environmental Impact Statement (Missoula, Mont., 2000); Sarah Van de Wetering, Bitterroot Grizzly Bear Reintroduction: Management by Citizen Committee?, in Across the Great Divide, supra note 5, at 150; Kemmis, This Sovereign Land, supra note 31, at 1–18. On the Bush administration's reversal of the Bitterroot grizzly bear reintroduction proposal, see U.S. Fish and Wildlife Service, Reevaluation of the Record of Decision for the Final EIS and Selection of the Alternative for Grizzly Bear Recovery in the Bitterroot Ecosystem, 66 Fed. Reg. 33623 (June 22, 2001); Grizzly Reintro Plan Dead, Endangered Species & Wetlands Report 3 (June 2001); Idaho Challenges Griz Reintro, Endangered Species & Wetlands Report 9 (February 2001); Green Groups Decry Grizzly Plan Reversal, 20(12) Land Letter 4 (June 25, 2001).

35. See Chapter 6 for a more detailed examination of new nature reserve and biodiversity conservation concepts, including discussion of the Northern Rockies Ecosystem Protection Act and Wildlands Project proposals.

36. For more information on organic legislation, see Robert L. Fischman, The National Wildlife Refuge System and the Hallmarks of Modern Organic Legislation, 29 Ecology L. Q. 457 (2002). An earlier version of this proposal appeared in Keiter, Beyond the Boundary Line, supra note 20, at 328-31. For other useful discussions of new biodiversity or ecosystem management legislation, see Bradley C. Karkkainen, Biodiversity and Land, 83 Cornell L. Rev. 1, 41–57 (1997); J. B. Ruhl, Biodiversity Conservation and the Ever-Expanding Web of Federal Laws Regulating Nonfederal Lands: Time for Something Completely Different?, 66 U. Colo. L. Rev. 555, 661-71 (1995); Julie B. Bloch, Preserving Biological Diversity in the United States: The Case for Moving to an Ecosystems Approach to Protect the Nation's Biological Wealth, 10 Pace Envir. L. Rev. 175, 217–22 (1992); Holly Doremus, Patching the Ark: Improving Legal Protection of Biological Diversity, 18 Ecology L. Q. 265, 318–24 (1991).

37. Ecological restoration policy and funding is addressed in Chapter 5.

38. See 16 U.S.C. §§ 79(k), (l) (Redwood National Park Organic Act amendments providing job displacement funds and employment opportunities for displaced local timber workers); see also Chapter 4 for a description of the economic assistance and transition program funds that accompanied the Northwest Forest Plan.

N I N E

Keeping Faith with Nature

1. See Roger N. Clark et al., Toward an Ecological Approach: Integrating Social, Economic, Cultural, Biological, and Physical Considerations, in Robert C. Szaro et al., eds., Ecological Stewardship: A Common Reference for Ecosystem Management (vol. 3) 297 (Oxford: Elsevier Science, 1999) (hereafter, Ecological Stewardship); James J. Kennedy and Jack Ward Thomas, Managing Natural Resources as Social Value, in Richard L. Knight and Sarah F. Bates, eds., A New Century for Natural Resources Management 311 (Washington, D.C.: Island Press, 1995) (hereafter, A New Century).

2. See Committee of Scientists, Sustaining the People's Lands: Recommendations for Stewardship of the National Forests and Grasslands into the Next Century (Washington, D.C.: U.S. Department of Agriculture, 1999); Robert Costanza, Ecological Economics: Toward a New Transdisciplinary Science, in A New Century, supra note 1, at 323; Gretchen Daily and Katherine Ellison, The New Economy of Nature: The Quest to Make Conservation Profitable (Washington, D.C.: Island Press 2002).

3. See Hanna J. Cortner and Margaret A. Moote, The Politics of Ecosystem Management 11–28 (Washington, D.C.: Island Press, 1999); Mark W. Brunson and James J. Kennedy, Redefining "Multiple Use": Agency Response to Changing Social Values, in A New Century, supra note 1, at 143.

4. See generally J. Barid Callicott and Michael P. Nelson, eds., The Great New Wilderness Debate (Athens: University of Georgia Press, 1998); Dave Foreman, The Wildlands Project and the Rewilding of North America, 76 Denver U. L. Rev. 535, 546–48 (1999); Michael E. Soule and John Terborgh, eds., Continental Conservation: Scientific Foundations of Regional Reserve Networks (Washington, D.C.: Island Press, 1999).

5. See Stephen Farber and Dennis Bradley, Ecological and Resource Economics as Ecosystem Management Tools, in Ecological Stewardship, supra note 1, at 383; John Gordon and Jane Coppock, Ecosystem Management and Economic Development, in Marian R. Chertow and Daniel C. Esty, eds., Thinking Ecologically: The Next Generation of Environmental Policy 37 (New Haven: Yale University Press, 1997). On the controversy surrounding federal mining law reform, see Christine Knight, A Regulatory Minefield: Can the Department of the Interior Say "No" to a Hardrock Mine, 73 U. Colo. L. Rev. 619 (2002); Sam Kalen, An 1872 Mining Law for the New Millennium, 71 U. Colo. L. Rev. 343 (2000).

6. See Vawter Parker, Natural Resources Management by Litigation, in A New Century, supra note 1, at 209; see also Charles F. Wilkinson, Crossing the Next Meridian: Land, Water, and the Future of the American West (Washington, D.C.: Island Press, 1992); Joseph L. Sax, The Legitimacy of Collective Values: The Case of the Public Lands, 56 U. Colo. L. Rev. 537 (1985).

7. See Richard D. Periman et al., Human Influences on the Development of North American Landscapes: Applications for Ecosystem Management, Ecological Stewardship (vol. 2), supra note 1, at 479; Oliver A. Houck, Are Humans Part of Ecosystems?, 28 Envir. Law 1 (1998); see also Mark J. McDonnell and Steward T. A. Pickett, eds., Humans as Components of Ecosystems: The Ecology of Subtle Human Effects and Populated Areas (New York: Springer-Verlag, 1993).

8. See Steven L. Yaffee, Regional Cooperation: A Strategy for Achieving Ecological Stewardship, in Ecological Stewardship, supra note 1, at 131; Bradley C. Karkkainen, Collaborative Ecosystem Governance: Scale, Complexity, and Dynamism, 21 Virg. Envir. L. J. 189 (2002); see also Richard L. Knight and Peter B. Landres, eds., Stewardship across Boundaries (Washington, D.C.: Island Press, 1998).

9. See Gordon and Coppock, supra note 5, at 44–46.

10. The Leopold quote is found at Aldo Leopold, A Sand County Almanac 239 (New York: Ballantine Books, 1966).

11. See Oliver A. Houck, On the Law of Biodiversity and Ecosystem Management, 81 Minn. L. Rev. 869 (1997); Robert B. Keiter, Beyond the Boundary Line: Constructing a Law of Ecosystem Management, 65 U. Colo. L. Rev. 293 (1994); see also Robert B. Keiter et al., Legal Perspectives on Ecosystem Management: Legitimizing a New Federal Land Management Policy, in Ecological Stewardship (vol. 3), supra note 1, at 9.

12. See P. Lynn Scarlett, A New Approach to Conservation: The Case for the Four Cs, 17(2) Nat. Resources and Envir. 73 (2002); James J. Kennedy and Michael P. Dombeck, The Evolution of Public Agency Beliefs and Behavior Toward Ecosystem-Based Stewardship, in Ecological Stewardship (vol. 3), supra note 1, at 85; Winifred B. Kessler and Hal Salwasser, Natural Resource Agencies: Transforming from Within, in A New Century, supra note 1, at 171; see also Cortner and Moote, supra note 3. On the analysis-paralysis point, see U.S. Forest Service, The Process Predicament: How Statutory, Regulatory, and Administrative Factors Affect National Forest Management (Washington, D.C., 2002); Brian Stempeck, Forest Service Explains Gridlock Costs, 21(10) Land Letter 5 (May 23, 2002); U.S. General Accounting Office, Forest Service: Appeals and Litigation of Fuel Reduction Projects (Washington, D.C., 2001).

13. See John D. Leshy, Shaping the Modern West: The Role of the Executive Branch, 72 U. Colo. L. Rev. 287 (2001). On western efforts to sabotage and reshape the Progressive Era conservation agenda, see Samuel P. Hays, Conservation and the Gospel of Efficiency: The Progressive Conservation Movement, 1890–1920, 271–76 (Cambridge, Mass.: Harvard University Press, 1959).

14. See John D. Leshy, The Babbitt Legacy at the Department of the Interior: A Preliminary View, 31 Envir. Law 199, 208–12 (2002).

15. See Daniel Kemmis, This Sovereign Land: A New Vision for Governing the West (Washington, D.C.: Island Press, 2001); Karkkainen, supra note 8.

16. See Gordon and Coppock, supra note 5; Bradley C. Karkkainen, Environmental Lawyering in the Age of Collaboration, 2002 Wisc. L. Rev. 555; see also Julia M. Wondolleck and Steven L. Yaffee, Making Collaboration Work: Lessons from Innovation in Natural Resource Management (Washington, D.C.: Island Press, 2000); Philip Brick et al., eds., Across the Great Divide: Exploration in Collaborative Conservation and the American West (Washington, D.C.: Island Press, 2001).

17. The quotation can be found at Lynton Caldwell, The Ecosystem as a Criterion for Public Land Policy, 10 Natural Res. J. 203, 208 (1970).

Abbey, Edward, 224

Active management, 147, 149–151, 291, 316–317

Adaptive management, 71, 86, 120, 166, 289, 292

Administrative Procedures Act of *1946*, 26–27, 32, 240, 251

Agencies. *See* Federal agencies; *and specific agencies*

Agriculture Department, 125, 129

Alaska National Interest Lands Conservation Act (ANILCA), 200–201, 334*n*10

Alligator National Wildlife Refuge, 159

Allowable sale quantity (ASQ), 82, 85

All-terrain vehicles (ATVs), 178, 262, 263

Anaconda Copper, 18

Animal units per month (AUMs), 63

Antiquities Act of *1906,* 23, 33, 121, 123, 184, 190, 196, 203, 205, 320

Applegate Partnership, 28, 253

Aquatic conservation strategy (ACS), 89, 99–100, 104, 110, 111

Aquatic habitat: loss of, 99; management standards, 163; in old growth forests, 84

Arches National Monument, 222, 223

Areas of critical environmental concern (ACECs), 39, 204

Association of Forest Service Employees for Environmental Ethics (AFSEEE), 68

Babbitt, Bruce, 39, 70, 96, 114, 116, 121, 127, 184, 185, 238, 246

Baca Ranch, 212

Bandelier National Monument, 140

Bears: management policy, 67, 70. *See also* Grizzly bear

Behan, Richard, 302

Biodiversity conservation, 5, 25, 80; aesthetic/recreational benefits of, 59; Clinton administration and, 97; defined, 52; economic value of, 58–59; ecosystem-based (coarse filter) approach to, 52–53, 284–285; enclaves and, 5, 186–187; enforcement by courts, 118–119; FWS mandate for, 69, 197; integrated strategy, 169–170; in nonequilibrium model of ecology, 51–52; shortcomings of national parks and wildlife refuges, 187, 197–198; species-based (fine filter) approach to, 52, 53. *See also* Ecological restoration; Spotted owl controversy

Biosphere reserve concept, 69–70

Bison restoration, 143

Bitterroot grizzly bear controversy, 161, 251, 257, 305, 323

Blackfeet tribe, 65

Bob Marshall Wilderness Area, 175, 177, 201

Bosworth, Dale, 298

Boundaries, ecosystem, 76, 151, 317

Bradley, Bill, 184

Brown, Ron, 96

Browner, Carol, 70, 96

Bureau of Land Management (BLM): aquatic conservation strategy and, 89; Clinton administration reforms, 39–40; collaborative process conservation and, 225–227; definition of ecosystem management, 72; devolution policy and, 27; ecosystem management initiatives, 68–69, 80, 116–117; ICBEMP restoration initiative and, 162, 163, 168; interagency relations, 197; land exchanges of, 212, 213, 214; legal challenges to, 69, 92–95; mandate and tradition of, 7, 22, 24, 39, 68, 204; motorized recreation and, 263; Northwest Forest Plan implementation and, 110, 111–

Bureau of Land Management (BLM)
(*continued*)
112; recreational access and, 270; Resource
Advisory Councils, 36; resource districts
of, 81; salvage logging harvests and, 106; in
spotted owl controversy, 84–85, 90, 91,
92–95; wilderness preservation and, 180,
181–186, 202, 203, 224
Bush, George H. W., 96
Bush, George W., 78, 116, 117; community-
based conservation endorsed by, 248, 301;
dismantling of Clinton era reforms, 125–
126, 179, 205, 207; forest health initiative
of, 157; ICBEMP restoration initiative
and, 167–168; in motorized recreation
controversy, 264, 266; Northwest Forest
Plan and, 109, 110, 111

Cabinet Yaak grizzly recovery region, 160, 161
Caldwell, Lynton, 79, 326
California Desert Conservation Area, 197,
202, 262–263
California Desert Protection Act, of *1994*, 202
Canyon Country Partnership, 220, 226, 227,
246, 255, 257
Canyonlands National Park, 223, 224
Carhart, Arthur, 198
Carson, Rachel, 57–58
Carter, Jimmy, 33, 196
Carver, Dick, 235
CASPO (California spotted owl), 277–278,
280, 285
Catron County (New Mexico), 227–234
Catron County Citizens Group, 246
Cerro Grande fire, 140
Chevron doctrine, 32, 34
Christianity, view of nature, 55
Church, Frank, 174
Civic republicanism, 36, 245
Civilian Conservation Corps (CCC), 137,
143, 144
Clean Air Act, 22, 60, 141
Clean Water Act, 11–12, 22, 60, 152–153, 154
Clements, Federic, 49, 51
Cleveland, Grover, 20, 33
Clinton, Bill: BLM reform under, 39–40, 95;

congressional response to preservation policy, 122, 123–124, 125; ecosystem management initiatives, 70–71, 113–117, 121, 163,
300–301, 320; Government Performance
Review program, 115; ICBEMP restoration
initiative, 167; land exchanges under, 212;
National Biological Survey created, 42; national monument designations of, 33, 121,
184, 196, 203, 205, 239; Northwest Forest
Conference, 96–97; private land regulatory strategy of, 216; Roadless Area Conservation initiative, 116, 121, 126, 179; salvage logging rider and, 105–106, 157. *See
also* Northwest Forest Plan
Coal mining, 61, 63
Coarse-filter approach to biodiversity conservation, 52–53
Collaborative conservation, 36, 220, 320–
321; administrative law reform and, 26–27,
120, 252–253; arguments against, 254–258,
304–306; benefits of, 246, 256; in Bitterroot grizzly bear restoration, 161, 251, 257,
305; Bush administration and, 248, 303;
courts and, 120–121; in critical habitat determinations, 228–232, 253; definition of
community, 246–247; in ecosystem management regime, 274, 287, 289, 299–302,
318, 322–323; expansion of, 26, 27–28,
245–246; federalism and, 25–26, 304; in
forest planning and management, 231,
233–234; grassroots nature of, 301; historical background of, 26–27; impact on
preservation, 206–208; in land use planning, 217–218; limitations of, 251–254;
livestock grazing advisory boards and, 26,
28, 39, 247, 251; national *vs* local interests
in, 303–306; origins of community-based
concept, 244–245; place-based initiatives,
302–303; principles of community-based
concept, 248–250; in recreation management area, 225–226, 257, 270–271; species
reintroduction and, 161, 217, 232, 257, 303,
305. *See also* Quincy Library Group
Colorado Canyons National Conservation
Area, 206
Colorado Plateau, 35, 180, 222

Columbia River, Native American fishing rights on, 65

Columbia River Basin, ecological restoration in, 115–116, 124, 162–169

Columbia River Gorge National Scenic Area, 215, 302

Community and the Politics of Place (Kemmis), 36

Community-based conservation. *See* Collaborative conservation

Condor restoration, 158

Congress, U.S.: disposal policy and, 17, 18; ecosystem management legislative proposal for, 307–310; ecosystem management movement and, 70, 80–81, 86, 122–125; environmental lobby in, 44; ESA reauthorization in, 239–240; ESA reform proposals in, 241; ICBEMP restoration initiative and, 165, 166, 167; land exchanges and, 212–213, 214–215; motorized recreation debate in, 266–267; national park creation by, 22, 23, 195–196, 197; oversight of federal agencies, 7, 32, 323–324; place-based legislative approach of, 302, 307–308; power over public land and resources, 11, 20, 28–33, 108, 321; preservation agenda of, 22–24, 195, 206–208; preservation laws in, 22, 38, 66–67, 124, 152, 324; private land regulatory powers of, 216; Quincy Library Group bill in, 278, 280–281, 296–297, 302; recreation fee program in, 269; salvage logging rider in, 105–106, 156–157; state/local interests and, 26, 35–36; timber harvest and, 90, 93–94, 95; wilderness preservation in, 175, 179, 182, 184, 198–203, 206; Wise Use movement allies in, 237, 241; wolf restoration and, 131, 134

Congressional Research Service, 122

Conservation biology, 54, 58, 74, 178, 192

Conservation and Reinvestment Act (CARA), 212

Conservation easements, 217

Constituency-based governance, 26–28. *See also* Devolution policy

Constitution, U.S., 11, 29, 31, 33, 210, 304

Cooney, James, 228

Cooperative Ecosystems Studies Unit, 115

Coronado National Forest, 106

County supremacy ordinances, 229, 234–236, 243

Courts: biocentric perspective and, 58; county supremacy ordinances in, 235–236; ecosystem-based policies and, 34–35, 118–122, 319; judicial review, 33–35; land exchanges in, 213; motorized recreation in, 266; Northwest Forest Plan in, 102–105, 252; preservation agenda and, 207; public participation and, 120–121; public *vs* private rights, 315–316; range restoration in, 154; salvage logging rider in, 106–108; spotted owl controversy in, 2–3, 87–95, 319; wilderness preservation in, 178, 179, 200, 201, 205; Wise Use movement in, 237–242; wolf restoration in, 132–133, 134, 136, 159, 216. *See also* Supreme Court, U.S.

Craighead brothers, 67

Critical habitat determination, 89, 121, 241

"Custom and culture" ordinance, 229, 236

Darwin, Charles, 55

DDT, 58

Defenders of Wildlife, 45–46, 132, 134, 135, 217

Defensible fuel profile zones (DFPZs), 277, 278, 280

Delisting process, 162, 253–254

Democratic principles, ecosystem management and, 77

Desert Citizens case, 213

Devolution policy, 25–28, 206–208, 253–254, 300, 302–304, 308. *See also* Collaborative conservation

DeVoto, Bernard, 311

Disposal policy, 17–19

Dombeck, Mike, 115

Douglas, William O., 58, 171

Downes, William, 133, 134, 136

Dwyer, William, 2–3, 90, 91, 92, 97, 98, 102, 103, 104, 108, 110, 118, 119

Eastern Wilderness Areas Act of *1975*, 200

Ecological restoration: active *vs* passive man-

Ecological restoration (*continued*)
agement, 147, 149–151, 316–317; collaborative process in, 161, 232, 251, 257, 305; comparison with ecosystem management, 146; comparison with remedial concepts, 145–146; comparison with traditional conservation policy, 146–147; condor, 158; defined, 145; under Endangered Species Act, 152, 154, 158–162; ferret, 158; forest (eastern), 143; forest (western), 155–157; goals for, 147–149, 151; grizzly bear, 160–161, 162, 251, 257, 303, 305, 323; large-scale projects, 151–152, 162–169; legal framework for, 152–153; opposition to predator recovery, 158–160; origins of, 144–145; rangeland, 153–155; spatial and temporal scales for, 151; strategies of, 145, 149–151; types of projects, 152; wildfires and, 5, 136, 138–141; wildlife refuges and, 142–143; wolf, 3–4, 45–46, 128–136, 150, 158–160, 216, 232, 238, 316–317

Ecological Society of America, 53–54, 71

Ecological succession theory, 49

Ecology: early development of, 49–50; ecosystem concept in, 49–51, 52; era of, 2–6, 12–14, 324–327; philosophical thinking about, 54–59; of preservation, 177–178, 187–190, 206–207; socioeconomic change and, 60–65; species extinction and, 49, 51; technological advances and, 53. *See also* Biodiversity conservation

Economic values, in policy agenda, 313–314

Ecosystem concept, 49–51, 52

Ecosystem ecology model, 51; and nonequilibrium theory, 51–52

Ecosystem management: biosphere reserve concept, 69–70; Bush administration and, 125–126; Clinton administration initiatives, 70–71, 113–117, 121, 300–301; collaborative processes in, 274, 287, 289, 299–302, 318, 322–323; congressional response to, 30–31, 80–81, 122–125; constituency-based governance, 26–28; criticism of, 13, 75–78; defined, x, 71–72, 76–77; ecological restoration and, 146; enclave-based system and, 5, 186–187; expansive reserve system and, 5–6, 25, 186–190; federal-state-local conflicts over, 10–11, 25, 300–301, 303–306; greater ecosystem concept, 190–191; historical background to, 16–28; introduction of, 66–71; judicial support of, 34–35, 118–122, 319; legal foundation for, 7, 10, 11, 13–14, 73–74, 78, 316; legislative models, 307–310; national *vs* local interests in, 303–306; principles of, 72–73; private landowners and, 215–218; regional, 67–68, 190–192, 317–318; scale issues, 66, 70, 76, 151–152, 256–257, 288–291, 317; scientific community endorsement of, 53–54; sustainability goal of, 71; *vs* traditional resource management, 73. *See also* Biodiversity conservation; Northwest Forest Plan; Quincy Library Group; Spotted owl controversy

Ecosystem Management for Parks and Wilderness, 70

Ecosystem Management Task Force, 71, 72

Elk restoration, 232

Elwha River dam removal, 169

Enclave strategy, 5, 186–187

Endangered American Wilderness Act of *1978,* 200

Endangered community argument, 242–243

Endangered Species Act of *1973,* 22, 24, 30, 34, 36, 37, 60, 81, 123, 320; collaborative conservation and, 228–232, 253–254; congressional reform proposals, 240–241; delisting process, 162, 253–254; ecological restoration under, 152, 154, 158–162; experimental population provision of, 4, 130, 131; federal-state conflict and, 11; FWS implementation of, 69, 74, 118–119; habitat conservation planning process, 114, 216; modern preservationist ethic and, 316; Northwest Forest Plan and, 103; private land under, 216, 316; reauthorization of, 239–240; snail darter case, 69, 88; spotted owl controversy, 87, 88–89, 93, 94–95; Supreme Court on, 41, 240; Wise Use movement litigation, 239–242

Endangered Species Committee (God Squad), 94–95

Endangered species registry, 3, 110, 117, 130, 253

Energy pyramid theory, 50

Enlibra model, 248, 300

Environmental impact analysis: cumulative effects, 119–120, 151; for ICBEMP, 164, 166; NEPA requirement for, 74, 240; for Northwest forests, 85, 93, 94, 104; for Quincy Group plan, 280–281, 282–283, 292

Environmental movement: collaboration with industry, 46, 276–278, 281–282, 288, 293; factors in emergence of, 60; influence on policy, 44–46; Northwest Forest Plan and, 103; in spotted owl controversy, 2–3, 84–85, 87; wilderness preservation and, 182, 184, 185–186, 200–201, 202; wolf reintroduction and, 131, 133, 217. See also Quincy Library Group

Environmental protection concept, ix, 10, 13, 22, 64

Errington, Paul, 130

ESA. See Endangered Species Act

Espy, Mike, 96

Ethics, ecology and, 54–59, 313–314

Everglades National Park, 187

Executive branch. See Federal agencies; President

Extractive industries: beneficiaries of disposal policy, 18; collaboration with environmentalists, 46; displaced by ecological priorities, 312; globalization trend and, 61–62; influence on public land policy, 43–44; opposition to wilderness bill, 174; regional influence of, 61; rural/urban conflicts over, 64–65; utilitarian perspective of, 42–43; Wise use movement and, 234. See also Mining industry; Ranching industry; Timber industry

Federal Advisory Committee Act (FACA), 252–253, 295, 321

Federal agencies: accountability of, 32, 323–324; centralized model and, 25; collaborative conservation and, 26–27, 120, 251–254, 304; concurrent jurisdiction and, 11, 26; discretionary authority of, 31–32; erosion of authority, 319–320; introduction of ecosystem management concepts, 66–71; judicial review and, 32, 34; legal mandates of, 7, 10, 37–41, 74, 204, 298; Northwest Forest Plan and, 98, 102; regulatory expansion of, 11–12, 22; scientific and technical expertise of, 21, 38, 314; single-resource agenda of, 7, 38, 39; utilitarian conservation model of, 21–22. See also specific agencies

Federal Land Exchange Facilitation Act, 212–213

Federal Land Policy and Management Act (FLPMA) of 1976, 68, 73, 74, 92–93, 151, 224; land exchanges under, 212; multiple use standard under, 39; preservation responsibilities under, 22, 24, 27, 39; presidential withdrawal powers under, 33; wilderness designation under, 181, 202

Federalism, public land and: Congress, 28–31; devolution policy and, 25–26, 304; executive branch, 31–33; judicial review, 33–35; states, 35–37

Federal Tort Claims Act (FTCA), 141

Feinstein, Diane, 280

FEMAT. See Forest Ecosystem Management Assessment Team

Ferret restoration, 158

Fine-filter approach to biodiversity conservation, 52, 53

Fires. See Wildfires

Fish and Wildlife Service, U.S.: aquatic conservation strategy and, 89; biodiversity conservation responsibilities of, 204, 208; critical habitat determination by, 89, 121, 241; definition of ecosystem management, 72; ecosystem management initiatives of, 117, 197; ESA implementation by, 69, 74, 118–119, 121, 158, 240, 241, 253–254, 283, 320; grizzly bear restoration plan of, 160–161, 251; mandate and tradition of, 7, 22, 24, 37, 41–42; in spotted owl controversy, 87, 88–89, 90, 91, 94, 229; wolf restoration plan of, 3, 130, 131–132, 133, 134–135, 159

Flora, Gloria, 236

FLPMA. *See* Federal Land Policy and Management Act
Foreman, Dave, 192
Forest. *See* National forests
Forest and Refuge Omnibus Act of *1976*, 175
Forest ecosystem, 83–84
Forest Ecosystem Management Assessment Team (FEMAT), 97–98, 99, 101, 252, 281
Forest fires. *See* Wildfires
Forestry, 21
Forest Service, U.S.: authority of, 31, 33–34; biodiversity conservation mandate of, 92, 204; definition of ecosystem management, 72; ecosystem management initiatives of, 68, 70, 80, 115–116; environmental analysis of, 119; interagency relations, 197; inventorying and monitoring obligations of, 120; land exchanges of, 213, 214–215; legal challenges to, 34, 38, 91, 92; livestock grazing program, 232; local interests and, 26, 229, 231, 232, 233–234, 247; mandate and tradition of, 7, 22, 37–38; motorized recreation and, 263, 266; in Northwest Forest Plan implementation, 110, 111–112; policy reform in, 38–39; Quincy Library Group plan and, 276, 277, 278, 280, 282–283, 292, 294–295; recreational access and, 270; salvage sale program of, 106, 107; scientific management in, 21–22; in spotted owl controversy, 84–85, 91, 92, 96, 277–278; timber harvest quotas of, 3, 68, 82; utilitarian conservation policy in, 20–21; wilderness management of, 174–175, 178, 179, 199–200, 201, 204–205, 239; wildfire policy of, 136–137, 138
Forsman, Eric, 83
Frye, Helen, 93
Fundamentals of Ecology (Odums), 50

Gallatin National Forest, 178, 190, 214
Gang of Four, 86
Gateway communities, 259–260, 261
General Accounting Office (GAO), 111–112, 122, 153, 213
General Mining Law of *1872*, 7, 123, 316
General Revision Act of *1891*, 33

Geographical information systems (GIS), 53
Gila National Forest, 198, 228–229, 232
Gila Wilderness Area, 232, 233
Glacier National Park, 69, 160, 175, 177
Glacier-Waterton International Peace Park, 192
Gold mining, 63
Golley, Frank B., 50, 311
Gore, Al, 70, 96
Gore Range Primitive Area, 199
Government Performance Review program, 115
Grand Canyon Trust, 44, 45
Grand Staircase-Escalante National Monument, 184, 185, 203, 214
Grassland banking system, 155
Great Basin, in Utah, 180
Greater ecosystem concept, 190–191
Greater Flagstaff Forests Partnership, 157
Greater Glacier-Bob Marshall Ecosystem, 190–191
Greater Grand Canyon Ecosystem, 191
Greater Yellowstone Coalition, 44, 45, 67
Greater Yellowstone Coordinating Committee (GYCC), 13, 67–68
Greater Yellowstone Ecosystem, 67, 169, 177, 190, 191, 254
Greater Yellowstone Vision Document, 43, 235
Greenworld, 87, 88
Grizzly bear: delisting debate, 254; habitat preservation, 179, 239; habitat size, 189; recovery program, 160–161, 162, 251, 257, 303, 305, 323

Habitat conservation area (HCA), 86
Habitat conservation plan (HCP), 216
Hage, Wayne, 238
Hansen, Jim, 182, 184
"Hard look" doctrine, 33
Harrison, Benjamin, 20, 33
Hays, Samuel P., 15, 60
Herger-Feinstein Quincy Library Group Forest Recovery Act of *1998*, 280, 282, 296–298
Hetch Hetchy controversy, 23, 40

Hogan, Michael, 107
Hoover, Herbert, 222
House Resources Committee, 30
Huckelberry Land Exchange, 213

ICBEMP. *See* Interior Columbia Basin
    Ecosystem Management Project
Ickes, Harold, 180
Industry: service-sector, 62, 64. *See also* Ex-
    tractive industries; Recreation and tourism
Interagency Coordination Working Group,
    102
Interagency Ecosystem Management Task
    Force, 117
Interagency Grizzly Bear Committee
    (IGBC), 160–161
Interagency Scientific Committee (ISC), 85–
    86, 90–91, 92, 188
Interest groups, 42–44. *See also* Environmen-
    tal movement; Wise Use movement
Interior Columbia Basin Ecosystem Manage-
    ment Project (ICBEMP), 13, 115–116, 124,
    162–169, 281, 285, 314, 318
Interior Department, 28, 90, 107, 114, 125,
    131, 241
Island biogeography, 51

Jackson, Henry, 174
Jamison strategy, 94
Jarbridge Shovel Brigade, 236–237
Jobs in the Woods program, 101, 112
Johnson, Lyndon Baines, 273
Judicial review, 32, 33–35, 36; public *vs* pri-
    vate rights, 315–316
Judiciary. *See* Courts; Supreme Court, U.S.

Kemmis, Daniel, 36, 219, 302
Kesterson National Wildlife Refuge, 67
Kitzhaber, John, 253, 255
Kleppe v. New Mexico, 29
Knudsen-Bandenberg Act of *1930,* 156
Kootenai National Forest, 106, 107

Lake Tahoe Regional Planning Agency, 302
Land and Water Conservation Fund Act of
    1965, 24, 63, 212, 269

"Land Ethic" (Leopold), 57, 58, 59, 210–211,
    318
Land exchanges, 212–215
Land grants, 17, 18
Laney, Kit and Sherri, 232
Leavitt, Mike, 182, 185
Leopold, Aldo, 47, 198, 228; ecological per-
    spective of, 56–57; ecological restoration
    initiatives of, 144; land ethic of, 57, 58, 59,
    210–211, 318; on wolf recovery, 130
Leopold Report, 40, 130, 137, 148
Lewis and Clark National Forest, 65, 178
Lindemann, Raymond, 50
Litigation. *See* Courts
Livestock depredation, wolf restoration and,
    4, 131–132, 135, 217
Livestock grazing: advisory boards, 26, 28, 39,
    247, 251; allotment levels, 232; incidental
    take permit, 241; overgrazing problems,
    153; property rights and, 238; rangeland
    reform regulations, 69, 116, 121, 123, 153;
    rangeland restoration, 153–155; rate of, 63;
    riparian habitat and, 231–232; Taylor Graz-
    ing Act of *1934,* 26, 39, 153, 247
Local control. *See* Collaborative conservation
Logging. *See* Timber harvest; Timber indus-
    try
Lolo-Kootenai accords, 177
"Lords of Yesterday," 17–18
Lubrecht group, 302
Lyons, Jim, 96

MacArthur, Robert, 51
Magic Pack, 131
Malpai Borderlands Group, 14, 28, 155, 217,
    220, 246, 253, 255
Man and Biosphere program, 69
Man and Nature (Marsh), 55–56
Marbled murrelet, 89, 110
Marsh, George Perkins, 55–56
Marsh, William Perkins, 127
Mech, David, 130
Melcher, John, 177
Memorandum of Understanding (MOU),
    231
Metcalf, Lee, 174, 175, 176

Mexican spotted owl, 228–229, 230

Migratory Bird Treaty Act of *1918*, 23, 30, 90

Mineral deposits, 6, 60

Mineral Leasing Act of *1920*, 7

Mining industry, 10, 39, 44, 61, 142, 199; attitude toward natural resources, 42–43; BLM regulations, 116–117; gold, 63; uranium, 222, 223–224

Mining laws, 20, 23, 34, 35, 39, 123, 316

Minnesota v. Block, 29

Mission 66 program, 63, 259

Moab (Utah), 221–227

Montana National Forest Management Act of *1991*, 177

Montana Natural Resources Utilization Act of *1988*, 177

Montana Wilderness bills, 176, 177

Montana Wilderness Association (MWA), 174, 175, 178

Montana Wilderness Study Act of *1977*, 175, 178

Motorized recreation, 117, 178, 194, 225, 262–268

Mountain biking, 224–225, 263

Mountain States Legal Foundation, 238

Muckleshoot Tribe case, 213

Muir, John, 1, 23, 45, 56

Multiple-use management, 7, 37–38, 68, 81, 92, 105, 308, 316, 327

Multiple Use-Sustained Yield Act of *1960*, 22

Murie, Adolph, 130

National Biological Survey (NBS), 42, 114, 123–124, 126

National Bison Range, 196

National Environmental Policy Act (NEPA) of *1970*, 22, 27, 34, 60, 68, 81, 201; cumulative effects analysis requirements of, 119–120, 151; fire policy and, 141; in ICBEMP, 164, 166; impact statement requirements of, 74, 240, 241; in Northwest Forest Plan litigation, 103, 104; in spotted owl litigation, 87, 90, 91, 92, 93

National Forest Management Act (NFMA) of *1976*, 7, 22, 27, 38, 68, 81, 123, 201; biodiversity provision of, 73, 90–92, 118, 119, 152; forest planning revisions, 115; jurisdictional boundaries of, 151; in Northwest Forest Plan litigation, 103, 104; salvage logging exceptions of, 156; in spotted owl litigation, 87, 90, 91, 92

National forests: authority over, 33–34; biological importance of, 6, 83–84, 205; concurrent jurisdiction over, 26; establishment of, 20, 24, 174; health of, 82, 83, 142, 155; in Montana, 176; multiple-use management of, 7, 37–38, 68, 81, 92, 105; restoration (eastern), 143; restoration (western), 155–157; road construction moratorium in, 116, 205; salvage logging rider and, 105–108, 156–157, 237; wildfires in, 137. *See also* Forest Service, U.S.; Northwest Forest Plan; Timber harvest; Wilderness preservation; *and specific names*

National Landscape Conservation System, 117, 126, 203, 315

National Marine Fisheries Service (NMFS), 88, 89, 110, 111, 163, 241, 253

National monuments: BLM responsibility for, 117; Clinton's designations of, 33, 121, 184, 196, 203, 205, 239, 320; presidential powers and, 23, 33, 208

National parks: biosphere reserve concept in, 69–70; as enclaves, 5, 186–187; ecology of, 187; environmental threats to, 66–67, 70; in greater ecosystem concept, 190–192; growth and development of, 196; political origins of, 22, 23, 195–196; predator eradication in, 3, 129, 143; preservation model and, 22–24; shortcomings of, 187, 197; species extinction in, 51, 197; in Utah, 180, 222–223; visitation growth, 259; wilderness designation, 202. *See also specific names*

National Park Service: collaborative conservation and, 251; definition of ecosystem management, 72; ecosystem management initiatives of, 70, 117; establishment of, 23; mandate and tradition of, 7, 40–41; Mission 66 program, 63, 259; motorized recreation and, 264, 266; scientific research agenda of, 124, 208; State of the Parks Report (*1980*), 67; wildfire policy of, 5, 137–138

National Parks Omnibus Management Act of *1998*, 124

National Recreation Trails Program, 266

National Trails System Act of *1968*, 63

National Wildland Recovery Corps, 191

National Wildlife Federation, 44–45, 174

National Wildlife Refuge Administration Act of *1966*, 24, 124

National wildlife refuges: ecological restoration in, 143; ecosystem management in, 41; mandate of, 124; political origins of, 41, 196–197; shortcomings of, 187, 197

National Wildlife Refuge System Improvement Act of *1997*, 7, 41, 124

Native Americans: ecological restoration goals and, 147–148; influence on public land policy, 65; in wolf restoration program, 134

Natural resource management. *See* Ecosystem management; Public land and natural resource policy

Natural Resources Defense Council, 44, 44–45

Nature Conservancy, 45, 216–217

Nature reserve design, principles of, 188–190

NEPA. *See* National Environmental Policy Act

"New West": collaborative conservation in, 225–226; environmental concerns in, 64, 226–227; Moab and, 224–228; recreation and tourism in, 63–64, 224–225; socioeconomic change in, 62–63, 221–224; urban-rural tensions in, 64–65

Nez Perce tribe, 134

NFMA. *See* National Forest Management Act

Niobrara National Scenic River, 251

Nixon, Richard, 265

North American Waterfowl Management Plan, 192

North Cascades Ecosystem, 191

Northern Rockies Ecosystem Protection Act (NRPA), 54, 178, 191

Northwest Economic Adjustment Initiative, 101

Northwest Forest Conference, 96–97

Northwest Forest Plan, 13, 45, 80, 89, 95, 96–113, 188, 204, 285; aquatic conservation strategy of, 99, 101, 104, 110, 111; development of, 96–102; ecosystem management principles of, 98; implementation process, 109–113; interagency coordination in, 102, 112; legal challenge to, 102–105, 106–108, 252; salvage logging rider and, 95, 105–108; socioeconomic provisions of, 101–102, 112; timber harvest and, 99, 101, 104, 108, 109

Northwest Forest Resource Council, 103

Northwest Forest Resource Council v. Glickman, 107

Northwest Timber Compromise, 90

Norton, Gale, 248, 305

Nye County case, 235–236

Odums, Eugene, 50

Off-road vehicles (ORVs), 262, 263, 264–267

Oil and gas exploration, 63

Oil and gas leasing, 201, 205, 237

"Old West": Catron county and, 228–232; custom and culture county ordinances, 229; economic problems in, 227–228, 233; extractive industries and, 234; opposition to environmental policies, 234–244; species reintroduction in, 232; underdog image of, 243

Olson, Sigurd, 130

Olympic National Park, 69, 169

Option *9*, 98–99, 101

Oregon and California Lands Act, 87, 92, 103

Oregon Plan, 253, 255

Outdoor Recreation Resources Review Commission, 63, 259

Outfitter Policy Act of *2001*, 268–269

Owens, Wayne, 182

Pelican Island, 23, 196

Pesticides, 58

Philosophy, ecology and, 54–59

Pinchot, Gifford, 20–21, 23, 26, 38, 56, 137, 314

Place-based initiatives, 302–303, 307–308

Plum Creek Native Fish HCP, 216

Plumas National Forest, 275, 276, 280

Population growth, 62
Portland Audubon Society v. Lujan, 93
Postindustrial economy, 62
Powell, John Wesley, 15, 17, 136
Predator extermination program, 3, 40, 129, 143
Predator recovery programs, 69; captive breeding, 150, 158; collaborative process in, 161, 251, 257, 305; compensation for livestock depredation, 132, 135, 217; grizzly bear, 160–161, 162, 251, 257, 303, 305, 323; opposition to, 158–160; wolf, 3–4, 45–46, 128–136, 158–159, 232, 238, 316–317
Preservation: devolutionary trend in (*See* Collaborative conservation); of ecosystem-level areas, 178, 190–192, 215–218, 326; enclaves, 5, 186–187; history of, 22–25; local opposition to, 192–194, 207, 224, 234–244; nature reserve design principles, 188–190; politics of, 195–198, 208; private ownership and (*See* Private land); shortcomings of, 187–188, 197, 206–207; strategic concessions debate, 194–195; *vs* utilitarianism, 314–315. *See also* Biodiversity conservation; Wilderness preservation; Wise Use movement
President: Antiquities Act powers of, 23, 33, 121, 123, 184, 190, 196, 320; national monument designations by, 23, 33, 121, 184, 196, 203, 208, 320; role in public land policy, 31–33, 320–321. *See also specific names*
Private land: in checkerboard pattern, 11, 209; collaborative process and, 217–218, 244, 289–290, 291; development on, 209–210; federal exchanges of, 212–215; federal purchase of, 211–212; grants, 17, 18; Leopold's view of, 210–211; market-oriented approach to ecological conservation on, 215–216; property rights and, 172, 210, 238, 242, 316; regulatory approach to ecological conservation on, 215–216
Privatization, 18–19, 24. *See also* Private land
Progressive Era, 19–22, 26, 56, 321
Property. *See* Private land
Property rights, 172, 210, 238, 242, 316
Provincial Advisory Committees, 112

Public land: acreage in west, 6–7; agencies, 37–42; in ecology era, 2–6, 12–14, 324–327; economic role of, 7–8, 10, 61, 63; exchanges, 212–215; industry on (*See* Extractive industries); jurisdictional authority over, 11; location and ownership of, 8–9; as political lands, 12; privatization of, 18–19, 24; purchases, 211–212; recreation on (*See* Recreation and tourism); timberlands, 81
Public Land Law Review Commission, 219
Public land and natural resource policy: agency authority over, 31–32, 319–320; boundary lines in, 5, 48, 66, 70, 76, 151–152, 256–257, 288–291, 317; centralized model, 25; Congressional authority over, 28–31, 108, 321; devolution, 25–28, 300, 303–306 (*See also* Collaborative conservation); disposal, 17–19; interest groups and, 42–46; judicial authority over, 32, 33–35, 319; Native American tribes and, 65; presidential authority over, 32–33, 320–321; in Progressive Era, 19–22, 56, 321; rural/urban conflict in, 64–65 (*See also* "New West"; "Old West"); science-values conflict in, 313–314; state and local influence over, 35–37. *See also* Ecological restoration; Ecosystem management; Preservation; Wilderness preservation
Public opinion, on environmental protection, 64
Public participation. *See* Collaborative conservation
Public Rangelands Improvement Act (PRIA) of *1978,* 153

Quincy Library Group, 13, 14, 28, 30, 124, 220, 233, 253, 257, 274–299, 320; active management strategy of, 291; attitude toward Forest Service, 276, 281; boundaries, 276–277, 288–291; congressional endorsement of, 278, 280–281, 296–297, 302; environmental critics of, 285, 287, 293, 295, 296; Forest Service role in, 276, 277, 278, 280, 282–283, 292, 294–295; geographical scale issues and, 289–291; industry-community collaboration in, 276–277, 281–

282, 288, 293–294; legal guidelines for, 295–296, 297–298; spotted owl contro-versy and, 275–276; spotted owl (CAPSO) guidelines, 277–278, 280, 285; sustainabil-ity principles of, 287; timber harvest and, 282–283, 286, 292

Ranching industry: collaboration with envi-ronmentalists, 46; devolution policy and, 26, 27; influence on public land policy, 42–43, 44; rangeland restoration and, 153–155; in regional economy, 6, 10, 62, 63; Sagebrush Rebellion, 10, 27, 35, 39, 43, 68–69; unclaimed lands and, 39; wolf restora-tion and, 131, 132–133, 135, 217. See also Livestock grazing

Rangeland: development on, 209; prescribed fire policy, 155; reform regulations, 69, 116, 121, 123, 153, 238; restoration, 153–155. See also Livestock grazing

Reagan, Ronald, 68, 69, 85, 131, 135–136, 201

Recreation fee program, 269

Recreation and tourism: attitude toward preservation, 194; benefits of biodiversity to, 59; collaborative process in, 225–226; gateway communities and, 259–260, 261; growth of, 63–64, 220–221, 258–259; in local economy, 6–7, 10, 62–63, 259–260, 261–262; motorized vehicle controversy, 117, 178, 194, 225, 262–268; mountain bik-ing, 224–225; regulatory authority over, 260–261, 268–272; strategic concessions to, 194

Redwood National Park, 66

Regional ecological restoration, 162–169

Regional ecosystem management, 67–68, 190–192, 317–318

Reich, Robert, 96

Resource Advisory Councils (RACs), 116, 153

Restoration ecology. See Ecological restora-tion

Riparian areas: livestock grazing and, 231–232; restoration of, 153–155

Rivers: restoration of, 150–151, 169; wild and scenic designations, 187

Rivlin, Alice, 96

Road construction: local opposition to, 224, 226–227; moratorium, 116, 205

Roadless Area Conservation initiative, 116, 121, 126, 179, 204, 239

Roadless Area Review Evaluations (RARE I and II), 175, 199–200, 201

Roadless rule, 179, 301, 306

Roosevelt, Franklin, 33

Roosevelt, Theodore, 23, 32, 33, 41, 129, 196, 203, 320

Rothstein, Barbara, 110–111

"Round River" (Leopold), 57

R.S. 2477 law, 267

Rural West. See "Old West"

Sacramento Bee, 277

Sagebrush Rebellion, 10, 27, 35, 39, 43, 68–69, 220

Salmon: on endangered species registry, 110, 163, 164; Native American fishing rights and, 65; restoration projects, 150–151, 163, 169, 253, 255

Salvage logging rider controversy, 95, 105–108, 156–157, 237

Sand County Almanac, A (Leopold), 56, 58

Sand Flats Recreation Management Area, 225–226, 255

San Francisco Bay Delta project, 169

San Juan Resource Area, 264

San Rafael Swell conservation area, 226, 267, 306

Santa Rosa and San Jacinto Mountains Na-tional Monument, 206

Sawtooth National Recreation Area, 135

Scapegoat Wilderness, 175

School trust land exchange, 190, 214

Science-values conflict, in public land policy, 313–314

Scientific Panel on Late Successional Forest Ecosystems (Gang of Four), 85, 86

Scientific utilitarian conservation model, 19, 21–22, 23, 314–315

Section 314 rider, 93

Section 318 rider, 90, 108

Senate Committee on Energy and Natural Resources, 30

Service-based economy, 62, 64, 259–260, 270–271

Shelford, Victor E., 171

Sierra Club, 23, 44

Sierra Club v. Morton, 58

Sierra Nevada Ecosystem Project (SNEP), 169, 281

Sierra Nevada Forest Plan Amendment, 115–116, 124, 281, 298

Sierra Nevada national forests, 274–276, 279, 280–281

Sierra Pacific Industries, 275, 276, 282, 289–290, 291

Silent Spring (Carson), 57–58

Siuslaw National Forest, 90

Ski areas, 214–215, 259–260, 261–262, 263

Sky Islands Wilderness Conservation Network Plan, 192

Smokey Bear, 4

Snail darter case, 69, 88, 95

Snowmobiles, 117, 262, 263, 264, 266

Society for Conservation Biology, 54

Society for Ecological Restoration, 145

Sonoran Institute, 45

Soule, Michael, 192

Species conservation. *See* Biodiversity conservation

Species extinction, 49, 51, 91, 128

Species recovery. *See* Ecological restoration

Spotted owl controversy, 45, 68, 83–126, 239; ecosystem management plan (*See* Northwest Forest Plan); federal agencies and, 84–85, 86, 228–229, 280–283; legal challenges in, 87–95, 240, 319; Quincy Library Group plan, 274–276, 278, 280, 286, 291; regional forest plan, 151–152; in Sierra Nevada forests, 277–278; in Southwestern forests, 228–229; threat to old growth forests in, 83–84, 85; timber harvest and, 2–3, 85, 86, 91, 92, 93–94, 95, 228, 275–276

Spotted owl habitat areas (SOHAs), 85

State and local government: county supremacy ordinances of, 229, 234, 235; influence on public land policy, 35–37; land use and property laws of, 210; in "New West," 226–227

State sovereignty, 29, 36. *See also* Devolution policy

Steen, Charlie, 222

Steens Mountain Cooperative Management and Protective Area, 206

Stegner, Wallace, 47

Stewardship concept, 56

Stone, Christopher, 58

Sun Valley Company, 214–215

Supreme Court, U.S., 11, 58; environmental litigation and, 33; on ESA mandate, 69, 88, 240; on president's withdrawal power, 32, 34; on the property clause, 29; on property rights, 210; on range reform regulations, 121; snail darter case, 69, 88, 95; on state sovereignty, 29, 36; on takings claims, 238; on timber harvest, 90

Surface Mining Control and Reclamation Act (SMCRA), 152

Sustainability principle, 71, 76, 250, 271, 287

Sustainable Biosphere Initiative, 53–54

Taft, William Howard, 20, 32

Takings claims, 238

Tansley, Alfred George, 49–50

Taylor Grazing Act of *1934,* 7, 26, 39, 153, 247

Tellico Dam, 88

Texas Gulf Sulfur company, 222, 223, 225

Thomas, Jack Ward, 79, 86, 97, 98, 115, 278

Thomas Report, 86

Thoreau, Henry David, 22

Timber harvest: decline in, 63; environmental damage from, 155–156; forest restoration and, 155–157; increased quotas, 68, 82; under Northwest Forest Plan, 99, 101, 104, 108, 109, 111; under Northwest Timber Compromise, 90; under Quincy Group plan, 282–283, 286, 292; salvage logging rider and, 105–108, 156, 237; spotted owl controversy and, 2–3, 85, 86, 91, 92, 93–94, 95, 228, 230, 275–276

Timber industry: collaboration with environmentalists, 46, 276–278, 281–282, 288, 293; ESA litigation of, 240; influence on public land policy, 42–43, 44, 82; Northwest Forest Plan litigation of, 103, 104; postwar

lumber demand and, 81–82; in regional economy, 10, 82, 83, 228; regional influence of, 61, 82; wilderness designation process and, 201. *See also* Forest Service, U.S.; Quincy Library Group; Timber harvest
Tourism. *See* Recreation and tourism
Trade organizations, 43
Trout Creek Mountains Working Group, 154, 246

Umpqua National Forest, 106
Umpqua River Basin, 108, 110–111
Unfunded Mandates Reform Act of *1995*, 123, 252
University of Wisconsin: Arboretum, 144; Society for Ecological Restoration, 145
Uranium mining, 222, 223–224
Urban growth, 62, 221
Urban-wildland interface, 221, 271, 281
Utah Schools and Lands Exchange Act of *1998*, 214
Utah Snowbasin Land Exchange Act of *1995*, 214–215
Utah West Desert Land Exchange Act of *2000*, 214
Utah Wilderness Coalition (UWC), 182, 185–186
Utilitarian conservation model, 19, 21–22, 23, 137, 142, 147, 314
Utilitarian view: of biodiversity, 58–59; of industry, 42–43; of nature, 55–56

Valles Caldera Preservation Act, 13, 125
Valles Caldera trust model, 302, 303, 307

Watt, James, 35, 182, 201
Weeks Act of *1911*, 143
Western Forest Health Initiative, 105
Western Governors' Association, 36, 64, 248, 300
White River National Forest, 263, 270
Wichita National Forest Reserve, 196
Wild and Scenic Rivers Act of *1968*, 24, 63
Wilderness Act of *1964*, 24, 63, 138, 172, 174–175, 198–199
Wilderness corridors, 178

Wilderness Inventory Handbook, 181
Wilderness preservation, 172–173; Clinton initiatives, 203–206, 207; in Congress, 175, 179, 182, 184, 198–203, 206; devolutionary trend in, 206, 207; ecology of, 177–178, 187, 206–207; ecosystem-level areas, 178, 191, 192; in Montana, 173–180; motorized recreation controversy and, 262–268; origins of, 24–25, 198; in Utah, 180–186, 224. *See also* National forests
Wilderness Society, 44, 56, 174
Wilderness study area (WSA), 224
Wildfires: Cerro Grande fire, 140; ecosystem approach to, 5, 136, 138–141; prescribed fire policy (forest), 137–138, 139, 157; prescribed fire policy (range), 155; salvage logging rider and, 105–106; suppression policy, 4, 5, 136–137, 140–141, 156; Yellowstone fires, 4–5, 138, 139
Wilkinson, Charles, 60, 273
Wilson, Edward O., 51
Wise Use movement, 10–11, 27, 35, 37, 65, 121, 220; congressional allies of, 237, 241; county supremacy ordinances in, 234–236, 243; endangered community argument of, 242–243; ESA litigation, 239–242; litigation strategy of, 237–239; property rights claims of, 238, 242; reform agenda of, 43–44
Withdrawal authority, presidential, 32–33, 34, 121, 123, 184, 190, 196, 320
Wolf extermination, 3, 129
Wolf restoration, 3–4, 45–46, 128–136, 150, 158–160, 216, 232, 238, 316–317
Worster, Donald, 1, 56
Wyoming Farm Bureau Federation, 132–133

Yellow Ribbon Coalition campaign, 275–276
Yellowstone National Park: bear management policy for, 67, 70; biosphere reserve, 69; creation of, 22–23, 195; ecological restoration goals for, 148, 149; greater ecosystem concept and, 67, 169, 177, 190, 191–192; grizzly bear restoration in, 160–161, 162; motorized recreation in, 117, 262, 264, 266; regional ecosystem management

Yellowstone National Park (*continued*)
  experiment, 67–68; visitation at, 259;
  wildfires in, 4–5, 138; wolf eradication in,
  3, 129; wolf restoration in, 3–4, 131–136,
  150, 238, 316–317
Yellowstone to Yukon (Y2Y) initiative, 54,
  178, 191–192, 193

Yosemite National Park, Hetch Hetchy controversy in, 23

Zilly, Thomas, 88, 89
Zion National Park, 214